W9-DGX-791

HEAT EXCHANGERS:
Design and Theory Sourcebook

A PUBLICATION OF THE INTERNATIONAL CENTRE FOR HEAT AND MASS TRANSFER

Founding Members:

American Institute of Chemical Engineers
American Society of Mechanical Engineers
Canadian Society for Chemical Engineering
Canadian Society for Mechanical Engineering
Institution of Chemical Engineers, London
Institution of Mechanical Engineers, London
National Committee for Heat and Mass Transfer of the
 Academy of Sciences of the U.S.S.R.
Societé Française des Thermiciens
Verein Deutscher Ingenieure
Yugoslav Society of Heat Engineers

Sponsoring Members:

Associazione Termotecnica Italiana
Egyptian Society of Engineers
Indian National Committee for Heat and Mass Transfer
Institution of Engineers of Australia
Israel Institute of Chemical Engineers
Koninklijk Instituut van Ingenieurs, Netherlands
Society of Chemical Engineers of Japan

HEAT EXCHANGERS:
Design and Theory Sourcebook

N.H. Afgan and E.U. Schlünder, Editors

SCRIPTA BOOK COMPANY
WASHINGTON, D.C.

McGRAW-HILL BOOK COMPANY
New York   St. Louis   San Francisco   Auckland   Bogotá
Düsseldorf   Johannesburg   London   Madrid   Mexico
Montreal   New Delhi   Panama   Paris   São Paulo
Singapore   Sydney   Tokyo   Toronto

Library of Congress Cataloging in Publication Data

Main entry under title:

    1. Heat exchangers—Congresses. I. Afgan, Naim,
ed. II. Schlünder, E. U., ed. III. International
Centre for Heat and Mass Transfer.
Lectures from the 5th seminar held by the
International Centre for Heat and Mass Transfer
TJ263.R4   621.4'022   74-8733
ISBN 0-07-000460-9

HEAT EXCHANGERS

6 – MAMM – 8 4 3 2 1

COMPREHENSIVE TOPICAL GUIDE

This sourcebook contains a wealth of design, performance, and rating data for a great variety of heat-exchange equipment. Tabular material is abundant, and each chapter has been profusely illustrated with graphical results and equipment drawings. Frequently, the data presented are new and heretofore unpublished.

The following guide will be helpful in directing the reader to the main topics, types of equipment, and systems described in the indicated chapter.

Special attention should be paid to the reference sections at the end of each chapter, these were carefully compiled to provide useful indications of sources for further and more detailed information.

CONTENTS

CONTRIBUTORS

Abadzić, E., Chapter 14, Linde Aktiengesellschaft, Munich, Germany

Baldus, H., Chapter 14, Linde Aktiengesellschaft, Munich, Germany

Bandel, J., Chapter 29, Institut für Thermische Verfahrenstechnik, Universität Karlsruhe (TH)—Federal Republic of Germany

Bardoux, H. I., Chapter 19, DYNAFLOW—INFONET, Amsterdam, Netherlands

Bird, G., Chapter 26, Postgraduate School of Studies in Chemical Engineering, University of Bradford, England

Carnavos, T. C., Chapter 17, Noranda Metal Industries, Inc., Newton, Connecticut, USA

Clark, A., Chapter 28, Department of Mechanical Engineering, University of Michigan, Ann Arbor, Michigan, USA

Donsi, G., Chapter 15, Istituto di Chimica Industriale e Impianti Chimici, Università Laboratorio di Ricerche sulla Combustione, C.N.R., Napoli, Italy

Dul, J., Chapter 32, Institute of Nuclear Research, Świerk, Poland

Forgó, L., Chapter 5, HÖTERV, Design Bureau for Industrial Power and Heat Supply, Budapest, Hungary

Gardner, Karl A., Chapter 2, Liquid Metal Engineering Center, P.O. Box 1449, Canoga Park, California, USA

Göğüs, Y., Chapter 24, Middle East Technical University, Ankara, Turkey

Granville, W. H., Chapter 26, Postgraduate School of Studies in Chemical Engineering, University of Bradford, England

Hausen, H., Chapter 9, Technische Universität Hannover, Federal Republic of Germany

Heinecke, E., Chapter 25, Kernforschungsanlage Jülich GmbH, Jülich, Germany

Hennecke, W., Chapter 31, Institut für Thermische Verfahrenstechnik, Universität Karlsruhe (TH), Federal Republic of Germany

Hewitt, G. F., Chapters 12 and 13, United Kingdom Atomic Energy Authority, Harwell, United Kingdom

Hickman, R. S., Chapter 30, Department of Mechanical Engineering, University of California, Santa Barbara, California

Hlavačka, V., Chapter 27, National Research Institute for Machine Design, Behovice, Czechoslovakia

Hoffman, T. W., Chapter 6, Department of Chemical Engineering, McMaster University, Hamilton, Canada

Kakaç, S., Chapter 24, Middle East Technical University, Ankara, Turkey

Kalinin, E. K., Chapter 8, Moscow Aviation Institute, USSR

Kolpashchikov, V. L., Chapter 35, Heat and Mass Transfer Institute, B.S.S.R. Academy of Sciences, Minsk, B.S.S.R., USSR

Korybalski, E., Chapter 28, Department of Mechanical Engineering, University of Michigan, Ann Arbor, Michigan, USA

Kostić, Ž., Chapter 23, Thermal Physics and Engineering Department, Boris Kidric Institute, Belgrade, Yugoslavia

Luikov, A. V., Chapter 35, Heat and Mass Transfer Institute, B.S.S.R. Academy of Sciences, Minsk, B.S.S.R., USSR

Marschall, E., Chapter 30, Department of Mechanical Engineering, University of California, Santa Barbara, California

Martynenko, O. G., Chapter 35, Heat and Mass Transfer Institute, B.S.S.R. Academy of Sciences, Minsk, B.S.S.R., USSR

Massimilla, L., Chapter 15, Instituto di Chimica Industriale e Impianti Chimici, Universitá Laboratorio di Ricerche sulla Combustione, C.N.R., Napoli, Italy

Mizushina, T., Chapter 16, Department of Chemical Engineering, Kyoto University, Kyoto, Japan

Moalem, D., Chapter 34, Department of Chemical Engineering Technion, Israel Institute of Technology, Haifa, Israel

Mirković, Z., Chapter 20, Research and Development Center for Thermal and Nuclear Technology, ENERGOINVEST, Sarajevo, Yugoslavia

Oka, S., Chapter 23, Thermal Physics and Engineering Department, Boris Kidric Institute, Belegrade, Yugoslavia

Özgü, M. R., Chapter 24, Middle East Technical University, Ankara, Turkey

Patankar, S. V., Chapter 7, Mechanical Engineering Department, Imperial College, London, S.W.7, England

Paules, B., Chapter 33, Centre de Cinétique et Physique Chimique C.N.R.S. Département de Génie Chimique, Nancy, France

Perrut, M., Chapter 33, Centre de Cinétique et Physique Chimique C.N.R.S. Département de Génie Chimique, Nancy, France

Pescod, D., Chapter 22, Division of Mechanical Engineering, Commonwealth Scientific and Industrial Research Organization, Highett Victoria, Australia

Petukhov, B. S., Chapter 11, Institute of High Temperatures, Moscow, USSR

Schlünder, E. U., Chapters 1, 29, and 31, Institut für Thermische Verfahrenstechnik der Universität Karlsruhe, Federal Republic of Germany

Schmidt, F. W., Chapter 18, The Pennsylvania State University, University Park, Pennsylvania, USA

Schneller, J., Chapter 27, National Research Institute for Machine Design, Behovice, Czechoslovakia

Semeria, R., Chapters 12 and 13, Commisariat à l'Energie Atomique, Grenoble, France

Sideman, S., Chapter 34, Department of Chemical Engineering Technion, Israel Institute of Technology, Hafia, Israel

Sikmanović, S., Chapter 23, Thermal Physics and Engineering Department, Boris Kidric Institute, Belgrade, Yugoslavia

Spalding, D. B., Chapter 7, Mechanical Engineering Department, Imperial College, London, S.W.7, England

Spigt, C. L., Chapter 19, DYNAFLOW–INFONET, Amsterdam, Netherlands

Taborek, J., Chapter 3, Heat Transfer Research, Inc. (HTRI) Alhambra, California

Tolubinsky, V. I., Chapter 21, Institute of Engineering Thermophysics, Academy of Sciences, Ukrainian SSR

van der Walle, F., Chapter 19, DYNAFLOW–INFONET, Amsterdam, Netherlands

von der Decken, C. B., Chapter 25, Kernforschungsanlage Jülich GmbH, Jülich, Germany

Willmott, A. J., Chapter 10, University of York, United Kingdom

Zozulya, N. V., Chapter 21, Institute of Engineering Thermophysics, Academy of Sciences, Ukrainian SSR

Žukauskas, A. A., Chapter 4, Academy of Sciences of the Lithuanian SSR, Vilnius, USSR

FOREWORD

The present HEAT EXCHANGER SOURCEBOOK combines recent developments in heat transfer with the practical applications of heat exchangers. The progress on basic research and scientific foundations of the science of heat transfer achieved at research institutions and universities is merged on these pages with the engineering advances made in the industries throughout the world in the design, operation, rating and performance of heat exchangers. The emphasis has been placed here on the industrial and practical heat exchange systems.

The HEAT EXCHANGER SOURCEBOOK has been assembled by the editors by inviting contributions from scholars and experts in their special fields from all over the world. The contributions were then reviewed by a special committee of distinguished specialists in the field of heat exchangers:

> E. U. Schlünder (F.R. Germany), Chairman
> K. A. Gardner (U.S.A.), Co-chairman
> T. Gregorich (Yugoslavia), Co-chairman
> R. Gregorig (F.R. Germany)
> D. F. Mattarolo (Italy)
> T. Mizushina (Japan)
> T. D. Patten (Great Britain)
> B. S. Petukhov (U.S.S.R.)
> R. Semeria (France)
> J. Taborek (U.S.A.)
> A. A. Žukauskas (U.S.S.R.)

Research and development in steady state and transient behavior are emphasized. Heat exchangers operating with two-phase systems, packed beds, direct contact heat exchangers as well as new concepts in design and calculation are presented. Computer methods are frequently employed and described. The editors feel that this HEAT EXCHANGER SOURCEBOOK presents an excellent insight into the state-of-the-art for both recent and future developments of heat exchangers and will be of great use to all engineers and scientists in the field of heat transfer.

The forum for the presentation of the work was the International Centre for Heat and Mass Transfer, specifically its Fifth Seminar, devoted to developments in heat exchangers. In contrast to earlier ones, which were oriented toward basic research and the scientific foundations in the field of heat and mass transfer, as stated above, this Seminar stressed the application of these foundations to the technical and engineering problems. The presentations were organized in 12 parts. Each part was opened by a lecture and followed by communications in the given field. All lectures and a special selection of the communications have been included in the HEAT EXCHANGER SOURCEBOOK. With representation of 26 countries, the effort is of a truly international diameter.

The International Centre for Heat and Mass Transfer received its support for this undertaking from UNESCO, Energoinvest (Yugoslavia), the Yugoslav Federal Council for Scientific Coordination, The National Science Foundation (USA), The Academy of Sciences of the USSR and the Boris Kidric Institute (Yugoslavia).

Special Acknowledgement is hereby extended to UNESCO for its financial support.

N. Afgan, E. U. Schlünder

Chapter 1

APPLICATION OF HEAT TRANSFER THEORY
TO HEAT EXCHANGER DESIGN

E. U. Schlünder (*)

The title of this chapter implies a provocation to comment on the relation between engineering and engineering science. Sometimes it appears that the practicing engineer is skeptical about applying science in solving his practical problems, while on the other hand the scientist reproaches the engineer with incomprehension.

In order to meet the aim of this volume we should accept this provocation and try to bridge this apparent gap for the benefit of both future heat transfer research and future heat exchanger design.

I would like to attack the problem by a critical review of the conditions governing and limiting heat transfer research activity.

The conditions might be listed in a set of theses which — by the way — not only refer to heat transfer but also to other engineering problems.

1. The subject of heat transfer research is the design, performance and development of all kinds of heat exchanging equipment. Accepting this, we have to concede that — deviating from the classical meaning of science — heat transfer research has no value in itself but only in connection with actual engineering problems.

2. The goal of heat transfer research is to predict design and performance data for heat exchanging equipment.

3. The method applied in heat transfer research is to investigate the engineering problems under well defined conditions thus enabling us to find out the ruling phenomena and laws.

4. The means for predicting design and performance data is the theory. The theory usually is based on hypotheses and combines mathematically a set of

---

(*) Institut für Thermische Verfahrenstechnik der Universität Karlsruhe, F.R. Germany.

variables and invariables. The latter ones usually are to be determined by experiments. The variables permit predictions beyond the range already known by experience.

5. The accuracy to be sought of the theory has to be of the same order as the data to be predicted. With technical problems usually the accuracy of the invariables to be determined by experiments as well as the data to be predicted are of the order of some percent. Therefore, the accuracy of theories for technical purposes in almost all cases is sufficient within a range of some percent.

6. The effectiveness of a theory is so much the better as it permits the reduction of experimental work. Refining a theory is useful only if this leads to a reduction of experimental work. This will be true if the number of the invariables could be reduced or if their general applicability could be extended. Refining a theory involving increasing experimental work is only to be justified if this leads either to new physical knowledge, finally reducing experimental work in a long range aspect, or if this leads to a higher degree of exactness which might be sought in a special case.

7. The application of theories in solving engineering problems is justified if the conditions in the practical case can be identified to be comparable to those well defined conditions under which the theory was derived.

8. The reliability of a theory is restricted because of two facts. On one hand the idealized conditions under which the theory has to be derived should be as close as possible to the practical conditions, but on the other hand away enough to guarantee the possibility for setting up a theory of general validity.

9. The limits in setting up a theory are given first by the possibility of analyzing the engineering problem at all and second by the chance to establish reasonable hypotheses.

10. The benefit of a theory is ambivalent. There are engineering problems whose actual conditions cannot be quantified completely. In this case we can expect at the best to achieve a physically based qualitative analysis. On the other hand, there are engineering problems for which the ruling physical laws are unknown. In this case we can expect at the best to achieve an empirically based quantitative correlation. To solve engineering problems, both qualitative and quantitative analyses are needed for. In a few cases both kinds of analyses can be presented by one unique theory. In the majority of engineering problems we have to be content with either one or the other kind of analyses. As far as possible both kinds should be tried, even if there will remain at the moment seemingly unbridgeable gaps between them encouraging future research.

In the following I would like to comment on these theses by use of some examples.

1. A recent development in the design of freeze dryers is the application of short contact vacuum dryers. They consist of a cascade of horizontal heated plates within a vacuum chamber. The frozen granular material, e.g. meat, vegetables etc., is moved on these plates along circular paths by a stirring device and falls down the cascade from plate to plate after each rotation.

From the theory of drying processes it is known that the drying rate under these conditions is entirely controlled by the rate of heat transfer. Therefore, the heat transfer coefficient between the heated plates and the moved granular bed determines the size of the dryer and thus must necessarily be known by the designer as a function of all variables involved.

2. To set the aim of research work, it is necessary to know the variables. This knowledge is to a certain extent based on arbitrary decisions. It is obvious that the heat transfer coefficient α might depend on the termal and mechanical properties λ, ρ, C of the granular material, the gas and the plate, the residence time t, the particle diameter d, the agitator speed n, the bed height H, the shape of the stirrer, the shape of the particles, etc.. Enumerating the variables, we usually stop if further ones obviously do not seem to be of significance. In this case we know further that the thermal properties of the plate are of no importance because the ratio of the heat penetrating coefficients $\sqrt{(\lambda\rho C)\text{Granular}} / \sqrt{(\lambda\rho C)\text{Metal}}$ goes to zero so that the local temperature of the plate remains unchanged during the short contact period of each particle. So the aim is to determine the function

$$a = a \, (\lambda_p, \; \rho_p, \; C_p, \; \lambda_G, \rho_G, C_G, \; t, \; n, \; d, \; H, \; \dots \text{ form, shape}).$$

3. The chosen method was to investigate the heat transfer coefficient between the granular material of spherical shaped particles of equal size brought into contact with a heated plate. The rate of heat supply was constant four times greater than zero. The rotations per minute of the stirrer were kept constant but were altered stepwise.

4. To set up a theory we first have to introduce one or more hypotheses. Because the theory should be valid in general, that means for all possible values of the variables, it would be useful to seek asymptotic laws first.

We choose the limiting cases of zero and infinite stirrer speed. In the first case we introduce the hypotheses that the granular material filled with high or low pressure gas can be treated like a homogeneous medium. Under the simplifying assumption that the isotherms are parallel we can calculate the mean heat conductivity λ_{so} of the granular bed by integrating over all local heat flux paths through gas and solid material lying in series. Regarding the fact that the heat conductivity of the gas goes to zero when approaching the contact point of the sphere with the plate and taking into account the influence of radiation, we obtain the following formulas (1):

$$\frac{\lambda_{so}}{\lambda} = \left(1 - \sqrt{1 - \Psi}\right)\left(\frac{1}{1 + Kn^* | \Psi} + \Psi Nu_r\right) + \sqrt{1 - \Psi}\ \frac{\lambda'_{so}}{\lambda}$$

$$\frac{\lambda'_{so}}{\lambda} = \frac{2}{N - M}\left\{\frac{\left[N - (1+Kn^*)\frac{\lambda_s}{\lambda}\right]B}{(N - M)^2}\ln\frac{N}{M} - \frac{B - 1}{N - M}(1 + Kn^*)\right.$$

$$\left. - \frac{B + 1}{2B}\frac{\lambda_s}{\lambda}\left[1 - (N - M) - (B - 1)Kn^*\right]\right\}$$

$$M = B\left[\frac{\lambda}{\lambda_s} + Kn^*\left(1 + Nu_r\frac{\lambda}{\lambda_s}\right)\right]. \qquad N = \left(1 + Nu_r\frac{\lambda}{\lambda_s}\right)\left(1 + Kn^*\right)$$

$$B = 1.25\left(\frac{1 - \Psi}{\Psi}\right)^{\frac{10}{9}}, \qquad Kn^* = \frac{2\sigma}{d}\left(\frac{2}{\gamma} - 1\right),$$

$$Nu_r = \frac{0,04\ C_s}{\frac{2}{\epsilon} - 1}\left(\frac{T}{100}\right)^3\frac{d}{\lambda}$$

Where: λ, λ_s = heat conductivities of the gas and the solid particles respectively, ψ = void fraction, d = particle diameter, σ = mean free path of the gas molecules, C_s = radiation coefficient, ϵ = emissivity, T = absolute temperature, γ = coefficient of accomodation.

Thus, if the thermal properties of the granular bed are known, it is easy to calculate the heat transfer coefficient as a function of the residence time t. The asymptotic solutions for short and long times t for the boundary condition of constant heat flux supplied to a perfect conducting plate are (2):

$$a = \frac{2}{\sqrt{\pi}}\frac{\sqrt{(\lambda\rho C)_{so}}}{\sqrt{t}} \qquad\qquad (1)$$

$$a = 3 \frac{\lambda_{so}}{H} \tag{2}$$

If the residence time goes to zero, experimental results have shown that there exists a limiting maximum value for α. Taking into account the fact that for very short residence times the whole heat transfer resistance is located in the gaseous gap between the particles and the plate, we can calculate this maximum value of α also. We obtain the formula (3):

$$a_{max} = 4 \frac{\lambda}{d} \left\{ (2\,Kn^* + 1)\,\ln\left(\frac{1}{2Kn^*} + 1\right) - 1 + \frac{1}{4}\,Nu_r \right\} \tag{3}$$

This equation gives also the limiting case if the residence time is large, but the contact time of the particles, due to the mixing motion caused by the stirrer, goes to zero. That means that formula (3) gives the maximum heat transfer coefficient, which cannot be exceeded at all even if the stirrer speed goes to infinity.

If we plot the heat transfer coefficient α versus the residence time t according to the formulas (1), (2) and (3), we see from fig. 1 that under the vacuum

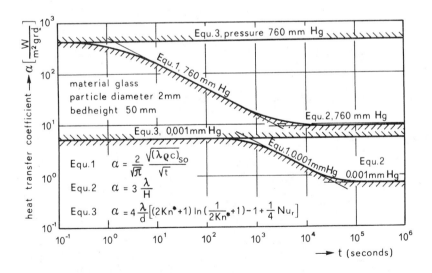

Fig. 1. Limiting laws controlling heat transfer coefficient α between a stirred granular bed and a heated plate.

conditions even with a residence time of a thousand seconds the stirrer speed has no influence, the value of α is always the maximum value. That means that there is no possibility of increasing the heat transfer by increasing the agitation of the granular bed under vacuum conditions. In contrast, under high pressure conditions the intensity of agitation might have a great influence on the heat transfer. To fill the gap between the limiting laws by a theory, we have to introduce another hypothesis on the contact time as a function of the mixing intensity. This concerns a problem of turbulent motion. Since we know from other problems of turbulent motion that in this area no satisfactory theories exist up to now, we should prefer to continue our investigations experimentally. The following figures show the experimental results.

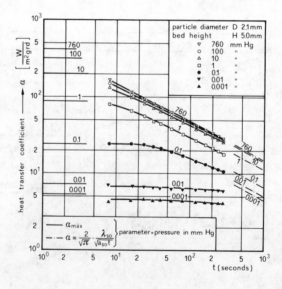

Fig. 2. Experimental values of heat transfer coefficient for zero stirrer speed and various pressures. Glass spheres 2.1 mm diameter.

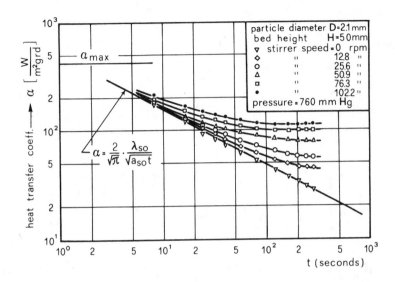

Fig. 3. Experimental values of heat transfer coefficient at 760 mm Hg pressure and various stirrer speeds. Glass spheres 2.1 mm diameter.

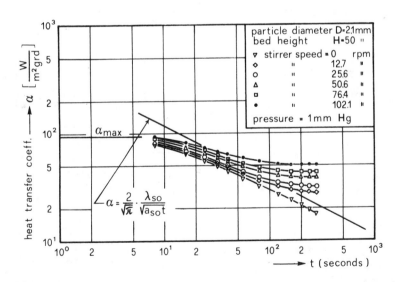

Fig. 4. Experimental values of heat transfer coefficients at 1 mm Hg pressure and various stirrer speeds. Glass spheres 2.1 mm diameter.

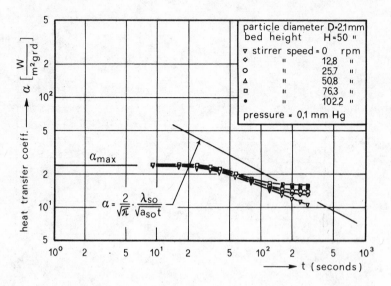

Fig. 5. as Fig. 4. but 0.1 mm Hg pressure.

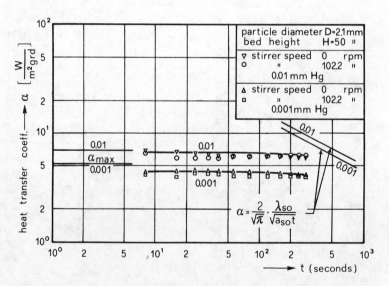

Fig. 6. as Fig. 4 but 0,01 mm Hg pressure and 0.001, resp.

These figures confirm the theory of the limiting laws. The data show further asymptotic values for the heat transfer coefficient for long residence times depending on the stirrer speed. Applying equation (2) to these asymptotic values, one could determine a hypothetical contact time as a function of the stirrer speed. But this would no longer be a theory but merely a correlation.

5.   The accuracy of the equations (1), (2) and (3) is of the same order as those of the variables, the constants, and the experimental data. The theoretical solutions in the intermediate range between equations (1) and (3) and (1) and (2) respectively can be found by graphical interpolation with sufficient accuracy so that no further theoretical work in this case is sought.

6.   The theory developed above shows that further experimental work is neither necessary nor useful under vacuum conditions. Thus, the time and costs for further research work is reduced.

7.   The theoretical and experimental results of this analysis can be applied to all kinds of apparatuses with heat transfer from plates to moved spherical shaped granular material if the thermal properties, the residence time, and the hypothetical contact time are known.

8.   The reliability is restricted by the difficulty of determining the hypothetical contact time, which can only be estimated. Since this time depends e.g. on the arbitrary shape of the stirrer, reliable data can only be obtained by measuring heat transfer coefficients applying those arbitrary stirrer shapes. No theoretical predictions but only empirical correlations of limited validity seem to be possible at the moment.

9.   The limitation of setting up a complete theory is due to two facts. First, we cannot quantify the actual conditions completely, e.g. all possible stirrer shapes. Second, even if we could do this, we have no satisfying hypothesis of turbulent motion under these conditions.

10.   The benefit of the presented analysis is to reveal the dominant physical laws ruling this kind of heat transfer process, thus enabling us to predict limiting laws and to acquire a better understanding of empirical formulas obtained from direct measurements at certain apparatuses. The remaining problem for the practicing engineer is to analyze a given apparatus under given operating conditions to find out how the theory can be applied. In this respect no further research can help him, and he has to rely on this imagination and discernment.

This was an example from the field of heat transfer. Now I would like to add another example from the field of mass transfer.

1. A recent development in the technique of removing organic compounds from water is the application of ion exchangers and active carbon filters. A vessel is filled with resin or carbon particles of about 1 mm diameter. While water is pumped through this packing, the organic compound is transferred to the particles. When the particles are saturated they are removed and regenerated. For the design of these apparatuses one must know the local and time dependent concentration distributions in the packed bed and in the water.

2. The aim is to set up a theory for predicting these concentration distributions. The experimental work to determine the invariables in this theory should be minimal.

3. The method could be to study the structure of the particles and to reveal the mechanism of the diffusive mass transport. In this case there would arise quite a lot of experimental work which might lead to very interesting fundamental results. But this would be a long range concept for an extended physicochemical research program. To solve our technical problem, we should follow another method which works faster and cheaper. This method is to establish hypotheses on the transport mechanism from which we can derive a theory. The theory will include limiting laws. These limiting laws will be checked by only a few systematical experiments. If they will be confirmed with the sufficient accuracy which is sought in this case, the theory can be applied to solve our technical problem even if the hypothesis introduced at the beginning do not really correspond to the actual physical phenomena.

4. The hypothesis is that diffusion only occurs inside the water-filled pores of the particles and that there is always equilibrium between the internal concentrations of the solid and the liquid phases. From the sorption isotherms we know that the capacity of the liquid phase is negligible compared with that of the solid phase. In the case of ion exchanging resin the sorption isotherms are nearly linear, while in the case of active carbon the equilibrium concentration y^* in the carbon is nearly independent of the liquid concentration. Under these circumstances the concentration profiles in an ion exchanging resin particle can be calculated by the Fourier theory of conductive processes. They will turn out to be steady curves belonging to the family of trigonometric functions. The profiles in a carbon particle however show an unsteady jump from saturation concentration to zero. This concentration jump is moving to the center of the particle. Due to this different behavior we find different equations for the sorption rates and the local and time dependent concentration distributions in the apparatus. If we introduce

dimensionless variables for the concentration in the solid phase η, in the liquid phase ξ, for the time τ and the length in flow direction ζ, we can compare both the ion exchanger and the active carbon filter in a short comprised mathematical form:

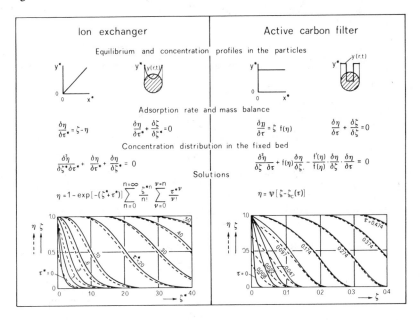

Fig. 7. Sorption kinetics and concentration distribution in ion exchanges and active carbon filters.

In these equations all quantities are known except the mass transfer coefficient β_1 between the liquid and the solid phase and the apparent diffusity δ_s in the solid phase. Both quantities can be determined by an adsorption experiment of laboratory scale. A small quantity of granules is brought into contact with the solution in a tumbler. From the measured adsorption rate at short contact times one gets the coefficient β_1, while from the adsorption rate at long contact times the coefficient δ_s can be determined. From these experiments [4], [5], it was found that for phenylic acid e.g. the diffusity in an ion exchanging resin is about 10^{-11} m²/s while in active carbon it is about 10^{-9} m²/s. The mass transfer coefficient β_1 was shown to be in good agreement with well known correlations for packed beds.

5. The accuracy of the theory is sufficient for many technical cases, although the sorption isotherms in both cases are idealized. If the actual isotherm deviates considerably from either the one or the other form, one would prefer numerical

solutions. At any rate, the typical form of the concentration profiles is between the two forms shown in Fig. 7—and these distributions are very similar.

6. The effectiveness of the theory is that only a few experiments at laboratory scale are necessary to know all the data required for the design of large scale apparatuses.

7. The results of the theory can be applied to large scale ion exchangers and active carbon filters. The conditions in the laboratory experiments are nearly the same as the technical conditions.

8. The reliability of the theoretical predictions may be limited if some other effects occur. If the cycles of loading and regenerating are relatively long, biological effects might influence the effectiveness of the sorption material.

9. The theory is limited to binary mixtures. This is a severe limitation because in all practical cases the polluted water includes a large number of organic compounds. Research must urgently be extended in this direction.

10. The benefit of this theory is the emergence of a hypothetically based quantitative analysis which permits prediction of design data as well as quick estimations of the influence of the various parameters for binary and quasi-binary mixtures.

One could continue to note examples for technical problems where the laws of heat and mass transfer play a dominant role. In any case we have to ask how to find and how to present these laws for the purpose of practical application. The form of the law is the theory. The theory is to be based on a consistent set of invariables. The question is to find out at what a level a certain number of physical quantities are to be considered constants. To illustrate this statement, we e.g. refer to the phenomenon of heat transfer through a stagnant gas at different pressures. The outstanding invariables are: at extremely low pressure the velocity of light radiation (Planck's law), at moderate pressure the velocity of sound (Knudsen's law), at normal pressure the velocity of diffusion (Fourier's law, Einstein's law resp.).

In our first example the theory could be based on a relatively small set of invariables, such as heat conductivity, heat capacity, density etc. as long as the granular bed was not agitated. The experimental results shown above were only to check the theory based on Fourier's law, while the fundamental experiments to determine the invariables of the theory as e.g. the heat conductivity of the particles and the gas had been carried out elsewhere. In the case of the agitated bed the heat transfer coefficients found in the experiments could not be explained by a theory the invariables of which were found in other experiments. The question is how to proceed in this case. We have to decide whether there is a chance of establishing a theory which includes more fundamental invariables from which the measured heat transfer coefficients can be derived or whether we presumably never can expect to find such a theory. In this case the measured heat transfer coefficients are to be considered the fundamental invariables themselves, which means to elevate the so called Newton's law of convection to the same level as Fourier's law of conduction.

In our second example there is a similar situation. The mass transfer inside the solid particles can be derived from Fourier's theory. The only invariable we need is the diffusity. But the mass transfer from the liquid phase to the particle surface under turbulent flow conditions cannot be derived from a theory a priori. So the experimentally found mass transfer coefficients are the invariables themselves. This point of view is thoroughly sufficient in order to set up a theory for predicting local and time dependent concentration distributions in ion exchangers or active carbon filters.

Let us consider another heat transfer phenomenon. The heat transfer coefficient α_v between a superheated surface and a boiling liquid usually is correlated by an equation of the type

$$\alpha_v = K \dot{q}^n \tag{4}$$

where $\dot{q}$ is the heat flux. The power n was found to be between 0.5 and 0.8. Much effort was spent during the last decades of heat transfer research to establish theories in order to predict the parameters K and n in this equation. A vast number of subtle investigations were carried out to study the influence of bubble nucleation, bubble growth, surface roughness, surface imparities, etc. Up to now nobody knows definitely whether these investigations will lead to definite predictions of the parameters in question. In particular it would be hard to decide whether further studies on bubble dynamics can deliver a contribution for deriving the parameters K and n from more fundamental physical invariables without increasing experimental work to determine the latter ones, or if we should consider instead the parameters K and n as invariables themselves. With respect to engineering problems, the latter point of view could be satisfying as far as a complete set of these parameters would be available. If we accept equation (4) as a fundamental equation, some consequences concerning the layout of evaporators are unavoidable. Usually, the size of evaporators operating under pool boiling conditions is calculated with the logarithmic mean temperature difference. We know that the logarithmic mean is only valid if the heat transfer coefficients are independent of the temperature difference. This does not hold if the heat transfer is governed by equation (4). In order to predict the size of an evaporator, we have to set up a theory the invariables of which are the parameters K and n. Such a theory leads to differential equation for the temperature distribution along the heat exchanging surface which has to be solved by numerical methods, see fig. 8. The result is presented in form of a correction factor for the heat flux $\dot{Q}/\dot{Q}_{\log}$, see fig. 9.

As we see from fig. 9 the error, when using the logarithmic mean, might be of some hundred percent [6].

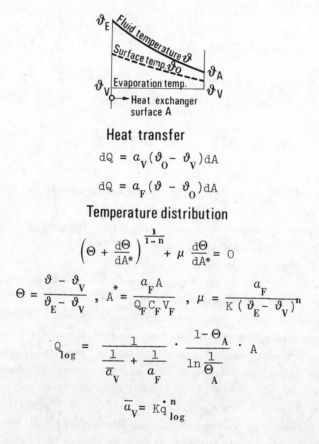

Heat transfer

$$dQ = a_V(\vartheta_0 - \vartheta_V)dA$$

$$dQ = a_F(\vartheta - \vartheta_0)dA$$

Temperature distribution

$$\left(\Theta + \frac{d\Theta}{dA^*}\right)^{\frac{1}{1-n}} + \mu \frac{d\Theta}{dA^*} = 0$$

$$\Theta = \frac{\vartheta - \vartheta_V}{\vartheta_E - \vartheta_V} \quad , \quad A^* = \frac{a_F A}{Q_F C_F V_F} \quad , \quad \mu = \frac{a_F}{K(\vartheta_E - \vartheta_V)^n}$$

$$\dot{Q}_{\log} = \frac{1}{\dfrac{1}{\bar{a}_V} + \dfrac{1}{a_F}} \cdot \frac{1 - \Theta_A}{\ln \dfrac{1}{\Theta_A}} \cdot A$$

$$\bar{a}_V = K\dot{q}_{\log}^{\,n}$$

Fig. 8. Temperature distribution in a evaporator operating under pool boiling conditions.

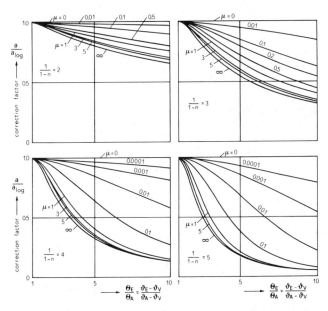

Fig. 9. Correction factor for the heat flux in an evaporator operating under pool boiling conditions.

As we can learn from this example, it is possible in solving practical problems to establish theories without knowing the fundamental physical laws ruling the involved phenomena, by introduction of certain summarizing invariables covering a variety of other fundamental physical constants without knowing their more or less complex interactions. For the scientist this is a very unsatisfying situation. This might be one of the reasons, that the heat transfer research on boiling phenomena during the last decades was so concentrated on microscopic details. Today, after having studied these details intensively, we have to confess that our original problem of predicting design data for heat exchangers seems to be more complicated than before. Now we have to answer the question whether these difficulties are only temporary and can be overcome or whether they might be of principal nature and thus any attempt to overcome them is expected to be useless. This question does not only concern boiling phenomena which were referred to only as an example, but concerns a great number of heat and mass transfer processes under technical conditions also, in particular transfer phenomena with all kinds of single and multiphase turbulent or randomly moving fluids.

At present we cannot decide definitely whether the difficulties in these cases are temporary or principal. But if we see that the progress in discovering the physical fundamentals is very slow, we should concentrate much more on the technical fundamentals of heat and mass transfer.

A quite similar situation exists in other disciplines too. In the field of metallurgy e.g. we find physical and chemical investigations about the internal structure of matter, but also a strong activity to determine technical parameters such as tensile strength, surface hardness, toughness, brittleness, ductility, fatigue tests etc. These data are determined by systematic experiments in standardized test apparatuses, and are compiled in special handbooks, since everybody knows that at least at present there is no chance of predicting these data from our knowledge about the internal structure of matter alone. On the other hand these data are urgently needed for setting up theories on material strength. In our field of heat and mass transfer the technical parameters as e.g. friction factors, heat and mass transfer coefficients etc. are determined primarily to obtain information about the physical fundamentals of which these parameters are composed. Thus, the experiments are performed more under the aspect of accepting or rejecting individual hypotheses and less under the aspect of a systematic compilation of technical data which can serve as invariables in theories for solving technical problems. That we should emphasize the latter aspect can be shown by another example, taken from the field of thermodynamics. The research on the virial coefficients to establish a general equation of state is one thing, and the international agreement in setting up standardized steam tables based on experimental data and an adapted empirical equation of state for this particular purpose only is another thing.

Realizing this kind of dualism as a typical feature of engineering science, we should stimulate a world wide agreement to strengthen research programs on heat and mass transfer by compiling reliable and systematic technical data which can serve as invariables in theories for solving engineering problems. In the first step we would have to define the problems, the goals, the methods, and then to standardize the test apparatuses. The second step would be to establish research programs and to distribute parts of them to those institutes which are willing to participate in such a program. Perhaps the International Centre for Heat and Mass Transfer in Belgrade might be a proper organization to originate a world wide concentration of heat and mass transfer research on a selected number of technical problems. If this would lead to an enlargement of our knowledge or technical invariables, we would be in a much better position to set up theories for the benefit of heat exchanger design.

Appendix

Ion exchanger

Equilibrium $\quad y^* = mx^*$

Reduced concentrations $\qquad \eta = \dfrac{y}{y^*(x_o)} \;;\; \xi = \dfrac{x}{x_o}$

Transfer from the liquid phase $\qquad \dot{n} = \rho_\ell \, \beta_\ell (x - x^*)$

Transfer in the solid phase $\qquad \dot{n} = \rho_s \bar{\beta}_s (y^* - y)$

Reduced time $\qquad \dfrac{1}{\tau^*} = \dfrac{\rho_s(1-\psi)}{a} \left\{ \dfrac{1}{\rho_\ell \beta_\ell} + \dfrac{1}{m} \dfrac{1}{\rho_s \bar{\beta}_s} \right\} \dfrac{m}{t}$

Reduced length $\qquad \dfrac{1}{\zeta^*} = \dfrac{\rho_\ell u}{a} \left\{ \dfrac{1}{\rho_\ell \beta_\ell} + \dfrac{1}{m} \dfrac{1}{\rho_s \bar{\beta}_s} \right\} \dfrac{1}{z}$

$$\dfrac{\bar{\beta}_s R}{s} \cong \sqrt{\left(\dfrac{1}{3}\pi^2\right)^2 + \dfrac{4}{\pi} \dfrac{R^2}{\delta_s t}} \quad , \quad \text{(mean value)}$$

Active carbon filter

Reduced concentrations $\qquad \eta = \dfrac{y}{y^*} \;;\; \xi = \dfrac{x}{x_o}$

Reduced time $\qquad \tau = \dfrac{\rho_\ell x_o}{\rho_s y^*} \dfrac{\delta_s t}{R^2}$

Reduced length $\qquad \zeta = (1-\psi) \dfrac{\delta_s}{R^2} \dfrac{z}{u}$

Biot-number $\qquad Bi = \dfrac{\beta_\ell R}{\delta_s}$

Adsorption rate function $\qquad f(\eta) = \dfrac{3\sqrt[3]{1-\eta}}{1 + \left(\dfrac{1}{Bi} - 1\right)\sqrt[3]{1-\eta}}$

Solution of the differential equation:

$$\varphi(\eta) = -(\rho - \rho_c(\tau)) = \frac{1}{3}\left(1 - \frac{1}{Bi}\right)\ln\eta + \frac{1}{6}\ln\frac{1 + \sqrt[3]{1-\eta} + \sqrt[3]{(1+\eta)^2}}{(1 - \sqrt[3]{1-\eta})^2}$$

$$+ \frac{1}{\sqrt{3}}\,\text{arctg}\,\frac{1}{\sqrt{3}}(2\sqrt[3]{1-\eta} + 1)$$

Inverse function $\qquad \psi(\rho - \rho_c(\tau))$

Saturation time for the first layer $\qquad \tau_1 = \frac{1}{2} - \frac{1}{3}\left(1 - \frac{1}{Bi}\right)$

Coordinate of saturation point ζc and liquid concentration ξ if $\tau < \tau_1$:

$$\zeta_c = \varphi(1) - \varphi(\eta[0,\tau])$$

$$\tau = \frac{1}{2}(1 - (1 - \eta(0,\tau)^{2/3}) - \frac{1}{3}\left(1 - \frac{1}{Bi}\right)\eta(0,\tau)$$

$$\xi = \frac{\eta(\ ,\tau)}{\eta(0,\tau)} \quad \text{and} \quad \lim_{\tau \to 0}\ \xi = e^{-3Bi\zeta}$$

For $\quad \tau > \tau_1$:

$$\zeta_c = \tau - \tau_1$$

$$\xi(\zeta,\tau) = \eta(\zeta,\tau)$$

$\dot{n}$ = rate of mass transfer
ρ = density
β = mass transfer coefficient
ψ = void fraction
a = specific particle surface per unit volume
x_o = concentration in the liquid phase at the entrance
δ_s = apparent diffusity in the particles
R = particle radius
u = liquid velocity in the empty cross section

REFERENCES

[1] P. Zehner
Experimentelle und theoretische Bestimmung der effektiven Wärmeleit-
fähigkeit durchströmter Kugelschüttungen bei mässigen und hohen Tempera-
turen.
Dissertation Universität Karlsruhe (FR) 1972 und VDI-Forschungsheft (im
Druck).

[2] H. S. Carslaw and J. C. Jaeger
Heat Conduction in Solids. Oxford, Clarendon Press.

[3] E. U. Schlünder
Wärmeübertragung an bewegte Kugelschüttungen bei kurzfristigem Kontakt.
Chem. Ing. Techn. 43 (1971) 11, S. 651-54.

[4] K. F. Ladendorf, E. U. Schlünder und H. Sontheimer
über die Austauschkinetik organischer Anionen an makoporöse Anionen-
austauschharzen.
Chem. Ing. Techn. 44 (1972) 5, S. 337-341.

[5] F. Schweiger
Diplomarbeit, Universität Karlsruhe (FR), Institut für Thermische Verfahrenstechnik (1972).

[6] E. U. Schlünder und H. Zemlin
Zur Berechnung der mittleren Temperaturdifferenzen bei überfluteten Ver-
dampfern.
Kältetechnik-Klimatisierung 23 (1971) 10, S. 292-95.

Chapter 2

ANTICIPATION OF OPERATING PROBLEMS IN THE DESIGN OF HEAT TRANSFER EQUIPMENT

Karl A. Gardner (*)

I. INTRODUCTION

It is the purpose of this chapter to consider possible problems in the operation of heat exchangers, and how to design to avoid them. Many are rather in nature and will be mentioned only in passing, or not at all. Others may be obvious, but difficult to solve, and still others may not be at all apparent; these will be discussed in greater detail except where more authoritative presentations are included in this volume.

It should not be, but sometimes is, necessary to point out that a heat exchanger is not only an apparatus for transferring heat from one medium to another, but is at the same time a pressure and/or containment vessel. More frequently it is two such vessels combined in an intricate way into a single piece of equipment, with diverse attendant problems. In the United States at least, and probably elsewhere, the rating — i.e., the determination of size and configuration of a heat exchanger — is frequently done by one specialized group of engineers, and the mechanical design assuring its safety and durability as a pressure vessel by another group. Too often members of these two groups are not well versed in one another's problems, and ill-considered compromises arising from failure of communication or understanding lead to inadequacy in service. This is really an organizational and procedural problem, not fundamentally one of design. It is hoped that the remarks which follow will be of some assistance in broadening the mutual areas of understanding between rating and design engineers, not only in industry, but also in academic circles.

The subject matter will be discussed under the broad headings of "Deficiencies in Heat Transfer and Pressure Loss Rating" and "Deficiencies as a Containment and Pressure Vessel."

(*) Liquid Metal Engineering Center, P.O. Box 1449, Canoga Park, Calif. 91304, U.S.A.

II. DEFICIENCIES IN HEAT TRANSFER AND PRESSURE LOSS RATING

A. Elementary Considerations

Since the rate of heat transfer, Q, in a heat exchanger is determined by the simple relationship,

$$Q = \int_{0}^{A_o} U_o \Delta t \, dA = A_o U_{om} \Delta t_m ,$$

(1)

the reason for failure of an exchanger to perform its heat transfer duty is to be sought in either of the two quantities U_{om}, the mean overall heat transfer coefficient, or Δt_m, the mean effective temperature difference between the heat exchanging streams. More properly, since both U_o and Δt may vary throughout the exchanger, the incorrectly integrated value of their product may be the culprit in a heat transfer deficiency. Nevertheless, it is convenient, and usually adequate, to consider each of U_{om} and Δt_m as separately determinable.

The overall coefficient U_o, may be expressed in the form:

$$\frac{1}{U_o} = \frac{1}{h_o} + \frac{1}{h_i}\left(\frac{A_o}{A_i}\right) + r_o + r_i\left(\frac{A_o}{A_i}\right) + r_w\left(\frac{A_o}{A_w}\right) ,$$

(2)

in which the h's represent heat transfer "film" coefficients, and the r's, heat transfer resistances, referred to the surfaces at which they occur as indicated by subscripts o (outside), i (inside), and w (wall). Multipliers (A_o/A_i) and (A_o/A_w) merely refer these values to the outside surface. The quantity, r_w, is the metal resistance of the tube wall in bare tube equipment, and is seldom susceptible to serious error, except possibly for combinations of boiling, condensing, or liquid metal streams, in which case it may be the major part of the total resistance to heat flow, and an error in tube wall thickness or thermal conductivity may be significant. The fouling resistances r_o and r_i, will be discussed later. Any one of the h's and r's may be a source of rating error.

The mean temperature difference is customarily expressed as a multiplier, F, times the logarithmic mean of the terminal differences; i.e., $\Delta t_m = F \Delta t_{log}$. The value of F has been determined for dozens of flow configurations, many of them by myself, based on the assumptions that U_o is constant throughout the exchanger, that the temperatures of both streams vary linearly with the rate of heat exchange, and that all elements of the fluid streams pass over proportional elements of heat transfer surface. One or more of these assumptions may be, and often is, unwarranted in actual practice.

With these elementary reminders of possible sources of error in rating heat transfer equipment, let us consider some of the most common causes.

B. Erroneous Estimation of Heat Transfer Coefficients

1. Inadequate Data

A common source of error in heat exchanger rating is the estimation of heat transfer coefficients under conditions for which experimental data are limited in range or completely lacking; e.g., the thermal conductivity of the fluids as a function of temperature, or a new flow configuration. Extrapolation beyond the verified range of data correlations or reasoning by analogy to some related conditions are usually resorted to under such circumstances. There is nothing wrong with this if it is done intelligently, preferably under the guidance of some theoretical approach which is supported by tests data over part of the range of variables involved. Extrapolation of empirical straight lines on arithmetic or log-log graph paper passing through a limited range of data is always hazardous, especially in the direction of increasing temperature for physical properties, or in the direction of decreasing Reynolds number for heat transfer or pressure loss correlations.

2. Erroneous or Misleading Correlations

Generalized correlations drawing on many sources of data without distinguishing properly between them by clear identifying symbols may give an unwarranted assurance of accuracy in some instances. A familiar example is the graph presented in many texts of the data points, and the curve through them, for the "law of the wall." In many cases the data points are not distinguished from one another and the best line thorugh them is quite well represented, in the turbulent region, by the familiar equation,

$$u^+ = 5.5 + 2.5 \ln y^+ \, , \tag{3}$$

which plots as a straight line on a semilog scale, as shown in Figure 1. If the data points are identified by Reynolds number, however, and appropriate curves drawn through them, it is seen that they are not straight lines at all but, rather, look like Figure 2. For many purposes, Equation (3) is sufficiently exact, but for others

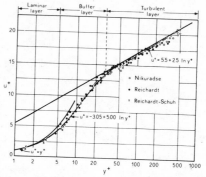

Fig. 1. Universal Velocity Distribution for Turbulent Flow in Circular Tubes (from H. Reichardt, NACA TM 1047, 1943).

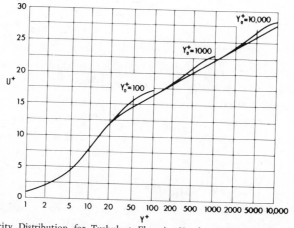

Fig. 2. Universal Velocity Distribution for Turbulent Flow in Circular Tubes—Typical Structure when Data Points Plotted with y_0^+ (centerline Value of y^+) as a Paramenter.

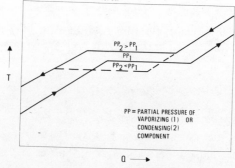

Fig. 3. An Impossible Temperature Difference Situation Not Recognizable from the Terminal Values Alone (dashed lines).

involving derivatives of u^+ it gives quite misleading results. This example is not direct-ly pertinent to heat exchanger rating, but it does illustrate the point that a compos-ite of undifferentiated data does not necessarily lead to the best correlation, parti-cularly when assembled from different investigators using different models of equip-ment over different ranges of variables.

C. Failure to Visualize Heat Versus Temperature Relationships

In an exchanger transferring both sensible and latent heat to or from one or both streams, particular care must be taken to determine the appropriate mean temperature difference by breaking the duty into a series of separate, easily calculated, portions. This is illustrated in Figure 3 by an extreme example encountered many years ago. The process involved heating a stream in an exchanger before it entered a reactor in which an exothermic reaction took place; after which the stream, with slightly altered composition, returned counterflow to itself through the same exchanger. With comfortable terminal temperature differences at both ends of the exchanger, the process engineer had apparently envisioned vs heat removal curves like those shown as solid lines. Unfortunately, the pressure loss through the reactor reduced the saturation temperature of the return stream below that of the entering stream as shown by the broken line. The exchanger could not have worked under the conditions specified to the heat exchanger designer because the central isothermal portion had a negative temperature difference between streams.

D. Maldistribution of Flow

If the flow is not uniformly distributed over the heat transfer surface or, still worse, if portions of the stream bypass the surface entirely, there are at least two adverse effects.

First, considering only the case of surface bypassing for purposes of example, the velocity and, hence, the film coefficient, h_o, are decreased by depar-ture of a portion of the total outside stream from the available flow area through the active tube surface.

Second, the bypass stream reaches the exchanger outlet uncooled (or unheated) in the worst conceivable case, and only there mixes with the active stream. The resultant outlet temperature, T_2, is given by

$$T_2 = xT_1 + (1 - x) \, T_2' = T_2' + x \, (T_1 - T_2') \; , \qquad (4)$$

where x is the fraction bypassed, T_1 is the inlet temperature, and T_2' is the outlet temperature of the active stream whose flow fraction is $(1 - x)$. Obviously, if the desired heat duty is to be accomplished, T_2' must be reduced (in the case of cooling) to

$$T_2' = T_2 - \frac{x(T_1 - T_2)}{(1 - x)} , \qquad (5)$$

and the effective mean temperature difference may be severely reduced, depending on the magnitudes of the bypass fraction, the cooling range, and the coolant inlet temperature.

These adverse effects on h_o and Δt_m may be drastic in the case of cooling viscous streams, as pointed out by Tinker [1] (*) more than 20 years ago, since the increasing viscosity of the active stream forces more and more of it to take the less viscous bypass route. It is appropriate to include Tinker's well known visualization of the various flow paths here as Figure 4.

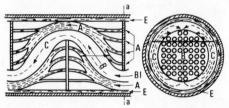

Fig. 4. Flow Paths of Various Streams through the Shell of a Cross-Baffled Exchanger (Courtesy of T. Tinker, and American Society of Mechanical Engineers)

The obvious cure for both these undesirable effects is to design so that bypass paths are minimized, especially those between baffle peripheries and the shell (axial bypass), and between the outermost tubes and the shell (cross-flow bypass). Untubed lanes through multi-pass tube bundles necessitated by pass-partitions in the heads may, depending on their orientation with respect to flow, also provide bypass paths. The means normally adopted to control bypass are anti-bypass strips or dummy tubes deliberately inserted into the tube layout, and judicious location of tie-rods at the tube bundle periphery.

Another cause of degradation of heat transfer capability due to flow distribution is the existence of stationary eddies (backwaters) which the main stream energizes but does not penetrate or sweep away. Such zones are probably not as prevalent as one thought except where sedimentary or particulate deposits may block the normal fluid escape routes through the annular orifices between baffles

---

(*) Superscripts refer to the References listed at the end of this treatise.

and tubes. Under such circumstances, not only the heat transfer coefficient and effective temperature difference suffer, but the pressure loss as well. While the heat exchanger designer may be helpful in minimizing such pockets, the system design should provide for necessary screens, filters, or traps.

E. Gas or Vapor Blanketing in Two-Phase Flow

Heat transfer surface may become insulated to varying degrees by accumulation of poorly conducting substances, gaseous or solid. The latter is discussed later under "Fouling."

1. Condensing with Noncondensable Gas

In condensing vapors in the presence of noncondensable gases, the gas concentration necessarily increases as condensation proceeds. The vapor can arrive at the heat transfer surface only by diffusion through a surface film at a rate dependent upon the difference between its partial pressure in the bulk of the stream and that at the tube wall, and dependent also on the effective thickness of the film, which varies inversely as a power of the velocity. The resultant gas blanketing toward the condenser outlet may reduce an initial high condensing coefficient almost to zero as the outlet is approached, unless appropriate design measures are taken to maintain a high flow velocity of the gas-rich vapor past the heat transfer surface. It is particularly important to integrate the $(U\Delta t)$ product carefully in such circumstances.

2. Vapor Blanketing in Boiling Equipment

Boiling equipment also may be subject to similar vapor blanketing and insulating phenomena, but for entirely different and less well understood reasons — sometimes with far more serious results than mere failure to provide the required rate of heat transfer.

One such phenomenon, known variously as "burn-out," "departure from nucleate boiling (DNB)," etc., results from generation of vapor bubbles from closely spaced nucleation sites at such a high volumetric rate compared to the surface on which they are generated that a vapor film is interposed between the surface and the liquid. The consequence is "film boiling," at a vastly reduced heat transfer coefficient as compared to nucleate boiling. This phenomenon is primarily associated with initially high heat flux and may occur even when the bulk temperature of the stream is below saturation temperature.

Film boiling may also be initiated by excessive metal temperature, as may be brought about by momentary deprivation of an adequate liquid supply to the heating surface. This "Leidenfrost effect" prevents a renewed liquid supply from wetting the surface. Dry zones thus created may either heal themselves, increase their extent, or remain stable, depending on the ability (or lack of it) of the surface wall to conduct heat away to adjacent nucleate boiling zones. Thus, unstable flow, whether excursive or periodic in nature, can lead to establishment of film boiling under average flow conditions which, if steady, would not cause departure from nucleate boiling, Once established, film boiling normally persists even though the heat flux, due to the poor coefficient, necessarily remains much lower than the critical value which may have caused it in the first place.

Another phenomenon, appropriately termed "dry-out," occurs as vaporization progresses to a point where much of the liquid content of the boiling stream exists in the form of droplets or mist, and the remainder is insufficient to wet the heating surface completely. This is primarily a function of vapor quality, although it varies also with many other parameters, including velocity, presence of centrifugal effects, pressure, and nature of fluid. The result of dry-out is, again, a very poor heat transfer coefficient which partakes of the nature of gas heating and, of course, reduces to a gas heating coefficient as complete vaporization is achieved.

Although much work has been done since the advent of the nuclear reactor, it cannot be said that a comprehensive knowledge exists on how to avoid vapor blanketing problems in boiling equipment for all liquids at all pressure, temperature, and flow conditions. In critical applications, recourse must be had to testing of model equipment or, since scaling up from models is not without its hazards, full-scale prototypes. In less critical equipment, liberal factors of safety on all parameters known to increase the likelihood of vapor blanketing are normally applied in the absence of actual test data. Examples are the restriction of heat flux to values of much lower than can probably be achieved, the deliberate introduction of high pressure loss into the water leg in boilers to avoid unstable flow, and the use of recirculating instead or "once-through" boilers to ensure against the problems of dry-out.

The major undesirable result other than loss of heat transfer capability, referred to earlier in this discussion on boiling, is tube failure caused directly or indirectly by overheat. Thermal stress fatigue caused by alternate local wetting and drying of the tube wall at the indeterminate mobile interface where dry-out occurs is also a matter of concern. This will be considered in greater detail later.

F. Fouling

Since Dr. J. Taborek, of Heat Transfer Research, Inc. (HTRI), will deal with fouling, in the next chapter, the following statements will be brief and historical rather than technical.

There is no need to expound on the fact that practically all nonmetallic solids are characterized by low thermal conductivity, and that if deposited on heat transfer surface by any mechanism whatsoever they constitute a heat transfer resistance, r_o and /or r_i, to the detriment of the overall heat transfer coefficient, U_o. Recognition of this effect first took the form of a "fouling factor," i.e., a multiplier less than unity applied to the overall coefficient for new, clean, equipment to provide some reasonable operating period before the apparatus was no longer capable of handling the specified heat load, and had to be shut down and cleaned. In those days, in view of the state of knowledge on heat transfer coefficients, a considerable component of the "factor" was necessarily a "factor of ignorance." It may be surprising that the phrase "in those days" was applicable into the 1950's for some classes of equipment—notably, surface condensers and feedwater heaters, as used in public utility installations.

Other more enlightened users and designers of heat transfer equipment in the chemical and petroleum fields recognized the problem for what it was in the early 1930's, or perhaps sooner — namely, a resistance that should be added in series to the other resistances in the heat flow path. The question then was how much to add for what fluid, at what temperature, at what velocity of flow. Individual users and designers set their own values based on their own experience, or lack thereof. Needless to say, the fouling resistances so chosen differed considerably, even for superficially identical conditions. At the time of my entry into the heat exchanger field in 1936, rating and design of such equipment was very similar to custom tailoring, depending as it did on the desires and whims of individual clients.

The Tubular Exchanger Manufacturers Association (TEMA) was founded in the late 1930's in an attempt to establish standards for high-quality shell-and-tube exchangers which would embrace as many as possible of the requirements of individual users and designers. The first edition of the TEMA Standards (1941) [2] included, among many other things, a listing of recommended fouling resistances for various services. Subsequent editions expanded the list based on such field experience as was available. The fourth (1959) edition, of which I was editor, made clear what should have been obvious from the beginning — namely, that the tabulated values of fouling resistance, constants in most cases, could not possibly be construed as taking into account the many physical and economic considerations necessary for optimization as outlined by Mueller [3] in 1954 in an invited lecture series at Purdue University. They are, rather, intended as guides to be followed in the absence of other more reliable data.

A semitheoretical approach to prediction of fouling was suggested in 1959 by D. Q. Kern and R. E. Seaton [4, 5] and developed further by Kern in 1966, [6] the basis being a deposition rate proportional to the mass flowrate and a suppression rate proportional to the shear of the flowing stream. A conclusion reached by Kern, et al., and also by C. H. Gilmour [7], was that it is frequently preferable to accept a high pressure loss, with its attendant higher pumping charges, to obtain the benefits of smaller equipment due to the combination of enhanced fluid film coefficient (h_o and/or h_i) and reduced fouling resistance (r_o and/or r_i).

The major scientific investigation of fouling sponsored by HTRI, and carried out in its laboratory and in portable field laboratories, will be dicussed by Dr. J. Taborek, as indicated previously. Suffice it to say here that this is, in this speaker's opinion, the most important work currently in progress from the stand-point of economic optimization of heat exchanger selection for specified services.

Fouling, like vapor blanketing, has other adverse effects than loss of heat transfer capabilities, especially when it occurs on the coolant side of high-temperature equipment. In this case, tube overheat and failure may be a hazard. Beyond this, porous or spongy deposits may act as concentration cells for corrosive agents which, in combination with a higher corrosion rate at elevated temperature, may also lead to tube failure. Some deposits, however, are beneficial if this, dense, and uniform. The thin, tightly adherent, magnetite film on steel boiler tubes is a good example in that it protects the underlying metal from the iron-water reaction which would otherwise proceed rapidly, and does so without any serious impedance to heat transfer. Finally, since pressure loss increases inversely as the fifth power of the diameter for flow inside tubes, fouling deposits there may, and often do have a severe adverse effect. The shell-side pressure loss may also be similarly affected.

The obvious cure for fouling problems, aside from that proposed by Kern, et al., is to eliminate the fouling agent from the stream, but this is not often practicable on process streams since the fluid itself is sometimes the fouling agent through coking or polymerizing reactions. With recirculating streams such as boiler water, jacket water, cooling tower water, and the like, however, there is no reason why fouling cannot be controlled by the user to standards imposed by the designer. For sedimenting deposits, the designer may frequently exercise some control over fouling by maintaining high velocity flow and avoiding abrupt changes in flow direction or velocity.

Brittle deposits such as calcium sulfate scale on tubes may be removed periodically in some types of specialized exchangers. If the scale is on the outside of bowed or coiled tubes, occasional thermal shocking will crack it off by flexure or differential thermal expansion. If it is on the inside of vertical tubes, a falling film type of design may be adopted which permits mechanical cleaning while operation continues. (The deposit need not be brittle in this case.)

Scraped surface designs with continuously rotating scraper blades are efficacious in the case of waxy or gelatinous deposits which would otherwise accumulate rapidly.

Beyond these cases, however, the designer can only rely on empirical fouling resistances recommended by TEMA and/or other sources, meanwhile attempting to learn more about the mechanism of fouling in specific circumstances. The knowledge so gained in laboratories and in the field will ultimately make it obvious how the system or component should be designed to minimize the fouling problem.

G. Closing Remarks on Rating

Little has been said here which is really new in principle. Leidenfrost's report on nonwetting of hot surfaces, for example, was written (in Latin) in 1756. Most deficiencies in exchanger rating are not due to ignorance of fundamental principles, nor even of lack of adequate data. They are, rather, the result of failure to apply what is known and to make conservative allowance for what is unknown. However, to the extent that provision of more surface than is required may be considered an economic deficiency in rating, these remarks do not apply.

III. DEFICIENCIES AS A CONTAINMENT AND PRESSURE VESSEL

A vessel designed for the purpose of containing (or excluding) fluids under pressure is quite obviously deficient in its function if any path opens between the interior of the vessel and ambient including, in the case of heat exchangers, paths between one fluid and the other. The result may range in seriousness from a minor nuisance to near disaster, depending upon the nature of the contained fluids and size and stability of the leakage path. Contrast, for example, the consequences of a leaky gasketed joint in a lubricating oil cooler with a tube rupture in a liquid- sodium-heated steam generator — an untidy spill of water or oil vs an almost explosive chemical reaction, which must be contained.

Various causes of failure and methods for their prevention are considered in what follows.

A. Mechanical Design

1. Pressure Vessel Codes and Standards

Pressure vessels are normally designed for an acceptably safe level of stress at the most severe combination of pressure and temperature anticipated in operation. It is customary to recognize that bending stresses and self-relieving stresses, such as those caused by restrained thermal expansion, may be permitted higher levels than stresses due to pressure and dead-weight loads, for which no self-relieving mechanism exists.

There are various standards and codes in existence throughout the world for the design of pressure vessels, in which allowable stresses are specified as a function of alloy, metal temperature, and nature of stress, and which also prescribe the design rules for specific vessel components. In the United States, at least, these codes are imposed by the separate states and some cities, and specified by most individual users. Almost all derive from the Boiler and Pressure Vessel Code of the American Society of Mechanical Engineers (ASME Code). [8] Practically all ordinary pressure vessel components are common with those of heat exchangers, so mechanical design for safety, in view of available codes and standards, is not normally a problem.

2. Heat Exchanger Supplements to Pressure Vessel Codes

There are exceptions or limitations to the design coverage for heat exchangers provided by the codes, mostly from the standpoint of ruggedness, operability, and maintainability, rather than that of safety. For example, a flat cover plate may deflect enough under pressure to permit pass-partition gasket bypassing between inlet and outlet compartments of the tube-side stream, even if it is in no danger of failure as a pressure vessel component. The TEMA standards, referred to before, provide design rules supplementing those of the ASME Code for these purposes; so also do the standards of the Heat Exchange Institute (HEI). [9] Both also set forth design rules for tubesheets; these are curiously lacking from the ASME Code, but not from other national and international standards, e.g., the British BS-1500, [10] the Dutch Stoomwezen rules, [11] and the forthcoming International Standards Organization (ISO) Pressure Vessel Standards. [11] (There are undoubtedly others with which I am not familiar.)

3. Differential Thermal Expansion

A fairly common failing in heat exchanger design is the specification of fixed tubesheets without sufficient thought to the possible consequences in operation. The economic temptation to this design (tubesheets integral with the shell) is considerable, since it is inexpensive compared to the floating head, removable tube bundle, alternatives with the same tube surface. If the shell-side stream is fouling, however, or corrosive, there is no access for thorough inspection or mechanical cleaning, although chemical cleaning may be used for some deposits. (In general, fouling streams should be directed through the tubes, not the shell, in any heat exchanger. Similarly, high-pressure streams should usually be through the tubes to avoid design for collapse. These two rules may occasionally conflict.) The main problem with fixed tubesheet exchangers, of course, is the restraint of free thermal expansion of both shell and tubes which may lead to buckled tubes or slipped tube joints, and ulimate leakage of failure. The use of an expansion joint in the shell will alleviate this situation, but at the risk of suffering a failure of this nonreplaceable part. More often than not, however, difficulties occur in fixed tubesheet exchanges which have no apparent need for expansion joints at the conditions for which they were calculated — normal operating conditions. The trouble usually is that a more severe condition has been overlooked — possibly startup, shutdown, or some process upset or emergency.

Even floating-head exchangers are not inherently immune to the differential expansion problems just mentioned since excessive temperature range on the tube side of multipass units may set up similar destructive stresses and/or tube joint loads between adjacent tube passes. Quadrant-type, four-pass tube layouts are particularly susceptible to this difficulty, since the hottest and coldest tube passes are next to one another. The same condition in fixed tubesheet exchangers, of course, may aggravate a situation which is already marginal. A "rule of thumb" frequently used is that the average metal temperature difference between adjacent tube passes should not exceed 50° F (28° C). Figure 5 illustrates the distinction between "ribbon" flow and quadrant flow, with some postulated average metal temperatures of the tube passes. While the ribbon flow tube pass arrangement does not violate the 50° F limitation, the quadrant layout exceeds it by a factor of three; the mixed flow arrangement exceeds it by a factor of two.

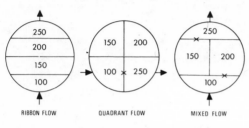

RIBBON FLOW   QUADRANT FLOW   MIXED FLOW

Fig. 5. Various Four-Pass Tube Layouts, Showing Undesirable Juxtaposition of Hot and Cold Passes for Quadrant and Mixed Quadrant and Ribbon Flow (x).

4. Vibration

A serious cause of heat exchanger failure as a pressure vessel is vibration of its components, most particularly of its tubes. This may lead to failure by fatigue due to the attendant cyclic stress reversals, or to thinning of component walls by fretting wastage as a result of impact or rubbing of parts in contact with one another, such as tubes within baffle holes. The vibration may be mechanically transmitted from other sources, as may happen during shipment or in shipboard or other compact machinery installations where the carrier may be a common supporting structure or the piping to rotating or reciprocating machinery. It may also be fluid transmitted, as in the pulsating stream from a reciprocating compressor. Finally, and most worrisome, vibration may be induced within the exchanger itself by a perfectly steady flow.

Flow-induced vibration is worrisome because it is not thoroughly understood, although much has been done within recent years, both theoretically and experimentally, to elucidate the phenomena involved, of which there are several. Most familiar is the alternate shedding of eddies from the downstream side of tubes and other obstacles in the flow stream, with resultant periodically alternating lift and drag forces on the obstacles. Whether the eddies have the nature of the von Karman vortices shed by single cylinders in a stream when multiple cylinders are involved, as in a tube bundle, is not so important as the demonstrated facts that:

1) Periodic forces are exerted on the tubes in tube bundles above some minimum Reynolds number.

2) The frequency of these foces increases linearly with the cross-flow velocity over a wide range of Reynolds numbers, thus implying the constancy of the dimensionless Strouhal number, $S = (\nu d_o/V)$, in which ν is the frequency, V the flow velocity, and d_o the outside diameter of the tubes. Although the Strouhal number is essentially independent of the Reynolds number up to approximately 200,000, it does depend on the tube arrangement and the spacing-diameter ratios in the transverse and longitudinal flow directions as shown by Chen [13] and Gregorig and co-workers. [14, 15, 16] The Strouhal number itself may range from 0.1 to 0.5, but for customary tube layouts the range is of the order of 0.2 to 0.3.

3) As the Reynolds number is increased above approximately 200,000, the energy associated with periodic eddy shedding becomes overshadowed by the increasing turbulent energy over a wide spectrum of frequencies and ultimately becomes imperceptible. The random buffeting of turbulent eddies near the natural frequency of the tubes, however, is still capable of exciting tube vibration and increasing its amplitude with increasing intensity of turbulence.

The vibrational amplitude, and hence the potential for causing damage, is a function of the forcing frequency, the tube natural frequency, the degree of damping, and the static deflection under a steady load of the same magnitude as the maximum of the periodically varying applied load. Figure 6, from den Hartog, [17] shows in dimensionless form the nature of the relationship between these variables. C is a measure of the degree of viscous damping, C_c being the value of critical damping above which vibration is not sustained. The ratio shown as abscissa is that of the exciting frequency to the natural frequency. The ordinate gives the ratio of maximum amplitude to static deflection.

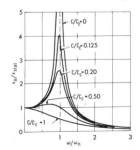

Fig. 6. Amplitudes of Forced Tube Vibration for Various Degrees of Damping (den Hartog).

Inspection of Figure 6 makes clear in a quantitative manner the well known fact that a resonant condition exists when the forcing and natural frequencies are nearly the same, and that such a situation is to be avoided at all costs unless a high degree of damping is available. For a heat exchanger subject to variable flowrate, it is also seen that it is desirable that the forcing frequency be less than, say, 75% of the tube natural frequency, and preferably less. This is not to say that operation at 125%, or more, would necessarily be any more hazardous, although a higher harmonic of the tube frequency might be excited. Nevertheless, to reach 125% it is necessary to pass through 100% as the flow is brought up to full load and, if steady operation at part load is required, it could readily fall at a resonant condition.

If the maximum cross-flow velocity is known, the forcing frequency is readily determined from the Strouhal number appropriate to the tube layout, thus:

$$\nu = SV/d_o \quad . \tag{6}$$

One difficulty is that V cannot be determined accurately without some sort of stream analysis. (Of course, if the full stream is assumed to go in cross-flow between adjacent baffles, and the resultant frequency is still sufficiently less than the tubes' natural frequency, this problem does not exist.)

The tube natural frequency is given by:

$$\nu_o = \frac{C_1}{P^2} \sqrt{\frac{EIg_c}{m}} \ , \tag{7}$$

where

P = baffle spacing
E = elastic modulus of tube material
I = moment of inertia of the tube cross section
g_c = gravitational constant
m = mass per unit length of tube wall, tube contents, and virtual mass of the shell fluid,

all in consistent units. This constant, C, is normally taken as 1.57 for the tube spans between adjacent baffles and 2.45 for the spans between end baffles and tubesheets, considered, respectively, as simply supported at both ends, and built in at one end and simply supported at the other. These values are for the first mode only.

Other things being equal, the cross-flow velocity in Equation (6) varies inversely as the baffle spacing, P, and so does the forcing frequency, ν. Since the natural frequency, ν_o, varies inversely as P squared, the frequency ratio of interest for use in Figure 6, ν/ν_o, varies directly as P. All other thinds are not equal, however because the fraction of the total shell stream which goes in corss-flow decreases as P decreases, but not in a directly proportional fashion. Thus, to decrease v/v_0 by a factor of two does not require halving the baffle spacing. Frequently the reduction of P is unacceptable because of the increased pressure loss which results; in such cases recourse must be had to other baffle arrangements, as will be discussed later.

With the foregoing rather oversimplified background information, consideration may be given to some of the options available to the designer to avoid vibration problems. For mechanically or fluid-transmitted vibrations, the choice of baffle spacing and clearances normally provide sufficient freedom in determining the natural frequency of the tubes and their maximum deflection. The latter can also be reduced by alternating the natural frequency of adjoining tube spans by using non-uniform baffle spacing so that each tube span exerts a snubbing influence on its neighbors. For flo-induced vibration, a brief historical comment will be of interest.

Whatever vibration problems that may have existed in the infancy of the heat exchanger industry had been essentially eliminated by restraints imposed on maximum unsupported tube spans and tube-to-baffle clearances. These were codified by the TEMA standards [2] at a time when shell diameters of exchangers rarely exceeded 40 in. (1000 mm). The rules adopted were basically empirical, and no velocity restraints were imposed. Nevertheless, the rules worked and tube vibration problems were very seldom encountered over a period of 20 or more years.

In the early 1950's, for the first time, the nuclear power industry demanded the transfer of vast flows of heat, not through water tube boilers as in the past, but through heat exchangers. As a result, in order to reduce the number of shells required, the diameters of the heat exchangers designed increased by a factor of 2 or more, i.e., to 80 to 100 in. (2000 to 2500 mm). The designers, as often as not, were architect-engineer employees without previous experience or intuitive sense of propriety in design. The trouble began in the late 1950's as these items of equipment finally came into operation at specified flowrates, whereupon a disturbing incidence of tube failures by vibration became evident.

Time and space do not permit pursuing this size effect on vibration in detail. The basic reason for the problem was, however, that years of essentially trouble-free operation of small to medium-sized exchangers — with little or no thought of velocity limitations — had led to an unwarranted sense of security in relying only on the TEMA-specified maximum unsupported tube spans. The greatly increased flows per shell necessitated considerably greater velocities with this constraint on baffle spacing, and the ratios of v/v_0 approached or passed the critical value of unity.

One method of reducing cross-flow velocities, and hence exciting frequencies, is shown in Figure 7. This, and its variants, effectively divides the shell stream into parallel paths within the exchanger, each having only a fraction (in this case, 1/2) of the cross-flow velocity it would have with ordinary segment-cut baffles. The net effect is to create a no-flow condition across a chordal (diametral, in this case) plane which has the same effect as an actual partition. This is even more effective in reducing pressure loss, since the number of tube layers crossed is also reduced.

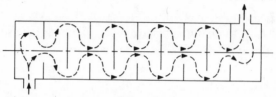

Fig. 7. Divided Flow Design to Reduce Cross-Flow Velocities.

B. Defective Materials and Workmanship

It might be thought that the designer has no control over shortcomings of a heat exchanger as a pressure vessel caused by defective materials or faulty workmanship — that this responsibility rests with the materials vendors and with the fabrication shop. While this may be true in a legalistic sense, the designer's conscience cannot be clear if he has imposed a configuration on the shop or the inspectors which strains the bounds of current technology or non-destructive testing. When design innovations are contracted for, engineering, shop, and inspection departments should all be in agreement that the design is feasible and the integrity of the product verifiable.

C. Suitability of Materials to Fluid Environment

It is obvious that the materials of construction of heat exchangers should retain their integrity in the fluid environment to which they are exposed, if they are to survive as pressure vessels. Parts such as tubes and floating heads are exposed to two fluids, and a material compatible with both sometimes does not exist. To the extent that oxidation is a problem, similar considerations may apply to external components such as the shell and heads. Loss of integrity may occur in various ways, as discussed in the following.

1. Wastage

Wastage is a generic term for actual loss of metal encompassing the mechanisms of corrosion, erosion, wear, and fretting, but not degradation of physical properties.

Corrosion reduces the thickness of pressure parts by chemical reaction with an aggressive agent in the ambient fluid(s). It may be uniform in nature, or of the pitting type. If uniform, and at a slow rate, it is often acceptable to use inexpensive materials such as carbon steel with excess thickness of parts, i.e., a corrosion allowance. With pitting corrosion, however, it is necessary either to use a nobler alloy, full thickness or as a cladding, or to treat the fluid to remove the aggressive agent as is done with boiler water treatment, for example. Cladding may of course be used as a substitute for uniform corrosion allowance. It should be understood that the cladding need not be metallic in all cases; Teflon coatings, or the like, are often used.

Erosion, sometimes abetted by corrosion, involves the physical removal of material by such mechanisms as scouring and cavitation. It is associated with areas of high flow velocity and/or sharp changes in flow direction, usually involving two-phase flows. Although some metals have superior erosion resistance, it is seldom warranted to use them for entire components, since erosive effects are so highly localized. I do not feel qualified to offer advice on erosion prevention beyond the obvious items of avoiding high velocities in two-phase streams, providing sacrifical excess material in areas of high velocity impingement, and streamlining entrance regions.

Wear and fretting have been discussed earlier from the standpoint of prevention by controlling vibration by proper design. It may be added here that wastage by these mechanisms is minimized by the use of materials of similar chemical composition and physical hardness for contacting parts susceptible to rubbing and impact. Admiralty metal tubes and commercial brass baffles are compatible from this standpoint; aluminum tubes and steel baffles are not.

2. Degradation

Degradation, as used here, refers to the loss of desirable physical properties, such as strength and ductility, without actual loss of material. This can come about by such mechanisms as corrosion embrittlement, mass transfer of alloy constituents, and fatigue, all affecting the metallurgical structure of components.

Embrittlement is known by many names, in American usage at least, depending upon the alloy affected. With brasses, it is called "season cracking," and with austenitic steels, "intergranular corrosion." Quite low concentrations of corrosive agents, differing from one matrial to another, are capable of reducing an originally sound ductile material to a mass of weakly joined crystals much like a lump of sugar, and with about as much strength. In ferritic steels, "caustic embrittlement" and "hydrogen embrittlement" have similar effects but, in this case, usually require some concentrating mechanism. Local dry-out in crevices in stressed areas such as tube joints in exchangers will serve this purpose in the case of caustic. Corrosion deposits in an acidic environment set up concentration cells for hydrogen attack.

"Stress corrosion cracking" is another mechanism for degradation, usually associated with austenitic materials but not necessarily confined to these alloys. The failure in this case is by extensive cracking networks, usually transgranular, originating at areas of high tensile stress in the presence of specific corrosive agents above some minimum concentration. The agent need not be present in the environment at this triggering concentration; crevices, again, often serve to provide the necessary strengthening mechanism. In the case of austenitic steels, chloride ion in the presence of oxygen, or hydroxyl ion, are the most troublesome agents.

As can be seen from what has been said to this point, material degradation as a result of corrosive agents in the environment is best attacked in the design phase by elimination of crevices and pockets where corrosion products can accumulate and, of course, by proper choice of materials.

Mechanical fatigue results from repeated cycling of strain through a range usually, but not necessarily, exceeding twice the strain at yield. The methods for control of vibrational amplitudes, and hence bending strain, have been discussed previously. Thermal fatigue results from repeated excessive strains due to restrained differential expansion within various components of a heat exchanger because of thermal transients in operation. Thermal shields or sleeves are effective in reducing the severity of transients, but obviously cannot be applied to tubes.

Although seldom encountered, and perhaps peculiar to liquid metal heat transfer systems, the mass diffusion of alloy constituents as a result of thermal gradients should be mentioned under metal degradation. If the system includes both austenitic and low-alloy ferritic material, even though the latter is at a lower temperature, there is a diffusion of carbon out of the ferritic zone into the liquid metal and, form there, into the austenitic zone. The consequence is reduction of strength of the ferritic material and reduction of ductility of the austenitic. The designer should make allowance for this in the permissible stresses or utilize higher alloy ferritic materials less susceptible to carbon transfer.

3. Crevices and Pockets

Although it should be amply evident from all that has been said so far, it is worth repeating that no-flow pockets and crevices are to be avoided in design for many reasons. One that has not been mentioned so far is the tendency of some organic streams to coke progressively in hot zones. The build-up of coke in hot tube joints has been known to collapse the tubes in many installations due to its exertion of constantly increasing external pressure. In any instance where tube joints do not require expansion throughout the full tubesheet thickness for the sake of holding power or leak tightness, consideration should be given to a "back-face" expansion to prevent access of the shell fluid to the crevices between tubes and tube hole walls. In many cases it is not a matter of concern; in others, it may be essential.

D.   Interleakage of Heat Exchanging Streams

If contamination or intermixing of one heat-exchanging stream with another must absolutely be avoided for any of various reasons, there are several design options available. Among these are double tubesheets and double wall tubes. The designer, here, attacks the consequences of leakage in addition to normal provisions against leakage.

1. Double Tubesheets

When tube walls are in little danger of failure, but tube joint reliability is mistrusted, the use of closely spaced tubesheets with tubes expanded into both is indicated. The space between them may be used to monitor for leakage of either fluid through the tube joints; in the case of floating heads, at least one tube must be used as communication between the intermediate space at the floating end and that at the stationary end. The possibility of differential radial thermal expansion between both tubesheets of a pair must be given special consideration to assure that this construction does not create more problems than it solves.

2. Double Tubes

If both tube walls and tube joints are considered interleakage risks, concentric tubes may be used, each connected to one of the tubesheets of a pair, with monitoring space between tubesheets as described before. Thermal coupling between inner and outer tubes may be by direct metallic contact (with leakage detection grooves), or by an intermediate fluid such as mercury or sodium-potassium eutectic. Needless to say, these constructions are not often adopted, but both have been used succesfully in nuclear steam generators.

One other application of double tubes does not require concentricity, but utilizes instead the conductivity of closely spaced metallic fins into which both tubes are expanded for thermal coupling.

Concentric bimetal tubes may also be the only solution for corrosion problems where no single metal has the necessary resistance to attack by both fluids in a heat exchanger. When used for this purpose, the tubes are drawn down to size together to provide compressive metal-to-metal contact. In this case double tube-sheets are not necessary; instead, the inner tubes are expanded into cladding material on the tube-side faces of the tubesheets.

IV. CLOSING REMARKS

In conclusion, it is hoped that this discussion of possible operating difficulties with heat exchangers has been instructive in the aggregate, even though no single topic has been treated in any depth. It should be evident that mastery of heat transfer and fluid mechanics alone is not sufficient to ensure a satisfactory design. Neither is expertise in any one of the several other disciplines involved. Since the individual who is master of them all is rare indeed, a close cooperation and mutual understanding among experts is necessary for all but the most routine design conditions.

REFERENCES

[1]   Tinker, T., "Proceedings of the General Discussion on Heat Transfer," Institution of Mechanical Engineers, London, and American Society of Mechanical Engineers, New York (1951), p. 84, 97, 110

[2]   "Standards of Tubular Exchanger Manufacturers Association," New York, N.Y. (First, 1941, through Fifth, 1968, Editions).

[3]   Mueller, A. C. "Thermal Design of Heat Exchangers," Research Series No. 121, Engineering Experiment Station, Purdue University, Lafayette, Indiana (1954), p. 47-50.

[4]   Kern, D. Q., and Seaton, R. E., "A Theoretical Analysis of Thermal Surface Fouling," Brit. Chem. Eng., Vol. 4, No. 5 (1959), p. 258.

[5]   Kern, D. Q., and Seaton, R. E., "Surface Fouling, How to Calculate Limits," Chem. Eng. Progress, Vol. 55, No. 6 (1959), p. 71.

[6]   Kern, D. Q., "Heat Exchanger Design for Fouling Services," Proceedings of the Third International Heat Transfer Conference, 1966, Vol. 1, p.170-178.

[7]   Gilmour, C. H., "No Fooling, No Fouling," Chem. Eng. Progress (July 1965), p.61.

[8]   "Rules for Construction of Pressure Vessels, ASME Boiler and Pressure Vessel Code, Section VIII," The American Society of Mechanical Engineers, New York, N. Y. (1971).

[9]   "Heat Exchange Institute Standards for Closed Feedwater Heaters," New York, N. Y., First Edition (1968).

[10] "British Standard 1500, Fusion Welded Pressure Vessels," British Standards Institution, London, England (1958).

[11a] "Regulation on Which the Assessment of Construction and Materials of Steam Apparatus, Vapor Apparatus on Pressure Vessels is Based," Staatsuitgeverij, The Hague, Netherlands.

[11b] "Calculation of Circular Tubesheets of Heat Exchangers With a Fixed and Floating Tubesheet and of Heat Exchangers With Two Fixed Tubesheets," Staatsuitgeverij, The Hague, Netherlands.

[12] First Draft Proposal, "Pressure Vessels," International Standards Organization, Document ISO/TC 11 (recr.-76) 248 EF (October 1968).

[13] Chen, Y. S., "Flow-Induced Vibration and Noise in Tube-Bank Heat Exchanger Due to von Karman Streets," Trans. ASME, Vol. 90, Series B (1968), p. 134- 146.

[14] Gregorig R., and Andritzky, H. K. M., "A Criterion for Vibration in Transverse Flow Over a Tube, Part 1," Chem. Ingr. Tech., Vol.39 (1967), p.894-900.

[15] Andritzky, H. K. M., and Gregorig, R., "A Criterion for Vibration in Transverse Flow Over a Tube, Part 2," Chem. Ingr. Tech., Vol. 40 (1968), p. 483-488.

[16] König, A., and Gregorig, R., "A Criterion for Vibration in Transverse Flow Over a Tube, Part 3, Vibration Experiments in the Nest of Tubes," Chem. Ingr. Tech., Vol. 40 (1968), p. 645-650.

Chapter 3

DESIGN METHODS FOR HEAT TRANSFER EQUIPMENT— A CRITICAL SURVEY OF THE STATE-OF-THE-ART

J. Taborek (*)

1. GENERAL COMMENTS AND OBJECTIVES

In this chapter we shall attempt to summarize some of the most important aspects related to the process of designing functional heat transfer equipment. By its very nature, this problem involves bridging the gap between a research type of analysis and the most practical level of engineering. If we consider the responsibilities facing a designer of heat transfer equipment, it is not surprising that the methods of design which are actually employed are inherently conservative and rarely reflect the latest status of academic research. In this sense, the tremendous worldwide effort in heat transfer research which is in progress produces results which rarely reflect directly the needs which a heat transfer equipment designer must face. Even in instances where research efforts are pertinent to practical design problems, the form of the results is often left in a state which the designer is reluctant or outright unable to employ.

We shall try to demonstrate that this does not need to be so and that research results can be meaningfully incorporated into the practice of heat transfer equipment design. Although limited by the allotted time and scope, examples have been selected with the particular hope of demonstrating the potentially effective results of such an attitude.

2. TUBESIDE FLOW

2.1 Tubeside Heat Transfer (Turbulent Flow)

This is the most fundamental problem in heat transfer and has been subject to research work since Reynolds recognized the basic principles in the

(*) Heat Transfer Research, Inc. (HTRI) Alhambra, California

1870's. Unfortunately, we are still searching for an adequate solution. Two methods of attack on this problem presently exist:

a) Analytical, attempting a fundamental solution of the basic equations describing the velocity and thermal profiles. Even though representing an ideal and ultimate solution, these attempts have been unsuccessful so far with respect to practical design considerations, especially if variable physical properties are considered.

b) Empirical solutions are generally accepted for industrial design practices which have to accomodate a wide variety of fluids. The basic correlation form can be expressed as

$$(Nu)_x = C(Re)_x^m (Pr)_x^n \phi^p \qquad (1)$$

where   C and m   are interrelated constants, particularly fitting a given data set (e.g., m = 0.7 to 0.9)

n   is usually presented as a constant varying between 0.33 and 0.6

ϕ_p accounts for the flow and thermal profile distortion between heating and cooling and is usually expressed as viscosity (or Prandtl number) ratio to a certain power for liquids or absolute temperature ratio for gases

x   is reference temperature to the physical properties used in evaluation of the dimensionless numbers. Bulk, film, wall and intermediate temperatures have been used in numerous published correlations.

In essence, the dilemma is between *accuracy* of a correlation and its ability to represent a *wide variety* of fluids under heating and cooling conditions. The large number of published correlations, usually based on limited data sets, makes the choice for an industrial designer difficult, with no authoritative evaluation offered anywhere.

For a limited range of physical properties, almost ideal solutions can be obtained. However, with spread of the physical properties representing a typical petro-chemical application, the error band produced by the best of this type of equations is ±40 percent, a rather unsatisfactory state of affairs considering the basic nature of the problem.

Newest developments suggest that the empirical equation form (Eq. 1) should be modified as follows:

$$Nu = C_1 \left[C_2 + f\,(Re)^m\,f\,(Pr)^n \right] \phi^p \qquad (2)$$

where $f(Re)^m$ reflects behavior of fluid group (such as water versus organics)

$f(Pr)^n$ becomes function of Pr itself as required by data and theory, assuming usually the form, $n = C_3 - C_4\,ln\,Pr$

ϕ_p becomes function (as exponent p) of Reynolds number and possibly the viscosity slope relations (for liquids)

This equation form accomodates a greater variation of physical properties and, therefore, more general applicability. The error range still remains at ±25 percent. Special fluid groups, e.g., water, still may require different correlations for best results.

A variety of solutions of the empirical-type tubeside equation is to remain in publications for many years to come because definite analytical solution acceptable for industrial applications is not in sight. What can be done in the meantime and what appears as a genuine challenge to the international community of heat transfer researchers can be summarized as follows:

a) Comprehensive evaluation of all the published empirical equations on a wide variety of data sets, establishing clear trends for future investigators.

b) Systematic attack on the outstanding problem of the heating-cooling correction factor, ϕ, which (originally derived by Sieder-Tate as the bulk-to-wall viscosity ratio from purely empirical data plot considerations) acts presently in correlational work as a catchall for the accumulated errors and is probably the least understood factor in the type of Eq. 2 if general validity to a wide quality of fluids is considered.

The importance of this work is further emphasized by the fact that tubeside derived correlational principles are adopted as a guide to other heat transfer geometry systems, such as flow across tube banks, flow in rectangular ducts, finned tubes and others. If we can not solve to reasonable perfection tubeside flow, the same gaps will obscure other systems correlations.

2.2 Tubeside Flow Heat Transfer (Laminar Flow)

Laminar flow inside tubes is of great industrial importance in the petro-chemical industry. While a large bulk of publications on this subject exists, it cannot be claimed that the problem has been resolved. Classical analytical solutions all failed to produce dependable design methods. Through numerical integration of the nonlinear partial differential equations, solutions may be possible with the help of extremely fast digital computers but will remain in the forseeable future impractical for industrial applications.

A number of empirical or semi-empirical equations were published and used in practice. Contrary to turbulent flow, the basis of the pertinent temperature difference must be observed. In the familiar equation

$$Q = h A \Delta T \tag{3}$$

the term ΔT may assume various definitions as long as it is consistent with the development and definition of h. The two most common cases for laminar flow are
 a) parabolic velocity profile and arithmetic mean temperature difference
 b) parabolic velocity profile and logarithm mean temperature difference (LMTD)

The two cases plotted as Nu versus Gz appear as shown in Figure 1. It

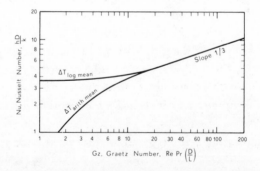

Fig. 1. Theoretical Average Ideal Laminar Heat Transfer Coefficients.

can be noticed that above $Gz \approx 20$ the two cases coincide. Below the critical Gz however, considerable differences may exist, as the log mean based solution transits into a constant Nu. This basic difference is rarely explained in text books.

The most popular and simple to solve correlation is due to Sieder and

Tate which (frequently unnoticed by the user) assumes the arithmetic mean ΔT

$$(Nu)_{arith\ mean} = 2.02 \left(RePr\ \frac{D}{L}\right)^{1/3} \left(\frac{\mu_b}{\mu_w}\right)^{1/4} \tag{4}$$

Here the difference between the theoretically derived constant 1.75 from the Graetz solution and 2.02 used was explained by natural convection which was otherwise neglected. As can be readily seen, for low Gz numbers (very large L) the equation if used with LMTD will break down.

Typical of the LMTD based solutions is that of Hausen [1]

$$Nu_{LMTD} = \left[3.66 + f\left(Re,Pr\ \frac{D}{L}\right)\right]\left(\frac{\mu}{\mu_w}\right)^{0.14} \tag{5}$$

which transits into the theoretically justified constant Nu if $Gz \to 0$. However, in this form, the equation does not account for natural convection and substantial errors will result under some conditions.

As a result of our extensive investigations, a new correlation type was developed, represented in principle as

$$Nu_{LMTD} = C + A\,(Re^*)^m\,(Pr)^n\left(\frac{D}{L}\right)^r\left(\frac{\mu_b}{\mu_w}\right)^p \tag{6}$$

where Re^* is a vectorial summation of Re and Gr.

Notice that the powers on Re^*, Pr and (D/L) are different, due to best results obtained by progressive regression analysis. About 85 percent of the available data was predicted within ±30 percent by Eq. 6. This compares with only 35 percent of the data within the same error level predicted by the Hansen equation. However, considerable further work remains to be done, especially as numerical solutions of the analytical forms of laminar flow will suggest improved combinations of correlational groups.

Somewhat different solutions will be obtained for vertical (up or down) flow and cooling or heating as the temperature profile changes for each case. Very little work has been done to date on these problems and errors will frequently reach ±100 percent levels.

The problems of laminar flow are further aggravated by the potential effects of maldistribution which can frequently cause poor functioning of multitube heat transfer equipment. In laminar flow *heating* of viscous liquids, a stable flow system usually results, but for *cooling*, the system is *inherently unstable*. If for whatever reason the flow or temperature in one tube becomes less than that of the remaining ones (possible effect of shellside stream maldistribution), a lower heat transfer coefficient will result with progressive decrease of the flow until a possible "freeze-up" results. Design of multitube viscous liquid coolers requires therefore some precautions as to equalization of flow distribution and potentially as a "maldistribution safety factor," a problem which is completely ignored by present research.

2.3 Pressure Drop Inside Tubes

Isothermal turbulent flow friction factor appears to be represented quite correctly in the customary forms. However, for nonisothermal applications, the use of the Sieder-Tate bulk-to-wall viscosity ratio factor to 0.14 power appears to be an empirical simplification worthy of further investigation.

In laminar isothermal flow, the classical solution of f-factor versus Reynolds number appears to hold quite well. The problem becomes complicated for *nonisothermal* conditions and effects of *natural convection*. Several different approach methods exist, none producing solutions of desirable accuracy. In a recent test against a large data bank, the writer determined considerable errors (up to 300 percent) especially in deep laminar flow cooling, using presently accepted method with the viscosity ratio to a constant power. A modified correction function was found necessary, expressed as a function of the isothermal friction factor, f_{iso}, and (ϕ^P, ψ^r):

$$f = f_{iso} \phi^P \psi^r \tag{7}$$

where ϕ^P = the μ/μ_w ratio, p *not* being a constant but a hyperbolic function

   ψ = the effect of natural convection expressed as a function of the group (Pr, Gr)

In transition flow, the ϕ function has to blend into turbulent correlations while the natural convection correction factor ψ has to decrease to 1.0. The solution of this problem is illustrated in principle in Fig. 2.

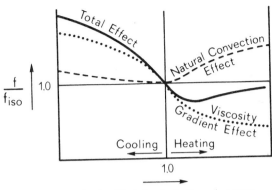

Fig. 2. Ratio of Actual Friction Factor to Theoretical Isothermal Friction Factor as Function of the Wall-to-Bulk Temperature Ratio.

2.4 Transition Flow Heat Transfer

Transition flow assumes considerable industrial importance as in many cases it can not be avoided because of other design considerations as well as that in many exchangers the flow transits from turbulent to laminar between the inlet and outlet of the exchanger. A survey of literature produced seven existing correlations for the transition region, represented by four different typical methods of approach:

a) Graphical connection of the laminar and transitional correlations plotted as Nu/Pr versus Re. This method is possible only if natural convection effects are neglected in the laminar correlation and both laminar and turbulent equations are fully compatible in all other terms. (Seider-Tate and Kern). [5]

b) Extensions of the turbulent correlation into transition range. (Hausen and Kuznetzov) [1, 22] The disadvantage is that no connection is established to the laminar range correlation, thus leaving a potentially disturbing step-function.

c) A unique, independent empirical correlation is developed for the transition region. (Metais and Norris) [23, 24] The disadvantage of this method is that a smooth connection is not established to either turbulent or laminar range equations.

d) Semitheoretical correlation is developed with validity for transition region only. (Gill) [25] (Same comments as for (c)).

As it is obvious that each of the presently existing methods shows some basic disadvantage, especially the disturbing step function at the boundaries, a new and much more promising method is proposed, defined as a proration function between the laminar and turbulent conditions

$$Nu_{trans} = \eta \, Nu_{lam} + (1 - \eta) \, Nu_{turb} \qquad (8)$$

where Nu_{lam} and Nu_{turb} are evaluated at Re = 2000 and 10000 respectively and η is an empirical function to be determined from data.

Thus, the natural convection effects and the L/D term typical to laminar flow blend systematically into the turbulent regime (where both effects are negligible) disregarding the type of equation used in either regime.

2.5 Flow Patterns in a Shell-and-Tube Exchanger

An additional problem in pressure drop prediction exists because of the design of a typical shell-and-tube exchanger. The flow path on the tube side of a shell-and-tube exchanger is rather complicated — a fact which is frequently not realized and not taken into proper account in the performance calculations. The figure below illustrates the main principles for a 4-tube pass internal floating heat construction.

Fig. 3. Flow Pattern in a Shell-and-Tube Exchanger (4-Tube Passes).

The following pressure drop stages are encountered:

(1) The inlet nozzle loss (stream expansion), function of design geometry
(2) Contraction loss at the entry into the first tube pass
(3) Expansion, turnaround and contraction in the floating head (if applicable) or turnaround only in U-tube bundles
(4) Expansion, turnaround and contraction in the channel
(5) Expansion from the last tube pass into the channel

(6) The exit nozzle loss (stream contraction)

(7) The frictional loss in the tubes proper

It is interesting to notice that the various pressure losses are never dealt with in heat transfer texts even though their magnitude can be considerable or even predominant with respect to the tubular friction loss proper, especially with low pressure gases and similar streams. In gross approximation, between one-half and two velocity head pressure loss is encountered at each of the above points. Data or dependable predictive methods are practically nonexistent outside of proprietary sources.

For shellside flow, the problems are even less understood as the flow from the nozzle enters a tube field with difficult to specify geometry. Nevertheless, the measured pressure drops in the nozzle-to-tube field transition can attain as much as 50 percent of the total pressure drop in the exchanger proper. Dependable data are nonexistent outside of proprietary sources and much more systematic work remains as a challenge to practically oriented researchers.

3. FLOW ON THE SHELLSIDE OF SHELL-AND-TUBE EXCHANGERS

We shall dwell on this subject at some greater length because of the universal importance to industry and the fact that this problem has been rather neglected lately in the literature. The shell-and-tube exchanger was introduced into industrial practice in the early 1900's when demands grew (particularly from the emerging process industry) for equipment which could handle relatively large heat loads which could not be satisfied by simple, single-tube devices. Multitubular equipment with its versatility was a natural answer. Early performance studies by U.S. and German sources go back to the 1910's and are usually based on *ideal tube banks*. The early heat transfer equations assumed the tubeside based form

$$Nu = C\, Re^m Pr^n \qquad (9)$$

with   Re   based on tube outside diameter and minimum crossflow area of the tube layout

m   0.6 for ideal tube banks (later confirmed for baffled shellside flow)

n   0.33, taken from tubeside correlations

C   0.6 for staggered tube layout on ideal tube banks

The inefficiencies connected with the nonideality of baffled flow suggested the need for adjusting the constant C which was typically reduced to 0.33 for staggered layouts (60 percent effectiveness of ideal tube bank). This equation type appears in all the early design methods.

The attempts to correlate pressure drop were much more difficult because of the greater sensitivity of ΔP to the actual flow patterns (compared to heat transfer) and not until systematic data sets on tube banks were available did correlations of respectable accuracy appear. An authoritative critical review of the developments on ideal tube banks was presented by Boucher and Lapple [2] (1948). Data in the laminar region were scarce until the work of Bergelin et al. [3] in 1952.

The correlational form was borrowed again from inside tube formulations and appears as,

$$\Delta P = C \ f \ \frac{G^2_{max} \ N_{rc}}{\rho} \tag{10}$$

where   f   =   friction factor derived from data and presented graphically as f(Re)

G_{max} = mass velocity based on minimum cross-sectional area in flow direction

N_{rc} =   total number of tube rows crossed

Subsequent developments concentrated on improving the correlations respecting the realities of baffled circular tube bundle flow. With the power of the Reynolds number confirmed as 0.6, the various approaches tried to adjust the constant C as a function of baffle geometry. Later, and mroe successful attempts concentrated on solving the actual flow relations in the baffled tube bundle. These evolutions are described in the following.

3.1 Methods of Calculation

A. Integral Methods

These methods, using the simple equation types described above, are termed "integral" as they consider the total flow as effective and introduce only simplified correction factor to account for the effects of bundle geometry. The two representative methods of this group are due to Donohue [4] and Kern [5], both of which became very popular in the 1950's [6] mainly because of the simplicity of calculations. However, the accuracy of the results is rather poor (especially for Kern's method) even for average geometry design configurations (see below) and errors can reach rather alarming magnitudes of hundreds of percent (usually on the safe side), especially for pressure drop and laminar flow.

B. Semianalytical Methods

It was recognized in the late 1940's [7] that better account must be taken of the *actual flow distribution* in the complex geometry of a baffled exchanger. The first practical application using this approach is due to Bell [8]. In an adjusted form his method can be represented as a modification of the "integral" type equation, using a number of correction factors which account for the non-ideality of baffled flow.

$$\frac{Nu}{RePr} = St = (j_i)[\phi \ \xi \ \gamma \ \lambda] \ Pr^{-0.66} (\mu/\mu_w)^{0.14} \qquad (11)$$

where j_i = ideal tube bank j-factor for any given tube layout

ϕ = the baffle window correction factor which accounts for the different effectiveness of crossflow versus window flow

ξ = the shell-to-bundle bypass stream correction factor, a function of the bundle bypass area (sealing strips) and, therefore, the type of bundle (fixed tubesheet, pull through, etc.)

γ = the tube-row number correction factor. In turbulent flow, the effectiveness of heat transfer increases with the number of tube rows up to about 10 and then stabilizes. In laminar flow, however, the effect of "adverse temperature profile" (similar to the L/D factor in tubeside flow) develops with the number of tube rows crossed and must be accounted for [10]

λ = the combined correction factor for baffle-to-shell and tube-to-tube hole leakage streams

The combined effect of these correction factors can range from almost the full effectiveness of an ideal tube bank (approximately 0.6) for a well-designed tube bundle to as low as 0.1 for poorly designed baffle geometry.

Pressure drop is treated in a similar way, i.e., use of correction factors for the crossflow portion of the tube bundle (between baffle tips) and the flow in the window and window turnaround. A bypass and leakage correction factor is used to calculate the "effective streams"which are related to ideal tube bank friction factors. Account is also taken of the fact that all streams combine in inlet and outlet baffle spacings.

In the final appraisal, the method has definite advantages and definite weaknesses, summarized as follows:

Advantages With a minimum amount of additional work, the efficiency factors of the individual streams account much more closely for effects of baffle geometry and manufacturing tolerances than can be accomplished by the pure integral methods. Despite the seeming complicacy, the method is, with appropriate graphical simplifications, relatively easy to use.

Disadvantages The individual correction factors are independent of each other and, therefore, their interaction cannot be reflected in a way as the theoretical flow model suggests. While additional refinements could be made the overall approach is considered less effective than the stream analysis type described below.

C. Stream Analysis Method

This method is based on the original work of Tinker [9] who recognized that the key to shellside flow solution is to divide the shell stream into a number of separate streams as shown in Fig. 4. In the order of decreasing efficiency, these are as follows:

B-Stream: This is the true crossflow stream and is considered fully effective for both heat transfer and pressure drop. Ideal tube bank correlations can be applied to the effective flow velocity based on an average integral cross-flow area.

A-Stream: This stream is created by the pressure differential driving force on the two sides of a baffle forcing fluid through the gap between tube and the baffle tube hole. Very high heat transfer coefficients were observed in the annular spaces and the A-stream can be considered fully effective for heat transfer but decreasing the overall pressure drop.

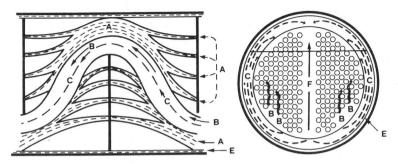

Fig. 4. Diagram of Fluid Streams through Shell Side of a Heat Exchanger.

C-Stream: The bundle bypass stream can attain very substantial values for pull through bundles without sealing devices. The stream is only partially effective for heat transfer as it contacts the heat transferring surface only on one side of the tubular field. Sealing devices, such as dummy tubes and strips, must be used for all designs with bundle-to-shell clearances larger than approximately one tube layout pitch.

F-Stream: Tube layout partitions of multitube pass bundles create open passages if placed in the direction of the main crossflow stream. Effectiveness of this stream is higher than for C-stream but blocking devices are again recommended.

E-Stream: The baffle-to-shell leakage stream is not only considered ineffective but can contribute to considerable distortion of the temperature profile (see below). The magnitude of this stream can reach substantial values, especially in laminar flow and is difficult to estimate as the tolerances of the shell diameter versus the baffle diameter are hard to establish on a predictive basis.

The relationship between the various streams, their associated resistances (K_j) and the resulting ΔP can be represented in the following schematic diagram (documented in greater detail in Ref. [10]).

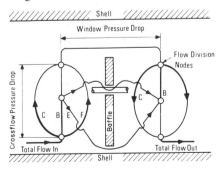

Fig. 5. Schematic Model of Shellside Flow Paths and Resistances across One Baffle Space.

Notice that the crossflow pressure drop, ΔP_x, must be equal for streams B, C and F while the window pressure drop, ΔP_w, serves as a driving force for the leakage streams A and E. For flow through the window, streams B, C, F and part of E combine.

The flow rate of any individual stream, j, can be shown to form the following relation with the flow resistance factor K_j.

for streams B, C, F:

$$Q_j = S_j \sqrt{\frac{1}{K_j}} \qquad (12)$$

for streams A, E:

$$Q_j = S_j \sqrt{\frac{1+z}{K_j}} \qquad (13)$$

with

$$K_j = f(Q_j)$$

where $z = \Delta P_w / \Delta P_x$ and S_j is the associated flow area.

The principle of the method is that the true flow rates of the individual streams are calculated first from the pressure drop relationships and then an "effectiveness" value is assigned to each stream for heat transfer calculations, defined separately for pure crossflow and window flow based on experimental values obtained under the various baffle geometry designs as shown in Fig. 6 below.

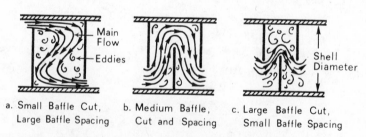

a. Small Baffle Cut,   b. Medium Baffle,   c. Large Baffle Cut,
   Large Baffle Spacing   Cut and Spacing   Small Baffle Spacing

Fig. 6. Flow Pattern as Function of Baffle Geometry.

$$Q_{\text{effective}} = \alpha\,(Q_B) + \beta\,(Q_A) + \gamma\,(Q_C) + \zeta\,(Q_F) \qquad (13a)$$

Coefficients α, β, γ, ζ are, in turn, functions of the baffle geometry.

An interesting side observation must also be made with respect to the distortion of the true mean temperature profile due to the effect of the bypass and especially shell-to-baffle leakage stream (E). Figure 7 illustrates the basic principle.

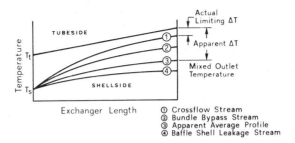

Fig. 7. Schematic Illustration of Temperature Profiles for Various Shellside Flow Streams.

The resulting design method is suitable to be solved only by means of digital computers. In spite of its relative complexity, calculation times on fast computers (such as the IBM 370) average around two seconds for a typical design case.

Of some interest is a comparison of the accuracy attributed to the various methods. Many of the previously published methods made obviously exaggerated claims to accuracy, mainly because of the limited data used. Evaluation of all the methods to the large HTRI data bank [10] produced the following interesting approximate picture.

Method	ΔP Data within $\pm 30\%$	Heat Transfer Data within $\pm 25\%$
Kern	10	30
Donohue	20	40
Bell	50	60
Stream Analysis	80	90

4. CONDENSATION

The field of condensation heat transfer is complicated by the interacting effects of the following operational parameters:

Geometry of Surface: Inside of horizontal or vertical tubes

Outside of horizontal or vertical baffled tube bundles

Vapor Hydrodynamics:	Condensate film laminar (no vapor shear)
	Gravity induced turbulency
	Vapor shear induced turbulency

Substance:	Single component
	Multicomponent fully condensable mixture
	Vapor with noncondensable gas present

The basic relationship can be best followed for condensation inside a vertical tube. The following figure illustrates the principles involved:

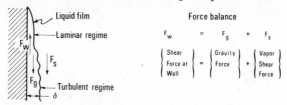

Fig. 8. Schematic Representation of Condensation on a Vertical Wall.

The condensate film is laminar until the laminarity is disturbed either by gravity or vapor shear forces. Completely different results are obtained for the three cases. In general, the Nusselt number for condensation may be expressed as

$$Nu_c = \frac{h_c \delta}{k_\ell} = \psi_c \tag{14}$$

where h_c is the heat transfer coefficient for condensation

k_ℓ is the thermal conductivity (liquid)

δ is the film thickness (the equation form for δ depends on the gravity force, F_g, or the shear force, F_s, is controlling)

ψ_c depends on the film flow regime, and is a constant for a laminar film and a function of the Reynolds and Prandtl numbers for a turbulent film

Various solutions of the specific cases were suggested, but a unified presentation still does not exist. The following is an attempt to summarize:

a) **Gravity Controlling Laminar Film (Nusselt solution, $F_s = 0$)** The term, ψ, becomes constant and analysis yields (with ρ_ℓ and ρ_v density of the liquid and vapor respectively)

$$\delta = C_1 (Re_c)^{1/3} \left[\frac{\mu_\ell^2}{\rho_\ell(\rho_\ell - \rho_v) g} \right]^{1/3} \tag{15}$$

b) **Gravity Induced Turbulent Film** As the condensate flow rate increases, turbulent film will eventually be established. Kirkbride [11] and Colburn [12] investigated this case, which resulted essentially in

$$\psi = C (Re_c)^{0.4} \tag{16}$$

$$\delta = \left[\frac{\mu_\ell^2}{k_\ell^3 \rho_\ell(\rho_\ell - \rho_v) g} \right]^{1/3} \tag{17}$$

A better solution of this problem appears to be obtained if we include also the effect of Prandtl number, such that

$$\psi = C \, Re^{0.47} \, Pr^{0.4} \tag{18}$$

c) **Vapor-Shear Controlling Turbulent Film** Under these conditions ($F_s \gg F_g$) an annular turbulent film exists and the solution assumes the type suggested by Boyko and Kruzhilin [13] with δ = inside tube diameter and

$$\psi = C (Re_c)^{0.8} (Pr)^{0.4} \xi \tag{19}$$

where ξ accounts for the vapor volume fraction and was determined empirically (for total condensation) as

$$\xi = 0.5 \left[\sqrt{\rho_\ell / \rho_v} + 1 \right] \tag{20}$$

Notice that the typical physical properties group pertinent to the Nusselt laminar film solution is not any more valid.

The three mechanisms described above can be represented graphically, plotting the average condensing coefficient, h_c, versus condensate flow rate, W_c, as shown in Figure 9 below.

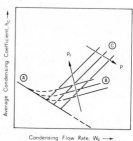

Fig. 9. Schematic Drawing Showing Average Condensing Coefficient, h_c, as Function of Condensate Flow Rate, W_c, for the Various Flow Regimes.

Transition regions obviously will exist, as both the dimensionless number groups as well as the physical properties group change between the regimes.

This figure represents graphically the three above discussed regimes

A) the Nusselt regime with laminar film
B) gravity induced turbulent film ($\approx Re^{0.47}$, $Pr^{0.4}$)
C) vapor shear induced turbulent film ($\approx Re^{0.8}$, $Pr^{0.4}$) and function of pressure through $\rho\ell/\rho_v$ effect

In the above figure, P indicates increasing pressure and Pr increasing Prandtl number.

The question of selecting the regime valid at any one given condition is difficult on an empirical basis. A rough rule-of-thumb (reasonably confirmed in comparison to data) indicates that in good approximation the *highest* value of h_c at any given W_c is valid even though much finer differentiation is desirable.

All the above deliberations were, of course, valid for average condensing coefficient over the entire length of the tube. Actually, the coefficient will vary considerably between entry and exit. Consequently, the true solution of the condensation problem is to determine *local* condensing coefficients. This leads invariably into iterative solution techniques, practical only with high-speed computers. Film thickness is predicted from the force balance in terms of shear. Heat transfer is then determined by relating the temperature gradient in the film to the universal velocity profile by a heat transfer-momentum analogy. Two solutions of this type are due to Rohsenow et al. [14] and Dukler [15].

There is no doubt that this type of approach will represent the ultimate solution of the condensation problem. However, at the present stage of development, they must be considered impractical in terms of general "design methods." Also, due to many assumptions inherent to their development, these methods are very difficult to adjust to experimental data and may be subject to unexpected extrapolation errors in nontested areas.

4.1 Condensation on the Outside of Vertical Tubes in Baffled Bundles

Condensation analysis on the outside of vertical tubes is identical to the form for inside tubes (or generally a vertical surface) described above. However, considerable difference enters the deliberation if the vertical tube is in a baffled bundle, as typically used in the industry. The baffles affect severely the condensate flow conditions as shown in Fig. 10 below so that many of the theoretical assumptions will be invalid. Solution of this problem will remain an empirical one, relying on experimental data taken under realistic operating conditions.

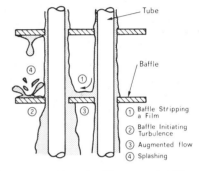

Fig. 10. Effect of Baffles on Vertical Film Condensation.

Stripping of the condensate layer by the baffles, accumulation of the condensate on the baffles and splashing of the condensate as it descends from baffle to baffle are the most obvious reasons. In general, scattered experience values indicate that the condensing coefficients will be lower than would be calculated for a single tube. Geometry of the baffles, tube-to-baffle clearances and other design criteria will have to be included in a generalized design method.

4.2. Condensation Inside Horizontal Tubes

This is the case of high industrial importance both for shell and tube units and particularly for air coolers. Correlation of the important mechanisms involved is complicated by the fact that both axial and circumferential flow of condensate must be considered, and by the many flow patterns which may exist. Also, in horizontal intube condensation, the vapor shear force acts at right angles to the gravity force and an analytical force balance is much more complicated than for vertical tubes. The most important mechanisms from the standpoint of condensation correlation are:

a) **Stratified flow** In which a laminar film drains down the walls forming a laminar pool flowing along the bottom of the tube.

b) **Wave flow** In which the condensate pool along the bottom of the tube is turbulent.

c) **Annular flow** Which occurs when vapor shear is the dominant force and the condensate film is maintained more or less evenly on the tube wall, producing a heat transfer mechanism identical to the vertical tube vapor shear-controlled case.

These are schematically shown in Fig. 11 below and explained as follows:

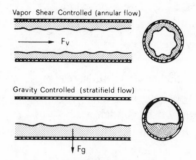

Fig. 11. Dhear and Gravity Controlled Condensation in a Horizontal Tube.

a) **Vapor-Shear Controlled Annular Flow** This case can be solved by the same methods as applicable to vertical flow as gravity effects are negligible.

b) **Gravity Controlled Stratified Flow** Assuming laminar flow of the stratified layer, the case was solved by Kern [5] in analogy to Nusselt's solution for vertical tube. However, the stratified condensate pool forms eventually ripples and internal turbulency, promoted by its own increasing flow velocity as well as vapor shear. This phenomenon (visually observed in a transparent tube at HTRI) is taken into account in a recent analysis of this problem by Chato [17], which produces more realistic values of the condensing coefficient (approximately 30 percent higher than Kern).

If the film flow regime applicable to any one correlation type is not properly recognized and respected, an extremely confusing picture can emerge, as shown in Fig. 12 below. [26]

4.3 Condensation on the Outside of Horizontal Tube Bundles

The classical solution of this case for single tube with laminar condensate flow is due to Nusselt and was later extended for N_{rv} tubes in a vertical row as follows:

$$h_n = 0.725 \, (\phi_c)^{1/4} \left(\frac{1}{\Delta T_c D_o} \right)^{1/4} (1/N_{rv})^{1/4} \tag{21}$$

where $\phi_c = \dfrac{k_\ell^3 (\rho_\ell - \rho_v) \rho_\ell g \lambda}{\mu_\ell}$

D_o = outside tube diameter, ft

ΔT_c = difference between average condensing fluid bulk temperature and average wall temperature

This analysis assumes that the condensate film remains laminar for the entire tube bank regardless of the value of N_{rv} and results in extreme penalties for large bundles.

e.g., for $N_{rv} = 20$, $(1/N_{rv})^{1/4} = 0.473$

Despite the high industrial importance of shellside condensation, virtually no progress has been made in this area and the conservative Nusselt solution is still used.

The actual flow conditions in a baffled tube bundle differ substantially from the idealized assumptions, as shown in Fig. 13a and 13b below.

The vapor is usually in crossflow to the tube bank due to the practice of using vertical baffle cuts to facilitate condensate drainage. Longitudinal flow will be observed in the vapor passage through the baffle window. The vapor usually enters with sufficiently high velocity to cause vapor-shear induced turbulency as shown in the above Figure and was actually observed on HTRI research exchanger with glass windows. Obviously, the Nusselt solution will not apply for this case.

Furthermore, it is logical to assume that the condensate film will remain laminar only for a given number of tube rows and then become turbulent as the condensate Reynolds number exceeds a critical value. This is analogous to the well-known relations for condensation on a vertical wall, where the theoretical

Nusselt relation applies only until a Reynolds number of about 2100 is reached.

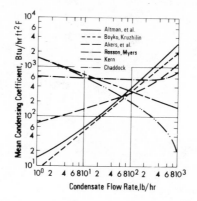

Fig. 12. Comparison of Correlations for 50 lb/sq in. abs Pentane, Condensing Inside Horizontal Tube.

In order to account empirically for an "average" turbulence effect, Kern suggests that the power on the $1/N_{rv}$ term be reduced from 1/4 to 1/6. Other researchers suggested that the penalizing tube row effect be neglected altogether — an approach which obviously reflects the combined effect of various inaccuracies.

A more rational and systematic analysis is suggested here.

a) **Gravity Controlled Condensate Film** Because of frequently imposed very low pressure drop limitations and particularly toward the outlet of the condenser, the limiting case of no vapor-shear effect must be considered.

Referring to the above Figure, the classical Nusselt relation, including the N_{rv} effect, will be valid until a critical Reynolds number is reached and the film becomes turbulent. The heat transfer mode changes at this point and can be expressed by an equation similar to that derived for vertical surface condensation under gravity induced turbulency

$$Nu = C \, Re^{m} Pr^{1/3} \tag{22}$$

where   m   will be somewhere between 1/3 and 0.4

b) **Vapor Shear Controlled Turbulent Film** In the vapor shear controlled region, such as at the entry to the tube bundle, the Nusselt solution is invalid. Very high coefficients (up to 20 times higher than the Nusselt solution) have been measured by HTRI in this region.

A suggested solution is again an approach similar to the vapor shear induced turbulency on a vertical surface (Boyko — Kruzhilin), with the exception that the Re-Pr effect must be adjusted to values obtained on tube banks

$$Nu = C \ Re^{0.6} Pr^{0.4} \ \xi \qquad (23)$$

where ξ is adapted from Eq. 20, accounting for the vapor volume fraction

C   is a constant, assumed as 0.6 for ideal tube bank

In an actual shell-and-tube exchanger bundle, the effects of vapor by-passing in the clearances between bundle and shell must be taken into account, adjusting the true flow velocity in Re and the constant C, according to data.

Combining the Nusselt condensation regime with the gravity induced and vapor-shear induced turbulent film regimes, relations as shown in Fig. 14 will be obtained (a graph similar to Fig. 9 for vertical surface condensation). Because of the potentially large variation in the value of the coefficients obtained under the various flow regimes as well as the fact that condensate drains to the bottom of the shell within each baffle spacing, an accurate design method must proceed stepwise from baffle to baffle, and is practically solvable only by digital computers.

4.4 Condensation on Finned Tubes in Horizontal Tube Bundles

Tubes with external radial fins may be used advantageously for condensation in horizontal tube bundles as long as the inside tube heat resistance is not controlling. Basic analysis of this problem was presented by Katz et al. [18] and Young and Ward [19]. It follows all the essential premises developed for plain tubes with the assumption that the fin surface behaves like vertical surface and the fin root area behaves like horizontal plain tube. The development was based purely on Nusselt's analysis and did not respect the differentiation of the flow regimes created by the various effects of gravity and vapor shear.

Essentially, the same premises will apply to finned tubes as they were discussed for plain tubes. The only additional effect particular to finned tube is that of "condensate retention." If the surface tension of the condensate is relatively high, the liquid has the tendency to be retained in the fin gaps, thus making the finned

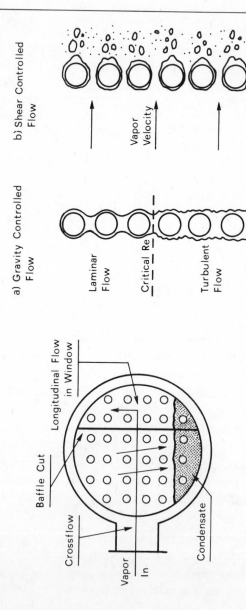

Fig. 13. a) Flow in Horizontal Baffled Condenser with Vertical Baffled Cut; b) Schematic Drawing Condensation on a Vertical Tube Row.

surface ineffective. This phenomenon was first reported by Katz et.al. in an unpublished report. It was independently confirmed at HTRI and from the parameters involved (surface tension, density and fin height to spacing ratio), dimensionless correction factors were developed which can account for this effect. In principle, this is illustrated in Fig. 15 below.

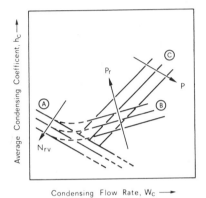

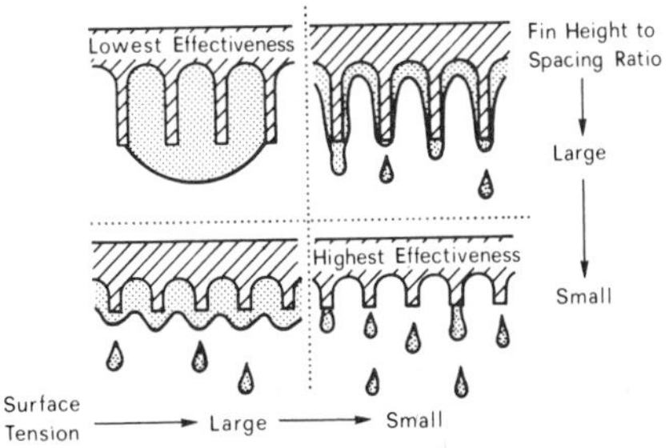

Fig. 15. Condensate Retention Effect as Function of Fin Geometry and Fluid.

Fig. 14. Schematic Relations for Condensation Outside Horizontal Tubes.

The effect can be very significant for fluids having large value of the σ/ρ group, such as water, and for tubes with high fin density. Close analysis of data on plain tubes indicated that the same surface tension phenomenon also takes place, but to lesser extent as the fin effect is not present. It has never been considered before, but it is suggested that its inclusion explains some existing discrepancies of data between water and hydrocarbons.

5. BOILING

No attempt is being made in this chapter to elaborate on the complex relations pertinent to boiling heat transfer and associated thermodynamic and hydrodynamic relations. The only brief mention warranted within this lecture is pertinent to quite basic experiments performed at the writer's Insitute [16], namely, boiling in tube bundles. Most past (and numerous) experiments were concerned only with boiling on single tubes or discs. The problem to be resolved was how these experiments relate to boiling in tube bundles. An extensive research project demonstrated the basic behavior as shown in Fig. 16 below.

a. Effect of Tube Bundle Geometry

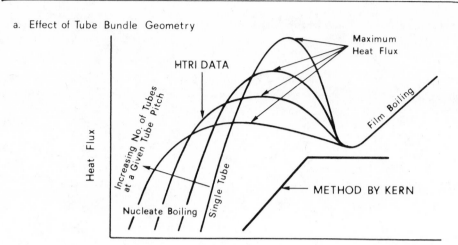

Fig. 16. Schematic Representation of Heat Flux vs. Temperature Difference for Single Tube and Multitube Bundles.

a) The heat flux was found to *increase* as function of increasing tube field density probably because of increased convection effect and reaches values up to 5 times that for an elementary surface under the same ΔT.

b) The maximum heat flux, on the contrary, was found to *decrease* as a function of the same parameter, probably because of "vapor choking."

c) For comparison, the method proposed by Kern [5] and widely accepted by the industry is shown to be extremely conservative for most cases.

d) Contrary to experiments on simple geometries, no pronounced effect of surface quality or tube material was found, apparently because of the convection influence.

These thoughts are presented here only as a documentation and suggestion for need of further detailed investigation.

6. SPECIAL PROBLEMS

There are a number of extraneous special problems which affect the accuracy of heat exchanger design. The most important ones are mentioned below.

6.1 Physical Properties

The dimensionless numbers associated with heat transfer and pressure drop correlations contain typically the following properties: viscosity, density, specific heat, and thermal conductivity. Analysis of the available data sources will show that for industrial streams, especially as used in the petro-chemical industry, the available values are completely inadequate and frequently inaccurate. [21] Errors in thermal conductivity alone can easily reach 100 percent. Predictive equations are not sufficiently dependable, especially for mixtures and polar molecules. Any further progress in accuracy of heat transfer and pressure drop methods will be conditioned by a better supply of physical property data. This subject, due to its general validity, would be especially suitable for a coordinated international cooperative effort. Too lilltle attention to this subject is alocated at present.

6.2 Fouling

Most heat transfer processes result in deposition of undesirable scale, commonly referred to as "fouling." The thermal resistance of such deposits can reach dominant magnitude, sometimes completely overwhelming the accuracy of heat transfer predictions. For example, a water-cooled condenser with overall coefficient of 500 Btu/hr-ft^2-F and typical fouling resistance of 0.002 hr-ft^2-F/Btu will use only one-half of the surface under clean conditions. While some progress has been reported recently [20], it is imperative that much more work be devoted to this neglected subject.

6.3 Effect of Geometric Tolerances

The shellside predictive methods based on the Stream Analysis principle require the knowledge of detailed manufacturing dimensions, especially
- shell-to-baffle clearances
- tube-to-baffle hole clearances
- tube-to-shell bypass clearance

These dimensions can be only estimated for any given exchanger size, being functions of shell diameter, bundle construction and the operating pressure. A computer design program which passes through the course of numerous solutions requires a dependable predictive system for all the pertinent dimensions. This establishes the need for a close tie between the pure research approach to the problem and the mundane facts of shop practices — a problem yet to be resolved.

6.4 Material Codes and Standards

There is a bewildering system of National Standards for construction of heat exchangers and pressure vessels in general, which are frequently overlapping and unsystematic. Large engineering companies operating on an international basis have to maintain special experts to clarify this unnecessarily obscured area. The need exists for uniform, internationally recognized codes and standards for the general benefit of all parties concerned.

REFERENCES

[1] Hausen, H., Zeit. V.D.K. Beiheft Verfahrenstechnick, No. 4, 91-98 (1943)

[2] Boucher, D. F. and Lapple, C. E., "Pressure Drop Across Tube Banks," CEP, 4, No. 2, 117-134 (1948)

[3] Bergelin, O. P., Brown, G. A. and Doberstein, S. C., "Heat Transfer and Fluid Friction During Flow Across Banks of Tubes — IV. A Study of the Transition Zone Between Viscous and Turbulent Flow," Trans. ASME, 74, No. 6, 953-960 (1952)

[4] Donohue, D. A., *Petrol. Refiner,* 34, No. 10 and 11 (1955)

[5] Kern, D. Q., *Process Heat Transfer,* McGraw-Hill, New York, 1950

[6] V.D.I., Wärmeatlas, VDI-Verlag (1957)

[7] Tinker, T., "General Discussion on Heat Transfer," 97-116, Inst. of Mech. Engrs., London, England (1951)

[8] Bell, K. J., "Exchanger Design Based on the Delaware Research Program," Petro. Engr., 32, No. 11, C26-36, C40a-C40c (1960)

[9] Tinker, T., "Shellside Heat Transfer Characteristics of Segmentally Baffled Shell-and-Tube Heat Exchangers," ASME Paper No. 47-A-130, N.Y. (1947)

[10] Palen, J. W. and Taborek, J., "Solution of Shell Side Flow Pressure Drop and Heat Transfer by Stream Analysis Method, " CEP Symp. Ser., 65, No. 92 (1969)

[11] Kirkbride, C. G., Ind. Eng. Chem., 26, 425 (1934)

[12] Colburn, A. P., Ind. Eng. Chem', 26, 432 (1934)

[13] Boyko, L. D. and Kruzhilin, G. N., *Int. J. Heat Mass Transfer,* 10, 361-373 (1967)

[14] Rohsenow, W. M. et al., Trans. ASME, 78, 1637 (1956)

[15] Dukler, A. E., CEP, 62,(Oct. 1959)

[16] Palen, J. W., Yarden, A. and Taborek, J., "Characteristics of Boiling Outside Large-Scale Horizontal Multitube Bundles," AIChE Symp. Ser., 68, No. 118

[17] Chato, J. C., *ASHRAE J.,* 4, 52-60 (1960)

[18] Katz, D. L., Young, E. H. and Balekjian, G., "Condensing Vapors on Finned Tubes," *Petrol. Refiner,* 33, No. 11, 175 (1954)

[19] Young, E. H. and Ward, D. J., "How to Design Finned Tube Shell-and-Tube Heat Exchangers — Part 2: Design of Finned Tube Condensers," The Refining Engr., **29**, 32 (1957)

[20] Taborek, J. et al., "Fouling — The Major Unresolved Problem in Heat Transfer, Parts I and II," CEP, **68**, Nos. 2 and 7 (1972)

[21] Nangia, K.K. and Taborek, J., "Physical Properties of Liquids-Status of Present Knowledge and Importance in Design Methods for Heat Transfer and Fluid Flow," presented at the AIChE-ASME Thirteenth Natl. Heat Transfer Conf., Denver, Colorado (1972)

[22] Kuznetsov and Leonenke, Khim Nauka i Prom, **4**, CA53:20948C, 406-407 (1959)

[23] Metais, HTL TR No. 51, Heat Transfer Lab., Univ. of Minn., Minneapolis, Minn. (1963)

[24] Norris and Sims, Trans. AIChE, **38**, 469 (1942)

[25] Gill and Lee, AIChE J., 8, 303-309 (1962)

[26] Bell, K.J., Taborek, J. and Fenoglio, F., "Interpretation of Horizontal In-Tube Condensation Heat Transfer Correlations with a Two-Phase Flow Regime Map," CEP Symp. Ser., **66**, No. 102 (1970)

Chapter 4

HEAT TRANSFER OF BANKS OF TUBES IN CROSSFLOW AT HIGH REYNOLDS NUMBERS

A. A. Žukauskas (*)

I. INTRODUCTION

In a previous article, I reviewed the data on heat transfer in a single tube and banks of tubes in crossflow of various fluids [1]. The influence of the properties of fluids, heat flux direction and flow pattern on heat transfer have been studied comprehensively. The influence of bank arrangement on the intensity of heat transfer was examined, particularly in the subcritical flow regime.

The present chapter is concerned mainly with heat transfer of banks of tubes at high Reynolds numbers. The fast growth of chemical and power industries and the emergence of some new branches of engineering demanded reliable equations for the calculation of heat exchangers working in the critical regime. Thus, the present chapter will deal, with problems of heat transfer of tube banks in crossflow in critical and supercritical regimes.

These problems are being investigated in many countries. My chapter presents a general review of investigations, performed at the Institute of Physical and Technical Problems of Energetics of the Academy of Sciences of Lithuanian SSR and in other research centers in recent years. It concerns heat transfer and hydraulic resistance of tube banks of various arrangement in crossflows of gas and fluids in the range of the Prandtl number from 0.7 to 500 and the Reynolds number from 10^4 to 2×10^6.

(*) Academy of Sciences of the Lithuanian SSR, Vilnius, U.S.S.R.

N.B. ALL FIGURES QUOTED IN TEXT ARE AT THE END OF THE CHAPTER

II. GENERAL FLOW PATTERN

Heat transfer is considerably influenced by the flow regime around the tube, while flow past banks of tubes is one of the most complicated problems of practical importance. The flow pattern around a tube in a bank is influenced by the surrounding tubes. In a contraction between adjacent tubes of a transverse row, the pressure gradient changes. This causes a corresponding change of velocity distribution in the boundary layer and of the flow pattern in the rear.

The flow pattern around a tube in a bank is determined by the arrangement and geometrical parameters of the bank. Banks of in-line and staggered arrangement are most common (Fig.1), and they are usually defined by the relative transverse $a = s_1/D$ and longitudinal $b = s_2/D$ center-to-center distances, called transverse and longitudinal pitches, respectively. In banks of both arrangements, flow around a tube in the first row is similar to the flow around a single tube, but the flow pattern in subsequent rows is different.

The velocity distribution around tubes in different rows of a staggered bank has a similar character. In an in-line bank inner rows are located in the circulation regions of the preceding rows, and the flow preceding one of the inner rows is vortical, with a nonuniform velocity distribution.

At low Reynolds numbers where the flow in a bank is laminar with large scale vortices in the circulation regions, their effect on the boundary layer of the front portion of a subsequent tube is eliminated by viscous forces and a negative pressure gradient.

With increasing Reynolds number the flow between tubes becomes vortical with a higher degree of turbulence. Although the front portion of a inner tube is influenced by vortical flow, a laminar boundary layer persists on the tube.

Such a flow regime covers a wide range of Reynolds numbers and alters its character only at $Re > 2 \times 10^5$, when flow in the bank becomes highly turbulent. The total drag of the bank varies like that of a single tube in the critical flow regime.

Thus, we may distinguish three flow regimes in tube banks with respect to Reynolds numbers: a predominantly laminar flow regime at $Re < 10^3$, a mixed or subcritical flow regime at $Re < 2 \times 10^5$, and a predominantly turbulent or critical flow regime at $Re > 2 \times 10^5$. We shall first consider the subcritical and the critical flow regimes.

III. INFLUENCE OF FLUID PROPERTIES ON HEAT TRANSFER

The last transfer of a tube in a bank is determined mainly by flow velocity, physical properties of the fluid, heat flux intensity, heat flux direction, and the arrangement of the tubes. The dimensionless relation is as follows

$$Nu = f\left(Re, Pr, \frac{\mu_f}{\mu_w}\; \frac{\lambda_f}{\lambda_w}\; \frac{Cp_f}{Cp_w}\; \frac{\rho_f}{\rho_w}\; \frac{S_1}{D}\; \frac{S_2}{D}\right). \tag{1}$$

To generalize experimental data, we use the following equation based on the functional relation (1).

$$Nu = cRe^m Pr^n . \tag{2}$$

For gases for which the Prandtl numbers are equal and constant, Eq. (2) becomes

$$Nu = cRe^m . \tag{3}$$

1. Influence of the Prandtl Number

Most of the fluids commonly used in practice have Prandtl numbers ranging from 1 to 1000. This means that in calculations of heat transfer a wrong choice of the exponent over the Prandtl number may lead to considerable errors. An exponent in the range 0.31-0.33 is still accepted by some authors. This value of n is suggested by theoretical investigations of heat transfer in a laminar boundary layer on a plate. However, later calculations and experimental measurements [2, 3] have revealed the dependence of the exponent over the Prandtl number on the flow regime in the boundary layer. For a laminar boundary layer on a plate n = 0.33, but for a turbulent boundary layer it amounts to 0.43. The above suggests that for mean heat transfer from a tube the value of n may be somewhere between 0.33 and 0.43.

The variations in the flow pattern around a single tube lead to changes of the exponent over the Prandtl number [4]. The exponent for Pr > 10 is equal to approximately 0.34 and 0.39, in the front and in the rear, respectively (Fig. 2).

Investigations of heat transfer in laminar and turbulent boundary layers suggest slight changes in the value of the exponent with large variations of the Prandtl number.

Our investigations of heat transfer [5] in 27 banks of tubes of different arrangements, in flow of various fluids in the range of Pr from 0.7 to 500, suggest that for the mean heat transfer of all sorts of banks the exponent over the Prandtl number has the value of 0.36.

2. Choice of Reference Temperature

In the process of heat transfer, the fluid temperature varies and causes variations of fluid physical properties. Thus, evaluation of the influence of the fluid physical properties on heat transfer is closely connected with accounting for the temperature variations in the boundary layer, in other words, with the choice of the so-called reference temperature, on the basis of which the physical properties are evaluated.

The influence of the variations of physical properties may be established by two methods. In the first method the physical properties are referred to the main flow temperature and an additional parameter is introduced in Eq. (2) to account for the properties variation. The second method is to choose a certain temperature value between that of the flow and that of the wall, which enables evaluation of the influence of physical properties on heat transfer. In this case, the relations for the heat transfer calculation remain the same as for constant physical properties.

We prefer to take the main flow temperature t_f as reference temperature in the range of moderate temperatures. This method is simple in practice and sufficiently accurate for practical purposes. In flows of viscous fluids, however, experimental results are higher for heating than for cooling, the discrepancy increasing with the increase of temperature difference.

Our investigations confirm that variations of viscosity and other properties in the boundary layer on a circular tube may well be accounted for by the ratio Pr_f/Pr_w, with a corresponding exponent. In this case, the results are referred to the main flow temperature.

Our investigations of changes of the exponent over Pr_f/Pr_w for the case of a laminar and a turbulent boundary layer on a plate give n = 0.25 for heating and n = 0.17–0.19 for cooling [2, 6]. For heat transfer of banks of tubes in crossflow of various fluids for most practical purposes, n = 0.25 is sufficiently accurate for both cases.

The ratio μ_f/μ_w is often used also. It should be noted that for viscous fluids, like water and oil, it is mainly viscosity that changes with temperature and therefore $Pr_f/Pr_w \approx \mu_f/\mu_w$.

Some other parameters have also been proposed to account for radical changes of physical properties [7].

We shall present experimental results of mean heat transfer from banks of tubes in the following form

$$Nu_f = cRe_f^m Pr_f^n (Pr_f/Pr_w)^{0.25} \tag{4}$$

The tube diameter and velocity in the smallest cross section will be chosen as characteristic parameters.

We are not concerned with the processes of convective heat transfer at high temperatures of gases and in the presence of chemical reactions in the boundary layer, and the formulas cannot be applied for those conditions.

IV. MEAN HEAT TRANSFER AND REYNOLDS NUMBER

In most cases heat transfer of tubes in the first row is considerably lower than in inner rows and becomes stable in the third or fourth row. The following discussion deals mainly with the mean heat transfer of a tube in an inner row of a bank.

In Fig.3 data of our experiments with in-line banks of pitches (1.30 X 1.30) [8] and (1.26 X 1.26) [9] in air flow are presented in form of :

$$Nu_f = f (Re_f) \tag{5}$$

The tube diameters are 19 mm and 23mm, respectively. The agreement of separate experiments is satisfactory. In the range of $Re < 2 \times 10^5$, the exponent over Re is m = 0.63. But for Re about 2×10^5, the exponent increases to m = 0.84, and a new law of heat transfer is established.

In Fig. 4 the results of in-line banks (2.06 X 1.37) in air are presented by the same relation. The experiments were performed in KFA Jülich. Tube diameter was d = 51 mm. Here, for $Re < 2 \times 10^5$ the exponent over Re is m = 0.65, and for Re 2×10^5 it increases to 0.82.

Fig. 5 presents heat transfer experiments by J. Stasiulevičius [11] with the staggered banks (2.2 X 1.3). Here for $Re < 2 \times 10^5$ the exponent over Re is m = 0.60, and for Re $> 2 \times 10^5$ it increases to 0.84.

In Fig. 6. the relation

$$K = Nu_f \, Pr_f^{-0.36} \, (Pr_f / Pr_w)^{-0.25} = f \, (Re_f) \,. \tag{6}$$

represents our earlier results [5] of heat transfer of the in-line bank (1.65 × 2.0) in flows of different fluids, together with the results of recent investigations of R. Ulinskas, P. Daujotas and myself on heat transfer of the in-line bank (1.5 × 1.5) in water. Tube diameters are 19 mm and 30 mm, respectively. The agreement both of separate experiments and of different fluids is satisfactory. Similar to experiments in air, for Re < 2 × 10⁵ m = 0.63, and for Re > 2 × 10⁵, m = 0.8.

Fig. 7 represents the same experiments on two staggered banks, of (1.5 × 1.5) in different fluids with t_w = const and q_w = const. For   Re < 2 × 10⁵, m = 0.60 and for Re > 2 × 10⁵ m = 0.8.

As we see, heat transfer is higher for q_w = const than for t_w = const. In heat transfer calculations and in the analysis of experimental results, a consideration of surface temperature variation is necessary. Different temperature distributions on the wall lead to different values of the heat transfer coefficient. Therefore, relations of heat transfer of a tube with constant surface temperature, as applied in the case of variable surface temperature, are only approximate, and in the case of a considerable surface temperature gradient they may be misleading.

In Fig. 8 heat transfer results of different authors are compared for banks of staggered arrangement in equilateral triangles with the transverse pitch from 1.3 to 2.0.

The continuous line corresponds the our results. Curve 1 – (1.5 × 1.3), from Bergelin et al. [12]; curve 2 – (1.5 × 1.5) and (2.0 × 2.0), from Grimison [13]; curve 3 – (2.0 × 2.0), from Kuznetsov and Turilin [14], and from Antuf'ev and Beletskij [15]; curve 4 – (1.3 × 1.5), from Lyapin [16]; curve 5 – (1.6 × 1.4), from Dwyer and Sheeman ; curve 6 – (2.1 × 1.4), from Hammeke et al. [10]. The results correlate well in the whole range of Re examined.

Thus, not only our results but also the results of other authors indicate that in a predominantly laminar flow, Re < 10³, the intensity of heat transfer of banks of tubes is proportional to Re to the power of 0.5. With the further increase of Re and the turbulization of the flow, the heat transfer becomes more intensive and proportional to Re to the power of 0.6 or 0.63. As follows from the results analyzed, at about Re > 2 × 10⁵ heat transfer increases suddenly, which is an indication of a new regime of flow and heat transfer. We have called it the predominantly turbulent or the critical regime.

The physics of heat transfer in the subcritical regime is studied in a number of references. Thus, in recent years this has been studied in detail in Boris Kidric Institute, Yugoslavia [17]. But the critical regime has only recently become an object of investigation, and the general flow pattern and the process of heat transfer still are obscure. I would like to concentrate here on the problems of the critical regime.

V. FLOW PATTERN AND HEAT TRANSFER IN THE CRITICAL RANGE OF Re.

1. Pressure Distribution.

The variations in the hydrodynamic conditions in the flow around the tube in a row of a bank are illustrated by the distribution of pressure. Fig. 9 shows the distribution of pressure coefficients in inner rows of staggered tubes in gas flow as a function of Re. Curve 1—second row, and curve 2—fourth row of the bank (2.0 X 1.4) at Re < 1.5 X 10^6 [18]; curve 3—second row, and curve 4—fourth row of the bank (2.2 X 1.5) at Re = 2.7 X 10^4 [19].

The pressure coefficient for a tube in a bank is given by

$$P = 1 - \left[\left(\eta_{\varphi=0} - \eta_{\varphi} \right) / \left(\rho u^2 / 2 \right) \right], \tag{7}$$

where u is the mean velocity in the minimum free cross-section.

Fig. 10 presents our recent measurements of pressure distribution in an inner row of the in-line bank (1.25 X 1.25) in water. Here, curve 1 corresponds to Re = 4 X 10^5, curve 2 to Re = 6 X 10^5, curve 3 to Re = 1.1 X 10^6.

Figures suggest a certain influence of Reynolds number on pressure distribution on an inner tube. The boundary layer separation point is at $\varphi = 150°$.

Fig. 11 presents measurements by S. B. Neal and J. A. Hitchcock [26] of velocity and turbulence distribution at several distances from the surface around a tube in an inner row of the staggered bank (1.33 X 1.17) at Re= 1.4 X 10^5.

The rapidly increasing velocities over the front portion of the tube (Fig.11) are accentuated by the presence of the adjacent tube in the previous row. This provides a local flow contraction followed by an expansion. Nearer to the point of separation, at $\varphi = 150°$, surface velocities decrease towards zero.

As presented by Achenbach [18], in a staggered bank at high Reynolds numbers the position of the separation point is somewhat different.

Our investigations of staggered and in-line banks (1.25 × 1.25) and (1.50 × 1.50) in the range of Re from $5 × 10^4$ to $2 × 10^6$ have revealed a determining effect of the pitch on the separation point, rather than of the Reynolds number.

2. Drag of a Tube in a Bank.

Fig. 12 shows the distribution of the skin friction coefficient $c_f = (\tau/\rho u^2)\sqrt{Re}$ around a tube in a staggered bank (2.0 × 1.4) [18]. Curve 1 corresponds to Re = $1.3 × 10^5$; curve 2 to Re = $8 × 10^6$; curve 3 to Re = $3.7 × 10^5$; curve 4 to Re = $1.4 × 10^6$. The effect of the Reynolds number on the friction is most evident in the region $60° < \varphi < 75°$.

3. Local Heat Transfer.

Fig. 13 gives variation of local heat transfer coefficient around a single tube and a tube in a bank, in the subcritical range of Re. In banks of both arrangements a higher turbulence intensity in the flow causes an increase in the heat transfer in the front portion as well as in the rear portion of a tube. Nevertheless, the maximum value of the heat transfer in the case of an in-line bank is observed at $\varphi = 50°$ because the impact of the stream on the surface occurs at this point.

On the tube in an inner row of the staggered bank, curve 3, the boundary layer is laminar initially. As it develops around the tube, along with thermal boundary layer, heat transfer coefficients decrease rapidly. At $\varphi = 90°$ the transition from laminar to turbulent boundary layer begins.

Immediately beyond the top of the tube there is a considerable thickening of the boundary layer. It is accompanied by an increased turbulence near the tube as the surface flow becomes unsteady. Although mean surface velocities are falling quickly, the movement of warm air from the surface together with the increased turbulence results in an improvement of local heat transfer coefficients [26].

Near the point of separation, surface velocities decrease towards zero, and the poorest heat transfer occurs at the point of separation or a little beyond it.

In banks of staggered arrangement a change of longitudinal and transverse pitches from 1.3 to 2.0 has hardly any effect on the character of heat transfer. In Fig. 14 we see local heat transfer at Re = 70.000. Curve 1 — (2.0 × 2.0) after Mikhaylov [20]; curve 2 — (1.5 × 1.5) after Mayinger and Schad [21]; curve 3 — (1.3 × 1.3) after Bortoli et al. [22]; curve 4 — (1.5 × 2.0) after Winding and

Cheney [23]. An increase in heat tranfer is observed at about $\varphi = 120°$ in the banks examined. Boundary layer separation occurs at $\varphi = 150°$.

A comparison of local heat transfer of staggered banks of different arrangement [24] shows a certain influence of pitch on the character of heat transfer in the subcritical range.

Fig. 15 illustrates the dependence of local heat transfer on Re in an inner row of a staggered bank. It presents our measurements in the flow of water. A distinct change in the character of local heat transfer is observed at the increase of Re from 1.6×10^5 to 9.3×10^5. We see essentially the effects of boundary layer turbulization on heat transfer.

With the increase of Re to 9.3×10^5, curve 3, the transition from laminar to turbulent boundary layer starts at $\varphi = 30°$, instead of at $\varphi = 90°$ at $Re = 8 \times 10^4$.

Fig. 16 presents analogous investigations of in-line banks.

It can be concluded from these data that in the critical regime flow through the bank becomes increasingly turbulent, together with a change in the velocity field in inner transverse rows.

The critical regime in the staggered and in-line banks examined is obvious at $Re > 2 \times 10^5$, expressed by the increase of the exponent m from 0.6 to 0.8-0.9. This means that with an increase of the Reynolds number, the heat transfer increases much faster than in a turbulent boundary layer along a plate. The rate of increase of heat transfer on a tube is somewhat similar to that of heat transfer increase on a plate in the transition from laminar to turbulent flow. Therefore, it should be supposed that in the critical regime the value of the exponent m is influenced not only by turbulence, but also by the tube arrangement, surface roughness, temperature difference, and the value of Pr. From this point of view, measurements of the interaction of velocity and temperature fields in the boundary layer of a tube would be of interest, similar to measurements at low Re numbers.

VI. THE EFFECT OF TUBE ARRANGEMENT

In the Institute of Physico-Technical Problems of Energetics, Academy of Sciences, Lithuanian SSR, we have thoroughly investigated the effect of pitch on heat transfer [5; 24]. The experiments were performed on 47 banks of different arrangements in crossflow of gas and fluids in the range of Re from 1 to 2×10^6.

Fig. 17 shows a comparison of the heat transfer from various banks of in-line arrangement at $\mathrm{Re} < 2 \times 10^5$. The exponent m over the Reynolds number varies from 0.55 to 0.73 for banks of different arrangements. It suggests an increase of m with constant longitudinal and decreasing transverse pitch. In fact, the value of m is influenced by changes in the ratio of longitudinal to transverse pitches.

In this flow regime m = 0.63 is acceptable for most banks of in-line arrangement. For most practical purposes, a mean heat transfer coefficient can be calculated with sufficient accuracy from

$$Nu_f = 0.27 \; Re_f^{0.63} \; Pr_f^{0.36} \; (Pr_f/Pr_w)^{0.25} \tag{8}$$

Fig. 18 shows a comparison of the heat transfer of different staggered banks in flow of viscous fluids at $\mathrm{Re} < 2 \times 10^5$. The exponent over Re is equal to 0.60 for all banks. The effect of pitch is clear—heat transfer increases with a decrease in the longitudinal pitch and, to a lesser extent, with an increase of the transverse pitch. The variation of c may be evaluated by the geometrical parameter a/b.

The generalized formulas for heat transfer of inner tubes in various staggered banks are

for a/b < 2

$$Nu_f = 0.35 \; (a/b)^{0.2} \; Re_f^{0.60} \; Pr_f^{0.36} \; (Pr_f/Pr_w)^{0.25} \tag{9}$$

and for a/b > 2

$$Nu_f = 0.40 \; Re_f^{0.60} \; Pr_f^{0.36} \; (Pr_f/Pr_w)^{0.25} \; . \tag{10}$$

Fig. 19 presents the heat transfer in inner rows of in-line banks in the flow of gas at $\mathrm{Re} > 2 \times 10^5$.

Fig. 20 shows a comparison of the heat transfer of different staggered banks in the flow of gas at $\mathrm{Re} > 2 \times 10^5$. Here, curve 1-fifth, and curve 2-first row of the bank (1.2 × 0.9), curve 3-fifth, and curve 4-first row of the bank (2.5 × 1.3).

As is seen from Fig. 20, with an increase of Re, the heat transfer of an inner row increases more than the heat transfer of the first row. Fig. 21 shows a comparison of the heat transfer of different in-line and staggered banks in the flow of water.

The influence of pitch on heat transfer in the critical regime is similar to that in the subcritical regime. The heat transfer of all staggered banks with widely spaced tubes, i.e. with large a/b ratio, is more intensive.

Similar results are obtained for in-line banks. Heat transfer is most intensive in banks with large transverse and small longitudinal pitches.

At the same time, as seen from the plots, in the range of $Re > 2 \times 10^5$ the law of heat transfer is not common to all the banks. Our investigations suggest some divergence in the intensity of heat transfer of banks in the flow of water.

Investigations of the effect of the pitch seem indispensable for the critical range of Re in flows of different fluids.

As a first approach, the following relations can be suggested for practical calculations of heat transfer on in-line banks of tubes in crossflow at $Re > 2 \times 10^5$:

$$Nu_f = 0.021 \, Re_f^{0.84} \, Pr_f^{0.36} \, (Pr_f/Pr_w)^{0.25} \, . \tag{11}$$

The next relation is recommended for calculating heat transfer from tubes in inner rows of staggered banks

$$Nu_f = 0.022 \, Re_f^{0.84} \, Pr_f^{0.36} \, (Pr_f/Pr_w)^{0.25} \tag{12}$$

VII. HYDRAULIC RESISTANCE OF TUBE BANKS

The resistance of a bank with viscous fluids of constant density is expressed by the following relation

$$\Delta p = f \, (u, \, s_1, \, s_2, \, D, \, z, \, \mu, \rho) \, . \tag{13}$$

The dimensionless form of this relation is

$$Eu = f \, (Re, \, s_1/D, \, s_2/D, \, z) \, . \tag{14}$$

Our measurements of the pressure drop across a tube bank, performed with 53 different banks in flows of air and other fluids, [5; 24; 25] suggest that resistance is mainly determined by the transverse pitch a and increases with a decrease of the transverse pitch. With the longitudinal pitch b increasing, the larger space between two neighboring rows permits the formation of vortices, which in many cases affects the resistance of the bank. The pressure drop across banks is proportional to the number of rows, and the entrance and exit conditions in the bank contribute more to the total loss of kinetic energy.

In Fig. 22 hydraulic resistance of the staggered bank (1.5 $\times$ 1.04) referred to one row [25] is presented. Digits denote the total number of rows. The resistance of a single row is similar to that of a single tube. It decreases at the critical Reynolds numbers, and increases slightly with a further increase of Re. The pressure drop coefficient of a staggered bank with many rows also decreases with an increase of Re, and at Re $> 2 \times 10^5$ it becomes independent of the Reynolds number.

In Fig. 23 the resistance characteristics of staggered banks with many rows are presented [25]. Fig. 24 presents the resistance of two in-line banks measured by R. Ulinskas and P. Daujotas in the flow of water. A similar character of hydraulic resistance of all banks in high Reynolds numbers is obvious.

As suggested by the analysis of experimental results, for simplicity of calculations a graphical interpretation of the data is most convenient. General graphs have been compiled from our results described previously, including the results of other authors on the pressure drop across banks in flows of gases and liquids. A satisfactory correlation has been achieved. [1, 24].

VIII. CONCLUSIONS

The preceding sections were devoted to the heat transfer of tubes in banks and the hydraulic resistance of tube banks. The main factors exerting an influence on the heat transfer process were analyzed. Calculation formulas were proposed, reflecting the general characteristics of the heat transfer of tubes in crossflow. The derivation of various equations and charts for various cases of banks of tubes in crossflow are out of the scope of this paper, constituting a separate problem. I hope the data presented here enables the derivation of the most efficient methods of calculation for crossflow heat exchangers.

NOMENCLATURE

a	relative transverse pitch, s_1/D
b	relative longitudinal pitch, s_2/D
c	constant
c_η	specific heat at constant pressure
D	diameter of tube, m
F	heat transfer surface, m^2
K_f	complex dimensionless terms, $Nu_f Pr_f^{-0.36}(Pr_f/Pr_w)^{-0.25}$
L	length, m
m	power index of Re
n	power index of Pr
P	pressure coefficient
η	pressure, N/m^2
$\Delta\eta$	pressure drop, $/N/m^2$
q	specific heat flux, W/m^2
S_1	transverse pitch of bank of tubes, m
S_2	longitudinal pitch of bank of tubes, m
S_2'	diagonal pitch of staggered bank, m
t	temperature, $°C$
u	fluid flow velocity, m/sec
z	number of tube rows in bank
α	heat transfer coefficient, $W/m^2 \ °C$
ζ	pressure drop coefficient, $2\Delta p/\rho u_0^2 z$
λ	thermal conductivity, $W/m°C$
μ	dynamic viscosity, $Nsec/m^2$
ν	kinematic viscosity, m^2/sec
ρ	density, kg/m^3
φ	angle measured from front stagnation point, deg
Nu	Nusselt number, $\alpha d/\lambda$
Pr	Prandtl number, $C_\eta \mu/\lambda$
Re	Reynolds number, $u_0 d/\nu$
Eu	Euler number, $\Delta\eta/\rho u_0^2$

Subscripts

f, o	conditions of the main flow
w	conditions on the wall

REFERENCES

[1]   A. Žukauskas, Int. Seminar. Herceg-Novi, Yugoslavia, (1969).

[2]   A. Žukauskas and J. Žiugžda, "Heat Transfer in Laminar Flow of Fluid", Mintis, Vilnius, 1969. (in Russian).

[3]   A. Žukauskas and A. Ambrazevičius, Int. J. Heat Mass Transfer 3, 305 (1961).

[4]   V. Katinas, J. Žiugžda, and A. Žukauskas, Lietuvos TSR MA Darbai, Ser. B. 4 (63), 209 (1970).

[5]   A. Žukauskas, V. Makarevičius, and A. Šlančiauskas, "Heat transfer in Banks of Tubes in Crossflow of Fluid", Mintis, Vilnius, 1968. (In Russian)

[6]   A. Šlančiauskas, R. Ulinskas, and A. Žukauskas, Lietuvos TSR MA Darbai, Ser. B 4 (59, 163 (1969).

[7]   R. Gregorig, Wärme Stoffübertragung 3, 26 (1970).

[8]   V. Makarevičius and A. Žukauskas, Lietuvos TSR MA Darbai, Ser. B. 3 (26), 231 (1961).

[9]   J. Stasiulevičius and P. Samoška, Lietuvos TSR MA Darbai, Ser. B. 4 (35); 77 (1963); 4 (55); 133 (1968).

[10]   K. Hammecke, E. Heinecke, and F. Scholz, Int. J. Heat Mass Transfer 10; 427 (1967).

[11]   J. Stasiulevičius and P. Samoška, Inzh. Fiz. J. 7, No. 11,10. (1964).

[12]   O.P. Bergelin, G.A. Brown, and S.C. Dorberstein, Trans. ASME 74, 953 (1952).

[13]   E.D. Grimison, Trans. ASME 59, 573 (1937).

[14]   V.I. Kuznetsov and S. Turilin, Izv. VTI. No. 11,23 (1952).

[15]   V.M. Antuf'yev and G.S. Beletsky, "Teploperedacha i aerodinamicheskiye soprotivleniya trubchatikh poverkhnostey v poperechnom potoke" Mashgiz, Moscow, 1948.

[16]   M.B. Lyapin, Teploenergetika No. 9, 49 (1956).

[17]   Ž. Kostič and S.N. Oka, Int. J. Heat Mass Transfer, 15, 279, (1972).

[18]   E. Achenbach, Wärme Stoffübertragung 2, 47 (1969)

[19]   Ž. Kostič and S. Oka, Int. Seminar, Herceg-Novi, Yugoslavia, (1968).

[20]   G.A. Mikhaylov, Sovietskoye kotloturbostroyeniye No. 12,434 (1939).

[21]   F.M. Mayinger and O. Schad, Wärme Stoffübertragung 1, 43 (1968).

[22] R.A. Bortoli, R.K. Grimmble, and J.E. Zerbe, Nuc. Sci. Eng. 1, 239, (1956).

[23] C.C. Winding and A.J. Cheney, Ind. Eng. Chem. 40. 1087 (1948)

[24] A. Žukauskas, Advan. Heat Transfer, 8 (1972).

[25] J. Stasiulevičius and P. Samoška, Lietuvos TSR MA Darbai, Ser. B. 4 (35), 83 (1963).

[26] S.B. Neal and J.A. Hitchcock, Heat Transfer, 1970, 3, FC7.8, Paris-Versailles, 1970.

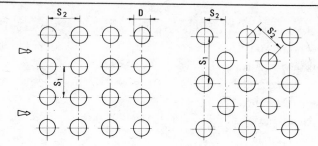

Fig. 1. Tube arrangement in banks.

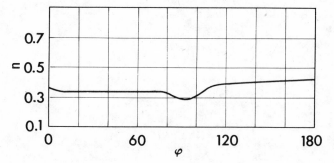

Fig. 2. Changes of the value of the exponent over Pr_f.

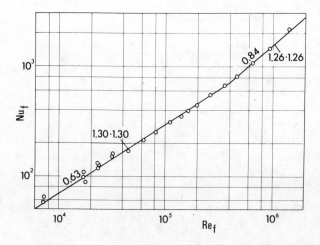

Fig. 3. Heat transfer of in-line banks in the flow of air at high Reynolds numbers.

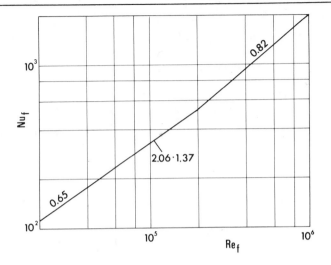

Fig. 4. Heat transfer of an in-line bank in the flow of air at high Reynolds numbers.

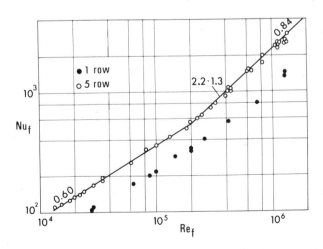

Fig. 5. Heat transfer of a staggered tube bank in the flow of air at high Reynolds numbers.

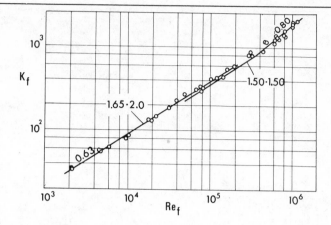

Fig. 6. Heat transfer of in-line tube banks in the flow of water at high Reynolds numbers.

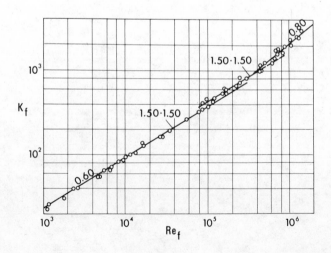

Fig. 7. Heat transfer of staggered tube banks in the flow of water at high Reynolds numbers.

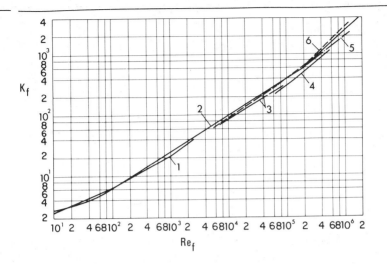

Fig. 8. Comparison of heat transfer of various staggered tube banks.

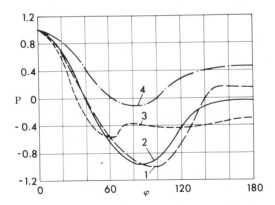

Fig. 9. Distribution of pressure coefficient in inner rows of staggered tube banks as a function of Re. 1-second, 2-fourth row of the bank (2.0 x 1.4) at Re = 1.5 X 10^6 [18].   3-second, and 4-fourth row of the bank (2.2 x 1.5) at Re = 4 X 10^4 [17].

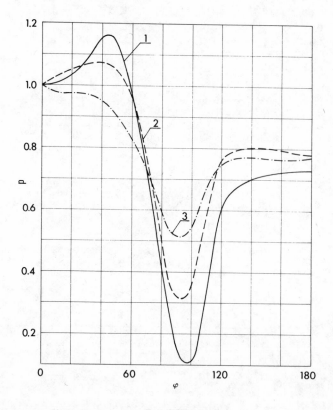

Fig. 10. Distribution of pressure coefficient in inner rows of the in-line tube bank (1.25 x 1.25) as a function of Re. 1-Re = 4 X 10^5; 2-Re = 6 X 10^5; 3-Re = 1.1 X 10^6.

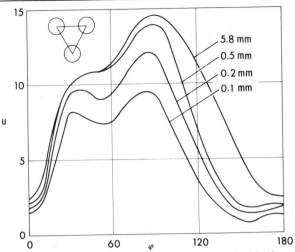

Fig. 11. Variation in local velocity around a tube in a bank [26].

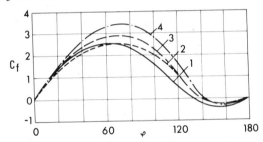

Fig. 12. Distribution of skin friction coefficient $C_f = (\tau/\rho u^2)\sqrt{Re}$ on a tube in a staggered tube bank, 2.0 X 1.4 [18]. 1-Re = 1.3 X 10^5; 2-Re = 8.0 X 10^6; 3-Re = 3.7 X 10^5; 4-Re = 1.4 X 10^6.

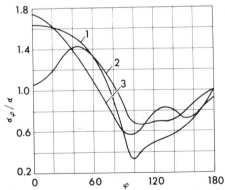

Fig. 13. Variation of local heat transfer of single tube and a tube in a bank. 1-single tube; 2-a tube in an in-line bank; 3-a tube in a staggered bank.

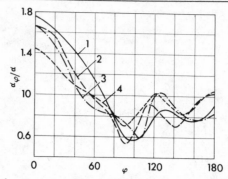

Fig. 14. Local heat transfer of a tube in various staggered banks at Re < 70,000. 1-(2.0 × 2.0), after Mikhaylov [20]; 2-(1.5 × 1.5), after Mayinger and Scad [21]; 3-(1.3 × 1.13)/after Bertolli et al. [22]; 4-(1.5 × 2.0), after Winding and Cheney [23].

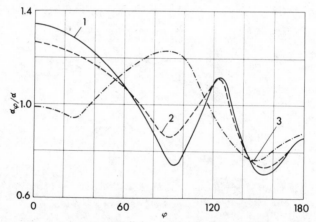

Fig. 15. Local heat transfer of tubes in staggered bank (1.5 × 1.5) in flow of water. 1-Re = 8 × 10^4; 2-Re = 1.5 × 10^5; 3-Re = 9 × 10^5.

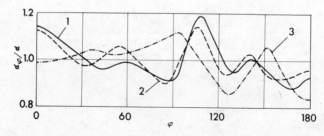

Fig. 16. Local heat transfer of tubes in in-line bank (1.5 × 1.5) in flow of water. 1-Re = 8 × 10^4; 2-Re = 1.5 × 10^5; 3-Re 9 × 10^5.

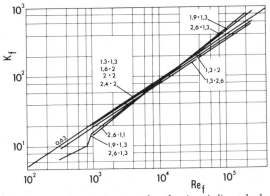

Fig. 17. Comparison of heat transfer of various in-line tube banks.

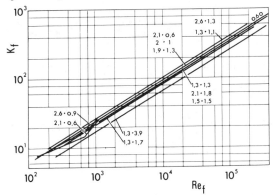

Fig. 18. Comparison of heat transfer of various staggered tube banks.

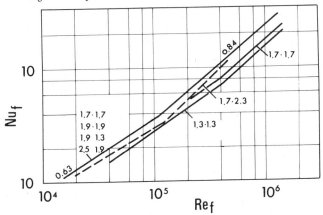

Fig. 19. Heat transfer of in-line tube banks at high Reynolds numbers [9].

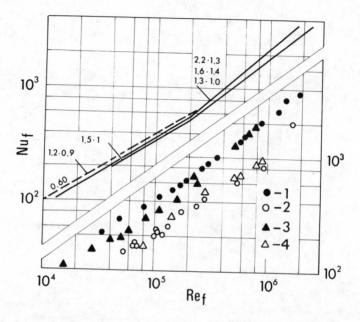

Fig. 20. Heat transfer of staggered tube banks at high Reynolds numbers [11]. 1-Fifth, and 2-first row of the bank (1.2 × 0.9); 3-fifth, and 4-first row of the bank (2.5 × 1.3).

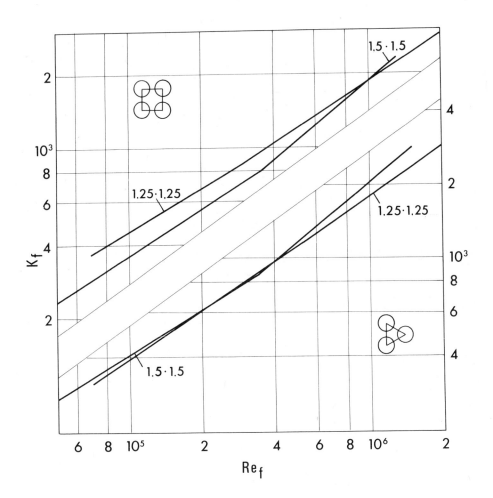

Fig. 21. Heat transfer of in-line and staggered tube banks in the flow of water.

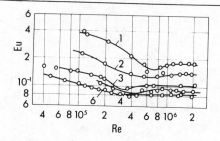

Fig. 22. Hydraulic resistance of the staggered tube bank, (1.5 × 1.04), referred to one row. Digits denote the total number of rows [25].

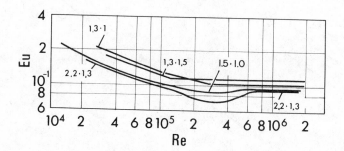

Fig. 23. Hydraulic resistance of staggered banks [25].

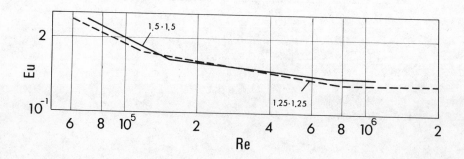

Fig. 24. Hydraulic resistance of in-line banks in the flow of water.

Chapter 5

SOME EXTRA-HIGH CAPACITY HEAT EXCHANGERS OF SPECIAL DESIGN

L. FORGÓ[*]

With regard to considerable recent technical advances, new requirements have to be met by heat exchangers, too, and constantly higher unit-capacities must be made available. In the field of power generation the unit capacity of turboalternators has been 150 MW to about 300 MW ten years ago, while at the present time their rating is as high as 1000 MW, units up to 2000 MW capacity are being developed, and turboalternators of even higher rating are already taken into consideration. Nuclear engineers require not only new constructions but also new heat transmitting mediums as well (CO_2, He, Na, etc.). The chemical industry, the production of fresh water from sea water, aviation, modern rocket techniques, as well as space investigations, etc. set new tasks, which must be performed by means of new solutions. Owing to the many aspects of this new technique, within the scope of a short lecture only some interesting examples, characterizing technical development, can be dealt with, in reference to a limited field only and without the aim of total coverage.

In the following I should like to deal with some up-to-date solutions illustrating the progress achieved in the development of heat exchangers, as well as with some suggestions which may be of interest in this field.

All these examples refer to methods where the requirements involving high thermal capacities necessitated the solution of particular problems.

---

[*]HŐTERV, Design Bureau for Industrial Power and Heat Supply, Budapest.

N.B. ALL FIGURES QUOTED IN TEXT ARE AT THE END OF THE CHAPTER

1. Air cooled condensing plant

It is known that heat discharge from condensers of thermal and nuclear power stations increases essentially in proportion to the unit-capacity of the installations, and the possibility of improving thermal efficiency is very limited. Under the present technical circumstances the upper limit for the unit-capacity of turboalternators is about 1200 MW. It is expected, however, that in the near future units of even higher capacity will be installed. Thus, in consideration that an amount of heat of 1.3×10^9 kcal/h has to be discharged from the condenser of a fossil fueled 1200 MW steam turbine, while in case of a light-water reactor this value amounts to around 2×10^9 kcal/h, it is obvious that problems arise which necessitate new solutions. This heat must be dissipated to the ambience at a low temperature level without contaminating the environment more than inevitable. Up to now the far smaller amounts of waste heat could be dissipated to natural water bodies by means of the so-called fresh-water cooling. In the case of the above-mentioned with much higher and further increasing capacities, this is generally no longer possible, because of the harmful warming of natural waters. Thus, this solution will not be applicable in the future. Another method to dissipate waste heat is the use of an evaporative type, the so-called wet cooling tower. In that case the natural waters are protected against water pollution. This has, however, the drawback that large amounts of water vapors would be discharged to the atmosphere.

The ambient air used in a fossil-fueled 1200 MW capacity turboalternator is about 2500 t/h vapor, which under certain weather conditions would favor fogging, or icing of the neighboring roads and airports.

From the viewpoint of environment protection, air cooled condensing installations have the advantage that these difficulties are not encountered at all. The waste heat is directly dissipated to the atmosphere (without the evaporation of water). It immediately rises to a great height and, consequently, contamination of the environment cannot occur.

The layout scheme of this air cooled condensing plant ("System Heller") is shown in Fig. 1.

The exhaust steam of the turbine is condensed in a jet condenser, the cooling water of which becomes recooled by means of the ambient air. The exceedingly high air flow required for cooling is usually produced by natural draft cooling towers. Though the layout of the installation is apparently simple, serious problems arise with regard to the economy of the whole system. These difficulties are due to the exceedingly great amount of heat, which has to be discharged at a small temperature difference, while considerable quantities of cooling water and air have to be kept moving. Thus, e.g. in case of the above mentioned fossil-fueled 1200 MW plant capacity the amount of heat to be discharged is around 1.3×10^9 kcal/h. This means that 130,000 m^3/h cooling water is to be circulated while the amount of circulating air is around 220×10^6 m^3/h. The pressure drop of flow on the water and air side must be kept within narrow limits because of economic reasons. Thus, e.g. the pressure drop of the air through the heat exchangers cannot exceed 3-4 mm w.g., and to maintain cooling water circulation only 0.5 - 0.7% of the generator rating can be used. Apart from these conditions which are difficult in themselves, other conditions are imposed by manufacturing, erection, corrosion resistance, etc. All these conditions, however, can be fulfilled with the following solution.

It is known that in the case of laminar-flow the heat transfer coefficient can be effectively increased — even if flow velocity is low — by decreasing the hydraulic diameter of the heat transferring surface. This circumstance ensures the efficiency of heat transfer even at a low flow resistance. Fig. 2 illustrates the design of a heat exchanger made according to this principle. The photo shows one section of the heat exchanger surface with the tubes carrying the cooling water and the fins along which the cooling air circulates. The slotted fins are arranged perpendicularly to the air flow. Thus, their dimensions are very small in the flow direction. With this design the heat transfer coefficient is relatively very high at a low air velocity. Consequently, the frontal area of these heat exchangers is very large. In order to keep space requirement within acceptable limits, the cooling elements are arranged zig-zagged. The tubes and fins shown on Fig. 2 are made of pure aluminum (99.5%). In order to ensure sufficient stiffness of the cooling elements, they are inserted into hot dip galvanized steel structures, which allows room for thermal expansion of the aluminum parts. The high thermal conductivity of the aluminum renders it possible to make the coolers with relatively thin and long fins and few tubes. The structural material and the applied special surface coating ensure adequate corrosion resistance. The fins are jointed to the tubes by means of cold pressing by which good heat conductivity is achieved. The standard cooling elements are 5 m long, 2.5 wide, and their thickness is 0.15 m (in the direction of air flow). Six, resp. 8, such cooling elements are assembled in a so-called "delta". Each delta of this type comes up to a power plant capacity of about 2 - 2.5 MW. Each delta is 15 resp. 20 m high; the two sides are made of the aluminum cooling elements; the third side is formed by steel stiffeners and, in case of certain weather conditions, with louvres for the regulation of air flow. The cooling deltas are completely factory assembled, pressure tested, and loaded in wagons. Fig. 3 shows a transport consisting of 10 deltas after arrival at the site of the power plant. In this case the 15 m long coolers were transported on wagons of 13 m length. Thus, it was necessary to place an unloaded wagon between two loaded wagons. The cooling deltas can be lifted from the wagon by a mobile crane and transported to the storage place or — under circumstances — immediately to the tower, as can be seen in Fig. 4. The cooling deltas can be transported by rail without any damage, even distances of several thousand km. Experience has allowed erection time at the site to be considerably reduced. In the case of adequate organization, 10 deltas can be mounted along the lower periphery of the cooling tower within 8 - 10 hours.

Fig. 5 shows the dry cooling tower for a turbine of 220 MW capacity. The dimensions of the tower are as follows: height 120 m, lower diameter 108 m, upper diameter 60 m. On Fig. 5 a steel framed cooling tower covered by corrugated sheet aluminum can be seen. This structure was prescribed at this site because of high seismic loading. Under normal circumstances such towers can be build also of reinforced concrete in the usual hyperbolic form.

In the case of very high capacities (800-1200 MW), the dimensions of the cooling tower are so large that difficulties arise when they are made of reinforced concrete. In such cases the steel structure offers better possibilities, as its static calculations include fewer uncertainties. Fig. 6 shows a steel-framed tower serving for a steam turbine of 800 MW capacity. In the instance of this tower — owing to the large amount of waste heat — the cooling deltas must be arranged along its lower periphery in two stages, each of 15 m height. With this arrangement even higher capacities can be installed in one tower, only the height of the cooling deltas must be increased to 20 m.

2. Multiple reheating cycle for steam turbines

An interesting suggestion for increasing cycle efficiency of large steam turbines has been made by L. Szücs [2]. Owing to the pressure of international competition, the efficiency of large turboalternators continuously increases. Under present technical circumstances, however, in order to achieve the slightest progress, intensive development has to take place. In the following, such research work will be dealt with.

It is known that the thermal efficiency of steam turbines can be increased by the multiple reheating of steam. This, however, imposes considerable surplus costs and, consequently, at the present time steam in the turbines is reheated only twice. At the same time multiple reheating of steam implies difficulties because the steam expanded in the turbine must be passed back to the boiler and subsequently returned to the turbine for further expansion, which makes it necessary to install pipelines of constantly larger diameter. This problem can be overcome if steam is not passed back to the boiler and reheated in it but instead some transmitting medium which transfers heat from the boiler to the turbine is subsequently used for reheating steam in heat exchangers, accomodated in the pipelines connecting the turbine casings. In selecting the transmitting medium mercury, resp. mercury vapor is suggested.

The main characteristics of this suggestion can be explained with reference to Fig. 7. The upper part of the figure shows the simplified scheme of the Big Sandy power station of 800 MW capacity [3]. As can be seen from the figure, the turbine operates with supercritical live steam and double reheat. The diameter of the pipeline behind the second reheater is 2 X 1200 mm, while operational pressure is 22 at and temperature 565°C. The surplus costs of further reheating would increase not only because of the long pipelines of large diameter, but also due to the necessary quick acting stop valves, which cut off the pipelines in case of eventual load drop of the generator thus preventing the runaway of the turbine.

The lower part of Fig. 7 shows the suggested new method. Here, the boiler generates superheated live steam as is usual at the present time. Subsequent reheating, however, is performed by using mercury vapor as a transmitting medium. For this purpose, a mercury vapor generator has to be installed at an adequate place in the boiler, where the generation of mercury vapor takes place at a pressure of 18.6 at and a temperature of 575°C. Mercury vapor thus generated is led through a pipeline to the turbine. The reheaters, which must have in this case very small dimensions, are accommodated in the pipelines connecting the turbine casings of different pressure and transfer heat between the condensing mercury vapor and the superheated steam. The mercury condensate is collected behind the reheaters and returned — by means of a pump — to the boiler. In this case the pipeline carrying mercury vapor would have a diameter of 2 X 1000 mm, and these dimensions can be further reduced behind each reheater. As the figure shows it is suggested that steam be reheated five times.

The applicability and economy of the suggested solution depends upon the following conditions:

a) The dimensions of the reheater must be reduced to such an extent that the reheater can be accomodated in the pipe connecting two casings;

b) Pressure drop on the steam side must be low;

c) The accommodation of the mercury vapor generator in the boiler has to be feasible technically and economically.

In reference first and foremost to the last item, it can be stated that mercury vapor generators were successfully tested many years ago and have been operating since that time (EMMET procedure). On the basis of experiences gained up to now, in principle no difficulties arise with these mercury vapor generators The same applies to the pipelines, pumps and fittings. As to the construction of the reheater, entirely new tasks have to be faced.

Strictly speaking, the reheater is a heat exchanger, which on the one hand condenses the mercury vapor and on the other hand reheats the already superheated steam to a higher temperature. Thus, it is evident that a device consisting of finned tubes must be selected.

In order to achieve a very high efficiency of heat transfer on the steam side in the case of a small pressure drop, the heat transferring surface is formed by thin plate fins and very small gaps in which the velocity of the laminar flow is low. Because of this low velocity flow between the gaps, the frontal area on the steam side is very large. Under these circumstances the space requirement of the reheater can only be reduced by arranging the transferring surface consisting of finned tubes in V-form and by installing as many of these elements as necessary. The upper part of Fig. 8 shows one section of the pipeline connecting the turbine casings, in the extension of which the reheater is accommodated. Though this figure is not exactly to scale, one can see that the pipeline has to be widened only slightly for accommodating the heat exchanger. The construction of the heat transfer surfaces with the finned tubes can be seen in the lower part of Fig. 8. These surfaces consist of flat tubes and very thin stainless steel strips joined by means of a special welding procedure. Attention has to be drawn to the dimensions: fin thickness 0.15-0.4 mm; gap 0.3-0.5 mm and distance between the tubes (thus, twice the height of the fins) 4-6 mm. The length of the fins in the direction of steam flow is about 20-60 mm. Apparently, this is a special construction realized by means of a special manufacturing technology. The strength of the flat tubes is ensured by the welded structure which makes the use of thin tube walls possible.

The heat transfer coefficient and the pressure drop in the case of laminar flow can be calculated on the basis of available technical literature. The question, however, arises in what measure flow resistance increases with the arrangement of the elements in V-form. To answer this question, model tests with air were carried out. They proved that under the above described circumstances it is possible to adjust the pressure drop on the steam side to about 0.5% of the absolute pressure. Another question is, whether the exceedingly narrow gaps do not become blocked by fouling. Obviously, this question can be answered only by long run tests carried out under normal operating conditions. Theoretically, it can be said that the formation of steam side deposits occurs only on the spots, where steam cools while expanding. In the mentioned case, however, neither considerable expansion nor cooling takes place, but on the contrary, steam becomes heated. If the Big Sandy power station were erected according to this suggestion, it would have been possible to achieve a 4.6% savings in specific heat consumption by means of reheating, repeated five times. It is evident however, that the realization of the above-mentioned suggestion necessitates the reconstruction of the boiler for accommodating the mercury vapor evaporator, and an entirely new turbine construction must be developed as well.

Another much simpler solution, which would be accordingly less advantageous, consists of effecting alterations only behind the second reheater. This could be realized by accommodating a reheater in the place marked with a dotted line in the upper part of Fig. 7. By way of experiment the electric heating of the reheater could be taken into consideration, since in this case the boiler ought not to be reconstructed, because one can assume that the mercury parts of the installation are known.

The "Lead technology"

Another method for heat transfer of manifold applicability which uses as a transmitting medium a liquid metal — preferably lead — has become known as the "lead technology" [4]. This process can be utilized — first and foremost — in chemical and petro-chemical plants, as well as in the processing of synthetic materials. It can further be used for waste heat recovery in the field of metallurgy and even for steam generation in nuclear power stations, etc. As this process usually works with liquid lead, with a melting point of $327°C$, processing above this temperature level can be taken into consideration. As a matter of interest, it has to be mentioned that it is relatively easy to put such installations, into operation since for melting 1 t of lead and heating it up to $400°C$ only 18000 kcals are required. The heat conductivity of fluid lead is 14 kcal/m h°C, sufficient for maintaining the whole metal bath practically at the same temperature. An advantage of the process lies in the relatively low cost of lead. Because of the manifold applicability of this process, different devices must be used in various fields. As an example, a steam generator of certain kind of nuclear power stations will be dealt with in more detail.

It is known that in the case of gas-cooled nuclear reactors one of the possibilities for generating electric power consists of producing with the hot gas high-pressure-superheated steam, which expands in a steam turbine. Usually, these steam generators are comprised of conventional structures, e.g. tubes, headers, etc. According to the suggestion made by F. Vollhardt, however, steam generation by means of lead technology can be realized in a more advantageous way, as it is shown in Fig. 9. This example refers to a helium-cooled high temperature reactor. The hot gas leaves the reactor at a temperature of 850°C and is led into a container partly filled with liquid lead. The inlet pipe dives into the liquid metal, and thus the gas comes into direct contact with the molten metal in the form of bubbles produced by means of an adequately perforated plate. Due to the large surface of the bubbles, the gas cools down very quickly and the released heat keeps the temperature of the liquid metal at about 650°C. As is shown schematically in the figure, the container is divided by means of a vertical partition wall into two parts. Consequently, this 650°C bath temperature prevails only in the left side of the tank. The superheater tubes of the steam generator are submerged into the liquid metal. Subsequently, the helium gas passes through the partition wall and comes again into contact with the fluid metal, in the form of bubbles. In this part of the tank the temperature of the metal is about 440°C. From here, the gas returns to the reactor through a droplet separator. The evaporator tubes of the steam generator submerge into the fluid metal in the second part of the tank. The helium gas leaving the tank can be further cooled by the economizer and afterwards is returned to the reactor by means of a compressor. Thus, continuous gas circulation is maintained. As has already been mentioned, the temperature of the lead-bath is in the first stage around 650°C, and here the evaporation of the fluid metal is not negligible. These metal vapors, however, enter — together with the helium gas — through the partition wall into the second stage, become condensed and retained by the here prevailing, lower temperature of the liquid metal. Thus, a certain amount of the liquid metal must be returned from the second stage to the first stage.

The separated metal droplets will be led back into the second stage as well. Evaporation in the second stage — at a temperature of 440°C — is already negligible. The entrainment of metal droplets by the gas can be kept within acceptable limits by maintaining surface loading of the liquid at a low level. Details concerning temperature control, etc. are not dealt with. It has, however, to be mentioned that the liquid metal is not sensitive to contamination. Foreign substances gather at the surface of the liquid and can be easily removed.

REFERENCES

[1]   J. Bódás: Dry cooling tower uses steel structure.
Electrical World, April 1, 1972   p.30.

[2]   L. Szücs: New Ways of Increasing Steam Turbine Cycle Efficiency.
Acta Techn. Scientiarum Hungaricae Vol. 58./1967/   p.111/132.

[3]   D. H. Williams, A. Hansprung: 800MW Expansion at Big Sandy Plant on the AEP System.
Proceedings of the American Power Conference 1968. Vol. 30.   p.491/499.

[4]   F. Vollhardt: "Lead technology", a new method of heat transfer.[Hungarian]. Energia és Atomtechnika XXIII. Jahrgang 1970. Nr. 10. p. 433/443.

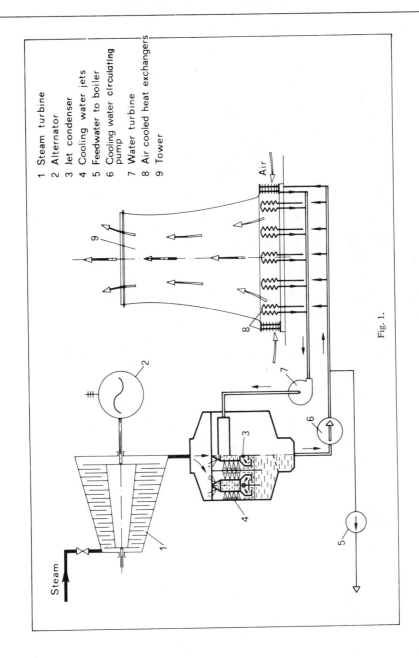

1 Steam turbine
2 Alternator
3 Jet condenser
4 Cooling water jets
5 Feedwater to boiler
6 Cooling water circulating pump
7 Water turbine
8 Air cooled heat exchangers
9 Tower

Air

Steam

Fig. 1.

Fig. 2.

Fig. 3.

Fig. 4.

Fig. 5.

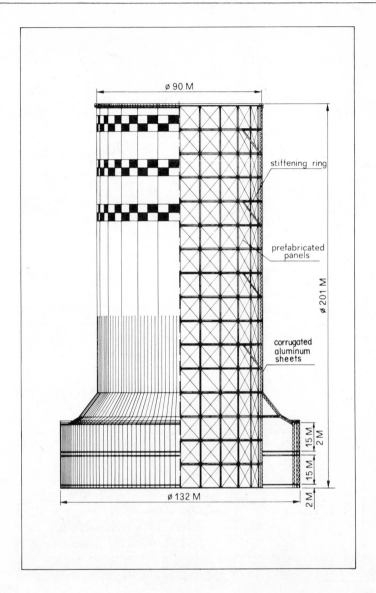

Fig. 6.

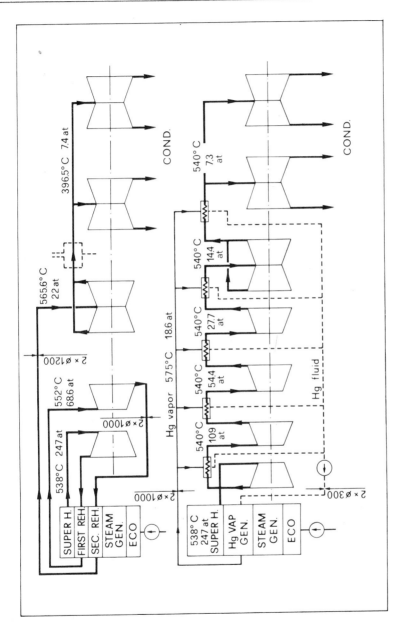

Fig. 7.

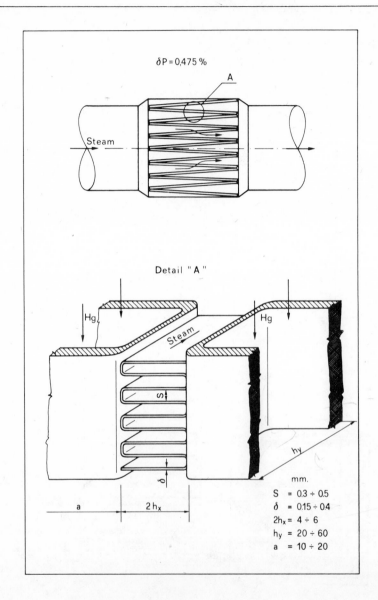

$\delta P = 0.475\%$

A

Steam

Detail "A"

Hg

Hg

Steam

S

δ

a

$2h_x$

h_y

mm.
S $= 0.3 \div 0.5$
δ $= 0.15 \div 0.4$
$2h_x$ $= 4 \div 6$
h_y $= 20 \div 60$
a $= 10 \div 20$

Fig. 8.

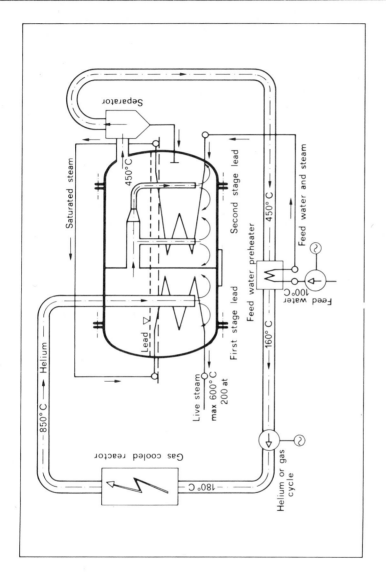

Fig. 9.

Chapter 6

THE OPTIMAL DESIGN OF HEAT EXCHANGER NETWORKS—
A REVIEW AND EVALUATION OF CURRENT PROCEDURES

T.W. Hoffman*

In most chemical and metallurgical processes, it is essential on economic grounds to recover and recycle energy. The most important part of this energy recovery is by way of heat transfer between outgoing hot and incoming cold streams. Moreover, temperature specifications on the hot and cold streams must be met and at the design stage a decision must be made regarding the use of a process stream or an external service (e.g., cooling water or steam) to accomplish the required heat duty. In even relatively simple situations, the problem of pairing and sequencing of exchanger streams becomes a large one and the use of the computer or short-cut techniques is necessary.

This chapter reviews and evaluates the essential details and evaluates some techniques which have appeared in the recent literature to solve this combinatorial problem.

THE GENERAL PROBLEM

In the design problem considered here there are s fluid streams, n of which are to be heated an m to be cooled to specified temperatures. The following specifications and/or assumptions pertain:

(i)   All information concerning flowrates, initial temperatures, heat capacities of the fluid streams is known.

(ii)   The final temperature requirements of the product, feed and intermediate process streams have been identified beforehand by process considerations. It is recognized that the optimal design of the overall process, of which the

(*) Department of Chemical Engineering, McMaster University, Hamilton, Canada.
N.B. ALL FIGURES QUOTED IN TEXT ARE AT THE END OF THE CHAPTER

heat exchanger network is only a small part, may be a strong function of these temperature specifications.*

(iii)   The availability of and costs for service heating or cooling (e.g., steam cooling water) are known. It is assumed that energy generated within the process under study cannot be "sold" outside it, so that recovery is only possible within it.

(iv)   Heat exchange is effected in a shell-and-tube heat exchanger and the overall heat transfer coefficient is known. This poses a problem since the overall coefficient depends on exchanger geometry and fluid properties (and hence temperature level) of the exchanging streams. Since the pairing of streams and temperature level are not known until the network has been established and further, the geometry of the heat exchanger is not known until its size is known, initial estimates of overall coefficients may have to be revised and the network reestablished.

This allows the economic design of any individual heat exchanger to be considered after the network has been established. Also, the interaction of the pressure drop with the overall process economics can be established later as well.

(v)   Heuristics, such as the minimum allowable approach temperature and maximum allowable temperature differences between hot and cold streams may be included.

(vi)   Operating cost data such as cooling water, steam, furnace fuel costs and capital cost data such as heat exchangers, coolers, heaters, furnaces are known. Capital costs of heat exchangers, furnaces etc. are correlated by a non-linear function of some characteristic, Z, by equations of the form:

$$C_{c_i} = a_i Z_i^{b_i} + k_i \tag{1}$$

where a_i b_i and k_i are constants depending upon the device i. Z_i is exchanger area for heat exchangers and heat load for furnaces, etc.

A constant amortization rate on capital equipment is specified.

(vii)   The controllability and cost of control and start-up which are related to process complexity are not considered in the economic objective function.

(*)Note that decomposition procedures for optimizing processes as suggested by Lasdon (1) still require the solution of the network optimization problem.

LITERATURE REVIEW

A brief "State-of-the-Art" review of previous work is presented below.

The early attempts at optimizing heat exchanger arrangements concentrated on the optimization of the exchanger arrangement and intermediate temperatures between heat exchangers, coolers and heaters for a single cold or hot stream. Bosnjakovic et al.[2], Mickley and Korchak [3], and Happel [4] consider the problem of optimizing the outlet temperature of a waste heat exchanger or exchanger trains. Ten Broeck [5] considered the optimization of a battery of waste heat exchangers. Westerbrook [6] applied dynamic programming to optimize a train of heat exchangers and a furnace. Bragin [7] employed the Discrete Maximum Principle to optimize the heat allocation and sequencing of hot streams to heat a cold stream. In each case however, the basic network was assumed known. Moreover, the methods are not applicable to interlocking networks where there are multiple hot and cold streams.

Hwa [8] was the first to consider the problem of optimal structure of multiple hot and cold streams. His formulation relies on piecewise linearization of the objective function and then solution by separable programming. A serious shortcoming of his method is that it requires the synthesis of possible network structures* beforehand.

Kesler and Parker [9] break up hot and cold streams into small finite elements of unit heat flow (e.g., 1×10^6 kcal./hr.) which they call exchangelets. They then consider the transfer of heat between exchangelets or groups of exchangelets. This allows a set of linear equations to be formulated along with the costs for each heat exchange combination and hence overcomes the problems associated with a nonlinear, nonconvex objective function. The optimal network is found by searching via alternate use of an assignment algorithm and a modified Linear Programming algorithm. The fact that finite elements are used may mean that global optimality has not been attained.

A similar linear programming approach has been reported by Kobayashi et al. [10]. They divide the process into an internal and external system. The internal system is comprised of process stream-to-process stream exchangers; the external

---

(*) The term network structure means the matching and sequencing of streams to achieve their desired temperatures.

system contains external sources and sinks. Similarly to Kesler and Parker, they break each stream up into finite heat elements and then apply an optimal assignment algorithm in linear programming to establish the optimum network structure of the internal system. Box's complex-method optimizing technique [11] is used to optimize the economic objective function of the combined internal and external system. Some iteration within this two-level approach may be necessary since some of the assumptions involved with the formulation of the internal structure may be invalid after the second optimization.

Nishida and coworkers [12] working like Kokayashi et al. with internal and external systems have proposed a graphical method or obtaining the best network structure. They then recommend that the optimum sizes and splitting of streams within this structure be found by use of Box's Complex method for constrained optimization problems. The rules for the graphical technique have been developed from the mathematical analysis of the special circumstance where all overall coefficients are equal and the heat capacity-mass flowrate product is the same for all streams. The method when applied to other cases does not guarantee optimality but does serve as a useful guide in structuring at least a near optimum network.

Masso and Rudd [13] use a heuristic method for building networks. It has a built-in learning capability which enables the system to move towards the optimal structure. The heuristics permit incorporation of very practical considerations but destory any guarantee of optimality.

King, Gantz and Barnes [14] have applied heuristic methods to an evolutionary approach to the design problem. Their method is quite general in that it can be applied to any process design problem; however, it does require a basic flowsheet at the outset.

Lee, Masso and Rudd [15] employ the branch and bound mathematical technique to reduce the amount of calculation required in this combinatorial problem. Their method guarantees optimality within the limitations of the heuristics that are included. This technique is ideal for heat exchanger networks where the same type of equipment is used in a relatively large number of places in the process.

Menzies and Johnson [16] have automated the branch and bound technique and applied it to optimize energy exchange networks where pressure and temperature effects are included. They also coupled this with a modular simulation program and lagrangian decomposition techniques to break large systems down into sub-systems of more manageable size. They applied it successfully to high and low pressure ethylene plants.

McGalliard and Westerberg [17] and Takamatou Hashimoto and Ohno [18] apply Lagrangian-based sensitivity analysis to the problem of optimizing a given exchanger network. This method thus serves as an alternative to the Complex method employed by the Japanese workers mentioned above. As will be shown later, except for the branch and bound approach, these methods should find use in optimizing the apparent optimum structure found by the other methods.

The purpose of this paper is to review the essential details of the methods of Kesler and Parker, Lee, et al., Nishida, et al. and Kobayashi et al. This will be done by applying these methods to a simple system. The predictions and limitations of the methods and some of the difficulties encountered in using the methods will be delineated by applying these techniques to obtain the optimal network of a real design problem. The techniques have been used as presented in the papers with little or no modification and hence results are presented from the standpoint of the user.

REVIEW OF METHODS

The Basic Problem.

The design problem to be used as a vehicle to demonstrate the methods is the one presented by Lee et al. and referred to in their paper as 4SPI. The details of the two hot and two cold streams are indicated in Table I; the basic economic and operating data are indicated in Table II.

The Graphical Method of Nishida, Kobayashi and Ichikawa.

Nishida et al. considered the heat exchange system to be broken into two subsystems (Figure 1): the interior subsystem where the process stream-to-process stream heat transfer occurs and the exterior subsystem where the heating and cooling by auxiliary services allows the temperature specifications to be achieved. This decomposition allows the interior and exterior subsystems to be synthesized separately. The key to the method, and indeed the link between the two subsystems, is the assignment of a total heat duty to the interior subsystem. This total interior heat duty determines not only the overall cost of the total heat exchange network but also the structure of the internal system. The initial step in applying this method is to determine the internal arrangement of exchangers and streams, given its total heat duty. The optimization of the whole system given this internal structure follows.

TABLE 1—DESIGN PROBLEM 4SPI

Stream No.	Flow rate	Input Temp.	Output Temp.	Heat Cap.
1 (S_{C_1})	20,643	140	320	0.70
2 (S_{H_2})	27,778	320	200	0.60
3 (S_{C_2})	23,060	240	500	0.50
4 (S_{H_1})	25,000	480	280	0.80

TABLE 2—DESIGN DATA

Stream (saturated) pressure		962,5 p.s.i.a. for problem 4SPI
		450 p.s.i.a. for problems 5SPI and 6SPI
Cooling water temperature	$t_w{}^i$	100°F
Maximum water output temp.	t_w	180°F
Minimum allowable approach	$\Delta T's$	
Heat exchanger	τ_{HE}	20°F
Steam heater	τ_H	25°F
Water cooler	τ_C	20°F
Over-all heat transfer coefficients		
Heat exchanger	U_{HE}	150 Btu/(hr) (sq ft) (°F)
Steam heater	U_H	200 Btu/(hr) (sq ft) (°F)
Water cooler	U_C	150 Btu/(hr) (sq ft) (°F)
Equipment down time	α	380 hr/yr
Heat exchanger cost parameters	a, b	350, 0.6
Cooling water cost	C_w	5×10^{-5} $/lb
Steam cost	C_s	1×10^{-3} $/lb

The basic problem in determining the optimum interior structure is defined as: For a given total heat duty in the interior subsystem and given process streams, synthesize a feasible exchanger network which will minimize the total area in this subsystem. This is stated as their defined optimum system.

In establishing this definition they have had to make the following additional limiting assumptions:

(i)   The cost of an exchanger is a linear function of heat transfer area; this means that the cost index, b = 1, and k = 0 in Equation (1)

(ii)   Only sensible heat is transferred.

(iii)   Each exchanger with its area calculated from the well-known expression.

$$A = \frac{q}{UF \, \Delta T_{l,m.}} \tag{2}$$

has the same effective overall heat transfer coefficient, UF.

In order to synthesize this optimum internal subsystem, they develop a number of rules based on analytically derived necessary conditions. An example of one of their necessary conditions with its inherent assumptions is: Given a system of equal number of hot and cold streams, all with equal hourly thermal capacity (weight flowrate-heat capacity product), assuming each stream is heat exchanged at most once and equal heat duty occurs in each exchanger, then it is shown that the streams should be matched.

$$(S_{h_1}, S_{C_1}), (S_{h_2}, S_{C_2}) \text{ --- } (S_{h_m}, S_{C_m}) \tag{3}$$

$$\text{where} \quad T_{h_1} \geqslant T_{h_2} \geqslant \text{ --- } T_{h_m} \tag{4}$$

$$\text{and} \quad T_{C_1} \geqslant T_{C_2} \geqslant \text{ --- } T_{C_m} \tag{5}$$

If the number of hot and cold streams is unequal, the matches are made in the same way with the coldest set matched with the hot streams if the number of hot streams is less than the number of cold streams and vice versa if the number of cold streams is the smaller.

Additional necessary conditions are summarized as follows:

- If multiple exchange is to occur between a hot and cold then it should be done countercurrently.

● If splitting of streams into substreams is to occur and each of these substreams if exchanged only once, then for optimal assignment of heat duty to each exchanger, the same fraction of heat should be assigned to each exchanger. Under these conditions the total area is independent of the number of exchangers.

These conditions become the rules for a graphical synthesis of the network on what they refer to as a heat content diagram (Figure 2). This is a plot of input and output temperatures of all streams in the system against their respectively hourly thermal capacity. Hot streams are shown above the horizontal axis, cold streams below. The origin of the horizontal axis is separate for each stream. The area of the block so formed represents the amount of heat to be removed from or added to each stream in order to satisfy its temperature specifications. This representation of the problem gives a very clear picture of any network design problem and is recommended to network designers.

Heat exchange is represented on the diagram by matching a hot block (or part thereof) with an equal area block in the cold section; this ensures heat balancing in an exchanger. Note that each horizontal division of a block corresponds to another heat exchanger for this stream; each vertical division corresponds to splitting a stream.

The three necessary conditions indicated above are used to establish the following rules for constructing the heat exchange network on this diagram:

(i)   Hot and cold blocks are matched (and numbered) consecutively in decreasing order of stream temperature (Theorem 1).

(ii)   For the hot blocks, the outlet temperature of exchangers in the i-th block is never lower than the inlet temperature of the $(i + 1)$-th block and similarly for the cold blocks the inlet temperature of the jth block is never lower than the outlet temperature of the $(j + 1)$-th block.

(iii)   If the total heat duty of the internal subsystem is smaller than both the total heat removed from hot streams or total heat added to cold streams, the highest temperature portion of hot blocks and coldest temperature portion of cold blocks are exchanged in the internal subsystem. The remainder is assigned to the heating and cooling external subsystem.

The application of these rules is demonstrated via the 4SPI problem. Figures (3) and (4) demonstrate the steps in solving this problem.

(i)   Figure 3 shows the heat content diagram for the streams listed in Table 1.

(ii)   An arbitrary total heat duty is assumed. This is chosen to allow a $20°F$. approach to be achieved between the inlet to hot stream S_{h1} and cold stream S_{C1}; hence a horizontal line is drawn at $460°F$ and the total heat duty on the hot streams provides an outlet temperature of $251.5°F$. on S_{h2}.

(iii)   From rule (ii), horizontal lines are drawn at the boundary temperatures as shown by the dotted lines on Figure 3.

(iv)   Boundaries between streams are ignored for the moment and the blocks of heat are separated as shown by the dotted lines on Figure 3.

(v)   Figure 4 shows the construction. By rule (1) the highest temperature hot stream is combined for exchange with the highest temperature cold stream (since the number of streams are equal). By the heat balance, the outlet temperature of the hot stream is $399.2°F$.

The remainder of the cooling of S_{h1} to $320°F$. is provided by part of the combined blocks 2 and 3, their outlet temperature by heat balance being $259°F$. Since there are two streams, S_{h1} must be split, the split (vertical line) or unknown weight flowrate in each stream is determined by the heat balance on blocks 2 and 3. The remainder of heat requirements on S_{C1} and S_{C2} between 240 and 259 are provided by the combined thermal capacity of S_{h1} and S_{h2} between 320 and 306.5. This leads to exchangers 4, 5 and 6 with an additional split on stream 1. The remaining heat in all of S_{h1} is exchanged with part of S_{C2} as shown. Exchanger 8 completes the network.

Figure 5 is the process flowsheet that arises out of the direct application of these rules. Obviously this network is much too complicated for a practical design since the control problems and costs may offset any expected saving.

A much simpler network results if one step in the rules is relaxed, namely rule (ii) concerning the minimum inlet and outlet temperatures. If rule (ii) is applied to the hot streams, but heat requirements of cold stream S_{C1} is allowed to be met entirely by S_{h1} and if the rules are applied rigorously for the remainder of the calculation, the heat content diagram as shown on Figure 6 results. This gives rise to the much simpler network, Figure 7.

This procedure may seem somewhat arbitrary but actually arises by considering ways to reduce the complexity of the rigorous network, as Nishida et al. have suggested. Note that the lower portion of the process flowsheet (Figure 5) shows almost a true countercurrent exchange between S_{h1} and S_{C1} except for the flow to exchanger 3 and 5.

The next step in the procedure is to determine the economic objective function for the entire process. Nishida et al. do not present details of their method but the GEMCS modular simulation routine has been found to be ideal for this purpose. The details of this technique are readily available* [19].

Basically, this simulation technique involves translating the process flow sheet into an information flow diagram in which information concerning the flow of material and thermodynamic state of any stream is transferred from one unit computation or module to another. All incoming information to the module is assumed known and this incoming information is modified in the unit computation; the outgoing information reflects these changes and becomes the incoming information for the next module in a specified calculation order. Here the unit computation describes heat exchangers and computes the heat and material balance when two or more streams are joined or split. Table III indicates a brief description of the modules. The procedure used is to size the exchangers in the network using the temperature specifications on each exchanger generated by the graphical procedure; this is done by the HEX2 modules (Figure 8a). The process is then simulated using the HEX1 modules in those situations where the outlet temperatures must not meet a temperature specification (Figure 8b). These exchanger areas are adjusted to ensure that all temperature specifications are met and the objective function evaluated. The optimization routine optimizes the entire network through manipulation of the fraction splits. If the heat duty of the internal subsystem is very much different from what was assumed in constructing the original network, the graphical procedure should be repeated to ensure that the internal network structure does not change. If it does, the entire procedure is repeated. In the example under consideration here the objective function was calculated to be $ 15,340 although the structure generated by the procedure was not optimized.

Global optimality is not ensured by this procedure but the simplicity of the procedure is its major advantage. Experience to date suggests that although the network structure is different from that generated by other techniques, the objective function is very close to that obtained by the other procedures to be discussed.

(*) A listing of the Fortran IV program and a user's handbook are available from the author.

TABLE III – DESCRIPTION OF MODULES USED IN NETWORK DESIGN, SIMULATION AND OPTIMIZATION

HEX1   calculates the outlet temperature given the heat transfer area and overall transfer coefficient by the effectiveness factor method (a simulator)

HEX2   calculates the area given incoming information and either the outlet temperature specification on one stream or the minimum approach temperature desired (a design model)

HEX3   calculates the area and stream requirements in order to meet a required temperature on the process stream

FURN1   calculates fuel requirements in a furnace to meet a required temperature on the process stream given the heating value of the fuel and furnace efficiency

COOL1   calculates the area and water requirements (given inlet and outlet temperatures) to meet a required temperature

JUNCO1   is a stream splitting module which requires specification of the fraction of the incoming flow going to each of the outgoing streams

JUNCO2   performs a heat and material balance on mixed streams

CONTL2   is a convergence tester. Since the networks usually involve recycle of stream information and all modules assume that incoming information is known, an iterative calculation is required on these streams. This module tests if the assumed and calculated information is within a given tolerance

CVRG1   is a convergence promotion module. Speed of convergence is increased over direct iterative calculations on the recycle streams. Model uses the method of Orbach and Crowe (20)

TEST1   tests if the temperature constraints have been met with the conditions of the simulation. If not it ensures appropriate increases/decreases in heat exchanger areas

OBJ 1   evaluates the objective function (cost/year) for any feasible network

OPT1   is a continuous optimizing routine to change the fraction split or other decision variable in the optimization

Branch and Bound Procedure

Lee, Masso and Rudd [15] were the first to suggest and demonstrate successfully the application of the branch and bound technique to the design of heat exchanger networks. Their application is reviewed below.

The mathematical foundation for this method may be simply stated as follows [15] [16]. If a given optimization problem A (maximization of an objective function, $O_A (D_A)$) is excessively difficult to solve, replace it by branching to a problem or set of problems, B, which are similar but much more easily solved than A. Problem B, however, must be selected to bound the original problem A. This means that if the optimal solution for problem A were available and inserted in the problem B, it must be a solution for problem B (satisfy all technical feasibility constraints) but not necessarily be optimal for B. This solution of B via the design A must indicate an equal or greater objective function $O_B (D_A)$ than the objective function for solution A, $O_A (D_A)$, in order to be a valid upper bound for A. This is expressed mathematically as

$$O_B \ (D_A) \ \geqslant \ O_A \ (D_A) \tag{6}$$

Furthermore, if the optimal solution of problem B is found and is feasible for problem A and gives equal values of the objective function when applied to both A and B, then it is also the required optimal solution to the original problem A.

The difficulty in using the method is in constructing the bounding problems for particular situations. Lee et al. have suggested the following

(ii)   Two approaches may be followed at this point: ·

(a) Combine the stream matches into stream processing paths for each primary stream, i.e., sequence the exchangers that have resulted from the stream matches of a particular primary stream. The cost of each processing path is easily determined; half the exchanger cost for a process stream/process stream match is assigned to any one path. For the 4SPI problem 34 = (10 + 5 + 7 + 12) processing paths are found. Note that although each path is feasible because streams are only used once, combination of paths to form the final network may not be possible because of multiple stream use.

TABLE IV EXAMPLES OF STREAM MATCHES

(NOTE: MAXIMUM HEAT EXCHANGE OCCURS IN
EACH HEAT EXCHANGER

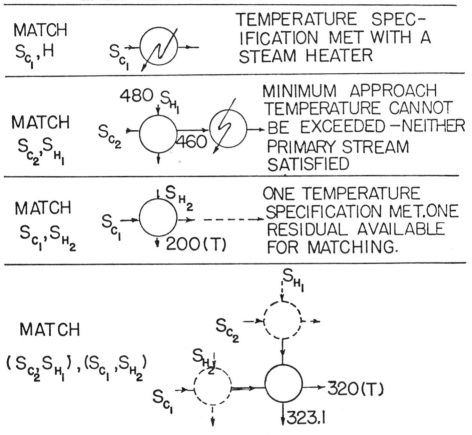

MATCH S_{c_1}, H		TEMPERATURE SPEC-IFICATION MET WITH A STEAM HEATER
MATCH S_{c_2}, S_{H_1}	480 S_{H_1} / S_{c_2} / 460	MINIMUM APPROACH TEMPERATURE CANNOT BE EXCEEDED –NEITHER PRIMARY STREAM SATISFIED
MATCH S_{c_1}, S_{H_2}	S_{H_2} / S_{c_1} / 200(T)	ONE TEMPERATURE SPECIFICATION MET.ONE RESIDUAL AVAILABLE FOR MATCHING.

MATCH

$(S_{c_2} S_{H_1}), (S_{c_1}, S_{H_2})$

RESIDUAL MATCHED WITH RESIDUAL
NO FURTHER MATCH POSSIBLE WITH
PROCESS STREAMS.

Unfortunately this method leads to a large number of possible networks: if there are m primary streams and n_i $(i = 1, m)$ possible paths the number of possible networks is

$$N = \prod_{i=1}^{m} n_i \tag{7}$$

Although most of these are infeasible, the computation task may be much too formidable even if automated.

This problem of evaluating all the feasible networks becomes the difficult design problem and the branching and bounding procedure is utilized again. In this case the difficult problem can be replaced by a set of problems that bound the one indicated above. Each set may be solved for the optimum network within it and then the optimum network is easily found from among the set.

In this example, a set of five problems is formulated as shown in Figure 9. This procedure leads to a drastic reduction in the number of primary stream processing paths in each problem. Similar branching can be initiated from any or all of these problems in the same way, although similar reductions with some eliminations will not result. The reader is referred to the original paper for further details.

This stream matching, path costing, branching procedure has been fully automated by Menzies and Johnson [16]. The 4SPI problem as outlined in Table I was solved using this program. The network is shown in Figure 10; its cost was found to be $ 15,454.*

SOLUTION BY LINEAR PROGRAMMING

Two independent solutions of this network optimization problem which utilize linear programming procedures have been reported. Kobayashi, Umeda, and Ichikawa [10] used the standard solution of the assignment problem in linear programming to determine the best internal subsystem network and then optimized the combined internal and external systems by Box's Complex optimization procedure. Kesler and Parker [9] formulated a special assignment algorithm procedure to determine the optimum total network. No further optimization was

---

(*) This figure is different from Lee's [15] since the amortization rate was assumed to be 25%; Lee did not report his value in his paper.

suggested.

Kessler and Parker's Solution

To cast the problem in the context of linear programming, Kesler defined the network problem in the following way.

Consider m hot streams and n cold streams as defined earlier. For the m hot streams, assuming each cold stream exchanges with each hot stream:

$$\sum_{j=1}^{n} Q_{ij} = \Delta H_i \qquad i = 1,2 \text{ --- } m \qquad (8)$$

where Q_{ij} is the heat duty in the ij^{th} exchanger and ΔH_i is the total heat removed in the ith hot stream.

Similarly for the n cold streams,

$$\sum_{i=1}^{m} Q_{ij} = \Delta h_j \qquad j = 1,2 \text{ --- } n \qquad (9)$$

These linear expressions satisfy the heat balance on each stream and the overall balance is indicated by

$$\sum_{i=1}^{m} \Delta H_i = \sum_{j=1}^{n} \Delta h_j \qquad (10)$$

The non-linearities are contained in the cost function, which is to be minimized:

$$\sum \sum \rho_{ij}(Q)Q_{ij} = C \qquad (11)$$

where the cost of removing heat, ρ_{ij}, is some non-linear function of the amount of heat removed, the physical properties of the streams and the temperatures of the streams.

The key to this method is the linearization of Equation (11) by fracturing each stream into heat elements or exchangelets, q, sufficiently small* that that objective function becomes a linear function of the elements.

(*) for example, 1×10^6 B.t.u./hr. or 3×10^6 kcal./hr. etc.

Mathematically this means,

$$\sum_{k=1}^{K_i} q_{ik} = \Delta H_i \tag{12}$$

for each of the i-1,2 --- m streams each of which has K_i elements and for the n cold streams, each with L_j elements

$$\sum_{\ell=1}^{L_j} q_{j\ell} = \Delta h_j \tag{13}$$

As before $\sum_i \Delta H_i = \sum_j \Delta h_j$ (14)

Each exchangelet can be defined uniquely by the parameter $X_{ikj\ell n}$ where i, is the hot stream number, k is the number of the element (hence defining the origin of the exchangelet), j is the cold stream number with which it is exchanging heat and ℓ is the number of element of stream j; n is the number of elements involved.

By this definition;

$$\sum_{j,\ell,n} X_{ikj\ell n} = q_{ik} \tag{15}$$

$$\sum_{i,k,m} X_{ikj\ell n} = q_{j\ell} \tag{16}$$

$$\sum_{i,j,k,\ell,n} \rho_{ikj\ell n} X_{ikj\ell n} = C \text{ (minimum)} \tag{17}$$

and

$$X_{ikj\ell n} = 0 \text{ or } 1 \tag{18}$$

The set of linear equations (12) through (18) in variables X fully defines the problem. The number of equations in the set to be optimized depends upon the number of streams in the problem and the number of elements into which each stream is fractured; this does not represent a large number (34 in the 4SPI problem to be discussed). The number of possible exchanges of single or multiple exchangelets, i.e., the number of X's, can be very large. By introducing certain limitations such as the minimum approach temperature or maximum difference in temperature between a hot and cold stream and the maximum size of a heat exchanger, this number can be significantly reduced. In the 4SPI problem (with $10°$ approach temperature and a $300°$ maximum temperature difference between heat source and heat sink) the number was 444.

Basically Kesler's procedure is to evaluate the cost of exchanging heat between each possible cold and hot exchangelet or group of exchangelets to provide a main matrix. This main matrix is then used to form an assignment matrix with each row representing one of the cold elements and each column representing one of the hot elements. This procedure ensures the integer character of the problem as expressed by Equation (18). Entries in this matrix other than zero indicate an exchange between the respective cold and hot elements. This procedure formulates the problem as a classical assignment problem in linear programming. Its solution is solved by standard methods although some modification was required to assure feasible solutions.

Kesler's Solution to the 4SPI Problem

The initial step in Kesler's method is to choose the size of a heat element. Kesler provides no guidance in this selection although its size cannot be greater than the smallest hourly heat capacity of any one stream. Since it is highly unlikely that each stream will form an integral number of exchangelets some heat imbalance will arise because of the integer requirements. Decreasing the size of the elements will ensure greater accuracy but will increase the dimensions of the main and assignment matrices significantly. In the 4SPI problem the element size was 4×10^5 B.t.u./hr. The fracturing of the streams is best represented on the heat content diagram, Figure 11. The horizontal lines indicate the temperature limits for each exchangelet. The area of each block should be the same. Note, however, that the cold streams really do not have an integer number of elements.

In this solution additional heating and cooling elements must be included to satisfy heating and cooling demands that cannot be met with the process streams. Furthermore, this method requires an equal number of hot and cold elements since each element must exchange heat with another element in order to meet the temperature specifications of all streams. Since an equal number of hot and cold elements arose with our choice of element size, there is a temptation to add a steam heater element (and hence a cooling water element to balance the element) as Kesler suggests and to search for a solution. No feasible solution could be found, however, since upon inspection it will be noted that the upper two elements of S_{C_1} cannot be exchanged with process fluids. The actual demands are clearly indicated by superimposing the cold part of the heat content diagram on the hot. Hence, it became necessary to modify Kesler's original program to allow for an unequal number of cold and hot exchangelets and to allow more than one external heat unit to be supplied. Unfortunately increasing the number of external hot units means that the number of external cold units must be increased accordingly. This then leads to different networks that are found by the other methods and represents a serious limitation of the method.

The optimal solution of the 4SPI network is shown as Figure 12. It's cost is appreciably greater than that obtained by the previous methods. This network was optimized using the GEMCS procedure indicated earlier and the 20° minimum approach temperature. The resulting cost was $ 15,977/year.

The Method of Kobayashi; Umeda and Ichikawa

As indicated earlier, like Nishida, Kobayashi et al. divide the system into the internal and external subsystems. Likewise, a total heat duty is specified for the internal system and then the internal network is determined through solution of the classical optimal assignment problem in linear programming.

Their linear programming formulation for the internal subsystem is almost identical to that of Kesler and Parker except that they require each exchanger to transfer only one unit of heat (one exchangelet); hence all exchangers have the same heat heat duty. Moreover, in this formulation they require an equal number of hot and cold exchangelets, although they have overcome this problem by introducing hypothetical streams which carry a high processing cost, so that they will not enter into the final structure. Similarly their solution allows for fewer exchangers than the number of exchangelets. A major difference arises in this method since it allows for each stream to be processed in more than one heat exchanger. This is achieved by splitting hot and cold streams into equal hourly heat capacities and hence doubling the number of streams to be considered. No guidance is given as to when a stream should be divided nor how large each exchangelet

Box's Complex method [11] is recommended for the final optimization. This method coupled with linearized modelling of the heat exchange network seems to be much more complex and restrictive than the simulation-optimization scheme suggested earlier in the paper.

No direct evaluation of this method has been made except to note that if the streams are split in the way suggested in this paper, Kesler's method can be utilized to find a solution. This was demonstrated with the refinery problem presented in their paper and utilized as a case study in the next section.

A CASE STUDY

An evaluation of these procedures would be incomplete without an application to a real problem. The case presented by the Japanese workers [10] [12]* involves the heat exchange system around a topping tower in a 80,000 bbl./day oil refinery. A cold crude oil stream is heated to 135°C at which temperature it enters a desalting unit. After leaving the desalter at 130°C it must be preheated to 355°C before it undergoes further processing. Seven hot streams of rather small hourly thermal capacity must be cooled to specified temperatures. A pipe still furnace is available for auxiliary heating and water coolers are used for auxiliary cooling. The inlet and specified temperatures along with their flows and heat capacities are indicated in Table V. Table VI presents the cost and operating data for the system. (The reported conventional system is shown on Figure 13.**); the synthesized networks along with their costs are presented on Figures 14 and 15.

The following points summarize this experience:

(1)   **Solution via Kesler and Parker's Method**

An exchangelet size of 3×10^6 kcal./hr. was used in this calculation which corresponded to the hourly heat capacity of the smallest hot stream, S_{h_2}. Considerable difficulty was experienced in expanding the original program to accept this problem. Approximately 100 k of memory on a CDC 6400 computer was required; solution was reasonably fast performing about 5 iterations/sec. If a smaller

---

(*) Both authors use the same example except that there are inconsistencies in the data both between and within the papers. Nishida's description is used here.

(**) The costs are not exactly as presented in their paper. This is partly because of inconsistent and incomplete data and also because the solution reported by Nishida et al. seemed to be in error.

TABLE 5   STREAMS PROPERTIES OF THE APPLICATION PROBLEM

(Hot streams)	T_h (°C)	T_{hi}^* (°C)	w_{hi} (ton/hr)	$C_p(T_{hi})$	$C_p(T_{hi}^*)$ (kcal/kg °C)
S_{h_1}	340	90	200	0.69	0.46
S_{h_2}	330	195	30	0.74	0.63
S_{h_3}	270	55	60	0.72	0.53
S_{h_4}	265	215	110	0.71	0.67
S_{h_5}	235	145	95	0.75	0.66
S_{h_6}	205	150	55	0.72	0.67
S_{h_7}	150	50	315	0.72	0.62

(Cold streams)	T_{cj} (C)	T_{cj}^* (C)	w_{cj} (ton/hr)	$C_p(T_{cj})$	$C_p(T_{cj}^*)$ (kcal/kg C)
S_c^1	25	135†	425	0.50	0.60
S_c^2	130†	355	425	0.60	0.79
S_w	30	55		1.0	1.0

† Input temperature to the desalting unit.

TABLE 6   NUMERICAL DATA OF THE APPLICATION PROBLEM

Operating conditions		
Thermal efficiency of furnace	0.80	
Heating value of fuel	9500	(kcal/kg)
Annual operating hours	8000	(hr/year)
Economic data		
Annual return of investment cost, δ	0.25	
(Operating cost)		
Unit cost of cooling water	0.83	(cent/ton)
Unit cost of fuel	1.94	(Cent/1)
Investment cost		
Shell and tube type heat exchangers	55.56 x A	($/unit)
Furnace	$2.50 \times Q^{0.7}$	($/unit)
Annual return of the investment cost	0.25	
Operating cost		
Unit cost of cooling water	0.83	(cent/ton)
Unit cost of fuel	1.94	(cent/1)

heat element were used it is doubtful whether the available 140 K of memory would be sufficient to solve the problem with the present program.

Only the solution indicated on Figure 14 was found. The program repeated the same calculations each time it found this solution, so that more iterations would have not effected another solution. This network could be optimized considerably however by the simulation method presented earlier, so that the cost/year could be reduced to $ 8.74 \times 10^5$ /year.

Many networks could be generated using Kobayashi's stream splitting technique; the best of these was not optimized so it is not reported.

(2)   **Solution by the method of Nishida et al.**

The ease with which a network can be synthesized and simulated (and hence optimized) even with a relatively difficult system makes this technique an extremely powerful one. Moreover, the computer requirements are only nominal, so that good networks can be synthesized even on small computers. The cost for control and implementation of the rather complicated networks that result should be evaluated and included in the objective function before a direct comparison with other networks can be made.

(3)   **The Method of Kobayshi et al.**

The complicated structure that results from this procedure is a serious drawback. Furthermore other than arbitrary rules for deciding stream splitting must be established before the method will prove useful.

(4)   **The Branch and Bound Method**

No solution could be obtained with the automated branch and bound procedure because of the high demands the method places on computer storage. Furthermore, although Lee et al. suggest that their method is amenable to hand calculation, this would not be possible in this case. The basic problem is in determining the stream matches. In the order of 650 separate exchangers were required (representing the feasible primary stream and residual matches). This meant that all the cost information of the heat exchangers and the stream history information associated with each of these matches had to be stored. This required in excess of the 140 K storage available on a CDC 6400 computer. Moreover since most of the storage was required for storage of information, overlaying of programs did not alleviate the problem. Considerable economy of storage could be effected by packing more than one number in each memory location. This modification is planned for the future since the branch and bound technique seems to be the most powerful optimization method available at present. Computer time to solve this problem would be less than 100 sec. on the CDC 6400 computer.

SUMMARY AND CONCLUSIONS

A review and an evaluation of the current methods of synthesizing heat exchanger networks has been presented. A number of shortcomings of these methods has been uncovered when they were applied to a real problem. In particular the high demand on computer memory of these methods mitigates against their use for designing large systems.

The method proposed by Nishida coupled with the optimization program presented here seems to be worthy of more evaluation mainly because of its nominal computer requirements. Since global optimality is not assured by this method only experience will dictate its real usefulness.

ACKNOWLEDGMENTS

This work was supported through a grant from the National Research Council of Canada. The Kesler program was programmed with some modification by Mr. J. Li on the CDC 6400 computer through a contract from the Atomic Energy of Canada. Dr. M.A. Menzies supplied his version of the branch-and-bound program and helped in modifying it to accomodate general heat exchanger networks. His help in receiving the current literature is also gratefully acknowledged.

REFERENCES

[1]   Lasdon, L.S. "Optimization Theory for Large Systems", MacMillan (1970).

[2]   Bosnjakovic, F., Vilicec M. and Slipcevic, B. VDI. Forschungsheft 432, Ausgabe B, Band 17 (1951).

[3]   Mickley, H.S. and Korchak, E.I., Chem. Eng., 69, No 23, 239, Nov. 1962.

[4]   Happel, J. "Chemical Process Economics", John Wiley (1969).

[5]   Ten Brock, H., Ind. Eng. Chem. **36** 64 (1944).

[6]   Westerbrook, G.T., Hyd. Proc. Pet. Ref. **40** 20 (1961).

[7]   Bragin, M.S. "Optimization Multistage Heat Exchanger Systems", Ph.D. Thesis, New York University (1966).

[8]   Hwa, C.S., A.I.Ch.E. – I.Ch.E. Symp. Ser. No 4, 111 (1965).

[9]   Kessler, M.G. and Parker, R.O., CEP Symp. Series, **65** No 92, 111 (1969).

[10]   Kobayashi, S, Umeda, T. and Ichekawa, A., Chem. Eng. Sc. **26**, 1367 (1971).

[11]   Box, M.J., Computer Journ. **8** 42 (1965).

[12]   Nishida, N., Kobayashi, S. and Ichikawa, A., Chem. Eng. Sci. **26** 1841 (1971).

[13]   Masso, A.H. and Rudd, D.F., A.I.Ch.E.J., **15**, No. 1, 11 (1969).

[14]   King, C.J., Gantz, D.W. and Barnes, F.J., Ind. Eng. Chem. Proc. Des. Dev. **11** 272 (1972).

[15]   Lee, K.F., Masso, A.H. and Rudd, D.F., I & EC Fundamentals, **9**, No 1, 48 (1970).

[16]   Menzies, M., and Johnson A.I., Paper 1C Presented at the Dallas Meeting of A.I.Ch.E. February 1972.

[17]   McGalliard, R.L. and Westerberg, A.W., Paper 2A presented at the Dallas Meeting of A.I.Ch.E. February 1972.

[18]   Takamatsu, T., Hasimoto, I. and Ohno, H., Ind. Eng. Chem. Proc. Des. Dev. 9, 368 (1970).

[19]   Crowe, C.M., Hamielec A.E., Hoffman, T.W., Johnson, A.I., Shannon, P.T., and Woods, D.R., "Chemical Plant Simulation", Prentice Hall (1971).

[20]   Orbach O. and Crowe, C.M., Can. Journ. Chem. Eng. **49** 509-13 (1971).

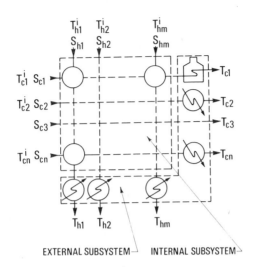

Fig. 1. Schematic Representation of the Combinatorial Heat Exchanger Network Design Problem.

HEAT CONTENT DIAGRAM

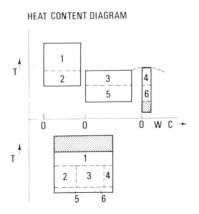

Fig. 2. Heat Content Diagram as Suggested by Nishida et al. [12].

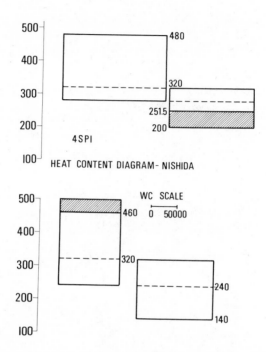

Fig. 3. The 4SPI Problem Represented on a Heat Content Diagram with Preliminary Construction.

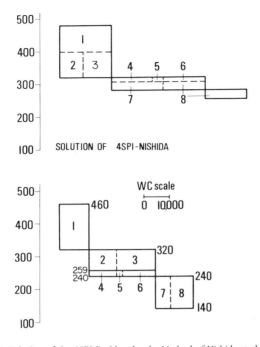

Fig. 4. Solution of the 4SPI Problem by the Method of Nishida et al. [12].

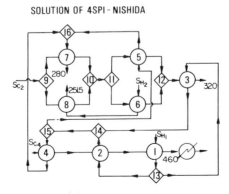

Fig. 5. Process Flowsheet for 4SPI Problem from Heat Content Diagram (Fig. 4).

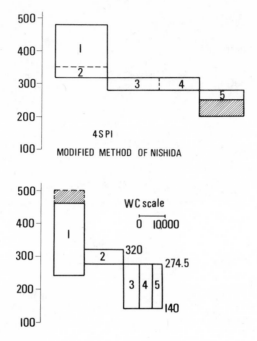

Fig. 6. Heat Content Diagram for Modified Procedure of Nishida et al. [12].

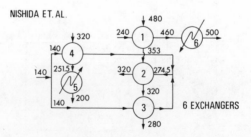

Fig. 7. Process Flowsheet Arising Out of Modified Method.

DESIGN MODE

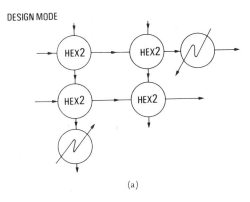

(a)

SIMULATION MODE

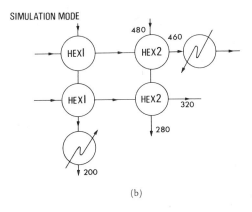

(b)

Fig. 8. (a) Using Modular Simulation Executive Computer Program in the Design Mode. (b) Using the Executive Program in the Simulation (Optimization) Mode.

CONSIDER THOSE NETWORKS ONLY WHERE:

1. NO STREAM MATCHING- STEAM OR WATER TO SATISFY T_{SPEC}

2. ALL NETWORKS COMPATIBLE WITH A MATCH OF S_{h_1} AND S_{c_1}
   E.G.

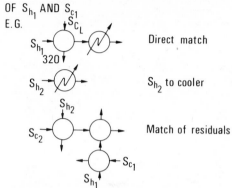

    Direct match

    S_{h_2} to cooler

    Match of residuals

3. COMPATIBLE WITH A $S_{c_1}- S_{h_2}$ MATCH

4. COMPATIBLE WITH A $S_{c_2}- S_{h_1}$ MATCH

5. COMPATIBLE WITH A $S_{c_2}- S_{h_2}$ MATCH

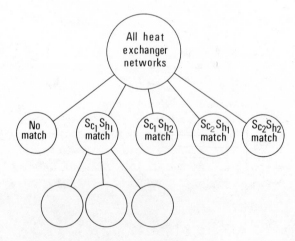

Fig. 9. Branch and Bound Representation of the 4SPI Problem Showing Branching to Smaller Size Problems where Networks are Considered Which are Only Compatible with Certain Specified Matches.

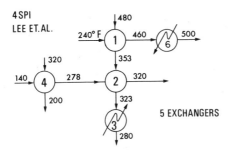

Fig. 10. 4SPI Process Flowsheet from Branch and Bound Method.

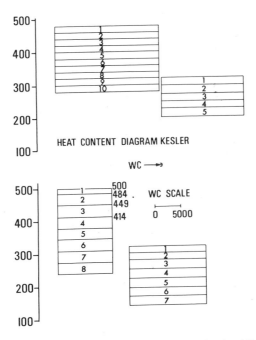

Fig. 11. The Solution of the 4SPI Problem by the Method of Kesler and Parker [9]. Exchangelets used in the solution Represented on a Heat Content Diagram.

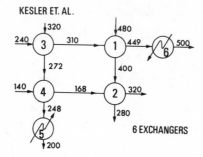

Fig. 12. 4SPI Process Flowsheet by the Linear Programming Method of Kesler and Parker [9] after further Optimization of Flowsheet.

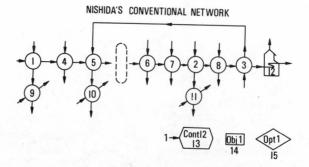

Fig. 13. The Conventional Process Flowsheet for the Refinery Heating/Cooling System (Cost = 11.02×10^5/year).

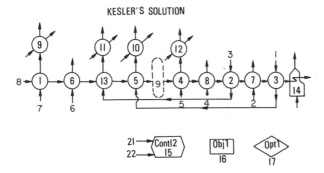

Fig. 14. Information Flow Diagram for Design/Simulation Program Showing Solution of the Refinery Problem by the Method of Kesler and Parker (Cost by Kesler Method = 13.87×10^5; Optimized Network Cost = 8.74×10^5/year).

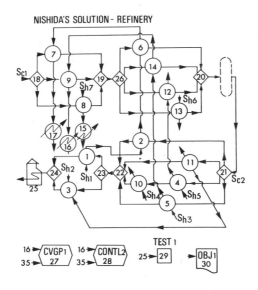

Fig. 15. Solution of the Refinery Problem by Nishida's Method. (Information Flow Diagram for Design/Simulation Program). (Cost = 7.33×10^5/year).

Chapter 7

A CALCULATION PROCEDURE FOR THE TRANSIENT AND STEADY-STATE BEHAVIOR OF SHELL-AND-TUBE HEAT EXCHANGERS

S.V. Patankar and D.B. Spalding (*)

1. INTRODUCTION

1.1 The problem considered

Fluid-dynamics theory, for reasons of history and fashion, has paid little attention to the phenomena which occur in industrial equipment such as heat exchangers, cooling towers and chemical reactors. Even now, when digital computers have removed the obstacles to numerical computation, the prized achievements of applied mathematicians concern more often the supersonic flow of air about a missile, or the development of an eddy behind a cylinder, than any phenomenon of interest to the process engineer.

As a consequence, the designer of process equipment must usually base predictions of its performance on assumptions about the flow patterns; departures from these assumptions are then allowed for by empirically-derived correction formulae. Thus a single-pass shell-and-tube heat exchanger may be designed by reference to the formula for the performance of an ideal counter-flow exchanger, modified by some performance-degradation factor which takes account of departures from the ideal flow pattern.

Unfortunately, although this practice can serve adequately for interpolation between equipment items of common type with only slight variations of geometry, it fails completely where the benefit of new geometrical configurations is to be explored; optimum configurations must therefore be determined experimentally as a rule.

The aim of the present chapter is to start a new fashion in process-equipment design, by applying to a shell-and-tube-heat exchanger a sophisticated numerical procedure for calculating fluid-flow distributions. The exchanger geome-

(*) Mechanical Engineering Department, Imperial College, London, S.W.7, England.

N.B. ALL FIGURES QUOTED IN THE TEXT ARE AT THE END OF THE CHAPTER

try has been deliberately chosen so as to necessitate a three-dimensional procedure: three velocity components have to be computed for the shell fluid at each point; and the lack of axial symmetry causes these velocities, and the temperature and pressure also, to vary throughout all three space dimensions.

It will be shown that the fields of velocity, temperature and pressure can indeed be calculated, in transient operation as well as in the steady state; and it will be argued that the increased realism of the predictions, as compared with more conventional calculation procedures, must often have such economic value to the designer as to render acceptable the greater cost of the computation.

1.2 Outline of the present paper

The present authors and their colleagues have reported elsewhere (Patanker and Spalding, 1972; Caretto et al., 1971) the procedure that they are developing for solving the differential equations which govern the flow of a continuous fluid. This can be applied to flow in the shell of a heat exchanger in a straightforward way, by construction of a finite-difference grid which is sufficiently fine to represent realistically the details of the tube-wall configurations. However there is another possibility: to regard the space within the shell as uniformly filled with fluid, through which however is distributed, on a fine scale, a resistance to fluid motion. It is the latter approach which is taken in the present paper.

This continuum approach has two distinct advantages; firstly, it allows use of a grid which is much coarser than that required for the straightforward approach, with consequent economy of computer time and storage; and, secondly, it provides an easy way of exploiting the extensive experimental data on the pressure-drop and heat-transfer performance of tube banks, without the errors which result from assuming that the flow through them is everywhere at the same velocity and angle. When computers are larger and our knowledge of turbulent flow more profound, the straightforward approach may prove preferable; at present, however, the continuum approach seems to have more to offer.

The organization of the paper is a simple one. In Section 2, we provide a brief resumé of the main features of the computational technique. Section 3 then illustrates the capabilities of the procedure by presenting the results of computations which have been made for the steady — and unsteady — state behavior of a shell-and-tube heat exchanger. No attempt is made to compare the results of the predictions with experimental data; but it is argued in the discussion, in Section 4, that such comparisons should now be made by the builders of heat exchangers. If, as seems probable, these comparisons are successful, designers may well decide that the present method of computation, or another of similar style, should be regularly used in design and optimization exercises.

2. THE SOLUTION PROCEDURE

2.1 The equations solved

2.1.1 Differential equations

The equations which are supposed to govern the distributions of the shell-fluid velocity components u, v and w in the three Cartesian coordinate directions x, y and z are:

$$\rho \, \frac{\partial u}{\partial t} + \rho u \frac{\partial u}{\partial x} + \rho v \frac{\partial u}{\partial y} + \rho w \frac{\partial u}{\partial z} = - f_x u - \frac{\partial p}{\partial x} \quad , \tag{1}$$

$$\rho \, \frac{\partial v}{\partial t} + \rho u \frac{\partial v}{\partial x} + \rho v \frac{\partial v}{\partial y} + \rho w \frac{\partial v}{\partial z} = - f_y v - \frac{\partial p}{\partial y} \quad , \tag{2}$$

$$\rho \, \frac{\partial w}{\partial t} + \rho u \frac{\partial w}{\partial x} + \rho v \frac{\partial w}{\partial y} + \rho w \frac{\partial w}{\partial z} = - f_z w - \frac{\partial p}{\partial z} \quad . \tag{3}$$

Here t stands for time, p for pressure, ρ for density and f_x, f_y, f_z for distributed-resistance coefficients. The latter may vary in an arbitrary manner with position, time and the magnitudes and directions of the velocities; but they are always greater than or equal to zero.

It will be noted that the usual terms representing viscous action are absent. Though their inclusion in the calculation procedure presents no difficulty, their influence is probably negligible inside such densely-filled spaces as the shell of a heat exchanger; we omit them entirely here so as to throw the special features of our problem into sharp focus.

The mass-conservation principle for the shell fluid provides another differential equation involving velocities, namely:

$$\frac{\partial \rho}{\partial t} + \frac{\partial}{\partial x} (\rho u) + \frac{\partial}{\partial y} (\rho v) + \frac{\partial}{\partial z} (\rho w) = 0 \quad . \tag{4}$$

The temperature T^s of the shell fluid obeys a differential equation which is similar in form to equations (1) to (3), namely:

$$\rho \left[C_v^s \frac{\partial T^s}{\partial t} + C_p^s \, u \frac{\partial T^s}{\partial x} + v \frac{\partial T^s}{\partial y} + w \frac{\partial T^s}{\partial z} \right] = a_s \, (T^w - T^s) \quad , \qquad (5)$$

where C_v^s and C_p^s are the specific heats at constant pressure and constant volume, α_s is a "volumetric heat-transfer coefficient", representing the ease with which heat can pass from the shell fluid to the tube wall at temperature T^w. Frictional heating is omitted from this equation, as well as the usual conduction terms, which here are negligible.

The tube-fluid temperature, T^t, obeys a similar equation; however, since tube fluid flows only in one direction, with mass velocity g_t, say, the equation has only one convection term. It is:

$$\rho_t \, C_v^t \frac{\partial T^t}{\partial t} + C_p^t \, g_t \frac{\partial T_t}{\partial z} = a_t \, (T^w - T^t) \quad . \qquad (6)$$

Here ρ_t, C_v^t, C_p^t are the relevant properties of the tube fluid, and the tubes are supposed to be aligned with the z direction.

Finally, there is a differential equation for the variation with time of the tube-wall temperature T_w. It is:

$$\rho_w C_v^w \frac{\partial T^w}{\partial t} = a_s \, (T^s - T^w) + a_t \, (T^t - T^w) \quad , \qquad (7)$$

where ρ_w and C_v^w are the relevant properties of the tube-wall material.

There is an important remark to be made about all these equations: it is that the densities ρ, ρ_t, ρ_w are to be interpreted as the mass of shell fluid, tube fluid and solid material respectively per unit heat-exchanger volume. Thus ρ is the density of the shell-side fluid multiplied by the proportion of the shell volume that is not occupied by the tubes and their contents. The ρ's can of course vary from place to place within the heat exchanger, to correspond with a non-uniform distribution of tubes.

2.1.2 Auxiliary relations

It is clear how the densities are to be calculated, but what about the f's and α's? For each of these, we need algebraic relationships connecting them with the prescribed geometry, the fluid properties, and the local velocity components. Dimensional analysis reveals that these relations must have the forms:

$$\frac{f_x \ell}{\rho u} = F_x\left(\rho \frac{u\ell}{\mu}, \frac{v}{u}, \frac{w}{u}\right), \qquad (8)$$

with similar expressions involving f_y and f_z; and:

$$\frac{a_s \ell}{\lambda} = F_s\left(\rho \frac{u\ell}{\mu}, \frac{v}{u}, \frac{w}{u}, \frac{c}{\lambda}\right), \qquad (9)$$

with corresponding expressions involving α_t. Here ℓ stands for a local length dimension, for example the tube diameter, λ represents the thermal conductivity of the shell-side fluid, and μ its viscosity. The functions may well possess more arguments than are shown above; thus the tube-wall thickness and thermal conductivity will normally enter the function F_s, in dimensionless form; and it will be the Reynolds and Prandtl numbers of the tube-side fluid which mainly influence the function F_t, from which α_t is obtained.

The functions F_x, F_s, etc. can be obtained from various fairly obvious sources. For example, if y is the direction normal to the tube axis, F_y can be deduced from experimental data on the pressure-drop performance of tube banks in cross-flow. Of course, it may be that data cannot be found for the particular tube arrangement in question, or for the influences of the velocity ratios u/v and w/v: then the F_y function must be supplied in part by way of a more detailed theoretical study, new experiments, or guesswork.

The F_t and F_s functions are obtainable from published or newly-developed correlations for heat transfer in tube banks and rod bundles. Ordinarily allowance must also be made for the heat-transfer resistances of the wall material and of layers of semi-insulating deposits.

Since our aim in this paper is to demonstrate a calculation method rather than put forward expressions for F_x, F_s, etc., quite crude and simple expressions for these functions are employed in our calculations in Section 3.

It is by way of the quantities f_x, f_y and f_z that the presence of the baffles is introduced into the mathematical system; for a baffle is nothing but an infinite resistance, which allows no flow in the direction aligned with the tubes, no matter how large a pressure gradient is exerted. A leaky baffle will be represented by a value which is large but finite for the direction normal to the baffle surface. Of course, since the baffles exert little resistance to flow along them, the f's in the other two directions will be scarcely influenced by their presence.

2.1.3 Boundary conditions

The boundary of a typical heat exchanger can be regarded as a nearly-closed surface which is impermeable both to heat and matter, penetrated by an entry and exit for shell-side fluid, and a number of entries and exits for tube-side fluid. In a multi-pass heat exchanger, several of the latter entries and exits are connected together.

Information about the rates of entry of shell-side and tube-side fluids, and about their temperatures, is easily expressible in terms of values of u, v, w, g_t, T^s and T^s for the appropriate parts of the heat-exchanger boundaries. For all the other parts, the values of the velocity components must be put equal to zero, and the temperature gradients likewise, in order to express the impermeability of the boundary.

The fluid-entry rates and conditions may vary, for an unsteady-state problem, as arbitrary functions of time. In such problems the initial temperatures and velocities prevailing within the heat exchanger must also be described.

2.2 The numerical procedure

2.2.1 Outline of the method

A brief outline will now be given of the general numerical procedure, which has been adapted for the solution of the above differential equations. Further information can be obtained from the already-cited references.

The starting point is the conversion of the differential equations into equivalent finite-difference equations, by way of integration over small sub-volumes of rectangular shape, each of which contains one of a three-dimensional array of points. The points at which the variables p, T^s, T^t and T^w are calculated all coincide: but the velocity components u, v and w are calculated for points lying between those of the above-mentioned grid.

The solution procedure is iterative, initial guesses for the variables being continually replaced by improved guesses derived from them. A special feature is the use of guessed values of the pressure distribution in the equations determining the velocity components, followed by correction of the pressures in a manner which tends to reduce the imbalances in the mass-continuity equations for each point.

Transient and steady-state problems are handled in essentially the same way. In transient problems, the distributions prevailing at the beginning of a time step are usually good approximations to the values for the end of the time step; so few iterations are needed before convergence is attained. Steady-state problems are formulated by setting the value of the time step to infinity; then about thirty iterations may be attained to ensure sufficient convergence.

2.2.3 Some special features

Two features of the method are worthy of note in the present context. First, because the distributed-resistance terms such as $f_x u$ are often non-linear in character, linearization is regularly employed. Thus

$$f_x u = a + bu \qquad , \qquad (10)$$

where a and b are constants chosen to satisfy both (10) and the equal-gradient relation:

$$\frac{\partial}{\partial u} (f_x u) = b \qquad , \qquad (11)$$

Secondly, the wall-temperature differential equation is not solved in the same manner as the equations for the other temperatures, because the absence from it of convection terms allows the following more advantageous practice. The value of T^w is eliminated from the finite-difference versions of both (5) and (6) by algebraic substitution; then (5) and (6) are solved by the usual iterative means; finally the values of T^w are deduced without further iteration from the values of T^s and T^t prevailing for the point and time in question.

3. RESULTS

3.1 The problem specification

Geometry. Fig. 3.1.1 shows the geometry of the heat exchanger for which we have performed the computations. The shell is a rectangular box with two baffles as shown. The inlet and outlet for the shell fluid are rectangular openings in the bottom and top walls of the shell. The tubes are fitted in a five-pass arrangement. Each pass is shown to consist of six tubes. At the end of each pass, the fluid from the six tubes is mixed in a header before it enters the six tubes of the next pass. (Our calculation of the shell-side flow is based on the assumption that we essentially have a continuum flow. In this connection, the representation of tubes in Fig. 3.1.1 and the mention of six tubes in the above description should be interpreted as follows: in a finite-difference calcualtion, the value of a variable at a grid point stands for the average value over the "grid cell" surrounding that point; similarly, one tube in Fig. 3.1.1 is supposed to represent a group of many tubes which would occupy the grid cell associated with that tube.) For clarity, tubes are not completely shown inside the shell. It can be seen that there is symmetry about the central xz plane; the computations are therefore performed over one-half of the heat exchanger.

Physical properties. The density and specific heats of the shell and tube fluids are taken as uniform; the specific heat C_v is taken to be equal to the specific heat C_p. In the calculation of the flow field in the shell, the distributed resistance is calculated from the values $f_x L/(\rho u) = f_y L/(\rho u) = 0.5$, and $f_z L/(\rho u) = 0.1$, where L is the x-direction width of the shell. The heat capacity of, and the heat transfer to, the shell walls and baffles are supposed to be negligible, The overall heat-transfer coefficient between the shell fluid and the tube fluid is taken as uniform. The following quantities are varied in the computations to be presented in this section: the overall heat-transfer coefficient, the heat-capacity rates of the shell and tube fluids, the heat capacities of the two fluids, the heat capacity of the tube walls, the ratio of the heat-transfer coefficients on the shell side and the tube side. Unless stated otherwise, the inlet temperatures of the shell and tube fluids are taken as 100 and 0 respectively (the temperature scale is immaterial here).

Computational details. All the computations reported in this paper were performed on a uniformly-spaced grid which used 5, 3 and 8 cells in the x, y and z directions respectively (in the y direction, the grid covered the region on only one side of the central xz plane). Thus, in the xy plane we had a grid node for every tube shown in Fig. 3.1.1; and in the z direction the inlet, the outlet and the baffles could be conveniently represented in 8 cells. The grid chosen is rather coarse, but is satisfactory for the present purpose of demonstrating the calculation procedure. The steady-state calculations needed about 20 iterations for convergence, and this required about 9 seconds on the CDC 6600 computer. For most of the unsteady-state calculations, 100 steps in time were performed and this needed 45 seconds of computer time when the transient behavior of both the flow flied and the temperature was computed. When the flow field was regarded as steady, and the unsteady behavior of the temperature field was computed, only 9 seconds of computer time were needed for 100 steps in real time. The computer storage requirement, including program was approximately 20 K.

3.2 Steady-state behavior

Flow field. For the problem described, the velocity field attained in the steady state is shown by velocity vectors in Figs. 3.2.1 and 3.2.2. Fig. 3.2.1 shows the vectors in the central xz plane; here one can clearly see how the flow is deflected by the baffles. Also the recirculations near the two ends of the shell can be noted. The velocity field in three different xy planes is shown in Fig. 3.2.2; from these figures, the three-dimensional nature of the flow can be imagined.

Temperature distribution. The heat-transfer coefficient used in the calculations can be expressed in terms of N, the number of transfer units based on the heat-capacity rate of the tube fluid. For equal heat-capacity rates of the shell and tube fluids, the temperature distributions in the two fluids is shown in Figs. 3.2.3, 3.2.4 and 3.2.5 for N = 1, 0.2 and 4 respectively. For presenting the shell-fluid-temperature distribution, we have chosen three lines (parallel to the z direction) on the central xz plane; the temperature variation on these lines is plotted in part (a) of the above-mentioned figures. In part (b), the variation of the tube-fluid temperature along the five passes is shown; the temperatures at the nodes nearest to the central xz plane are used.

A number of interesting points can be noted in these figures. Firstly, the shell-fluid temperature varies in a manner quite different from that of a conventional counter-flow heat exchanger. Secondly, there are locations where the shell-fluid temperature drops below even the outlet temperature of the shell fluid; this happens in the recirculation regions near the end wall and the baffles. Understandably, the temperature non-uniformity is most striking for large values of N. The tube-fluid temperature mostly rises continuously as it goes through the five passes; but the later passes do not seem to be as effective as the earlier ones. This is because large portions of the fourth and fifth passes go through the region of low shell-fluid temperature near the outlet in the shell. For N = 4, the effect is so severe that for certain lengths of the fourth and fifth passes, the tube-fluid temperature actually decreases. The merit of the present calculation method mainly lies in revealing such surprising behavior, which cannot be predicted by conventional design methods.

Influence of the heat-capacity rates. Fig. 3.2.6 shows the variation of the temperature effectiveness of the heat exchanger with the number of transfer units for three different ratios of heat-capacity rates of the two fluids. Also plotted are two dashed curves which show the variation of effectiveness for a counter-flow and a parallel-flow exchanger when $\dot{C}_s/\dot{C}_t$ equals 1. It can be seen that our predictions for the same ratio lie between the two dashed lines. This is expected because our heat exchanger can be considered to be a counter-flow one in some parts and a parallel-flow one in others.

3.3 Unsteady behavior: starting up from rest

Now we consider the transient behavior of the heat exchanger during the starting-up operation. Initially both the fluids are at rest and are at equal temperature of zero. At time $t = 0$, both fluids are admitted through their respective inlets; the shell fluid comes in at a temperature of 100, while the tube fluid enters at 0. The heat-capacity rates are taken to be equal, and N is taken as unity. The transient behavior will depend on the heat capacities of the two fluids; the heat capacity of the tube walls is taken to be zero for these calculations. The results are plotted in Fig. 3.3.1 for three different heat-capacity ratios. The abscissa is t/t_{res} where t_{res} is the residence time of the tube fluid. With the heat-capacity rates taken to be equal, the ratio of heat capacities also stands for the ratio of the residence times of the two fluids. Thus, for H_s/H_t equal to 0.1, the residence time of the shell fluid is one-tenth that of the tube fluid, and hence the steady state is reached faster. When H_s/H_t is 10, there is a corresponding slow approach to the steady state.

For the heat-capacity ratio of unity, the nominal residence times of the two fluids are equal. Even then the outlet temperatures of the two fluids seem to start rising long before one residence time elapses. The apparently surprising behavior can be traced to two reasons: firstly, because of the non-uniformities of the shell-side flow, some of the shell fluid can travel from the inlet to the outlet in less than one (nominal) residence time; secondly, the five-pass arrangement can cause, through the agency of the tubes, the temperature effects to transverse the shell in one fifth of the time. Once again, it is our detailed calculation of the three-dimensional velocity and temperature fields that has enabled us to reveal such effects.

3.4 Unsteady behavior: variation of inlet temperatures

The situation considered in this sub-section is the one in which the flow field is steady, but the inlet temperatures of the two fluids are varied with time. For this situation, the equations become linear and the results for a complex variation of the inlet temperatures can be obtained by the superposition of the solutions for two basic problems. In these problems, we start from equal temperatures of the two fluids; in one problem, the inlet temperature of the shell fluid is suddenly changed and maintained at the new value; in the other, the inlet temperature of the tube fluid is given a step change. For the calculations to be presented below, the conditions used are: $N = 1$, equal heat-capacity rates, equal heat capacities of the two fluids, and zero heat capacity of the tube walls.

Two basic cases. Fig. 3.4.1 shows the variation with time of certain outlet temperatures for the two basic cases. For the first one, the two fluids are initially at a temperature of zero; then, at $t = 0$, the inlet temperature of the shell fluid is raised to 100. In the second case, we start with a temperature of 100 everywhere; then at $t = 0$, the tube-fluid inlet temperature drops to zero. In both cases, the outlet temperatures of the individual passes and that of the shell fluid are plotted. The values refer to the central xz plane. It can be seen that the earlier passes attain the steady state quicker than the later passes.

Periodic variation of the tube-fluid inlet temperature. Although the solutions for the above two cases suffice to yield, by super-position, the solution for any arbitrary variation of the inlet temperatures we can directly handle an arbitrary variation of inlet temperature almost as easily as the step change. Since we march forward in time, all that is necessary is to reset the inlet temperatures to their new values after each time step. To demonstrate this possibility, we now present the results for the case in which, after a steady state has been reached with the inlet temperatures of the shell and tube fluids as 100 and 0 respectively, the inlet temperature of the tube fluid is varied in a periodic fashion about the mean value of zero, with an amplitude of 20 and a period equal to one residence time of the tube fluid. The variation of some of the outlet temperatures with time is plotted in Fig. 3.4.2. It can be seen that a steady periodic state has been reached after about two residence times. The amplitudes diminish as we proceed from an earlier pass to a later pass; and there is a phase difference between the various outlet temperatures plotted.

3.5 Unsteady behavior: effect of the heat capacity of the tube walls

In the calculations presented so far, the heat capacity of the tube walls was taken as zero. In this subsection we describe some computations which include the heat capacity of the tube walls as an additional parameter. With the inclusion of this parameter, it becomes necessary to distinguish between the heat-transfer coefficients on the shell side and on the tube side of the tube wall. The thermal resistance of the tube wall itself is taken to be negligible.

Various heat capacities of the tube walls. Fig. 3.5.1 shows the transient behavior of the outlet temperatures of the two fluids for different heat capacities of the tube walls. The initial conditions are the same as those of "starting-up from rest" (Section 3.3). The heat-capacity rates are equal, the heat capacities of the two fluids are equal, and the shell-side and the tube-side heat-transfer coefficients, α_s and α_t, are taken to be equal; N based on the overall heat-transfer coefficient is taken as unity. It can be seen from Fig. 3.5.1 that the inclusion of the tube-wall heat capacity slows down the approach to the steady state.

Influence of different heat-transfer coefficients on the shell and tube side. Lastly, in Fig. 3.5.2, we show, for the same problem as above, the variation of the outlet temperatures for three different values of α_s/α_t. The heat capacity of the tube walls is taken to be equal to that of the tube fluid. The number of transfer units N is unity based on the overall heat-transfer coefficient. At the start, the inlet temperature of the shell fluid is suddenly increased to 100. When α_s is large, the shell-fluid and the tube-wall temperatures must rise together, and this slows down the approach to the steady state. The value of α_s/α_t does not however seem to affect the rise of the tube-fluid temperature. This is because the initial step change in inlet temperature is made on the shell side: then the tube walls either lower the available temperature difference by slowing down the rise of the shell-fluid temperature (for $\alpha_s/\alpha_t \gg 1$) 0 or increase the effective heat capacity of the tube fluid by reducing the tube-side thermal resistance (for $\alpha_s/\alpha_t \ll 1$); in either case, the tube-fluid temperature rises at about the same rate.

4. CONCLUDING REMARKS

4.1 Achievements

The calculations presented in this papaer have demonstrated that, once the appropriate computer program is available, it is possible to make predictions of the performance of shell-and-tube heat exchangers that appear to be both plausible and interesting. It is true that no successful comparisons with the experiment have been displayed; indeed, no special care has been taken to ensure that the inputs to these calculations correspond to any particular realizable situation. Nevertheless, it should be clear that the method is flexible enough to allow the input of geometrical, physical-property and internal-resistance data of arbitrary complexity; and, since no earlier method of analysis has actually provided any predictions at all for the complex recirculating flow of shell-side fluid, the present method can hardly fail to represent an advance in predictive accuracy and usefulness to designers.

The heat-exchanger which has been the subject of study here has been of relatively simple geometry, with only two baffles and five passes. Partly for this reason, the grid chosen was a coarse one, allowing speedy and economical calculation. However, all that is needed to permit the analysis of heat-exchangers of greater geometrical complexity is an appropriately enlarged grid and somewhat greater computer time. No difficulties of principle are involved, even when the shell has multiple entries and exits, or when the heat exchanger has more than two fluids passing through it. It therefore seems that a tool of considerable practical utility is in embryonic existence.

4.2 Future tasks

It is easy to produce interesting and plausible demonstrations of the ability to perform predictions; to turn the predictive scheme into computer program that designers can use in their day-to-day work is a task of greater magnitude, demanding as much ingenuity but more varied knowledge. Tasks of two distinct kinds must be performed.

First, the predictive scheme must be validated, i.e. demonstrated to give realistic quantitative predictions over a sufficiently wide range of conditions to provide confidence that predictions for other conditions will also be valid. It is not only heat-exchanger experiments which are here in question; for example, the general phenomena of three-dimensional flows through bundles of rods and tubes have been left unexplored by fluid-dynamicists, quite apart from the heat-transfer aspects of the situation. A program of research validating the prediction procedure should therefore concern itself with some quite elementary experiments on the way a fluid distributes itself through a volume partially packed with a regularly-arranged solid. Such studies, it might be mentioned, would be useful also in connection with process equipment of different types, for example cooling towers, regenerators and packed beds.

The second task is to streamline and supplement the computational procedure so that it can accept problem specifications of the kind and in the form that it is convenient to designers to provide, and can express its results in a form that corresponds to the designers need. This means that the computer program must contain sequences to choose the optimum grid configuration to suit the specified geometry, must select the expressions for f's and α's by reference to the geometrical information, must choose the appropriate property values of the fluids when only their names are specified, and must, on occasion, compute the geometry that is needed in order to provide a specified performance rather than the performance that will result from a specified geometry.

A question that naturally arises is: when all this has been done, will the resulting computer program be so expensive to run that the designers will, after all, be constrained by lack of funds to continue the use of their existing methods? No final answer is yet possible. At present, it appears that the cost of a performance-prediction "run" on the computer will cost some tens of pounds; an optimization run may cost some hundreds. Probably this expenditure will be small in comparison with the benefits; but only time can prove this expectation.

5. ACKNOWLEDGMENTS

Thanks are due to Mr. A. Nigam and Mr. D. Sharma for the assistance in the preparation of the figures.

6. REFERENCES

[1]   Caretto, L.S., Gosman, A.D., Patankar, S.V. and Spalding, D.B. "Two calculation procedures for steady-three-dimensional flows with recirculation". Third International Conference on Numerical Methods in Fluid Dynamics, Paris, July 1972.

[2]   Patankar, S.V. and Spalding, D.B. "A calculation procedure for heat, mass and momentum transfer in three-dimensional parabolic flows". To be published in Int. *J. Heat Mass Transfer,* Vol. 15., 1972.

NOMENCLATURE

c	specific heat
c_p, c_v	specific heats at constant pressure and constant volume
$\dot{C}_t$	heat capacity rate of the tube fluid ($\equiv g_t c_p^t$)
$\dot{C}_s$	heat capacity rate of the shell fluid
f_x, f_y, f_z	distributed-resistance coefficients
g_t	mass-flow rate of the tube fluid
H	heat capacity per unit volume
ℓ	a local length dimension
L	the x-direction width of the shell
N	the number of transfer units based on the overall heat-transfer coefficient and the heat-capacity rate of the tube fluid ($\equiv \alpha_0/\dot{C}_t$)
p	pressure
t	time
t_{res}	the residence time of the *tube fluid* ($\equiv H_t/\dot{C}_t$)
T	temperature
u, v, w	velocity components of the shell fluid
x, y, z	co-ordinates
α_s, α_t	volumetric heat-transfer coefficients on the shell side and tube side
α_0	the overall volumetric heat-transfer coefficient
η	the temperature effectiveness ($\equiv (T_{out}^t - T_{in}^t)/T_{in}^s - T_{in}^t)$)
λ	thermal conductivity
μ	viscosity
ρ	the effective density (when used without subscript, it refers to the shell fluid)

Subscripts and Superscripts

in	inlet condition
out	outlet condition
s	the shell fluid
t	the tube fluid
w	the tube wall

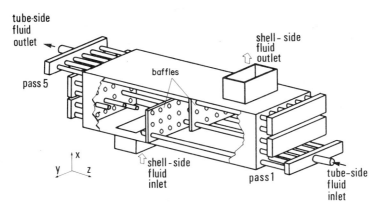

Fig. 3.1-1 The geometry of the heat exchanger.

Fig. 3.2-1 Velocity vectors in the central xz plane.

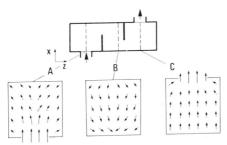

Fig. 3.2-2 Velocity vectors in the three xy planes.

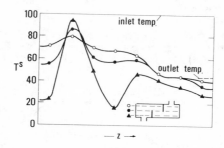

a) Variation of the shell-fluid temperature.

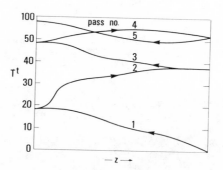

b) Variation of the tube-fluid temperature.

Fig. 3.2-3 Temperature distribuition on the central xz plane for N = 1 and equal heat-capacity rates.

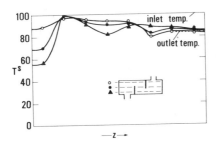

a) Variation of the shell-fluid temperature.

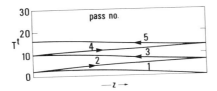

b) Variation of the tube-fluid temperature.

Fig. 3.2-4 Temperature distribution on the central xz plane for N = 0.2 and equal heat-capacity rates.

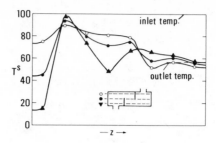

a) Variation of the shell-fluid temperature.

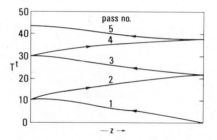

b) Variation of the tube-fluid temperature.

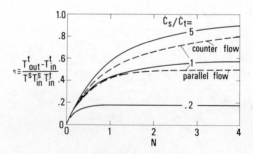

Fig. 3.2-6 Variation effectiveness for different heat-capacity-rate ratios. (The dashed lines represent one-dimensional counter-flow and parallel-flow solutions for $C_s/C_t = 1$.)

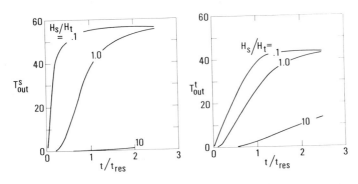

Fig. 3.3-1 Variation of the outlet temperatures of the two fluids in the "starting-up from rest" operation.

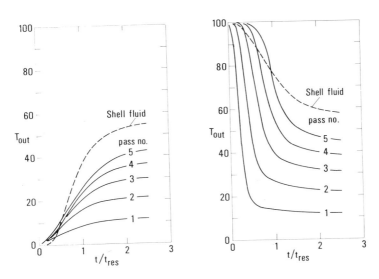

a) Step change in the shell-fluid inlet temperature.   b) Step change in the tube-fluid inlet temperature.

Fig. 3.4-1 Variation with time of the outlet temperatures for the two basic cases

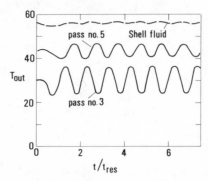

Fig. 3.4-2 Variation on the outlet temperatures created by a periodic variation of the inlet temperature of the tube fluid

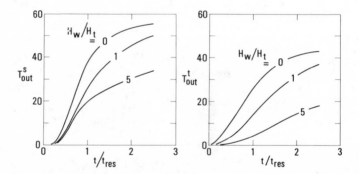

Fig. 3.5-1 Variation of the outlet temperatures for different heat capacities of the tube walls

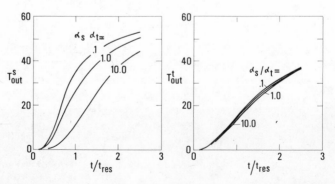

Fig. 3.5-2 Variation of the outlet temperatures for different ratios of the shell-side and the tube-side heat-transfer coefficients

Chapter 8

TUBULAR HEAT EXCHANGERS WITH BILATERAL HEAT TRANSFER AUGMENTATION AND CALCULATION OF A HEAT EXCHANGER UNDER UNSTEADY OPERATING CONDITIONS

E. K. Kalinin (*)

This chapter summarizes the results obtained by the author and his colleagues, G. A. Dreitser, S. A. Yarkho, V. A. Kuzminov, A. S. Neverov, and B. S. Baibukov, at the Moscow Aviation Institute.

The first part of this chapter contains: the experimental results of heat transfer intensification in tubes and in longitudinal flow past tube bundles; the design relations for tube bundle heat exchangers, and; the description of the technology of tube rolling. Then, a comparative analysis of the calculation of heat exchangers with bilateral heat transfer intensification and smooth tubes is given. Recommendations on a choice of optimum rolling sizes are presented as well.

The second part of this chapter presents: the results of calculating unsteady gas heat transfer in tubes; the results of calculating unsteady heating and cooling of tubes, and; a description of methods for calculating a heat transfer apparatus under unsteady operating conditions.

I.

Heat transfer intensification may serve as an effective means of essential improvement of heat exchanger characteristics. We shall show that with the help of heat transfer intensification it is possible: 1) to decrease a surface, volume and weight of a heat exchanger under invariable heat power, flow rates of coolants and hydraulic losses, 2) to increase heat power with invariable volume, flow rates of coolants) hydraulic losses, 3) to diminish hydraulic losses (power for rolling coolants) under invariable volume, heat power and flow rates of coolants, or 4) to accomplish compromised improvement of all these parameters.

A comparison will be made of tubular counter-flow heat exchangers

(*) Moscow Aviation Institute, USSR
N.B. ALL FIGURES QUOTED IN TEXT ARE AT THE END OF THE CHAPTER

with smooth tubes and tubes with special devices for artificial flow turbulization.

Usually, artificial flow turbulization is related not only to a heat transfer increase (Nusselt number, Nu) but also to an increase of the hydraulic resistance coefficient ξ.

The simple analysis shows that intensification allows the above improvement of heat exchanger characteristics to be achieved if

$$\frac{\xi}{\xi_s} < \left(\frac{Nu}{Nu_s}\right)^{3.5} \tag{1}$$

The greater the inequality (1), the more effective the method of heat transfer intensification.

The results of the investigation of this method were reported in detail at the 1969 Seminar of The International Centre for Heat and Mass Transfer and were published [1,2,3,5,7,9,12,13,14,15,16]. Briefly the essence of this method is as follows. The smooth tube is rolled by screw rollers, as is shown in Fig. 1. Such a simple device allows simple and cheap security of industrial production of rolled tubes. As a result of rolling (Fig. 2), at the inner surface of a tube smooth projections are formed, and at the outer surface the cross grooves are formed. In the flow past smooth projections inside tubes the cross eddies may not be formed, but the system of longitudinal screw eddies is formed. This leads to additional generation of turbulence at a height of a projection comb, increases the turbulence rate and turbulent thermal conductivity near a wall. When this turbulence starts decreasing, the next projection is set up, etc.

The results on the effect of the height and pitch of projections upon Nu and ξ are presented in Fig. 3, from which it is seen that with an increase in Re and Pr the effective height of a projection decreases and comparatively small pitches t/d = 0.25-0.5 are more profitable.

In the flow past cross grooves eddies are formed at external boundaries where there occurs additional generation of turbulence.

The effect of a pitch and relative depth of grooves upon heat transfer is shown in Fig. 4.

At the Moscow Aviation Institute a great number of investigations of this method of intensification in tubes and in longitudinal gas and liquid flow past tubes is being carried out.

The basic design relations for heat transfer and hydraulic resistance obtained in these experiments are of the form:

Heat transfer of gases $Re_b = 10^4 \div 4.10^5)$ and liquids at $10^4 < Re < Re_2$

for tubes

$$\frac{Nu}{Nu_s} = \left[1 + \frac{\ell g \, Re - 4.6}{35} \right] \left\{ 3 - 2 \, exp \left[\frac{-18.2 \, (1 - d/D)^{1.3}}{(t/D)^{0.326}} \right] \right\}$$

at $d/D = 0.88 \div 0.98$; $t/D = 0.25 \div 0.8$

$$\frac{Nu}{Nu_s} = \left[1 + \frac{\ell g \, Re - 4.6}{30} \right] \left[\left(3.33 \frac{t}{D} - 16.33 \right) \frac{d}{D} + \left(-3.33 \frac{t}{D} + 17.33 \right) \right]$$

at $d/D = 0.88 \div 0.98$ $t/D = 0.8 \div 2.5$

The hydraulic resistance of liquids and gases at $Re = 10^4 \div 4.10^5$ $Pr = 0.7 \div 50$ for tubes

$$\frac{\xi}{\xi_s} = \left[1 + \frac{100 \, (\ell g Re - 4.6) \left(1 - \frac{d}{D} \right)^{1.65}}{exp(t/D)^{0.3}} \right] exp \left[\frac{25 \, (1 - d/D)^{1.32}}{(t/D)^{0.75}} \right]$$

for $d/D = 0.9 \div 0.98$ $t/D = 0.5 \div 10$

$$\frac{\xi}{\xi_s} = \left[1 + \frac{\ell g \, Re - 4.6}{3.45} \right] \left(1.2 - 0.4 \frac{t}{D} \right) exp[9.8 \, (1 - d/D)^{0.74}]$$

for $d/D = 0.88 \div 0.98$; $t/D = 0.25 \div 0.5$

Heat transfer of rolled tube bundles at $S/D_a = 1.1 \div 1.5$;

$ha/d_{э\infty} = 0 \div 0.1$; $t/D_{е\infty} = 0.25 \div 2$

$$\frac{Nu}{Nu_s} = 1 + 0.6 \frac{\ell g \, Re - \ell g \, Re_1}{\ell g \, Re_2 - \ell g \, Re_1} \left[-exp \left(-35.8 \frac{ha}{d_{э\infty}} \right) + 1 \right] \cdot$$

$$\cdot \left\{ 1 - \exp\left[- \frac{0.62}{(t/d_{\epsilon\infty})^{0.982}} \right] \right\}$$

at $Re_1 < Re_L < Re_2$

$$\frac{Nu}{Nu_s} = 1 + 0.6 \left[1 - \exp\left(-35.8 \, \frac{ha}{d_{\mathfrak{z}\infty}} \right) \right] \left\{ 1 - \exp\left[- \frac{0.62}{(t/D)^{0.982}} \right] \right\}$$

at $Re_2 < Re_L < 10^5$

Hydraulic losses of rolled tube bundles at $ha/d = 0 \div 0.1$; $t/d_{\mathfrak{z}\infty} = 0.25 \div 2$ and $s/d_{\mathfrak{z}\infty} = 1.1 \div 1.5$

$$\frac{\xi}{\xi_s} = 1 + \left\{ 1.55 \, \frac{ha}{d_{\mathfrak{z}\infty}} \, (\lg Re - 3.5) - 0.035 \sin\left[\left(1 - 22.44 \cdot \frac{ha}{d_{\mathfrak{z}\infty}} \right) \pi \right] \right\} x$$

$$\left\{ 1 - \exp\left[-\frac{0.62}{(t/d_{\mathfrak{z}\infty})^{0.982}} \right] \right\}$$

at $Re = 3.10^3 \div 2.10^4$

$$\frac{\xi}{\xi_s} = 1 + \left\{ 3.21 \, \frac{ha}{d_{\mathfrak{z}\infty}} \, (\lg Re - 2.27) + 0.09 \, (\lg Re - 4.3) \, x \cdot \right.$$

$$\left. \cdot \, x \, \sin\left[\left(1 - 22.44 \, \frac{ha}{d_{\mathfrak{z}\infty}} \right) \pi \right] \right\} \left\{ 1 - \exp\left[- \frac{0.62}{(t/d_{\mathfrak{z}\infty})^{0.982}} \right] \right\}$$

at $Re = 2.10^4 \div 10^5$

Special studies have shown that rolled tubes do not lose strength.

The technology for assembling heat exchangers consisting of such tubes does not differ from that for assembling exchangers with smooth tubes.

In order to evaluate the effectiveness of the intensification method under consideration, numerical calculations were made of heat exchangers with the same grid of a tubular bundle with water flow in an intertube space and with gas flow in tubes. Thus, heat transfer is limited by gas heat transfer in tubes.

Calculations are made for two Re numbers equal to 4.10^4 and 4.10^5 with a different combination of rolling parameters t/D and d/D.

The calculated results at prescribed values of ΔP, Q and the flow rates of coolants are given in Figs. 5.6 and 7. As is seen, the effectiveness of rolling grows with an increase in Re, since in this case the hydraulic resistance of smooth tubes increases considerably. Intensification allows the volume of a tubular bundle to be decreased twice.

Fig. 8 illustrates the results of the same calculation of a change in the relation of heat power of heat exchangers made of rolled and smooth tubes at the same volumes, Δp and flow rates of coolants.

In the case of the optimum relation of rolling parameters, the heat power increased 1.7 times.

Finally, in Fig. 9 it is shown how the relation of total hydraulic losses $\Delta p/\Delta ps$(powers for gas pumping) depends on the rolling parameters and on the Re number if such calculation is conducted, provided that $V = V_s$, $Q = Q_s$ and $G = G_s$. It is seen that intensification allows hydraulic losses to be decreased more than 5 times. The best results in this particular case are obtained in the case of rolling with d/D = 0.92 and t/D = 0.25.

The results of calculating intensification effectiveness when the heat transfer coefficient of a heat exchanger limits heat transfer in the intertube space are presented in the table (Fig. 10).

The effectiveness of rolling increases with a decrease in a bundle pitch S/D_H. The best results are obtained in the case of rolling with $t/D_H = 0.22 \div 0.25$ $d_H/D_H = 0.93 \div 94$.

The calculations have shown that with commeasurable values of the heat transfer coefficients inside and outside tubes, the effect of intensification from two sides is noticeably pronounced.

Our investigations have shown that the intensification method considered has greater effect with film boiling in tubes. N. V. Zozulya's investigations have shown its effectiveness of the intensification method for condensers, especially with vapor condensation in tubes.

All the aforesaid allows us to conclude that the intensification method described is technological and very effective for different types of tubular heat exchangers.

II. CALCULATION OF HEAT EXCHANGERS OPERATING UNDER UNSTEADY CONDITIONS

Engineering methods for calculating heat exchangers and other heat exchanging devices which have chambers for flow of coolants are usually based on a one-dimensional description of the flow in channels using the concepts of heat transfer and hydraulic resistance coefficients. If the heat transfer coefficient depends on a distribution of boundary conditions along the length, these conditions unknown beforehand, then a conjugated problem is to be solved, i.e. the system of equations for flows and a wall.

If this relation is very weak, as for turbulent flows, then by introducing the concept of the heat transfer coefficient the problem is reduced to simultaneous solution only of one-dimensional energy equations for flows.

In the case of unsteady operating conditions the problem is, as a rule, conjugated. Investigations [10] have shown that for turbulent flows the heat transfer coefficient α depends, in real cases, not only on the law of a change in wall temperature T_w or flow rate of a coolant G in time but also on the rate of this change $\partial T_w/\partial\tau$. Consequently, the generalized relations for the number Nn may be obtained in principle. This means that the one-dimensional description of the flow still remains effective for constructing the engineering calculations.

Thus, the conjugated problem of channel-wall flow may be formulated in the following way:

1) heat conduction equation for a wall

$$\rho Cp \, \frac{\partial T}{\partial \tau} = \text{div} \, (\lambda \, \text{grad} \, T) + q_v \tag{1}$$

or if

$$\lambda = \text{const and } \alpha = \lambda/\rho Cp$$

$$\frac{\partial T}{\partial \tau} = a \, \nabla^2 T + q_v \,/\rho Cp \tag{1a}$$

2) motion equation for flow

$$\frac{G}{W} \frac{\partial W}{\partial \tau} + G \frac{\partial W}{\partial Z} = F\rho g_z - \frac{\partial p}{\partial Z} F - \xi \frac{\rho W^2}{2d} \tag{2}$$

3) continuity equation

$$\frac{\partial \rho}{\partial \tau} F + \frac{\partial G}{\partial Z} = 0 \tag{3}$$

4) energy equation for flow (assuming di = cpdTb)

$$\frac{GCp}{W} \frac{\partial T_b}{\partial \tau} + GCp \frac{\partial T_b}{\partial Z} = Ua (T_w - T_b) \tag{4}$$

The conjugation conditions assume the equality of temperatures and heat fluxes on both sides of a heat transfer surface. This system may be closed if the equations for Nn and ξ are known.

For a number of years the investigations of unsteady heat transfer in turbulent gas flows [4,6,8,10,11,13] have been carried out at the Moscow Aviation Institute, and soon investigations of such a problem regarding liquid flows will be commenced.

In these investigations the heat transfer coefficients $\alpha = q_w/T_w - T_b$ and ξ are found when solving the system of equations (1) ÷ (4), which is closed with the aid of measuring a necessary number of qualities in experiments. It was shown that with unsteady heating and cooling in the case of a constant flow rate $G = \rho wf = const$ heat transfer may be generalized as follows:

$$K = \frac{Nu}{Nu_o} = f (Re, \psi, K_{T_g}) \tag{5}$$

where

$$\psi = \frac{T_w}{T_b} ; \quad K_{Tg} = \frac{\partial T_w}{\partial \tau} \cdot \frac{d}{T_w} \sqrt{\frac{\lambda}{gCpG}}$$

No is the stationary value of the Nusselt number

$$Nu_o = f\left(\frac{z}{d} \; ; \; Re; \; \psi\right)$$

The number K_{Tg} takes into account the relation of heat transfer versus a change in a wall temperature, which for gases is mainly related to the effect of $\partial T_w/\partial \tau$ upon generation of turbulence near a wall. The experimental data on gas cooling over a range $Re = (0.32 \div 2)10^5$ $\psi = T_w/T_b = 1 \div 0.6$, $K_{Tg} = (0 \div 20).10^5$ were generalized by the equation

$$K = 1 + \left[(14.97 \, \psi^3 - 16.07 \, \psi^2 - 0.526 \, \psi + 3.193)\cdot(Re\cdot10^{-5})^{1.85-3\psi} + \right.$$
$$\left. + 46.7 \, \psi^3 - 119.1 \, \psi^2 + 99.09 \, \psi - 27.08)\cdot(K_{Tg}\cdot10^5)\right]$$

$$(6)$$

The comparison of equation (6) with the experiment and the effect of different parameters on heat transfer are shown in Fig. 11.

With gas heating over a range $Re = 8.10^4 \div 5.10^5$ $\psi = 1 \div 1.5$; $K_{Tg} = 0 \div 10^{-4}$ the experimental data are correlated by the equation:

1) lift of heat load;

$$K = 1 + \left(0.185 \, \frac{T_w}{T_b} + 0.038\right)\cdot \exp\left[\left(2.875 - 2.9 \, \frac{T_w}{T_b}\right)\cdot Re_b\cdot10^{-5}\right]\cdot$$
$$\cdot\left[\exp\left(1.06\cdot10^4\cdot K_{Tg}\right) - 1\right]$$

2) drop of heat load

$$K = 1 - \left(1.41 \, \frac{T_w}{T_b} - 0.97\right)\left[1 - \exp\left(A\cdot K_{Tg}\right)\right] \qquad (7)$$

where

$$A = \begin{cases} 793 \, |K_{Tg}|^{-0.177} & \text{at } -0.4\cdot10^{-4} < K_{Tg} \leqslant 0 \\ 1.47 \, |K_{Tg}|^{-0.8} & \text{at } -2.2\cdot10^{-4} \leqslant K_{Tg} \leqslant -0.4\cdot10^{-4} \end{cases}$$

The comparison of this equation with the experiment and the nature of the effect of Re, ψ, K_{rg} upon K are given in Fig. 12.

The presence of empirical equations (6) and (7) allows the system of equations (1)-(4) to be closed and used as engineering methods for calculating unsteady heating or cooling of a channel with a gas.

Such calculations were made and compared with those in which the heat transfer coefficient is found by quasi-stationary relations.

The first calculation was made for heating a tube by a turbulent gas flow when its inlet temperature increases in jumps or according to some prescribed law.

For $Bi = \alpha \delta_w / \lambda_w \ll 1$ the calculation was so simplified that it was reduced to the use of monograms to a considerable extent.

The calculated results of one particular case are given in Figs. 13 and 14.

As is seen, calculation involving the use of quasi-stationary relations gives deviations in the values of T_b and T_w.

The second calculation was conducted under gas heating conditions in the case of cooling of a tube (inlet temperature of a gas decreased in jumps or smoothly) and heating of a tube (heat evolution in tube walls was measured in jumps or smoothly). The condition $Bi \ll 1$ was also used. The calculation was conducted on an electronic digital computer.

The calculation results are given in Figs. 15 and 16. As we see, here in the majority of regimes calculation involving the use of quasi-stationary relations gives a noticeable deviation, although in some regimes the difference is not great when heat evolution is increased.

In conclusion, the method of thermal calculation of a tubular counter flow heat exchanger under unsteady operating conditions will be presented assuming that flow rates of coolants and their inlet temperatures are prescribed.

III. CALCULATION OF UNSTEADY OPERATING
CONDITIONS OF A HEAT EXCHANGER

The processes, during which flow rates of coolants (due to a pressure or velocity) and their inlet temperatures change with time, will be mainly referred to unsteady operating conditions of a heat exchanger. In these processes a very essential fact is that a wall temperature varies with time and consequently the wall either accumulates some of a heat flux from a "hot" coolant (with wall heating) or transmits additional heat to a "cold" coolant (wall cooling). It is therefore necessary to write down a system of three energy equations (for both coolants and a wall). Equations (4), for a wall assuming the equality of heat fluxes received and given by wall material, will be used for coolants. Then, the system of equations may be presented as:

$$G_1 \frac{Cp_1}{W_1} \frac{\partial T_1}{\partial \tau} + G_1 Cp_1 \frac{\partial T_1}{\partial Z} = U_1 q_1 \tag{8a}$$

$$G_2 \frac{Cp_2}{W_2} \frac{\partial T_2}{\partial \tau} + G_2 Cp_2 \frac{\partial T_2}{\partial (\ell - z)} = U_2 q_2 \tag{8b}$$

$$(\rho CF)_w \frac{\partial T_w}{\partial \tau} = U_1 q_1 + U_2 q_2 \tag{8c}$$

Here, subscript "1" refers to a liquid flowing in the intertubal space along the tubes, subscript "2" refers to a gas flowing in the opposite direction in tubes, and subscript "w" refers to the wall material. F_w is the cross-section of a tube wall $F_w = [\pi(d_1^2 - d^2)]/4$; V is the heated wall perimeter; $V_1 = \pi d$; $V_2 = \pi d$; d_1 and d are the outer and inner tube diameters, respectively; q_1 and q_2 are the heat fluxes between a wall and liquid and between a wall and gas, respectively. Without restrictions on the group of the solution, q_1 and q_2 will be considered to be the known functions of time and a coordinate z. Then, the replacement

$$Z = \frac{z}{d} \; ; \; Ho_1 = \int_0^\tau \frac{W_1(\tau) d\tau}{d} \; ; \; Ho_2 = \int_0^\tau \frac{W_2(\tau) d\epsilon}{d} \tag{9}$$

$$\theta_1 = \frac{T_1}{T_o} \; ; \; \theta_2 = \frac{T_2}{T_o} \; ; \; \theta_w = \frac{T_w}{T_o}$$

(where $T_0 =$ coust is the characteristic temperature) at $C_p =$ coust and $\overline{w}/w = 1$ at the sections of partition along the channel length reduces system (8) to the form:

$$\frac{\partial \theta_1}{\partial Ho_1} + \frac{\partial \theta}{\partial Z} = \frac{U_1 \, d}{G_1 \, Cp_1 \, T_o} \quad q_1 \, (Ho_1 ,Z) = 4 \; Sto_1 \qquad (10a)$$

$$\frac{\partial \theta_2}{\partial Ho_2} - \frac{\partial \theta_2}{\partial Z} = \frac{U_2 \, d}{G_2 \, Cp_2 \, T_o} \quad q_2 \, (Ho_2 ,Z) = 4 \; Sto_2 \qquad (10b)$$

$$\frac{\partial \theta_w}{\partial Ho_1} = \frac{- \, U_1 \, q_1 \, (Ho_1 ,Z) \; + \; U_2 \, q_2 \, (Ho,Z)}{\rho_w C_w F_w W_1 T_o} \; d \qquad (10c)$$

The relationship between Ho_1 and Ho_2 at any time moment may be obtained by their differentiation with respect to

$$\tau : \quad \frac{dHo_1}{\partial \tau} = \frac{W_1}{d} \quad \text{and} \quad \frac{\partial Ho_2}{\partial \tau} = \frac{W_2}{d}$$

Dividing the first expression by the second gives

$$\frac{dHo_1}{dHo_2} = \frac{W_1 \, (\tau)}{W_2 \, (\tau)} = \frac{W_1 \, (Ho_1)}{W_2 \, (Ho_1)} = \frac{W_1 \, (Ho_2)}{W_2 \, (Ho_2)} \qquad (11)$$

If the replacement is made in (10)

$$q_1 = a_1 \, (Ho_1 ,Z) \; [\theta_w (Ho_1 ,Z) - \theta_1 \, (Ho_1 \, Z)] \; To \qquad (12a)$$

$$q_2 = a_2 \, (Ho_2 ,Z) \; [\theta_w (Ho_2 ,Z) - \theta_2 \, (Ho_2 \, Z)] \; To \qquad (12b)$$

then the system of partial equations of the first type is obtained. Its numerical solution on an electronic computer is rather cumbersome.

Therefore, system (10) is first reduced to the system of integral equations. The Cauchy problems for the domains $Ho \leqslant z \leqslant z_2$ and $z \leqslant Ho < \infty$ may be solved for equations (10a) and (10b). Since consideration is made of counterflow, these domains and the statement of the Cauchy problems for equations (10a) and (10b) will then be somewhat different.

For equation (10a) the solution to the Cauchy problem for the first domain $Ho_1 \leqslant z \leqslant z_c$) will be

$$\theta_1 (Ho_1 Z) = 4\int_0^{Ho_1} Sto_1 (Y\ Y + Ho_1 + Z)\ dY + \varphi_1(X - Ho_1) \qquad (13)$$

and for the second domain $(Z \leqslant Ho_1 < \infty)$

$$\theta_1 (Ho_1 Z) = 4\int_0^Z Sto_1 (Y + Ho_1 - ZY)dY + \psi_1 (Ho_1 - Z) \qquad (14)$$

The solutions of the Cauchy problem for equation (10b) are obtained from the solutions of equation (10a) by replacing z by $z_e - z$ since the second coolant (in our case, a gas) enters into the channel at $z = z_e$ and moves to the side of a decreasing z.

Mathematically, such replacement is caused by the fact that it is convenient to consider the phenomena in different coordinate systems Ho_1, Z and Ho_2, Z with different direction of abscissa axes in the system with a common abscissa axis and common coordinate origin but with different scales along the dimensionless time axis (coordinate axis). This may be achieved by replacing $z_1 = z_e - z$. On physical grounds the above becomes obvious, if we recall that during time $\Delta\tau_1$ the particle of the first coolant covers a distance $z_1 = w_1 \Delta\tau/d$ from the inlet and will have a coordinate z_1, and the particle of the second coolant moves in the opposite direction from $z = z_e$ and during the same time covers a distance $z_e - z_2 = w_2 \Delta\tau/d$ and at a moment $\tau_0 + \Delta\tau$ will have a coordinate z_2. Thus, for equation (10b) for the first domain $(Ho_2 \leqslant z_e - z \leqslant z_e)$ the solution of the Cauchy problem $[Ho_2 = 0; \psi_2 = \varphi_2 (z_e - z)]$ will be of the form

$$\theta_2 (Ho_2 Z) = 4\int_0^{Ho_2} Sto_2 (YY - Ho_2 + Z_e - Z)dY + \varphi(Z_e - Z - Ho_2) \qquad (15)$$

and for the second domain $(z_e - z \leqslant Ho_2 < \infty)$ the solution of the Cauchy problem assumes the form $[(Z_e = Z\ ;\ \theta_2 = \psi_2(Ho_2)]$

$$\theta_2 \, (\text{Ho}_2 \, Z) = 4 \int_{Ze}^{Ze-Z} \text{Sto}_2 \, (Y + \text{Ho}_2 - Z_e + ZY) dY +$$

$$\qquad\qquad + \Psi_2 (\text{Ho}_2 - Z_e + Z) \qquad\qquad (16)$$

If total dimensionless time is the case, for example, Ho_1 is introduced by means of equation (11), then the system of equations will be written in one coordinate system Ho_1, Z. It is obvious that in this system the first and second domains in a general case are divided not by a straight line but by a curve, since in the equation $\text{Ho}_1 - b(Z_e - Z) = 0; b = f(w_1/w_2)$.

The first domain for a gas will be much smaller because the flow velocity for gases is usually much greater than for liquids.

In practice, for gases in the majority of cases it is possible to consider only the second domain of a process, and for liquids with small flow velocities both domains should be examined. With regard to the transformations made, system (10) may thus be considerably simplified by replacing partial equation (10a) by equation (13) for the first liquid domain, and by equation (14) for the second domain and by replacing equation (10b) by equation (15) for the first domain of a gas or by equation (16) for the second domain of a gas. Since in the checking calculation it is necessary as a rule to know not the temperature distribution of coolants but only their outlet values, these replacements allow additional simplification of a system. Then, for example, for the second domain instead of equations (14) and (16) it is possible to use simpler equations:

$$\theta_1 \, (\text{Ho}_1 Z_e) = 4 \int_{o}^{Ze} \text{Sto}_1 \, (Y + \text{Ho}_1 - Z_e Y) dY + \Psi_1 (\text{Ho}_1 - Z_e) \qquad (17a)$$

$$\theta_2 \, (\text{Ho}_2 \, 0) = 4 \int_{Ze}^{o} \text{Sto}_2 \, (Y + \text{Ho}_2 , Y) dY + \Psi_2 (\text{Ho}_2) \qquad (17b)$$

and equation (8c) may be presented also in the integral form

$$\theta_w \, (\text{Ho}_1 Z) = \int_{o}^{\text{Ho}_1} \frac{- \, U_1 \, q_1 \, (\text{Ho}_1 \, Z) + U_2 \, q_2 \, (\text{Ho}_1 \, Z)}{\rho_w \, C_w \, F_w \, W_1 \, T_o} \, d \cdot d\text{Ho} \qquad (17c)$$

where q_1 and q_2 are determined by equations (12a) and (12b). If it is taken into account that Ho_1 and Ho_2 are determined by equations (9) and (11) and

$$Sto_1 (Y + Ho_1 - Ze_1 Y) =$$

$$\frac{U_1 d}{Cp_1} \frac{a_1 (Y + Ho_1 - Ze) [\theta_w Y + Ho_1 - ZeY) - \theta_1 (Y + Ho - Ze_1 Y)]}{G_1 (Y + Ho_1 - Ze)}$$

$$(18a)$$

$$Sto_2 (Y + Ho_2 Y) = \frac{U_2 \cdot d}{Cp_2} \cdot \frac{a_2 (Y + Ho_2 Y)[\theta_w (Y + Ho_2 Y) - \theta_2 (Y + Ho_2 Y)]}{G_2 (Y + Ho_2)}$$

$$(18b)$$

then the system of integral equations (17) may be obtained. This system is considerably simpler than (10) for solution on electronic computers.

In this system, as well as in (10), the heat transfer coefficients α_1 and α_2 may be taken from experiments on similar unsteady processes.

In the case of variable heat capacity the enthalpy of coolants should be introduced into consideration instead of a temperature. It is not difficult to see that numerical calculation requires a greater number of iterations even for the simplified system of integral equations (17) and, consequently, an electronic computer with great quick-action and storage. Numerical calculation is more complex than the calculation of a heat exchanger operating under stationary conditions. Nevertheless, such calculations of heat exchangers are necessary.

The aforesaid allows the formulation of problems of future unsteady heat transfer investigations as follows:

1. Development and unification of methods for calculating basic types of heat exchangers and devices when solving conjugated problems in the one-dimensional description of a flow. These methods should be presented in the form of electronic computer programs written in basic computer languages.

2. Study of an unsteady heat transfer mechanism in turbulent flows, experimentation and theoretical calculations, and accumulation of design relations for the Nusselt number and the hydraulic resistance coefficient in the case of flow of coolants in different channels.

3. Development of numerical calculation methods for heat exchangers in laminar flow of coolants and unification of these methods in the form of prepared electronic computer programs.

REFERENCES

[1] Kalinin E.K., Jarkho S.A., On the effect of nonisothermity upon the hydraulic resistance coefficient in a turbulent water flow in tubes with artificial flow turbulizers, Teplofiz. Vysok. Temp. No. 5, 1966.

[2] Kalinin E.K., Jarkho S.A., The effect of the Reynolds and Prandtl numbers on effectiveness of heat transfer intensification in tubes, Inzh. Fiz. Zhurn. vol. XI, No. 4, 1966.

[3] Kalinin E.K., Jarkho S.A., Flow intermittance and heat transfer under conditions of artificial flow turbulization in tubes, Izv. AN BSSR, Ser. Fiz. - Tekhn. Nauk No. 2, 1966.

[4] Kalinin E.K., Koshin V.K., Danilov Y.I., Dreitser G.A., Galitseysky B.M., Isosimov V.G., Unsteady Heat Transfer in Tubes Resulting from changes in Heat flow, Gas Mass flow rate and Acoustic Resonance, Proceedings of the Third International Heat Transfer Conference, August 7-12, 1966, v. III, Chicago III, pp. 57-70.

[5] Kalinin E.K., Dreitser G.A., Yarkho S.A., The Experimental Study of Heat Transfer Intensification under Conditions of Forced Flow in Channels, Proceedings of ISME 1967, Semi-International Symposium 4-8, Sept. 1967, v 1, Tokyo.

[6] Kalinin E.K., Koshkin V.K., Galitseisky B.M., Dretser G.A., Izosimov V.G., Unsteady heat transfer in a tube when applied a heat flux and gas flow rate, Teplofiz. Vysok.-Temp. 1967.

[7] Kalinin E.K., Dretser G.A., Study of heat transfer intensification in longitudinal air flow past a narrow tube bundle, Inzh. Fiz. Zhurn. vol. XV, No. 3, 1968.

[8] Kalinin E.K., Dreitser G.A., Izosimov V.G., Correlation of experimental data on unsteady convective heat transfer with a change in a heat flux. Teplofiz. Vysok. Temp. vol. 7, No. 6, 1969.

[9] Kalinin E.K., Dreitser G.A., Yarkho S.A., Kusminov V.A., Local Separation in Turbulent Flow in Channels as a Method of Heat Intensification, Paper of International Symposium "Heat and Mass Transfer in Flows with Separated Regions and Measurement Techniques", Herceg-Novi, Yugoslavia, September, 1-13, 1969.

[10]   Kalinin E.K., Dreitser G.A., Unsteady Convective Heat Transfer and Hydrodynamics in Channels. "Advances in Heat Transfer", v. 6, Academic Press, 1970.

[11]   Kalinin E.K., Koshkin V.K., Dreitser G.A., Galitseisky B.M., Isosimov V.G., Experimental Study of Unsteady Convective Heat Transfer in Tubes, Int. J. Heat Mass Transfer, v. 13, p. 1271-1281, 1970.

[12]   Kalinin E.K., Dritser G.A., Yarkho S.A., Kusminov V.A., The Experimental Study of Heat Transfer Intensification Under Conditions of Forced Single and Two-Phase Flow in Channels, "Augmentation of Convective Heat and Mass Transfer", ASME, United Engineering Center, 1970, 80-90.

[13]   Kalinin E.K., Koshkin V.K., Coolants and Heat Exchangers, Mashgiz, 1971.

[14]   Kalinin E.K., Dretser G.A., Kuzminov V.A., Neverov A.S., On the effect of height and pitch of location of turbulizers upon heat transfer intensification in a tube, Izv. AN BSSR, Ser. Fiz.-Tekhn. Nauk No. 3, 31-35, 1971.

[15]   Kalinin E.K., Jarkho S.A., Study of heat transfer intensification in flow of gases and liquids in tubes, Inzh. Fiz. Zhurn. vol. XX, No. 4, 1971.

[16]   Kalinin E.K., Dreitser G.A., Kozlov A.K., Study of heat transfer intensification in longitudinal flow past tube bundles with different relative pitches, Inzh. Fiz. Zhurn. vol. XXII, No. 2, 1972.

Fig. 1. Rolling tube process

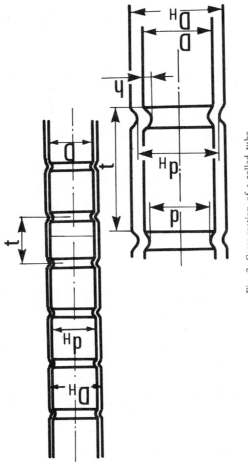

Fig. 2. Cross-section of a rolled tube

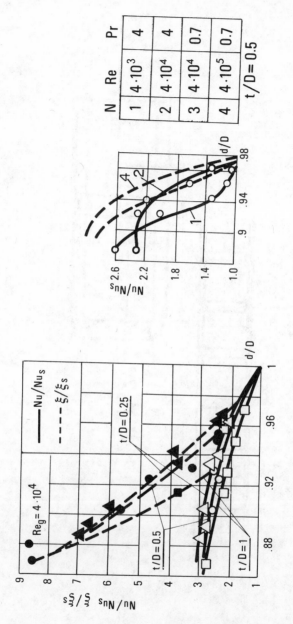

Fig. 3. Effect of a height of turbulizers on effectiveness of heat transfer intensification in tubes.

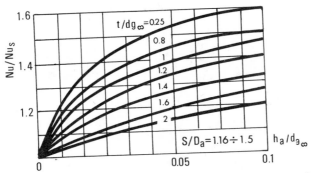

Fig. 4. Effect of depth of grooves of rolling $h_a/d_{\epsilon\infty}$ and rolling pitch $t/D_{\epsilon\infty}$ on heat transfer intensification in longitudinal flow past tube bundles for $S/Da = 1.16 \div 1.5$ at $Re > Re_2$.

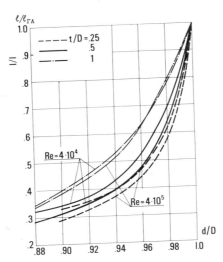

Fig. 5. Length ratio of tubular heat exchangers $1/1_{sm}$ versus rolling depth d/D.

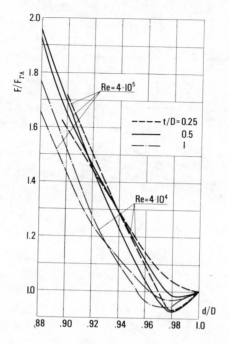

Fig. 6. Area ratio of cross sections of tubular heat exchangers F/F_{sm} versus rolling depth d/D.

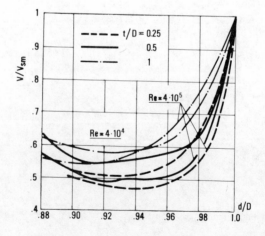

Fig. 7. Volume ratio of tubular heat exchangers V/V_{sm} versus rolling depth d/D.

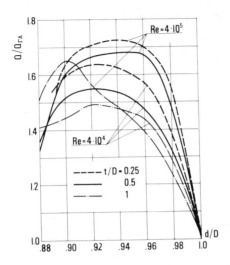

Fig. 8. Heat power ratio of tubular heat exchangers Q/Q_{sm} versus rolling depth d/D.

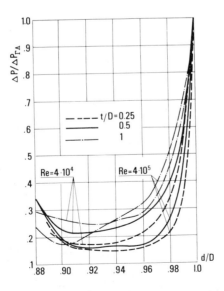

Fig. 9. Pressure drop ratio of tubular heat exchangers $P/\Delta P_{sm}$ versus rolling depth d/D.

t/D	0.225				0.45				0.9			
d/D	0.9	0.925	0.95	0.97	0.9	0.925	0.95	0.97	0.9	0.925	0.95	0.97
$V/V_{ГЛ}$												
Re = $4 \cdot 10^4$	0.715	0.7	0.709	0.75	0.748	0.731	0.739	0.815	0.83	0.81	0.815	0.875
Re = 10^5	0.733	0.71	0.724	0.77	0.766	0.74	0.754	0.796	0.842	0.818	0.823	0.864
$Q/Q_{ГЛ}$												
Re = $4 \cdot 10^4$	0.267	1.299	1.289	1.18	1.235	1.254	1.243	1.15	1.15	1.16	1.16	1.1
Re = 10^5	1.245	1.275	1.26	1.2	1.211	1.24	1.22	1.175	1.13	1.155	1.14	1.111
$\Delta p/\Delta p_{ГЛ}$												
Re = $4 \cdot 10^4$	0.433	0.406	0.42	0.546	0.485	0.456	0.472	0.6	0.62	0.589	0.606	0.715
Re = 10^5	0.463	0.424	0.446	0.518	0.513	0.467	0.5	0.569	0.65	0.602	0.625	0.695

Fig. 10. Table of calculation results for heat exchangers in which heat transfer is limited by a gas in the intertube space.

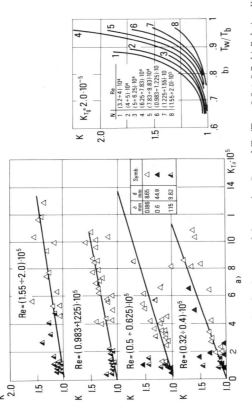

Fig. 11. a) Relation $K = f(K_{Tg})$ for fixed ranges of the number Re and $T_w/T_b = 0.8 \div 0.7$. b) Relation $K = f(T_w/T_b)$ for fixed value of $K_{Tg} = 2.0 \cdot 10^{-5}$ and different ranges of Re.

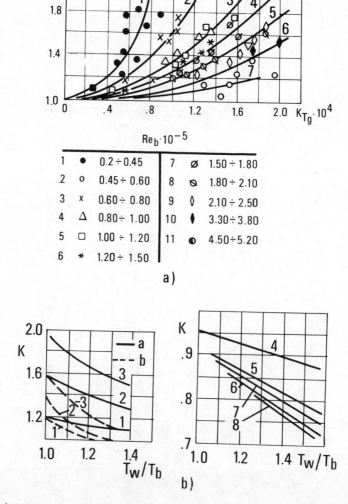

		Re$_b \cdot 10^{-5}$			
1	●	0.2 ÷ 0.45	7	∅	1.50 ÷ 1.80
2	○	0.45 ÷ 0.60	8	◐	1.80 ÷ 2.10
3	×	0.60 ÷ 0.80	9	◊	2.10 ÷ 2.50
4	△	0.80 ÷ 1.00	10	♦	3.30 ÷ 3.80
5	□	1.00 ÷ 1.20	11	◕	4.50 ÷ 5.20
6	✳	1.20 ÷ 1.50			

a)

b)

Fig. 12. a) Relation $K = f(K_{Tg})$ for different ranges of the number Re with a stepwise increase of heat evolution in tubes. $T_w//T_b = 1.1 \div 1.2$. b) Relation of K versus T_w/T_b at different K_{Tg}.

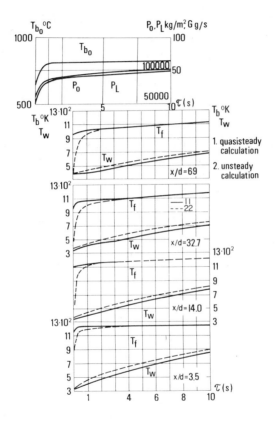

Fig. 13. Change of T_w and T_b in time with gas cooling in a tube (δ = 0.6 mm d = 44.8 mm).

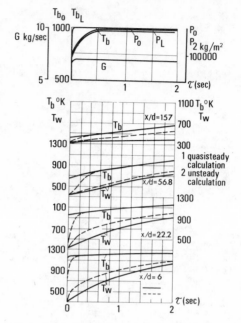

Fig. 14. Change of T_w and T_b in time in time with gas cooling in a tube ($\delta = 0.186$ mm $d = 8.65$ mm).

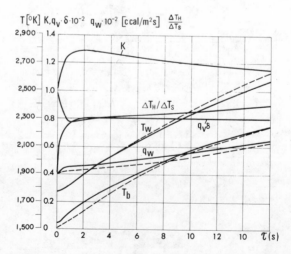

Fig. 15. Change of T_w, T_b, $\Delta T_H/\Delta T_s$, K, Q_v, δ, q_w in time with load increase.

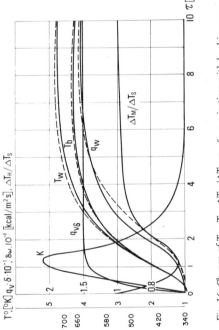

Fig. 16. Change of T_w, T_b, $\Delta T_H/\Delta T_s$, q_v, δ, q_w, in time with load increase.

Chapter 9

SURVEY OF THE HEAT TRANSFER THEORIES
IN REGENERATORS

H. Hausen (*)

After much experience, I have always found that it is a very difficult and lengthy task to explain the different theories of heat transfer in regenerators. Therefore in such a short survey, only the main ideas of these theories can be indicated. I want first to discuss an almost exact theory, which enables the determination of an overall heat exchange coefficient for regenerators. If the value of this coefficient is known, the heat transferred in a regenerator can be calculated in the same simple manner as in a recuperator.

For practical purposes, however, numerical methods of solution have been developed, which allow any required approximation to be calculated to the exact values and which can be applied in cases for which exact solutions are not known. I myself want to discuss some of those methods, which can be regarded as a solution to the integral equation governing the temperature distribution in a regenerator.

A regenerator differs from a recuperator in the following way. Through a recuperator the fluids flow continuously without any interruption. Usually the fluids are separated by walls, mostly of tubular shape, through which the heat is transferred. Contrary to this, a regenerator has a heat string porous mass, called matrix or chequerwork (fig. 1). The stream of the fluids is periodically reversed, so that at one time only one fluid is passing the regenerator. During the hot period the hot fluid gives up its heat to the matrix and is thus cooled. During the cold period the cold fluid flows in the opposite direction and is heated by taking up heat from the matrix. To effect a continuous heat transfer, at least two regenerators are necessary, so that at the same time one fluid is cooled in the first regenerator and the other fluid is heated in the other regenerator (Fig. 2).

An exact definition of a regenerator must include the fact that even in

---

(*) Technische Universität Hannover, F.R. Germany

N.B. ALL FIGURES QUOTED IN TEXT ARE AT THE END OF THE CHAPTER

the steady state the temperatures of the matrix and the fluids depend not only on local coordinates but also on time. This dependance on time, which is periodic in the steady state, is the reason why the bahavior of a regenerator and the theories describing it are much more complicated than those of a recuperator.

The lowest part of Fig. 3 represents the distribution of temperature in the matrix of a regenerator during the cold period after the steady state is reached. For simplicity it is assumed that the heat capacities of both fluids flowing during the respective periods are equal and do not depend on temperature. The heat transfer coefficients are also regarded as constant. In such a case the temperature distribution in a recuperator is purely linear. The temperature distribution in a regenerator differs from this by two essential features. The temperature lines change with time, for example downwards in the cold period, as represented in Fig. 3. Secondly, the lines are linear only in a middle part of the regenerator. Near the ends of the regenerator they are curved.

This behavior of a regenerator can be understood if one considers the exact solutions of the differential equation which describe the transfer of heat in the regenerator. There exists an infinite number of partial solutions, called eigenfunctions, which fulfill the boundary condition for the reversals of gas flow in regenerator operation. These correspond principally to the eigenfunctions of a stretched string, which are vibrations of different frequencies. As a real vibration of a string consists of a basic vibration and of higher vibrations or eigenfunctions, so the periodic behavior of a regenerator can be regarded as a vibration of temperatures consisting of many eigenfucntions. The basic vibration or zero-eigenfunction is linear in space and time (see upper part of Fig. 3). The middle part of Fig. 3 shows the sum of the higher eignefunctions, each of them multiplied by an adequate factor. By adding this sum of eigenfunctions to the zero-eigenfunction the real temperature distribution as represented in the lowest part of Fig. 3 is obtained.

From the middle part of Fig. 3 one sees an essential feature of the higher eigenfunctions. They have their greatest values at both ends of the regenerator and decrease with increasing distance from these ends. In regenerators of sufficient length they become so small that in a middle part the zero-eigenfunction alone governs the changes of temperature. Near the entrance of the fluid, i.e. on the left side of Fig. 3 the higher eigenfunctions are temperature vibrations of different frequencies. Near the exit of the fluid they are aperiodic (Fig. 4).

By combining all eigenfunctions, each multiplied by a suitable factor, the constant entrance temperature of the gas can be represented as illustrated in Fig.

5. The linear time variation of the zero-eigenfunction is the more counterbalanced, the more higher eigenfunctions are added, The approximation, which is attained by using the higher eigenfunctions up to eigennumber 4, may be seen from line 4.

I want to emphasize, that such a calculation of the periodic temperature only involves the mean temperatures in the cross section of the matrix. The temperature distribution over a cross section has been determined separately for the zero-eigen function. The result is shown in Fig. 6 for the cold period. While the fluid temperature decreases and is smaller than the temperature of the elements, the temperature in a cross section decreases principally as in fig. 6, in which the lines represent the temperature distribution at different times. At the beginning of the cooling period the curve has a parabolic form. The first effect of the cooling is that this parabola is transformed into another one of equal shape but open in the opposite direction. Finally, the parabola slides down without changing its form.

If one includes these temperature differences in the matrix elements, one is able with help of the theory we have discussed to establish an overall heat exchange coefficient for regenerators. Rummel [1] has defined such an overall heat transfer coefficient k by the equation

$$Q_{Per} = k \, F \, (T + T') \, \Delta \vartheta_M , \qquad (1)$$

in which Q_{Per} denotes the quantity of heat transferred in one regenerator during one period, F the surface of the heat storing matrix of one regenerator, T and T' the durations of the two periods and $\Delta \vartheta_M$ the mean temperature difference between the two fluids. $\Delta \vartheta_M$ is to be determined in the same way as for recuperators.

As the temperature distribution in a regenerator is composed of the zero-eigenfunction and the higher eigenfunctions, the overall heat exchange coefficient k, also has to be calculated in two steps. The first step yields a coefficient k_0 determined for the zero-eigenfunction. The next step is to calculate the ratio k/k_0, which corresponds to the influence of the higher eigenfunctions. From the detailed theory of the zero-eigenfunction there follows the equation

$$\frac{1}{k_0} = (T + T') \left[\frac{1}{hT} + \frac{1}{h'T'} + \left(\frac{1}{T} + \frac{1}{T'} \right) \frac{\delta}{\lambda_\beta} \phi \right] \qquad (2)$$

in which h and h' are the heat transfer coefficients in both periods respectively, δ and λ_β thickness and heat conductivity of the elements of the matrix. ϕ is a function which indicates the degree to which the heat enters into the interior of the matrix elements. It depends on $\delta \vartheta / \alpha$ $(1/T + 1/T')$, where a is the thermal diffusivity of the matrix. The value of ϕ is determined by a complicated theoretical equation but can easily be taken from Fig. 7 for matrices, which consists of plates, cylinders or spheres. The ratio k/k_0 is represented by Fig. 8. The parameters used in this figure are the so called "reduced length" Λ and the "reduced period" Π . Λ and Π are determined by the equations

$$\Lambda = 2 \frac{k_o (T + T') F}{C_{Per}} \qquad (3)$$

$$\Pi = 2 \frac{k_o (T + T') F}{C_s} \qquad (4)$$

in which C_{Per} denotes the mean value of the heat capacities of the fluids passing the regenerator during the heating period and the cooling period respectively, and C_s denotes the heat capacity of the heat storing matrix of one regenerator By. multiplying k_0 from equation (2) by k/k_0 from Fig. 8, the sought overall heat exchange coefficient K is determined as defined by equation (1) of Rummel.

Fig. 8 for k/k_0 differs a little from the old one, which I published in 1942. Three months ago I drew the new diagram presented here using values given me privately by Willmott and by Schellmann and other values published by Lambertson [2] and by Sandner [3]. All these values, which have been determined with help of an electronic computer, coincide very well with each other.

In view of the complexity of the theory, the method described to determine k seems to be relatively simple. Once the value of k is found, a regenerator can be calculated in the same way as a recuperator for the technical purposes. Using equation (1), one cannot only determine the quantity of heat being transferred in given regenerators but also the essential dimensions of regenerators to be built, in the first place the surface area F, which is necessary for a required amount of heat to be exchanged.

Schofield, Butterfield and Young in their 1961 paper [4] have shown that this method, combined with values of h and h' determined by the experiments of Böhm [5], can successfully be applied to the calculation of hot blast stoves.

However, with a few exceptions the eigenfunctions themselves, because of a special difficulty in determining their coefficients, have never been used to calculate the temperature distribution in a regenerator and its efficiency, as well as the ratio k/k_0. Even before this method was known, a search had begun for other methods. At first methods were proposed which to a great extent are empirical in character and include those of Rummel [1] and Shack [6] . Very useful finite difference methods have been developed by Nusselt [7] and Hausen [8] and adapted to the use in electronic computors mostly by Willmott. These methods will be discussed by Willmott himself.

A second group of methods is based on an integral equation, which governs the steady state behavior of a regenerator and has been published by Nusselt [9] . Nusselt has given an exact solution of this equation, which consists of an infinite series of integrals. This solution is too complicated for practical purposes, and several other solutions of an approximate kind have been developed. One, called the heat pole method, had been proposed by myself and later refined by Ilffe [10] and myself.

The idea of the heat pole method is as follows. The main part of the regenerator matrix may have the same temperature as the gas entering on the left side of the regenerator. This temperature may be set equal to zero. Only in one stripe of breadth $\Delta\epsilon$ the temperature may be higher by the amount 1. The gas flowing to the right takes up a part of the heat contained in the heat pole and carries it to the other stripes of equal breadth $\Delta\epsilon$, which lie on the right side of the heat pole. So, after some time the heat pole will be distributed over the different stripes as shown on the right part of fig. 8. The function Δw representing this distribution is called the heat pole function. It is related to Green's function used by the mathematicians. Δw indicates the influence of an amount of heat in one stripe on the temperatures in the other stripes at a later time. So, if at the beginning the four stripes in Fig. 8 have the temperatures f_1, f_2, f_3, and f_4 respectively, for example the temperature of stripe 4, after the considered time will be found

$$f_1 \Delta w_4 + f_2 \Delta w_3 + f_3 \Delta w_2 + f_4 \Delta w_1 \tag{5}$$

In this way, for a given initial temperature distribution of the matrix the temperature distribution at a later time can easily be calculated if the values of the heat pole function are known. These values can be determined from theoretical equations but are not simple.

An advantage of the heat pole method is that the temperature distribution in the steady state can be determined directly without the necessity of calculating foregoing periods or intermediate temperatures in a period. The result of this procedure is an approximate solution of the integral equation mentioned. In the refined heat pole method the mean values of each of the stripes are replaced by the values at the limits of the stripes, and the summing after equation (5) is replaced by a numerical integration with help of the Simpson rule.

The method of Nahavandi and Weinstein [11] consists in representing the initial temperature distribution $f(\xi)$ by the power series

$$f(\xi) = \sum_{j=0}^{m} a_j \xi^j \tag{6}$$

and inserting this into the integral equation. Finally, one gets $m + 1$ linear equations, which determine the coefficients a_j for the steady state. But before establishing these linear equations, one has to evaluate for each value of ξ numerically $m + 1$ integrals of the form

$$\int_0^{\xi} \epsilon^j K(\xi - \epsilon) \, d\epsilon \tag{7}$$

where $K(\xi - \epsilon)$ is the so called nucleus of the integral equation.

The required temperatures of the matrix are finally determined by solving the linear equations for a_j and inserting their values into equation (6).

Finally, I want to mention the method of Sandner [3] . Sandner starts from the exact solution of the differential equations, as developed by Nusselt. He has transformed the integral terms contained in this solution and finally replaced them by asymtotic and partially empirical expressions. The approximation of these expressions to the exact solution is so good that great distances between the values of ξ can be chosen. Thus, in the method of Sandner the complexity of his final expression is counterbalanced by the small number of values of ξ, for which the calculation is to be carried out. The results he obtained using his method are extremely accurate.

REFERENCES

[1]   K. Rummel: Die Berechnung der Wärmespeicher auf Grund der Wärme-durchgangszahl. Stahl und Eisen, Vol. 48 (1928), pages 1412-1414.

[2]   T.J. Lambertson: Performance factors of a periodic-flow heat exchanger. Trans. ASME, A, Vol. 80 (1958) p. 586-592.

[3]   H. Sandner: Beitrag zur linearen Theorie des Regenerators. Dissertation Technische Universität München 1971.

[4]   J. Schofield, P. Butterfield, and P.A. Joung: Hot blast stoves. Journal of the Iron and Steel Institute Vol. 199 (1961) p.229-240.

[5]   H. Böhm: Versuche zur Ermittlung der konvektiven Wärmeübergangszahlen an gemauerten engen Kanälen. Arch. Eisenhüttenwesen. Vol. 6 (1933) p. 423-431.

[6]   A. Schack: Die Berechnung von Regeneratoren. Arch. Eisenhüttenwesen. Vol. 17 (1943/44) p. 101-118.

[7]   W. Nusselt: Der Beharrungszustand im Widerhitzer. Z. VDI Vol. 71 (1927) p. 85.

[8]   H. Hausen: Nährungsverfahren zur Berechnung des Wärmeaustausches in Regeneratoren. Z. angew. Math. Mech. Vol. 11 (1931) p. 105-114.

[9]   W. Nusselt: Der Beharrungszustand im Winderhitzer. Z. VDI Vol. 72 (1928) p. 1052-1054.

[10] C.E. Iliffe: Thermal analysis of the contra-flow regenerative heat exchanger. Proc. Inst. Mech. Engrs. Vol. 159 (1948) p. 363-372.

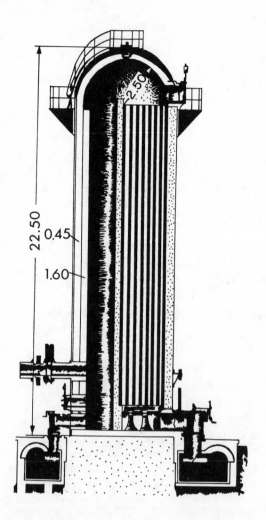

Fig. 1. Cowper stove.

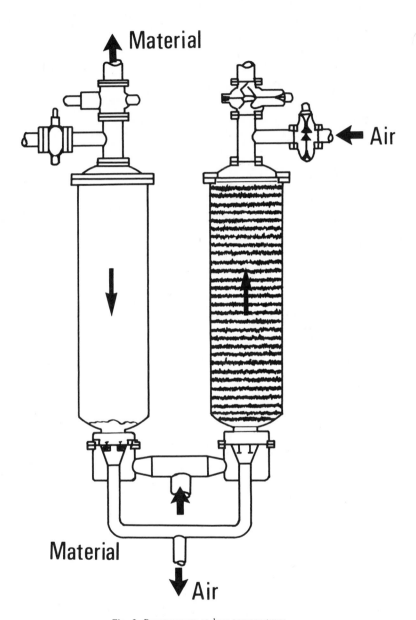

Fig. 2. Regenerators at low temperatures.

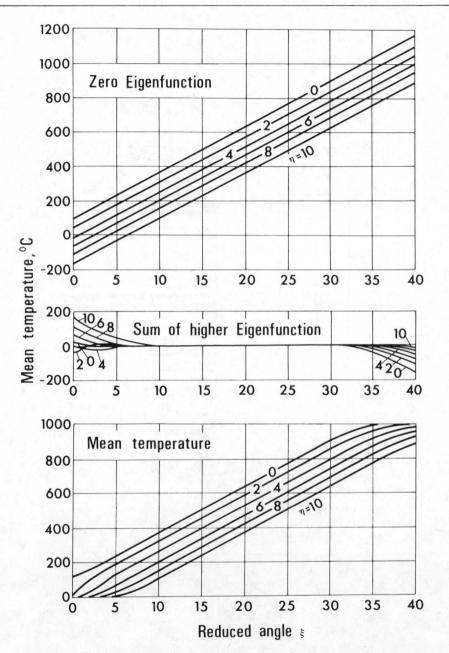

Fig. 3. Temperature distribution in the matrix of a regenerator.

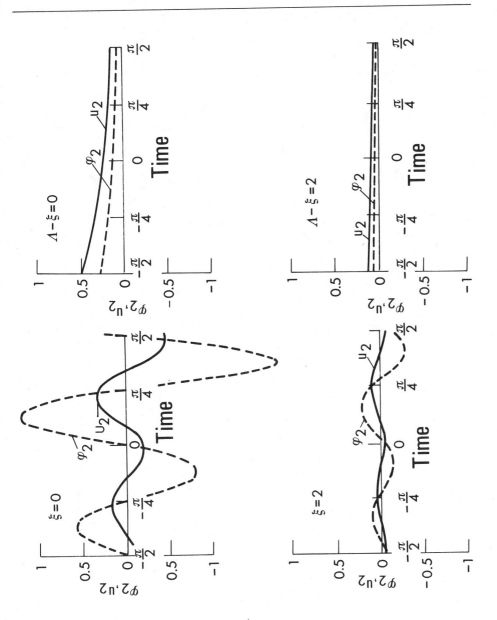

Fig. 4. Eigenfunction for K = 2.

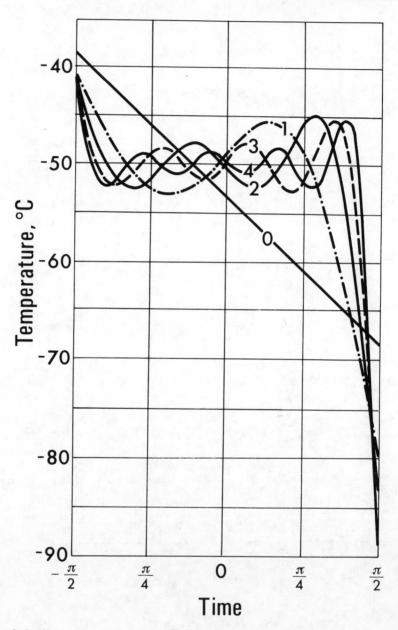

Fig. 5. Combining the engeinfunctions to represent the constant entrance temperature of the gas.

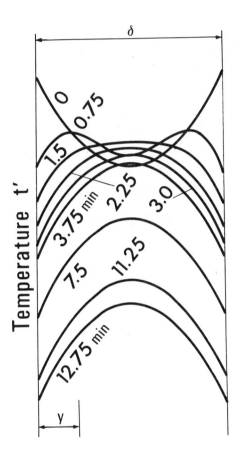

Fig. 6. Temperature changes in a cross-section of a matrix element.

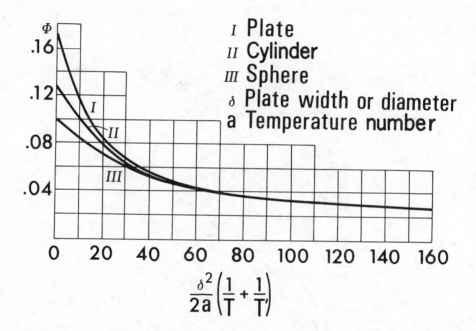

Fig. 7. Factor ϕ on the equation for k_o.

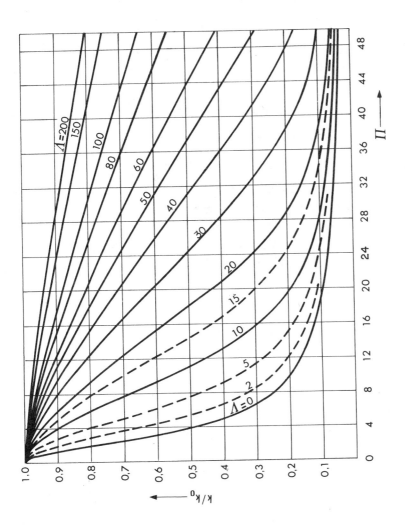

Fig. 8. Ratio k/k_0 of the overall heat transfer coefficient k to the coefficient k_0 after the zero-eignfunction.

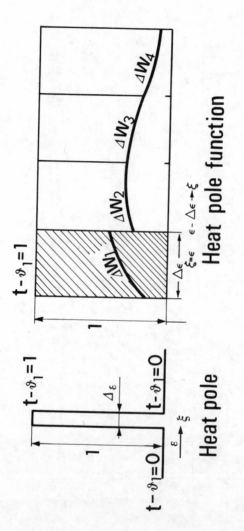

Fig. 9. Heat pole and heat pole function.

Chapter 10

DEVELOPMENTS IN REGENERATOR THEORY SINCE THE ADVENT OF THE DIGITAL COMPUTER

A. J. Willmott (*)

Prior to 1945, the theory of regenerators was progressed in two directions. Methods were proposed principally by Nusselt [1] and Hausen [2] , and later in 1948 by Iliffe [3] to solve the differential equations

$$\frac{\partial t}{\partial \xi} = T - t \tag{1}$$

$$\frac{\partial T}{\partial \eta} = t - T \tag{2}$$

describing the temperature behavior of the gases, $t(\xi,\eta)$ and the heat storing matrix, $T(\xi,\eta)$ in the regenerator. The methods involved lengthy calculations which were not practicable in an era prior to that of the digital computer. Professor Hausen has mentioned alternative methods which were developed by Rummel, Shack and himself among others, where the regenerator was represented by an equivalent recuperator.

In this chapter, I restrict my considerations to methods involving the solution of the differential equations.

These methods of solution of the regenerator problem incorporating the reversal boundary condition fall into two distinct classes. The "open methods" are those in which some arbitrary, but if possible well chosen distribution of temperature is imposed upon the solid matrix. Subsequently, the temperatures at the end of successive periods are evaluated repeatedly until the mathematical model achieves cyclic equilibrium. The solving of equations (1) and (2) is regarded as an initial value problem and typical of these methods are those of Lambertson [4]

(*) University of York, U.K.

N.B. ALL FIGURES QUOTED IN THE TEXT ARE AT THE END OF THE CHAPTER

(1958) and Willmott [5] (1964).

Once the digital computer was available, the speed of calculation became such that the continuous cycling of the model to equilibrium was no longer a deterrent for practical calculations.

It is interesting to note however, that immediately before digital computing machines became generally available, open methods of solution of the differential equations were attempted using analogue computers.

Heindlhofer and Larsen [6] described in 1945 an electrical analogue of the flow of heat in a regenerator system. This device, developed in 1935, was used in connection with the design of regenerators for open hearth furnaces and coke ovens. A very similar device was developed by Tipler [7] with a view to calculating the effectiveness of the regenerator applied to the Gas Turbine cycle.

Essentially, the thermal capacity of the regenerator packing at equally spaced positions down the length of the regenerator was represented by a bank of electrical capacitances. The voltage on each capacitance corresponded to the temperature of the packing at the equivalent position in the regenerator.

This bank of condensers was scanned first by an initially charged bank of condensers representing the hot gas, in series with an electrical resistance representing the resistance to heat transfer between the hot gas and the solid. Then, in the reverse direction, the condenser bank was scanned by a second initially discharged system of condenser-resistance units representing the flow of cold gas through the regenerator. See Fig. 1.

This electric analogue of the regenerator was cycled repeatedly until equilibrium was achieved.

Hendlhofer and Larsen measured the voltages using an electrostatic voltmeter while Tipler arranged for the measured "temperatures" to be displayed on a cathode ray oscilloscope.

These electrical analogues were special purpose devices used to represent the temperature behavior of the regenerator. In the late 1950's and early 1960's attempts were made to use general purpose analogue computers. These consisted of summing and integrating amplifiers which could be interconnected with potentiometers in a variable manner enabling the same analogue computer to be applied to different problems. Schuerger and Agarwal [8] from the United States Steel Corporation and Dancy and Meyer [9] at the Jones and Laughlin Steel Corporation in Pittsburgh used such general purpose analogues to simulate Cowper Stove behavior.

In the middle 1950's, digital computers became more readily available for scientific calculations and in 1958, Lambertson [4] described his method for representing the regenerator. He did not solve the differential equations in an explicit manner. In a similar way to that adapted by Heindlhofer and Larsen, he considered the temperature behavior of successive equally spaced sections of regenerator packing. Whereas the electrical analogue could compute the temperature variations in a continuous manner, Lambertson adopted a step-by-step procedure suitable for digital calculations.

Although Hausen [10] had described a finite difference method of solving the differential equations, I myself [5] implemented another method for the Ferranti Pegasus computer. In my method, the two equations are represented in difference from using the trapezoidal rule.

In this method, equation (1) is integrated using:

$$t_{r+1,s} = t_{r,s} + \frac{\Delta\xi}{2}\left\{\left(\frac{\partial t}{\partial\xi}\right)_{r+1,s} + \left(\frac{\partial t}{\partial\xi}\right)_{r,s}\right\} \tag{3}$$

Similarly, equation (2) is integrated in the following manner

$$T_{r,s+1} = T_{r,s} + \frac{\Delta\eta}{2}\left\{\left(\frac{\partial T}{\partial\eta}\right)_{r,s+1} + \left(\frac{\partial T}{\partial\eta}\right)_{r,s}\right\} \tag{4}$$

where the subscripts r and s refer to distance and time positions on a finite difference grid, and where $\Delta\xi$ is the distance step length and $\Delta\eta$ is the time step length. Equations (3) and (4) are expanded using the replacements

$$\left(\frac{\partial t}{\partial\xi}\right)_{r,s} = T_{r,s} - t_{r,s} \quad ; \quad \left(\frac{\partial T}{\partial\eta}\right)_{r,s} = t_{r,s} - T_{r,s} \tag{5}$$

In Hausen's method, the differential equation (1) is rewritten in the form

$$\frac{\partial^2 t}{\partial\xi\partial\eta} + \frac{\partial t}{\partial\xi} + \frac{\partial t}{\partial\eta} = 0 \tag{6}$$

The following finite difference substitutions are then made:

$$\frac{\partial t}{\partial \xi} = \frac{t_{r+1,s} - t_{r,s} + t_{r+1,s+1} - t_{r,s+1}}{2\Delta\xi} \tag{7}$$

$$\frac{\partial t}{\partial \eta} = \frac{t_{r,s+1} - t_{r,s} + t_{r+1,s+1} - t_{r+1,s}}{2\Delta\eta} \tag{8}$$

$$\frac{\partial^2 t}{\partial\xi\partial\eta} = \frac{1}{\Delta\xi} \left\{ \frac{t_{r+1,s+1} - t_{r+1,s}}{\Delta\eta} - \frac{t_{r,s+1} - t_{r,s}}{\Delta\eta} \right\} \tag{9}$$

A similar treatment can be effected with regard to a rearranged form of equation (2).

An exactly similar method to that of Hausen was described by Saunders and Smoleneic, [11] later reported by de Allen [12].

The importance of these finite difference methods lies in the fact that they can be readily adapted to problems in which variable mass flow of the gases and temperature dependent thermal properties of both gas and solid are incorporated. In 1968, I [13] described a solution to the problem where a variable mass flow of air through a Cowper Stove occurs when a by-pass main is included in the regenerator system.

In representing the temperature behavior of a regenerator by equations (1) and (2), two important assumptions are made.

Firstly, it is assumed that the thermal conductivity of the heat storing matrix is zero in a direction parallel to that of the gas stream. Secondly, in a direction perpendicular to gas flow, either

(1) the thermal conductivity is considered to be infinite, in which case the solid matrix is isothermal in this perpendicular direction, or
(2) it is considered to be finite and a bulk heat transfer coefficient, h is developed which incorporates the surface resistance to heat transfer between gas and solid and the resistance internal to the heat storing solid.

In this latter case, Hausen proposed a form:

$$\frac{1}{\overline{h}} = \frac{1}{h} + \frac{d\phi}{3\lambda} \tag{10}$$

where h is the surface transfer coefficient, d is the semi-thickness of the walls of heat storing material surrounding the channels through which the gases flow, λ is the finite thermal conductivity and ϕ is a factor built in to account for the effect of the hot-cold period and cold-hot period reversals.

Professor Hausen has indicated in his paper that the function ϕ is developed for the zero-eigenfunction. It is assumed that such a ϕ factor is applicable when the effect of higher order eigenfunctions is incorporated into the model of the regenerator.

In order to investigate the acceptability of this simplifying assumption, a fuller model of the regenerator has been developed in which the heat transfer within the heat storing matrix is calculated directly using the Fourier equation. The new model is expressed mathematically by the equations:

$$\frac{\partial t}{\partial \xi} = T_0 - t \tag{11}$$

$$\frac{\partial T}{\partial W} = \frac{\partial^2 T}{\partial Z^2} \tag{12}$$

where ξ is again a "dimensionless distance" down the length of the regenerator, Z is the corresponding perpendicular distance and W is the reduced time. These two equations are linked by a surface boundary condition

$$\frac{\partial T}{\partial Z} = N (T - t) \text{ at } Z = 0 \tag{13}$$

I [14] have described an open method for solving these equations, using finite difference techniques. The importance of this work lies not in the methods of solution of equations (11) and (12), but in the comparison between the temperature behavior of the same regenerator calculated using equations (11) and (12) and that computed using the simplified equations (1) and (2).

The "3-D calculations" involve the dimensions ξ, W and Z using equations (11) (12) and (13) while the "2-D calculations" only involve ξ and η in equations (1) and (2). Whereas the 2-D model in general predicts a linear variation of exit gas temperature, the 3-D calculations yield a non-linear variation; this non-linear effect becomes more marked the lower the value of ϕ, that is the more severe the effect of the reversals in regenerator operation.

In Figure 2 this comparison is illustrated graphically. A measure ψ of the difference between the two computed exit gas temperatures has been defined to be

$$\psi = \frac{A_2 - B_2}{A_3 - B_3}$$

The closer ψ to unity, the closer the 2-D model represents the fuller 3-D model of the thermal regenerator. In Figure 3 the variation of ψ with the reduced length Λ of the regenerator and with the correction factor ϕ is displayed, where

$$\Lambda = \frac{hA}{WS}$$

As Λ diminishes, so the effect of the higher order eigenfunctions increases over the whole length of the regenerator. For each value of ϕ, the value of ψ decreases as the reduced length decreases. On the other hand, as reduced length increases, the value of ψ approaches an asymptotic value, suggesting that under such circumstances, the 3-D calculated exit gas temperature can be estimated using only the zero-eigenfunction. Indeed Professor Hausen has suggested how this might be done.

The value of ψ is most sensitive to the value of the factor ϕ. Professor Hausen has shown how the temperature changes with time in the cross-section of a matrix element.

At a reversal, the parabolic temperature profile is inverted. The smaller the value of ϕ, the greater is the proportion of the hot or cold period which is devoted to inverting the parabolic profile. This inversion process is reflected in the exit temperature calculated using the 3-D method.

In coming to the end of this paper, I want to turn to the second group of methods used for solving the equations describing regenerator behavior. These are the closed methods.

In the closed methods, the reversal condition, which specifies that the solid temperature distribution at the beginning of a period is identical to that at the conclusion of the previous period, is incorporated directly. By embodying this condition for both hot and cold periods simultaneously, within the mathematical method for the solving of the differential equations, one specifies implicitly the cyclic equilibrium condition that the solid temperature distribution at the beginning of a complete cycle of operation, a cycle consisting of a hot/cold period followed by a cold/hot period, is identical to that at the beginning of the previous cycle.

For equations (1) and (2), Professor Hausen has mentioned a group of methods which is based on a pair of integral equations which govern the steady state behavior of a regenerator which has been published by Nusselt [1]. A number of methods have been proposed to solve these integral equations.

For equations (11) and (12), Collins and Daws, as reported by Evans and Edwards, [15] also obtained an integral equation. In the Nusselt method for equations (1) and (2), the integral equations are solved for the solid temperature distributions at the beginning of the hot and cold periods. In the Collins and Daws method for equations (11) and (12), the integral equations are solved for the time variations of the exit gas temperature for the hot and cold periods.

Now both methods essentially involve the evaluation of a dependent variable y as a function of x, that is $y(x)$. In the Nusselt method, y is the solid temperature at the beginning of a period and x is reduced distance down the regenerator. In the Collins and Daws approach, y is exit gas temperature in a particular period and x is time.

Two possible approaches have been made. Hausen and Iliffe for the Nusselt integral equation replace the integrals

$$\int_0^\xi y(\epsilon) \, K(\xi - \epsilon) \, d\epsilon$$

by numerical quadrature formulae so that the integral equations are reduced to a set of simultaneous linear equations which are solved for the values of $y_0, y_1, y_2, \ldots$ where

$$y_j = y(j\Delta x)$$

where Δx is the distance between the equally spaced positions where $y(x)$ is evaluated. In a similar way, Collins and Daws reduce their integral equations to a set of linear algebraic equations using Newton-Cotes formulae for integral evaluation.

An alternative approach has been to represent the initial temperature distribution $y(x)$ by a power series

$$y(x) = a_0 + a_1 x + a_2 x^2 + \ldots + a_m x^m$$

Again the integrals

$$\int_0^\xi y(\epsilon) \, K(\xi - \epsilon) \, d\epsilon$$

are replaced by numerical quadrature formulae. This time a set of simultaneous linear equations is generated which is solved for the coefficients $a_0, a_1, a_2 \ldots$. This method has been developed by Nahavandi and Weinstein [16].

For 3-D models, [equations (11) and (12)]. Tomeczek [17] has developed this approach as an alternative to that presented by Collins and Daws. The power series

$$a_0 + a_1 x + a_2 x^2 + \ldots + a_m x^m$$

represents the time variation of the exit gas temperature of the regenerator and the derived linear algebraic equations are solved for the coefficients $a_0, a_1, a_2 \ldots$.

In both approaches, it is required to solve a set of linear algebraic equations of the general form

$$A\underline{z} = \underline{h}$$

This method will break down if the determinant $|A|$ of A becomes very small, that is if the matrix A becomes singular.

When the temperature behavior of the gas/solid is almost linear, that is when the zero-eigenfunction dominates the situation, it would seem reasonable to use few levels in the regenerator for the Hausen-Iliffe approach to determine its overall behavior. Similarly, one might expect to solve the equations (1) and (2) with a few elements of the power series in the Nahavandi and Weinstein method.

In fact the opposite is true.

The discretisation of the integrals yields a set of simultaneous linear equations which becomes increasingly ill-conditioned the larger the ratio of the reduced length Λ to the reduced period Π, which parameters Λ and Π describe the regenerator and its operation, and have already been mentioned by Professor Hausen. The larger Λ/Π, the more linear the temperature behavior of the regenerator, in both distance and time.

The effect of this ill-conditioning is only relieved by taking small step lengths $\Delta\xi$ in the quadrature formulae. In Figure 4, for a fixed value of reduced period, the change in the thermal ratio calculated by Iliffe's method as the number of ordinates taken in the ξ-direction is increased, is displayed for increasing values of reduced length. It will be observed that quite meaningless results are computed unless a sufficient number of ordinates is employed. This is caused by the ill-conditioning of the linear equations.

One may respond by saying that the Iliffe method is just applicable to such large Λ/Π ratios, and indeed the thermal ratio η_{REG} approaches the value:

$$\frac{\Lambda}{\Lambda + 2}$$

Alternative methods of calculation are available if more precise values are required.

However, once a program has been written for a digital computer implementing, for example, the Iliffe method of solving equations (1) and (2) there always exists the temptation to use this program for data parameters outside the ranges for which the computer program was designed.

The weakness I have described in the Iliffe method sounds a note of caution to those who seize the enormous advantages of the digital computer to solve this and other scientific problems.

REFERENCES

[1]   W. Nusselt, Die Theorie des Winderhiters, Z. Ver. Dt. Ing., **71**, 85-91, (1927). (R.A.E. Library Translation No. 269).

[2]   H. Hausen, Naherungsverfahren zur Berechnung des Wärmeaustausches in Regeneratoren, Z. angw. Math. Mech., **2**, 105-114, (1931). (R.A.E. Library Translation No. 98).

[3]   C. E. Iliffe, Thermal analysis of the counterflow regenerative heat exchanger, J. Inst. Mech. Engrs., **159**, 363-372, (War emergency issue 44), (1948).

[4]   T. J. Lambertson, Performance factors of a periodic-flow heat exchanger, Trans. Amer. Soc. Mech. Engrs., **159**, 586 - 592, (1958).

[5]   A. J. Willmott, Digital computer simulation of a thermal regenerator, Int. J. Heat Mass Transfer, **7**, 1291-1302, (1964).

[6]   K. Heindlhofer and B. M. Larsen, An electrical analogue of the flow of heat in a regenerator system, AIME Metals Technology, (August 1945).

[7]   W. Tipler, An electrical analogue to the heat exchanger, Proc. VII Int. Congr. App. Mech., **3**, 196-210, (1948).

[8]   T. R. Schuerger and J. C. Agarwal, Limitations of blast furnace stoves, Iron Steel Eng., 143-156, (October 1961).

[9]   T. E. Dancy, H. W. Meyer and H. F. Ramstead, Principles of stove operation, Iron Steel Eng., 67-73, (July 1963).

[10] H. Hausen, Wärmeübertragung in genestrom, Gleichstrom und Kreuzstrom, Springer Verlag, Berlin, (1950).

[11] O. A. Saunders and S. Smolienc, Heat regenerators, Proc. VII Int. Congr. App. Mech., **3**, 91-105, (1948).

[12] D. N. de. G. Allen, The calculation of the efficiency of heat regenerators, Q. Jl. Mech. Appl. Math., V(4), 455-461, (1952).

[13] A. J. Wilmott, Simulation of a thermal regenerator under conditions of variable mass flow, Int. J. Heat Mass Transfer, **11**, 1105-1116, (1968).

[14] A. J. Willmott, The regenerative heat exchanger computer representation, Int. J. Heat Mass Transfer, **12**, 997-1014, (1969).

[15]   J. V. Edwards, R. Evans and S. D. Probert, Computation of transient temperatures in regenerators, Int. J. Heat Mass Transfer, **14**, 1175-1202, (1971).

[16]   A. N. Nahavandi and A. S. Weinstein, A solution to the periodic-flow regenerative heat exchanger problems, Appl. Scient. Res., **A10**, 335-348, (1961).

[17]   J. Tomeczek, The periodical heat transfer in an asymmetrical heat regenerator, Private communication, (1971).

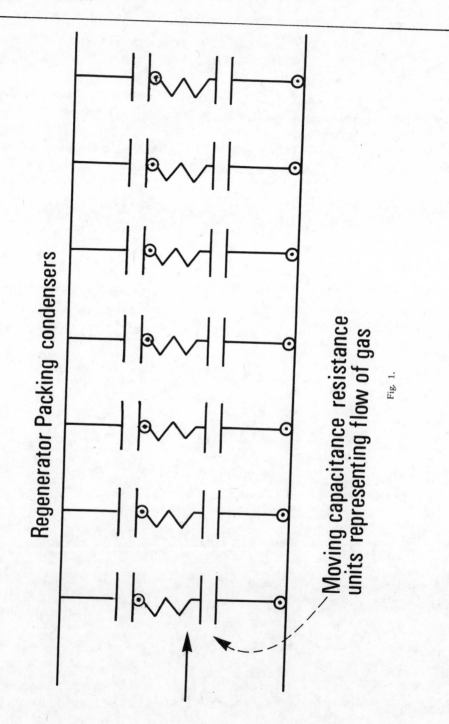

Fig. 1.

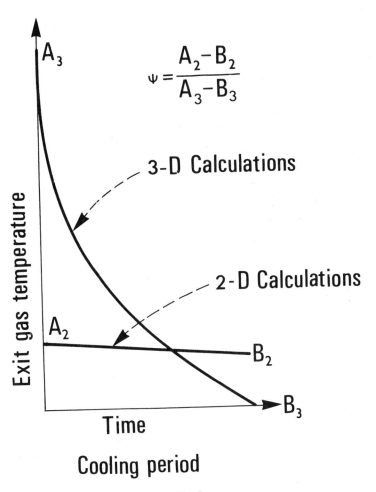

Fig. 2.

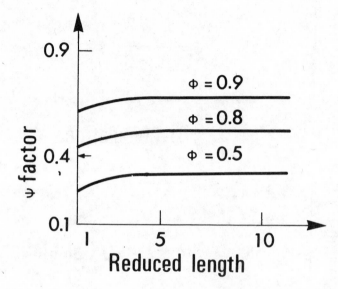

Fig. 3.

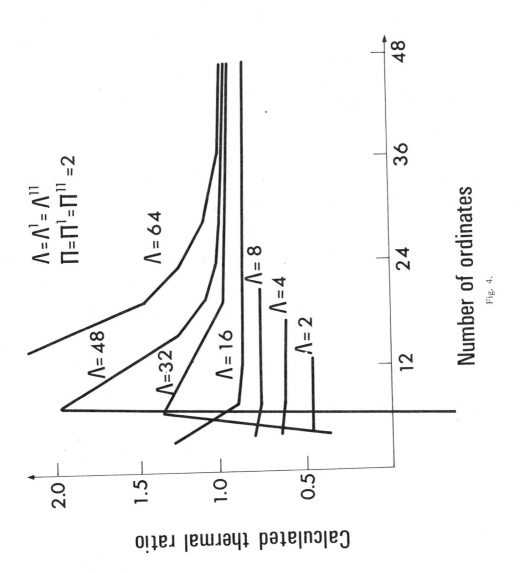

Fig. 4.

$\Lambda = \Lambda^{1} = \Lambda^{11}$
$\Pi = \Pi^{1} = \Pi^{11} = 2$

$\Lambda = 64$

$\Lambda = 48$

$\Lambda = 32$

$\Lambda = 16$

$\Lambda = 8$

$\Lambda = 4$

$\Lambda = 2$

Calculated thermal ratio

2.0

1.5

1.0

0.5

Number of ordinates

12

24

36

48

Chapter 11

TURBULENT HEAT TRANSFER IN TUBES WITH VARIABLE FLUID PROPERTIES

B. S. Petukhov (*)

Introduction

Modern technological development is characterized by the use of a wide variety of heat-transmission media in heat-transfer systems (heat exchangers included), as well as by a considerable expansion of the range of temperature drop and heat flux per unit area. At the same time, requirements placed upon the accuracy in calculating heat-transfer systems tend to become more stringent.

In the case of great temperature drop and high heat flux per unit area and sometimes (for example, in the single-phase near-critical region) in the case of their moderate values, dependency of the physical properties of heat-transmission media on temperature and pressure affect considerably the hydrodynamics and heat transfer. An inadequately accurate consideration of the effect of variable physical properties is likely to result in considerable errors, especially when calculating stressed heat-transfer systems. Therefore, the problem of heat transfer at variable physical properties has recently acquired considerable practical importance.

The complexity of the problem and inadequacy of the information accumulated in the field of hydrodynamics and heat transfer in turbulent flow of a liquid with variable physical properties have made analysis of this problem in the most general form difficult up to now.

The analysis can be carried out to cover individual flow conditions and specific groups of heat-transmission media. In the case of turbulent flow of a liquid with variable physical properties, one should distinguish between:

   a) a viscous-inertia-gravity flow; and

   b) a viscous-inertia flow.

(*) Institute of High Temperatures, Moscow, U.S.S.R.
N.B. ALL FIGURES QUOTED IN TEXT ARE AT THE END OF THE CHAPTER

In the former, thermogravitational forces have a considerable effect on hydrodynamics and heat transfer in the case of a forced flow of liquid. In the latter, the effect of these forces is rather considerable.

From the viewpoint of the nature of dependencies of physical properties on temperature and their effect upon hydrodynamics and heat transfer, one should distinguish between:

a) liquid-metal heat-transmission media;
b) non-metal liquid heat-transmission media (condensed media);
c) gaseous heat transmission media;
d) single-phase heat-transmission media at near-critical parameters of state.

The temperature dependence of the physical properties of heat-transmission media has an effect on both the time mean and fluctuating characteristics of the flow, that is, upon turbulent transfer processes. The latter problem has not been really investigated as yet, which hinders the development of theoretical calculation methods. Although the theoretical methods enable a qualitatively correct evaluation of the effect of variable physical properties on heat transfer and skin friction, a quantitative agreement of the theoretical calculation and experimental results is, by no means, constantly observed. It is, therefore, desirable that the engineering calculations should be based upon the results of experimental investigations.

The objective of the present lecture is to briefly outline the results of new investigations in heat transfer at variable properties and show their application to the calculation of heat-transfer devices.

Heat Transfer at Constant Properties

In order to carry out an analysis and calculation of heat transfer at variable physical properties, one should have an adequately accurate equation for heat transfer at constant properties.

Petukhov, Kirillov and Popov [1,2] have performed a theoretical calculation of heat transfer in a smooth, round pipe under conditions of turbulent flow of a liquid with constant properties and fully developed velocity and temperature profiles. Based on this calculation, the following interpolation equation for heat transfer has been obtained in [1]:

$$Nu = \frac{\frac{\xi}{8} \, Re \, Pr}{K + 12.7 \sqrt{\frac{\xi}{8}} \, (Pr^{2/3} - 1)}$$

where   (1)

$$K = 1.07 + \frac{900}{Re} - \frac{0.63}{1 + 10 \, Pr} \quad {}^{*}$$

ξ — is the skin friction coefficient in turbulent flow of liquid in a smooth, round pipe. Thus, for instance, according to the formula by Filonenko

$$\xi = (1.82 \, \ell g Re - 1.64)^{-2}$$ (2)

Eq. (1) agrees with the most reliable experimental data on heat and mass transfer to an accuracy of ±5 percent. It is true in the Reynolds number range from 4.10^3 to 5.10^6 and in the Prandtl (or Schmidt) number range from 0.5 to 10^6.

Widely used power dependencies of the type

$$Nu = CRe^m Pr^n$$

at constant values of C, m and n are not sufficiently accurate. This can be seen from Fig. 1 the ratio of the Nusselt number calculated from the Dittus and Bolter formula (C = 0.023, m = 0.8, n = 0.4) to the Nusselt number according to Formula (1). At certain values of Re and Pr the error can reach 30 percent.

The Effect of Thermogravitational Forces on Heat Transfer

The effect of thermogravitational forces on hydrodynamics and heat transfer at forced turbulent flow is more pronounced the greater the Grashof number and the smaller the Reynolds number and, at the same time, depends considerably upon the relative direction of the vector of forced flow velocity and of

---

(*) This more accurate expression for k has been recently obtained by the authors of [3].

the vector of thermogravitational forces.

Shown in Fig. 2 for the case of an upward flow in a vertical heated pipe (far from the entrance at q_w = const) are the limiting curves corresponding to the Nusselt number by one percent deviation from its value at viscous-inertia flow, under the effect of thermogravitational forces. The curves calculated theoretically by Poliakov are in good agreement with the measurement results of [4]. The region above the curve in question corresponds to the viscous-inertia flow, the region below the curve corresponds to the viscous-inertia-gravity flow.

Presented in Figs. 3 and 4 are the results of our and Sryghin's [4] joint measurements of heat transfer in upward and downward flow of water in vertical heated pipes at a constant heat flux density on the wall. The experimental data in Figs. 3 and 4 are presented by the coordinates Nu/Nu_0, Ra_A/Re^n (n = 2 in the case of upward flow and n = 1 in the case of downward flow) where $Nu = \frac{\alpha d}{\lambda}$ is the measured local Nusselt number far from the pipe entrance (that is, in the region of fully developed heat transfer), $\alpha = q_w/(t_w - t_b)$ is the local coefficient of heat transfer, q_w is the specific heat flux on the wall, t_w and t_b are the wall temperature and bulk temperature of liquid in the pipe section under consideration, Nu_0 is the Nusselt number at constant properties, calculated from Eq. (1) at the same values of Re and Pr as Nu, $Ra_A = Cr_A$ Pr is the Rayleigh number, $Gr_A = \frac{g\beta d^4 A}{16\nu^2}$ $A = dt_b/dx$ is the axial gradient of the bulk temperature. It should be noted that the parameter Ra_A/Re^n may be written differently

$$\frac{Ra_A}{Re^n} = \frac{Gr_q}{4\, Re^{n+1}} \quad \text{where} \quad Gr_A = \frac{g\beta d^4 q_w}{\nu^2 \lambda} \,, \quad Re = \frac{\bar{w}d}{\nu} \,.$$

In the case of upward flow in a heated pipe with an unheated portion at the entrance (cf., Fig. 3), the Nu/Nu_0 ratio at first decreases with an increase of Ra_A/Re^2, and then increases at $Ra_A/Re^2 > 10^{-4}$. The observed decrease of heat transfer is due to the decrease of turbulent transfer under the effect of thermogravitational forces. The growth of heat transfer at $Ra_A/Re^2 > 10^{-4}$ is due to the dominating influence of thermogravitational forces on the time mean flow. The function of forced flow in this region mainly consists of the supply and removal of the heat transmission medium to and from the pipe.

The experimental data on heat transfer under conditions of upward flow in a vertical heated pipe are described with an accuracy of ±10 percent by the equations:

if $Ra_A/Re^2 \leqslant 10^{-4}$

$$\frac{Nu}{Nu_o} = \left(1 + 1.15 \cdot 10^4 \frac{Ra_A}{Re^2}\right)^{-1} \tag{3}$$

if $Ra_A/Re^2 > 10^{-4}$

$$\frac{Nu}{Nu_o} = 10\left(\frac{Ra_A}{Re^2}\right)^{1/3} \tag{4}$$

In the case of downward flow in a heated pipe (Fig. 4), the ratio Nu/Nu_0, following an insignificant decrease, grows with an increase of Ra_A/Re. Some decrease of Nu/Nu_0 is probably due to a small decrease of the velocity gradient near the wall under the effect of thermogravitational forces. The growth of Nu/Nu_0 is caused by an intensive mixing of liquid in the wall layer under the influences of the same forces.

The experimental data on heat transfer under conditions of downward flow in a heated pipe are described with an accuracy of ±10 percent by the equation:

$$\frac{Nu}{Nu_o} = \left(1 + 0.5 \frac{Ra_A}{Re}\right)^{1/3} - 0.15 \exp\left[-8\left(\frac{Ra_A}{Re} - 0.5\right)^2\right] \tag{5}$$

At $Ra_A/Re > 1$, the second term in Eq. (5) becomes negligibly small and can be omitted.

The physical properties of liquid (λ, β, ν) in Eqs. (3), (4) and (5) have been taken at the bulk temperature in the given pipe section. The equations are valid at $x/d \geqslant 40$, $300 \leqslant Re \leqslant 3.10^4$, $300 \leqslant Ra_A \leqslant 8.10^5$ and $2 \leqslant Pr \leqslant 6$.

In the case of flow in horizontal pipes, the circumferential distribution of the heat transfer coefficient becomes inhomogeneous due to the effect of thermogravitational forces. The measurements taken in water at Re values from 8.10^3 to 35.10^3 and Cr values from 10^7 to 5.10^9 show that on the lower generatrix of the pipe the heat transfer coefficient increases 1.15 to 1.20 times, and on the upper generatrix it decreases 2-3 times as compared to α in the case of a viscous-inertia flow. While so doing, the average circumferential heat transfer coefficient decreases only 10-15 percent. The inhomogeneity in the circumferential distribution of α is greater than the greater Gr and the smaller Re. More detailed information is contained in [5].

Heat Transfer and Skin Friction in the Flow of Liquid with Variable Viscosity

In the case of a viscous-inertia flow of liquid non-metal heat-transmission media, the principal effect is produced on heat transfer and skin friction by the variation of viscosity with temperature, the other properties of such heat-transmission media depending only slightly on temperature.

The analysis shows that the dependency of Nu on Re and Pr in the case of variable viscosity remains about the same as in the case of constant properties. Therefore, the Nu/Nu_0 ratio is essentially independent of Re and Pr. The variable viscosity effect can be accounted for by making the Nu/Nu_0 ratio dependent on the parameter μ_w/μ_b, i.e., the ratio of viscosity coefficients at a wall temperature t_w and bulk temperature of the liquid t_b. This dependency, according to the experimental data of a number of authors, is presented in Fig. 5. Groups of experimental points are well observed near the curves, to which corresponds an equation from [1,6]:

$$\frac{Nu}{Nu_o} = \left(\frac{\mu_w}{\mu_b}\right)^n \tag{6}$$

where $n = -0.11$ when heating the liquid and $n = -0.25$ when cooling the liquid. Nu_0 is found from Eq. (1). The physical properties of the liquid (with the exception of μ_w) in Eq. (6) are taken at a temperature t_b in the given pipe section, and μ_w - is taken at a temperature t_w in the same section.

Eq. (6) is valid in the range of values of μ_w/μ_b from 0.08 to 40, Re from 10^4 to $1.25.10^5$ and Pr from 2 to 140.

As seen from Fig. 5, heat transfer to a liquid whose viscosity is temperature-dependent is, in the case of heating, greater and, in the case of cooling, lower than at constant viscosity.

The effect of variable viscosity on the skin friction coefficient is shown in Fig. 6 with the aid of measurements by Lafay [7]. The experiments were conducted in the flow of water in a round, essentially smooth, pipe far from the entrance at a constant specific heat flux on the wall. The experiments have shown the ratio of the coefficients of skin friction at variable and constant viscosity, ξ/ξ_0, to be dependent on both μ_w/μ_b and Re and to be practically independent of Pr (on the variation of the latter from 2 to 6). As seen in Fig. 6, ξ is more dependent upon Re the more μ_w/μ_b differs from unity. With an increase of μ_w/μ_b the ratio ξ/ξ_0 increases and, consequently, when heating the liquid ξ should be higher than when cooling the liquid.

Lafay [7] has described his experimental data and the data by other authors obtained when heating the liquid, by an empirical equation

$$\frac{\xi}{\xi_o} = 1 - \frac{1}{2}(1 + M)^n \lg(1 + M)$$

where

$$M = \left(\frac{\mu_b}{\mu_w} - 1\right)\left(\frac{\mu_b}{\mu_w}\right)^{0.17} \tag{7}$$

$$n = 0.17 - 2 \cdot 10^{-6} Re + \frac{1800}{Re}$$

For the case of cooling the liquid in a round pipe far from the entrance one can recommend the equation of [6]

$$\frac{\xi}{\xi_o} = \left(\frac{\mu_w}{\mu_b}\right)^{0.24} \tag{8}$$

The coefficient of skin friction in the isothermal flow of liquid, ξ_0, is calculated according to Eq. (2) or analogous to other dependencies. The physical properties of liquid in (7) and (8), with the exception of μ_w, are taken at a temperature t_b in the given pipe section; and μ_w is taken at a temperature t_w in the same section.

Eqs. (7) and (8) are valid in the range of values of Re from 10^4 to 3.10^5, Pr from 1.3 to 10, and μ_w/μ_b from 0.3 to 1 [Eq. (7)] and from 1 to 2 [Eq. (8)].

In the viscous-inertia flow of liquid metals the temperature drop through the flow is usually not so great. The physical properties of metals in the working temperature range depend only slightly on temperature. Therefore, when calculating heat transfer and skin friction, the physical properties of liquid metals can be regarded as practically constant and equal to their values at the bulk temperature of liquid in the given section.

Heat Transfer in the Flow of Gas with Variable Properties

In the viscous-inertia flow of gaseous heat-transmission media the effect of variable gas properties on heat transfer and skin friction is usually calculated with the aid of the temperature ratio parameter $\psi = T_w/T_b$. However, such an approach is inadequate inasmuch as the individual peculiarities of gases are not taken into account. Such peculiarities manifest themselves in the different nature of temperature dependencies of the physical properties (c_p, μ, λ) of gases. Fig. 7 illustrates the relative variation of physical properties of monatomic, diatomic and polyatomic gases as a function of temperature.

If the dependency of ρ, c_p, λ and μ on T is described by power functions with indexes n_ρ, n_c, n_λ and n_μ and if the fact is taken into consideration that for a gas with variable properties the dependency of Nu upon Re and Pr is about the same as for the case of constant properties, one can readily obtain a system of dimensionless numbers reflecting quite fully the effect of variable properties. In a viscous-inertia flow of gas with low subsonic velocities this system has the form

$$\frac{Nu}{Nu_o} = F\left(\frac{x}{d}, \psi, n_\rho, n_c, n_\lambda, n_\mu\right) \tag{9}$$

Here, Nu and Nu_0 are the values of Nusselt numbers at variable and constant properties over the entire length of the pipe, the thermal entry region included.

In accordance with Eq. (9), the present author, together with Kurganov [3], has processed experimental data on heat transfer to argon, nitrogen, air, carbon dioxide and ammonia. All the experimental data refer to the case of heating the gas at a constant specific heat flux on the wall. As a result, an equation has been obtained which correlates experimental data with a root-mean-square error of 2.5 to 5 percent (for different gases). This equation has the following form:

$$\frac{Nu_b}{Nu_{ob}} = \psi^{0,53\, n_\rho + \frac{1}{3} n_\lambda + \frac{1}{4} n_c - \phi\left(\frac{x}{d}\right) n_\mu \lg \psi} \tag{10a}$$

If we take into account that $n_\rho = -1$ and use power dependencies for properties, Eq. (10a) can also be written in the form:

$$\frac{Nu_b}{Nu_{ob}} = \left(\frac{\lambda_w}{\lambda_b}\right)^{1/3} \left(\frac{cp_w}{cp_b}\right)^{1/4} \psi^{-\left[0.53 + \phi\left(\frac{x}{d}\right)\lg \frac{\mu_w}{\mu_b}\right]} \tag{10b}$$

The values of $\phi(x/d)$ are taken from the following table:

x/d	10	20	30	40	50	60	70	80	90	100	∞
$\phi\left(\dfrac{x}{d}\right)$	0.11	0.24	0.38	0.55	0.73	0.89	1.02	1.13	1.21	1.27	1.50

The indexes "b" and "w" show that the appropriate physical properties are selected at temperatures T_b and T_w in the given pipe section. Nu_{ob} is calculated from Eq. (1) with the introduction of a correction for the thermal entry region.

Eq. (10) is true at Re_b 7.10^3, $1 \leqslant 4 \leqslant 4$ and $\dfrac{q_w}{\overline{\rho_w} c_{p1} T_{b1}} < 0.007$ (c_{p1} is the value of c_p at a gas temperature T_{b1} at the entrance).

Shown in Fig. 8 is the dependency of Nu_b/Nu_{ob} of ψ according to Eq. (10) at $x/d = 70$ for some technically employed gases. The curves for polyatomic gases at $\psi > 1$ are positioned substantially higher than for monatomic and diatomic gases. This is an essential difference which has to be taken into consideration in thermotechnical calculations. It is worthy of note that the extrapolation of Eq. (10) with respect to the gas cooling region to at least the values of $\psi \geqslant 0.5$ (cf., Fig. 8) is not at variance with the available experimental data. Indeed, the experimental data on heat transfer when cooling diatomic gases show, in accordance with Eq. (10), that $Nu_b/Nu_{ob} \simeq 1$ is within the measuring accuracy (± 10 percent). In the case of cooling polyatomic gases, Eq. (10) predicts the growth of Nu_b/Nu_{ob} with an increase of ψ. However, this regularity needs to be experimentally proved.

Heat Transfer in a Single-Phase Near-Critical Region

When speaking of heat transfer in a single-phase near-critical region, we refer to heat transfer processes occurring at higher-than-critical pressure and close-to-critical or pseudo-critical temperature (i.e., temperature corresponding to the maximum of heat capacity at constant pressure).

The specific nature of heat transfer in the single-phase near-critical region lies in the fact that the physical properties of heat-transmission medium in this region are subject to strong and specific variation due to temperature and depend considerably on pressure. The nature of variations of physical properties can be seen in Fig. 9, which presents the data for carbon dioxide at a pressure of 100 bar.

Up to now, a considerable amount of experimental work has been carried out on heat transfer in the single-phase near-critical region [*]. The experiments have been conducted mostly with water and carbon dioxide and, to a lesser extent, with other heat-transmission media. Heat transfer was studied mostly in the flow in round pipes under conditions of heating at a constant specific heat flux on the wall. However, the complexity of the problem and the absence of information on the regularities of turbulent transfer on strong variations of properties have made it impossible so far to develop adequately accurate and general methods for calculating heat transfer in the single-phase near-critical region. Therefore, the recommandations provided below are necessarily of an approximate nature.

In the case of single-phase heat-transmission media at near-critical parameters it is of special importance to distinguish between the viscous-inertia and viscous-inertia-gravity conditions of flow and heat transfer. The latter conditions are observed rather frequently due to the strong temperature dependency of density. In addition, because of practical reasons one should distinguish between normal conditions and conditions of deteriorating heat transfer. The normal conditions are characterized by a smooth (usually monotonic) variation of the wall temperature over the pipe length (at q_w = const) without strongly pronounced maxima of t_w (cf., curves 1 and 2 in Fig. 10). On the contrary, the conditions of deteriorating heat transfer are characterized by the presence of strongly pronounced maxima ("peaks") in the wall temperature distribution over the pipe length (cf., curves 3 and 4 in Fig. 10). The normal conditions and conditions of deteriorating heat transfer are observed both in viscous-inertia and in viscous-inertia-gravity flow of liquid.

In the case of a viscous-inertia flow, thermogravitational forces produce no effect on the flow and heat transfer. In this case, conditions of deteriorating heat transfer are observed with all orientations of the pipe in the field of gravity. In the case of a viscous-inertia-gravity flow, the conditions of deteriorating heat transfer observed in an upward flow in vertical heated pipes are rather weakly pronounced in the flow in horizontal pipes and are absent in a downward flow in vertical pipes. In all of the aforementioned cases, the conditions of deteriorating heat transfer only occur as a result of a certain combination of mean mass velocity and heat flux per

---

(*) Cf., review papers [6,8,9].

unit area and, even in this case, only when the temperature of liquid in the given pipe section is $t_b < t_m$ while the wall temperature in the same section is $t_w < t_m$, where t_m is the pseudocritical temperature at a given pressure.

In heat-transfer systems featuring a present specific heat flux on the wall the conditions of deteriorating heat transfer are very undesirable, for they involve the danger of an inadmissible increase of the wall temperature. It is, therefore, of great importance when evaluating the conditions to rule out the possibility of deteriorating heat transfer. According to the data of [10], this condition, following the introduction of some reasonable margin, has the appearance:

$$\frac{q_w}{\overline{\rho w}} \leqslant 0.02 \left(\frac{c_p}{\beta}\right)_{tm} \sqrt{\frac{\xi}{8}} \tag{11}$$

where q_w is the specific heat flux on the wall,

$\overline{\rho w}$ is mean mass velocity,

c_p and $\beta = 1/\rho(\partial\rho/\partial t)\rho$ denote, respectively, specific heat and the coefficient of volume expansion at constant pressure (the values of c_p and β are taken on isobars at points corresponding to pseudo-critical temperature t_m), and ξ, is the skin friction coefficient calculated from Eq. (2). The relationship (11) is valid for water and carbon dioxide at $P/P_{cr} = 1.02 + 1.3$ (here, P_{cr} is critical pressure), $\rho w = 400 + 2,500 \text{ kg/m}^2$ sec, $q_w = (0.1 + 5) \cdot 10^6 \text{ w/m}^2$ and $d = 2 + 20$ mm.

Several empirical formulae have been suggested for calculating heat transfer in viscous-inertia flow. Only a few of them satisfy the minimum accuracy requirements and, if so, only for normal heat transfer conditions. The formula by Krasnoschekov and Protopopov [11], which allows calculations to an accuracy of about ±20 per cent, offers some advantages. This formula has the appearance:

$$Nu_b = Nu_{ob}\left(\frac{\rho_w}{\rho_b}\right)^m \left(\frac{\overline{c_p}}{c_{p_b}}\right)^n \tag{12}$$

where Nu_{ob} is the Nusselt number at constant properties calculated from Eq. (1), $\overline{c_p} = h_w - h_b/T_w - T_b$ is the mean integral heat capacity in the temperature range of from T_b to T_w.

For water at $1.02 \leqslant P/P_{cr} \leqslant 1.49$ m = 0.3, for carbon dioxide at $1.02 \leqslant P/P_{cr} \leqslant 5.3$ m = 0.35 − 0.05 P/P_{cr}.

The exponent n depends on T_w/T_m and T_b/T_m. This dependency is shown in Fig. 11.

Eq. (12) is valid for normal conditions when heating the liquid at q_w = const, $2.10^4 \leqslant Re_b \leqslant 8.10^5$, $0.85 \leqslant Pr \leqslant 55$, $0.09 \leqslant \rho_w/\rho_b \leqslant 1$, $0.02 \leqslant c_p/c_{pb} \leqslant 4$.

The problems of heat transfer in viscous-inertia-gravity flow have been studied rather inadequately thus far. Some idea of the effect of thermogravitational forces on heat transfer can be gathered from Fig. 12 which presents the results of measuring heat transfer in an upward flow of carbon dioxide in a vertical heated pipe ca. ∼30 mm in dia. [12]. In the Figure, the measured Nusselt number has been related to the value calculated from Eq. (12) and presented as a function of Gr/Re. At Gr/Re $\leqslant 10^5$ thermogravitational forces produce almost no effect on heat transfer, whereas at high values of the parameter the heat transfer grows with an increase of Gr/Re. Heat transfer grows considerably even at relatively high values of Re. Thus, at Re $\simeq 10^5$ heat transfer doubles, and at Re $\simeq 6.10^4$ it triples. A qualitatively similar picture is observed in the case of a downward flow in heated pipes.

Conclusion

The problem of heat transfer in the turbulent flow of liquid with variable physical properties presents one of the major problems of the heat transfer theory nowadays. A comprehensive study of this problem brings about the development of more general methods for calculating flows and heat transfer, which are in better conformity with the actual conditions. In order to make progress in this field, attention should be paid to the study of the turbulent transfer processes and to the development of more perfect models of turbulent transfer on this basis.

REFERENCES

[1]   B. S. Petukhov and V.V. Kirilov, Teploenerghetika (Journal of Thermal Power Engineering), No. 4, 1958.

[2]   B. S. Petukhov and V.N. Popov, Teplofizika vysokikh temperatur (High Temperatures), vol. 1, No. 1, 1963.

[3]   B. S. Petukhov, V.A. Kurganov and A.I. Gladuntsov, Teplo- i massoperenos (Heat- and Mass Transfer), vol. 1, 1972, Minsk.

[4]   B. S. Petukhov and B.K. Stryghin, Teplofizika vysokikh temperatur (High Temperatures), vol. 6, No. 5, 1968.

[5]   B. S. Petukhov and A.F. Poliakov, 4th International Heat Transfer Conference, vol. 3, Rept. 7, Versailles, 1970.

[6]   B. S. Petukhov, Advances in Heat Transfer, vol. 6, Academic Press, Inc., New York, 1970.

[7]   J. Lafay, Mesure du coefficient de frottement avec transfert de chaleur en concection forcée dans un canal circulaire, Centre d'Études Nucléaires de Grenoble, Rapport-CEA-R-3896.

[8]   B. S. Petukhov, Teplofizika vysokikh temperatur (High Temperatures), vol. 6, No. 4, 1968.

[9]   W. B. Hall, J.D. Jackson, A. Watson, The Inst. Mech. Engin. Proc., 1967-1968, vol. 182, part 31.

[10] B. S. Petukhov, V.S. Protopopov, V.A. Silin, Teplofizika vysokikh temperatur (High Temperatures), vol. 10, No. 2, 1972.

[11] E. A. Krasnoschekov, V.S. Protopopov, Teplofizika vysokikh temperatur (High Temperatures), vol. 4, No. 3, 1963; vol. 9, No. 6, 1971.

[12] N. P. Ikriannikov, B.S. Petukhov, V.S. Protopopov, Teplofizika vysokikh temperatur (High Temperatures), vol. 10, No. 1, 1972.

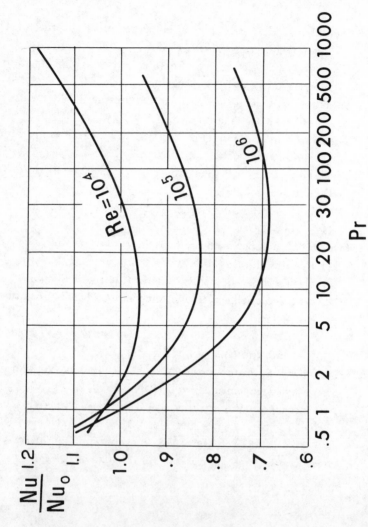

Fig. 1. The ratio of the Nusselt numbers according to the Dittus and Bolter formula and Formula (1).

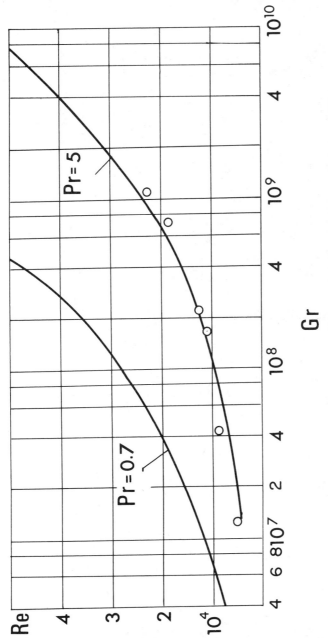

Fig. 2. The limits of the effect produced by thermogravitational forces on heat transfer under conditions of upward turbulent flow in a vertical heated pipe.

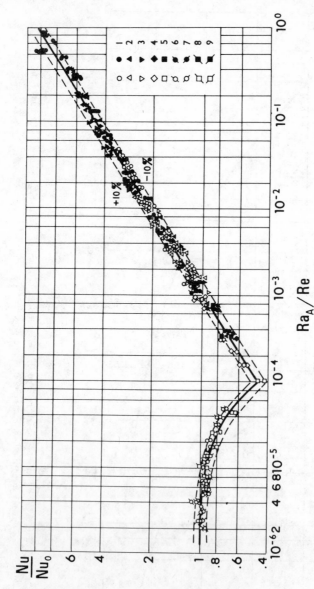

Fig. 3. Heat transfer under conditions of upward turbulent flow in a vertical heated pipe. Solid curves correspond to Eqs. (3) and (4).

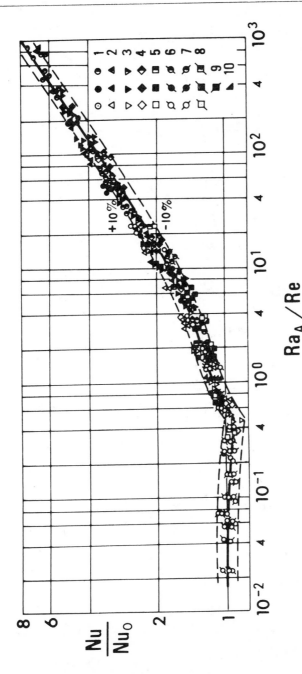

Fig. 4. Heat transfer under conditions of downward turbulent flow in a vertical heated pipe. Solid curves correspond to Eq. (5).

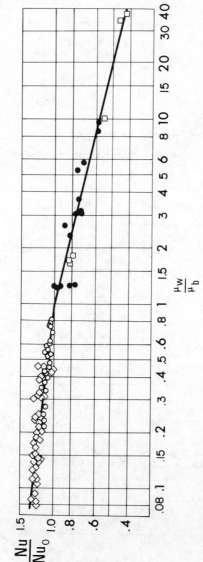

Fig. 5. The effect of variable viscosity on heat transfer under conditions of turbulent flow of various liquids.

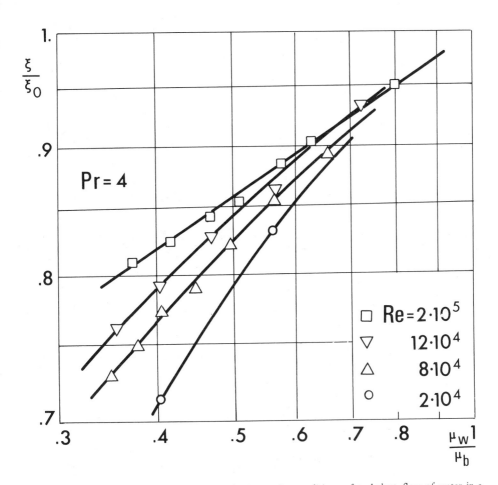

Fig. 6. The effect of variable viscosity on skin friction under conditions of turbulent flow of water in a heated pipe.

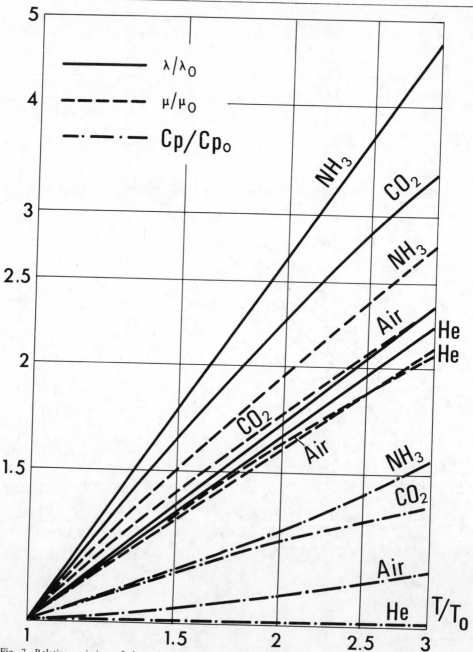

Fig. 7. Relative variation of thermal conductivity, viscosity and heat capacity with the temperature for some gases at $P = 1$ atm and $T_0 = 373.2°K$.

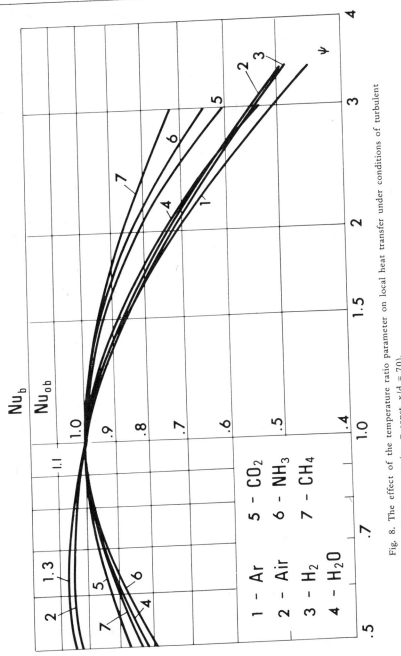

Fig. 8. The effect of the temperature ratio parameter on local heat transfer under conditions of turbulent flow of various gases (q_w = const, x/d = 70).

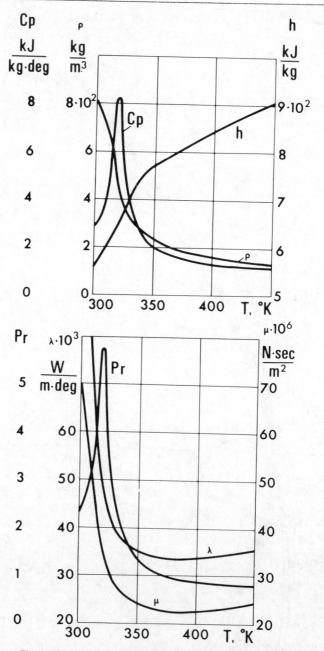

Fig. 9. Physical properties of carbon dioxide at P = 100 bar.

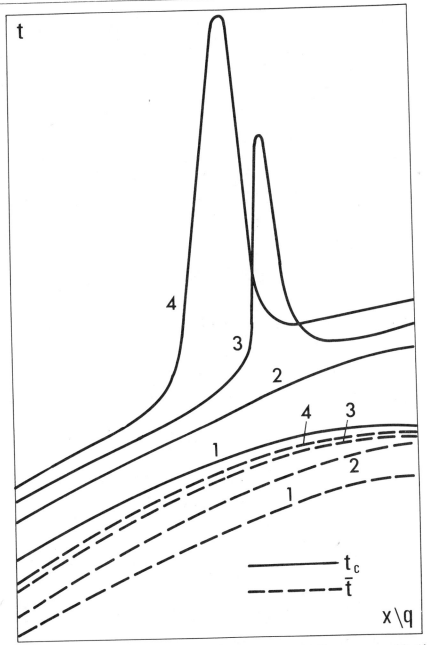

Fig. 10. Typical curves of the variation of wall temperature (t_w) and bulk temperature of liquid (t_b) along the pipe length for the normal conditions (curves 1 and 2) and conditions of deteriorating heat transfer (curves 3 and 4).

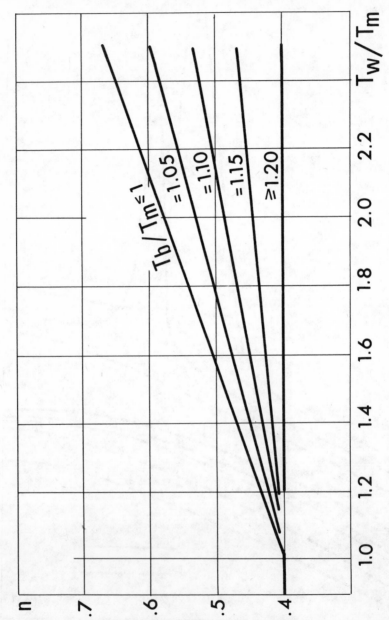

Fig. 11. The exponent *n* for the equation (12).

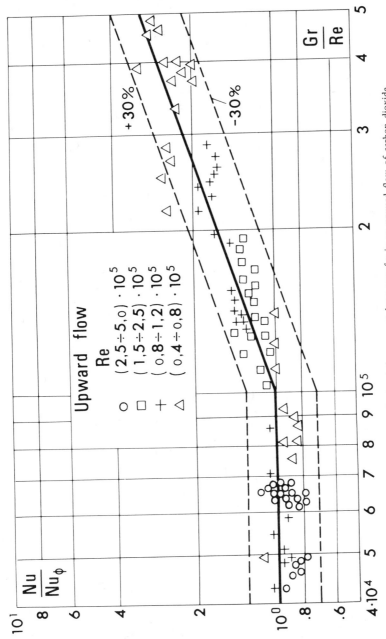

Fig. 12. The effect of thermogravitational forces upon heat transfer in an upward flow of carbon dioxide in a vertical heated pipe ($P/P_{cr} = 1.1 - 1.4$).

Chapter 12

ASPECTS OF TWO–PHASE GAS–LIQUID FLOW

G.F. Hewitt (*) and R. Semeria (**)

1. INTRODUCTION

This chapter is intended to give the designer of equipment in two-phase flow an appraisal of various aspects of the subject on which work is proceeding and where new information relevant to design is emerging. It is also aimed at indicating to the research worker those areas in which further work might be profitable.

Two-phase flow is of very great industrial importance and is also fascinating from a basic point of view. These two motives - not always well connected - have led to the publication of many thousands of papers. We have, therefore, had to be extremely selective in our choice of material and references.

The field of two-phase flow received great impetus from the needs of nuclear reactor designers. In reactors, a far more accurate knowledge of the phenomena is required than was considered necessary for industrial plant. In the light of the results obtained, however, it seems likely that great improvements and economies are possible in non-nuclear equipment by having a better knowledge of the heat transfer and flow behavior.

A close link exists between heat transfer and hydrodynamics in two-phase and it is impossible to consider heat transfer except in the context of hydrodynamics. Aspects of heat transfer are discussed in a second paper by ourselves at this Seminar.

In discussing two-phase flow hydrodynamics, we shall start with a consideration of flow regimes, then go on to define the variables involved and indicate their relevance. The general equations of two-phase flow are then discussed and the prediction methods for pressure drop and void fractions are described.

(*) United Kingdom Atomic Energy Authority, Harwell, U.K.

(**) Commisariat à l'Energie Atomique, Grenoble, France

N.B. ALL FIGURES QUOTED IN THE TEXT ARE AT THE END OF THE CHAPTER

Finally, we shall discuss the types of equipment used in two-phase flow and indicate the types of problem involved.

Readers wishing to have further information are refered to the list of books and major reviews which are listed at the beginning of the reference list at the end of this paper.

2. FLOW PATTERNS

The main flow patterns encountered in two-phase flow are as follows:

Bubble flow. Here the gas phase is distributed in discrete bubbles within a liquid continuum. Typically, this regime exists for volumetric gas concentrations (void fractions) less than 30%.

Plug flow. In this type of flow some of the gas bubbles have nearly the same cross-section as that of the channel and move along in the characteristic bullet-shaped bubbles. The bubbles of gas are separated by lengths occupied mainly by liquid which may or may not contain a dispersion of smaller gas bubbles. In heat exchangers, plug flow is more complex due to interaction with the tubes.

Churn flow. If the velocity of a two-phase mixture flowing in plug flow in a vertical channel is increased (by, for example, adding heat to the system and generating more vapor), the structure will eventually become unstable. In wide-bore tubes, this instability eventually results in the complete destruction of the slug flow with consequential "churning" or oscillatory action. In narrow-bore tubes, the transition from slug to annular flow (see below) is smoother but it is still accompanied by the characteristic instabilities in the vapor bubbles.

Annular flow. In this type of flow there is a liquid layer or film on the channel walls which presents a more or less continuous interface to a stream consisting mainly of gas, which flows in the center of the channel. The liquid film may or may not contain gas bubbles and the gas core may or may not contain entrained liquid droplets or agglomerates. Many authors have distinguished between annular and "annular dispersed" flow (i.e. annular flow with entrainment) but the existence of annular flow without droplet entrainment is rare and it is safer not to make a distinction.

The following regimes are characteristic of horizontal flow:

Stratified flow. In this case the separation of the liquid and gas phases is complete, the liquid flowing at the bottom of the channel and the gas at the top.

Wavy flow. As the gas velocity is increased in stratified flow, large surface waves begin to build up on the liquid layer. This region is called "wavy flow."

Slug flow. As the gas velocity is further increased in the wavy flow region, the waves eventually become big enough to reach the top of the channel. These are propagated at high velocity and wet the whole of the channel surface leaving a liquid film covering the surface in between the bridging waves or "slugs."

The general nature of these regimes is indicated in Figure 1 for various configurations and flow directions.

The prediction of flow regime for a given situation is very difficult. The effects of upstream geometry (bends, etc.) and rate of formation of the phases are usually significant. However, the use of empirical correlations (flow regime maps) is often useful, if not taken too seriously. Examples of flow pattern correlations for the various configurations are as follows:

(1) Horizontal flow: Baker (1958) modified by Scott (1964).

(2) Vertical flow: Hewitt and Roberts (1969).

(3) Heat Exchangers: Grant and Murray (1972).

Information is lacking on flow regimes in large diameter pipes, at large velocities where the flow patterns are difficult to characterise and in complex geometries.

3. THE SYSTEM VARIABLES

3.1 Independent Variables

In two phase flows the variables which can be regarded as independent are:

1. The channel geometry: cross-sectional area, shape of the channel in cross-section and in length, inlet geometry, especially with a two phase inlet (mode of initial mixing of the phases), roughness, nucleation properties on the wall (if degassing, boiling or condensation occurs), thermal properties of the wall (coupling in instability, etc.).

2. Flow rates of each phase: not only the inlet flow rates but also the rates of change along the channel are important.

3. Physical properties of the phases: this includes surface or interfacial tension, wettability of the wall, internal nucleation properties.

4. Heat flux (for heat exchangers): this includes the total thermal power and the distribution of the heat flux (axially and radially).

3.2 Classification of Two Phase Flow Parameters

It is possible to classify two-phase system parameters (dependent variables) as follows:

(a) First order parameters: these are of direct relevance in design and can be divided into primary design factors such as pressure drop and heat transfer coefficient and secondary design factors such as flow distribution and stability. It is information on these parameters that the designer needs in assessing a proposed system.

(b) Second order parameters: in this category one may classify time-averaged observations which are not of primary relevance in design, but which assist in the development and testing of analytical models of the system. Examples of this type of parameter would be liquid film thickness, radial mass flow, distribution and particle size.

(c) Third order parameters: in this class are included parameters in which the time-varying nature of the parameter is important, examples being interfacial structure velocity fluctuations and temperature fluctuations. Measurements of these parameters are aimed at providing a qualitative understanding and a quantitative description of the system and in assisting in building up more detailed models. Most two-phase systems are unsteady in so far as the local velocity is varying with time (due to turbulence, say) and, very often, the local distribution of the phases is also varying.

The calssification of the various parameters is as follows:

First Order Parameters: These can be divided into primary and secondary design factors:

Primary design factors: pressure drop, heat transfer coefficient, mass transfer coefficient, mean phase content (void fraction), maximum heat flux.

Secondary design factors: vibration (transient momentum flux), flow distribution, stability, flow or quality metering.

Second Order Parameters: The following parameters are considered: flow pattern, time averaged film thickness and amplitude distribution, mass flow distribution, phase distribution (local void fraction), velocity and local momentum flux distribution, concentration distribution, mixing characteristics (including phase mass transfer), drop, bubble or particle size and distribution of sizes, shear stress, temperature distribution, entrainment, contact angle.

Third Order Parameters: These are taken to include time variations in film thickness (waves), phase and component concentration, velocity, pressure and temperature. This grouping also includes photographic observation of local phenomena.

4. MODELS FOR PARALLEL FLOWS

4.1 Conservation equations

The most common approach to setting up equations for two-phase flows (see for example Hewitt and Hall Taylor (1970), Wallis (1970)) is to use a one-dimensional flow concept closely allied to similar models for single phase flow. Using this approach, a force-momentum balance for two phase flow in a channel of constant cross sectional area A and constant perifery S, taken over an element δ_z in the direction of flow, gives the following result:

$$\int_A \left[p - \left(p + \frac{dp}{dz} \cdot \delta z \right) \right] dA = \int_S \tau_o \delta z \, \delta S + \int_A \rho g \, \sin\theta \, \delta z \, dA$$

$$+ \int_A \frac{d}{dz} (G_L u_L + G_G u_G) \, dA$$

where p is the pressure, τ_0 the wall shear stress, ρ the local density, θ the angle of inclination of the channel to the horizontal, and G_L and u_L the local mass fluxes and velocities of the respective phases. An energy balance and a mass conservation equation can be similarly derived. Assuming constant pressure across the cross section and constant shear stress around the wall, the above equation reduces to

$$- \frac{dp}{dz} = \frac{S}{A} \tau_o + \int_A \rho g \, \sin\theta \, dA + \int_A \frac{d}{dz} (G_L u_L + G_G u_G) \, dA$$

The three terms on the r.h.s. of this equation can be regarded respectively as the frictional, gravitational and accelerational components of the pressure gradient.

Although commonly used, the above approach contains a number of hidden assumptions which can sometimes be invalid. An alternative approach is to start from the complete equation for the conservation of mass, momentum and energy and to carry out the appropriate space and time averaging, with a critical examination of the assumptions involved. Such an approach has been taken by Delhaye (1969) and is further developed by Boure (1971), and Boure and Reocreux (1972). The following description is derived from this work.

The procedure adopted was to start with the laws of conservation, to establish local differential equations and interface jump conditions. The local instantaneous equations are then averaged over a surface or over a segment and are then time averaged.

With reference to Figure 2, let us consider a portion of interference between the two phases 1 and 2 and a fixed control volume V such that it is divided by the interface into two subvolumes $V_1(t)$ and $V_2(t)$ belonging respectively to the phases 1 and 2. Let $A_i(t)$ be the portion of interface located inside the volume V, $\xi(t)$ the intersection of the boundary of V with the interface $A_1(t)$ and $A_2(t)$ the non-material surfaces bounding the fixed volume V in phases 1 and 2. The surfaces $A_1(t)$ and $A_2(t)$ are non-material surfaces and mass can be transferred across them. They are fixed in space, but their areas vary with time, when the interface is moving in such a way that the sum $A_1(t) + A_2(t)$ is constant.

The conservation equation can be written in a general form based on the above description. For example the momentum equation is given as:

$$\frac{d}{dt} \int_{V_1(t)} \rho_1 v_1 \, dV + \frac{d}{dt} \int_{V_2(t)} \rho_2 v_2 \, dV = - \int_{A_1(t)} \rho_1 (v_1 v_1) \cdot n_1 \, dA - \int_{A_2(t)} \rho_2 (v_2 v_2) \cdot n_2 \, dA$$

$$+ \int_{V_1(t)} \rho_1 F \, dV + \int_{V_2(t)} \rho_2 F \, dV + \int_{A_1(t)} n_1 . \pi_1 \, dA + \int_{A_2(t)} n_2 . \pi_2 \, dA$$

if we can neglect the effect of surface tension.

Where the subscripts 1 and 2 refer to the respective phases, v is the vector velocity, n_k the unit normal vector directed outside phase k, F is the external force per unit mass and π is the stress tensor. Delhaye transformed the conservation equations to local equations plus interface jump conditions, and performed space averaging over the cross section of each phase. For a cylindrical tube of constant cross section A, the momentum equations for the respective phase, projected on the tube axis, were as follows:

$$\frac{\partial}{\partial t} (R_k <\rho_k v_k>) + \frac{\partial}{\partial z} (R_k <\rho_k w_k v_k>) - R_k <\rho_k F> - \frac{\partial}{\partial z} (R_k < \pi_k . n_z >)$$

$$+ \frac{1}{A} \int\limits_{C\,(z,t)} P_k \left[v_k (v_k - v_i) \right] \cdot n_k \, \frac{dC}{n_k n_k C} - \frac{1}{A} \int\limits_{C\,(z,t)} \pi_k \cdot n_k \, \frac{dC}{n_k n_k C} -$$

$$- \frac{1}{A} \int\limits_{C_K(z,t)} \pi_k \cdot n_k \, \frac{dC}{n_k n_k C} = 0 \qquad\qquad (k = 1 \text{ or } 2)$$

Where ξ represents the intersection curve of the interface with the cross section plane, C_K the intersection curve of the cross-section area occupied by phase k with the tube wall; R_K the ratio AK/A; w the velocity component along the z axis; n_{kC} the normal unit vector to C located in the cross-section plane and directed outside phase k; and $v_i \cdot n_k$ the displacement velocity of the interface. The interface condition:

$$\sum_{k\,=\,1,2} \left\{ \rho_k \left[v_k (v_k - v_i) \right] \cdot n_k - n_k \cdot \pi_k \right\} = 0$$

has also to be satisfied. Delhaye then went on to make the following assumptions:

(a) The space average pressures across the cross-section are the same for the two phases.

(b) There is no mass transfer at the wall.

(c) The variation of the stress tensor along the axis can be neglected. This would be invalid with large pressure gradients.

(d) The correlation coefficients relating the mean values of quantity to the average of the instantaneous local values are all unity. This should not be so if there is, for instance, a very peaked velocity distribution!

The following expression is then obtained for the momentum equation projected along the axis:

$$\frac{dp}{dz} = - \frac{d}{dz} (R_1 \rho_1 W_1^2 + R_2 \rho_2 W_2^2) - (R_1 \rho_1 + R_2 \rho_2) F_z$$

$$+ \frac{1}{A} \left[\overline{\oint\limits_{\xi_1} (n_1 \cdot \pi_1)_z \, d} + \overline{\oint\limits_{\xi_2} (n_2 \cdot \pi_2)_z \, d\xi} \right]$$

Where R_K is the volumetric concentration of phase k, η_k is the unit normal vector perpendicular to the wall, π_k the stress tensor and ξ_k the part of the perifery in contact with the phase k. This equation is similar to the first equation in this section confirming the limits and validity of the usefulness of the three terms.

4.2 Homogeneous model

The most drastic simplification is the use of the homogeneous model in which it is assumed that the phases are homogeneously mixed and flow with a uniform velocity. The mean density is given by

$$\frac{1}{\rho_H} = \frac{x}{\rho_G} + \frac{(1-x)}{\rho_L}$$

and the momentum equation simplifies to

$$-\frac{dp}{dz} = \frac{S}{A}\tau_o + G^2 \frac{d}{dz}\left(\frac{1}{\rho_H}\right) + \rho_H g \sin\theta$$

the three terms on the right hand side being respectively frictional, accelerational and gravitational components of the total pressure gradient. The usual procedure for calculation of the frictional term is to use a two phase friction factor f_{TP} in the equation

$$\frac{S}{A}\tau_o = \frac{1}{2}\frac{S}{A}\frac{f_{TP}G^2}{\rho_H}$$

f_{TP} being calculated from single phase friction factor data using the two phase Reynolds number:

$$Re_{TP} = \frac{4GA}{S\mu_{TP}}$$

A number of definitions are available in the literaure for μ_{TP}, an example being that of Dukler et al. (1962):

$$\mu_{TP} = (1 - x) \, \frac{\ell_H}{\ell_L} \cdot \mu_L + x \, \frac{\ell_H}{\ell_G} \cdot \mu_G$$

Where μ_L and μ_G are the viscosities of the gas and liquid phases respectively.

In general, the homogeneous model, though simple, does not give very accurate predictions of two phase flow and it is usually necessary to use one of the distributed flow models

4.3 Distributed flow models
4.3.1 Continuous distribution models

S. G. Bankoff (1960) proposed a slip ratio correlation based on having different distributions of void fractions (α) and velocity (without local slip between the phases) over the cross sections of the channel.

Zuber and Findlay (1964) introduce a local slip ratio and, by integration, in the section, obtain the following general relation:

$$\frac{Q_G}{\alpha A} = C_0 \, \frac{Q_G + Q_L}{A} + u_0$$

where C_0 is a distribution coefficient which is calculated from the profiles of void and velocities:

$$C_0 = \frac{\displaystyle\iint_A \alpha u_s \, dA}{\displaystyle\iint_A \alpha \, dA \cdot \iint_A u_s \, dA} \qquad \text{with } u_s = \alpha u_G + (1 - \alpha) u_L$$

in this formula α, u_s, u_G, u_L are local values.

Similarly:

$$u_0 = \frac{\displaystyle\iint_A \alpha(1 - \alpha) \, (u_G - u_L) \, dA}{\displaystyle\iint_A \alpha \, dA}$$

In fact C_0 and u_0 are empirically determined from experimental data by plotting $QG/\alpha A$ versus $(QG + QL)/S$: for a given flow pattern a linear relationship is generally established and the following expressions are usually proposed for C_0 and u_0 :

Flow pattern (upward vertical flow)	C_0	u_0
BUBBLE FLOW		
Circular tube (d < 5 cm)	$1.2 \ (p/p_c \leqslant 1)$	
Circular tube (d > 5 cm)	$1.5 - 0.5 \dfrac{p}{p_c}$	
$- \ a < 0.1$		$u_0 =$ terminal vecolity of one independent bubble.
$- \ a > 0.1$		
$d_B < 0.5$ mm		$u_0 = g \ \dfrac{d_B^{\ 2} (\rho_L - \rho_G)}{18 \mu_L} (1 - a)^2$
$1 < d_B < 20$ mm		$u_0 = 1.53 \left[\dfrac{g \ (\rho_L - \rho_G)}{\rho_L^2} \right]^{\frac{1}{4}} (1 - a)^{3/2}$
Rectangular channel	$1.4 - 0.4 \dfrac{p}{p_c}$	idem
SLUG FLOW	1.2	$u_0 = 0.35 \left[\dfrac{g \ (\rho_L - \rho_G) \ d}{\rho_L} \right]^{\frac{1}{2}}$
CHURN FLOW	1	$u_0 = 1.53 \left[\dfrac{\sigma g \ (\rho_L - \rho_G)}{\rho_L^2} \right]^{\frac{1}{4}}$
ANNULAR FLOW	1	$u_0 = 23 \left[\dfrac{\mu_L \ (1 - x) \ G}{\rho_L \rho_G d} \right]^{\frac{1}{2}} \dfrac{\rho_L - \rho_G}{\rho_L}$

The direct use of such expressions in computer codes for the calculations of pressure drop can lead to some numerical difficulties due to the abrupt change of the equations from one flow pattern to another one. Strictly speaking, expressions for C_0 and u_0 allow prediction of hydrostatic pressure drop only. Acceleration pressure gradient is approximatively obtained by using the separated flow expression and, if need be, the slip ratio K calulated from C_0 and u_0 :

$$K = \frac{Q_G}{Q_L} (C_0 - 1) + C_0 + \frac{u_0 A}{Q_L}$$

4.3.2 Separated Flow Models

An alternative simplifying assumption is that the two fluids flow separately in the channel. The velocity of each phase is assumed constant, but different; a new variable is often defined: the slip ratio K

$$K = \frac{u_G}{u_L}$$

where :

$$u_G = \frac{xW}{\rho_G} \cdot \frac{1}{aA}$$

$$u_L = \frac{(1 - x) W}{\rho_L} \cdot \frac{1}{(1 - a) A}$$

$$K = \frac{x}{1 - x} \cdot \frac{1 - a}{a} \cdot \frac{1}{\rho^*} \quad \text{with } \rho^* = \frac{\rho_G}{\rho_L}$$

This relation shows that it is necessary to know two of the three variables α, x, K for defining a separated liquid-gas flow.

The momentum equation gives the expression for the different pressure drop terms:

$$\frac{dp_g}{dz} = \left[a\rho_G + (1 - a)\rho_L \right] g \sin \theta$$

$$\frac{dp_a}{dz} = G^2 \frac{d}{dz} \left[\frac{x^2}{a\rho_G} + \frac{(1-x)^2}{(1-a)\rho_L} \right]$$

$$\frac{dp_F}{dz} = \frac{S}{A} \tau_o$$

For calculating the different terms, one needs:
- a correlation for α (or K)
- a method for obtaining τ_0

In the case of annular dispersed flows, more sophisticated separated flow models have been used: for instance a pure liquid film and a homogeneous mixture in the core (droplet flow). In this case, a supplementary correlation is needed for the determination of the percentage of the liquid mass flow entrained in the gas core E. Comparison between the acceleration pressure drop calculated with this model and the homogeneous and the separated flow models respectively is given in Figure 3.

4.4 Flow Pattern Models
4.4.1 Bubble flow

This flow pattern is the most probable when the void fraction is less than 0.3 and if the tube diameter is much greater than the bubble dimensions. The size distribution of the bubbles at the beginning of their life is controlled by the mode of formation: injection of gas, degassing, boiling. The departure diameter of the bubble leaving the birth place is the result of a balance between gravitational force, drag, inertia and surface tension; for instance, at atmospheric pressure the order of magnitude of diameter is 2 mm. In stagnant liquid and low void fraction, bubble dimensions are stable and the terminal velocity of independent bubbles u_∞ can be determined.

In air-water mixture, the terminal velocity is:

$$u_\infty = \sqrt{\frac{4}{3} \frac{\rho_L - \rho_G}{\rho_L} \frac{g d_B}{C}}$$

C is the drag coefficient, a function of the Reynolds, Froude and Weber number as follows:

d_B < 0.3 mm Stokes law for solid spheres is valid

$$C = 24 \; Re^{-1} \qquad Re = \frac{u_\infty \, d_B}{\nu_L}$$

0.3 < d_B < 2 mm Circulation in the bubble

$$C = 18.2 \; Re^{-0.684}$$

2 mm < d_B < 7 mm Ellipsoidal shape

$$C \simeq 0.366 \; We.Fr^{-1}$$

d_B > 7 mm Bubble cap

$$C \simeq 2.61$$

In a bubble swarm, the terminal velocity is lower [Wallis (1970)]. In a turbulent flow of liquid the size distribution is controlled by the coalescence and breakup phenomena, and an "equilibrium" condition can be reached. At CEN Grenoble, R. Roumy has observed that this dimension is reached rapidly after a sudden change of section, after a change of flow rate, and after flow through obstacles.

In vertical upward flow, bubbles are moving more rapidly than the liquid and the same result is observed in negative pressure gradients irrespective of the orientation (for instance downward flow at high velocity).

The Zuber-Findlay model is well adapted for the bubble flow with values of C_0 and u_0 indicated above. Frictional pressure drop at high flow rates (Q_L/A > 2m/s) can be calculated by the simple formula:

$$\frac{\Delta p_{f_{TP}}}{\Delta p_{f_L}} = (1 - a)^{-1.8}$$

At lower velocities, the interphase friction becomes preponderant and at a given α, the measured friction pressure drop is higher than the calculated value.

4.4.2 Slug flow

For slug flow Griffith and Wallis (1959) proposed the use of a formula very similar to the Zuber-Findlay expression. The most probable value of C_0 is given above (C_0 = 1.2). Greskovich and Shrier (1971) have confirmed recently that this figure is valid for horizontal or slightly inclined tubes. The value of u_0 , which is zero in horizontal flow, increased rapidly with the inclination and is equal or greater than u_0 in vertical position for inclinations greater than 2 deg.

Greskovich gives the method for calculating the frictional pressure drop using an empirical correlation for the slug flow friction factor:

$$\frac{dp}{dz}_f = \frac{Q_L}{Q_G + Q_L} \cdot \frac{2\rho_L f \left(\frac{Q_L + Q_G}{A}\right)^2}{d} \qquad \text{in horizontal flow}$$

with
$$f = 0.0048 + 3980 \text{ Re}^{-1.285}$$

$$\text{Re} = \frac{(Q_G + Q_L) \cdot d \cdot \rho_L}{A \cdot \mu_L}$$

In inclined tubes (2-10 deg.) the same model applies but the determination of the liquid slug core holdup leads to a more complicated set of equations.

An example of the application of slug flow is the gas lift pump. In this apparatus (Fig. 4) gas (or another light fluid) is injected at the bottom of a tube immersed in liquid: the driving force due to the buoyancy of the gas is used to pump the liquid. In spite of the poor efficiency (0.1 - 0.7) compared to a mechanical pump, this system is widely used for erosive or corrosive liquids and suspensions. Stenning and Martin (1967) have proposed a simple theory assuming a slug flow in the tube. Typical results are indicated in Fig. 5.

4.4.3 Annular Flow

In annular flow, the independent variables of the system can be regarded as the channel geometry (including mode of phase entry), the flow rate, the physical properties and the heat flux (if any). The dependent variables include critical heat flux, heat transfer coefficient, pressure gradient, liquid film flow rate and liquid film thickness. The last three of these are the prime ones from which the others can be calculated; since these three variables are inter-dependent, a simultaneous solution for all three is required using three simultaneous relationships.

Two such relationships are now moderately well established:

(1) The triangular relationship. Basically, this relationship allows the calculation of pressure drop, film thickness or film flow rate from a knowledge of the other two. Briefly, this relationship is based on evaluation of the shear stress profile in the liquid film, determination of the velocity profile from the shear stress profile and the effective local viscosity (with due allowance for the effect of turbulence if appropriate) and integration of this velocity profile. A review of this relationship is given by Hewitt and Hall Taylor (1970) and by Hewitt (1970). The performance of the relationship is surprisingly good.

(2) Interfacial roughness correlation. To a first approximation at least, it would seem that the shape of the interface is such that, for a given mean film thickness, the interface presents a fixed effective roughness to the gas core (Gill et al. (1963), Hewitt (1970), Wallis (1970)).

The closure of the annular flow prediction problem requires a further relationship which must necessarily involve a prediction of the liquid entrainment phenomenon. This could either take the form of a correlation for the fraction of the total flow which is entrained or could be in the form of a relationship for entrainment dynamics. The latter form is preferable since it can cope with non-equilibrium situations. An extensive discussion of the mechanisms of entrainment and of the development of entrainment relationships is given by Hewitt (1972).

The most promising approach at the present time appears to be to consider the processes of droplet deposition, and liquid entrainment from the surface of the liquid film as being independent. The process of droplet deposition is much better understood than is entrainment. Cousins and Hewitt (1968) show that the rate of deposition is proportional to the concentration of droplets in the gas core and that the constant of proportionality was relatively insensitive to the flow velocity and tube diameter. The mechanism if deposition appears to be turbulent diffusion and the stochastic behavior of droplets in a turbulent gas has been investigated by Hutchinson et al. (1970) who show that the calculated deposition rate is in good agreement with that observed.

The rate of entrainement is therefore, the remaining unknown and Hutchinson and Whalley (1972) have recently proposed a correlation of entrainment rate as a function of the group (τ_im/σ) where τ_i is shear stress at the interface, m the film thickness and σ the surface tension. The form of the correlation is illustrated in Figure 6 and the fit to existing data is encouraging, though the scatter is still unacceptable for accurate calculation. However, preliminary calculations of closed solutions of annular flow using Figure 6, are being carried out at Harwell and are showing considerable promise.

It is clear that the problem of annular flow prediction is far from satisfactory solution though there have been some encouraging developments recently. Attention needs to be paid to the further development of the interfacial roughness and entrainment relationships.

4.4.4 Drop Flow

"Drop flow" can be defined as the case where the flow consists of a dispersion of droplets in a gas. In adiabatic flow, pure drop flow does not occur since there is almost invariably a liquid film on the wall, in which case the flow should be categorised as "annular." Drop flow can exist, however, in heated systems where the wall is kept dry by evaporation. Such a situation would exist, for instance, in the "post burnout" region. The properties of the drop flow are of considerable importance in determining wall temperature distribution in this region. The vapour is cooled by the evaporation of the droplets and is heated from the channel wall. The balance between these two processes determines the bulk vapor temperature and then the wall temperature. The rate of evaporation of the droplets depends on the difference in temperature and velocity between them and the vapor and the acceleration of the droplets depends on the drag force which in turn depends on the velocity difference. Also, if the velocity difference is too high, the droplets will split. All of these processes can be represented by simple differential equations which can be integrated along the channel. This approach has been adopted, for instance, by Bennet et al. (1968) who show an encouraging degree of success in predicting post-burnout wall temperature distribution.

4.5 Empirical Models
4.5.1 Lokhart-Martinelli and Baroczy models

The correlations most widely used as a first approach for the evaluation of the frictional pressure drop in gas-liquid flows are those of Lockhart and

Martinelli (1949) and (for steam-water flows) Martinelli and Nelson (1948). There is no physical basis for these correlations which assume that the ratio of the two phase pressure gradient to the pressure gradient for one or other of the phases flowing alone, is function only of the ratio of the pressure gradients for the respective phases. Thus:

$$\frac{\left(\dfrac{dp_F}{dz}\right)_{TP}}{\left(\dfrac{dp_F}{dz}\right)_L} = \Phi_L^2 \qquad \frac{\left(\dfrac{dp_F}{dz}\right)_{TP}}{\left(\dfrac{dp_F}{dz}\right)_G} = \Phi_G^2 \qquad \frac{\left(\dfrac{dp_F}{dz}\right)_L}{\left(\dfrac{dp_F}{dz}\right)_G} = X^2$$

$$\Phi_L^2 = f(X^2) \quad or \quad \Phi_G^2 = g(X^2)$$

In reality Φ_L^2 and Φ_G^2 are functions not only of X^2, but also of G, d_0, physical properties of the fluids (thus pressure), flow pattern, heat flux, etc., so that the prediction is sometimes far from the experimental results ($\pm 60\%$ and higher) especially for steam-water mixtures. Martinelli et al. proposed a classification of the two phase flows based on the laminar or turbulent regime of each phase with different graphical functions $\Phi_L^2(X^2)$ for each combination. Curves in the original papers allow the calculation of the different components of the two phase pressure drop (friction, gravitational and acceleration) either for a given quality or in an integral form for a boiling channel with uniform axial heat flux.

Many authors have published corrections of the Martinelli curves taking into account the influence of G, pressure etc.: one correlation which seems rather good for a large range of parameters and number of fluids is that of Baroczy (1966). Like Martinelli–Nelson he used the ratio

$$\Phi_L^2 = \frac{\left(\dfrac{dp_F}{dz}\right)_{TP}}{\left(\dfrac{dp_F}{dz}\right)_{LO}}$$

where $(dp_F/dz)LO$ is the frictional pressure drop gradient for the total flow flowing as if it had the liquid properties. Baroczy gives graphs of Φ_{LO}^2 as function of a property index $\rho^*(\mu_L/\mu_G)^{0.2}$ with mass flux and quality as parameters. The complex shape of the curves, which cross each other, is not very satisfactory though the accuracy of the correlation is much greater than that of the original Martinelli models.

Such models cannot take into account the abrupt change in pressure drop gradients due to flow pattern transition (bubble-slug-annular-dispersed) which are sometimes the origin of oscillations and instabilities.

4.5.2 The Chisholm and Sutherland model

A model which has a more satisfactory physical basis than do the Martinelli type models, is that of Chisholm and Sutherland (1969). These authors use the equation:

$$\Phi_L^2 = \frac{\left(\dfrac{dp_F}{dz}\right)_{TP}}{\left(\dfrac{dp_F}{dz}\right)_L} = 1 + \frac{C}{X} + \left(\frac{1}{X}\right)^2$$

where X is the Lockhart-Martinelli parameter and C is a constant. Chisholm and Sutherland suggest that the most suitable form of equation for C for general engineering purpose was:

$$C = \left[1 + (C_2 - 1)\left(\frac{v_G - v_L}{v_G}\right)^{0.5}\right]\left[\left(\frac{v_G}{v_L}\right)^{0.5} + \left(\frac{v_L}{v_G}\right)^{0.5}\right]$$

for the condition $v_G = v_L$, C has the required value of 2. The constant C_2 was assigned various "recommended" values for the different situations considered in steam-water flow. Those values used in the analysis are given below:

Flow condition	C_2	Comments
Rough tubes		
(1) $G > 1,500 \ Kg/m^2 \, s$	1	
$= 1.1 \times 10^6 \ lb./ft.^2 \, h.$		
or $30 > (v_G/v_L)^{0.5} > 9$		

(2) $G < 1{,}500 \; Kg/m^2 \, s$ 1500   If $C_2 > 4$

       for $(^{V}G/v_{L})^{0.5} < 9$ $G \; (Kg/m^2 \, s)$ let $C_2 = 4$

Smooth tubes

(3) $G > 2{,}000 \; Kg/m^2 \, s$ 1

       $= 1.5 \;\; x \; 10^6 \, lb./ft.^2 \, h.$

       or $30 < (^{V}G/v_{L})^{0.5} > 9$

(4) $G < 2{,}000 \; Kg/m^2 \, s$ 2000   If $C_2 > 4$

       for $(^{V}G/v_{L})^{0.5} < 9$ $G \; (Kg/m^2 \, s)$ let $C_2 = 4$

4.5.3 Effect of surface tension

None of the empirical models mentioned above take explicit account of the effect of surface tension. This property has a pronounced effect on the behavior of two-phase flows but it is only rarely considered. However, the CISE group at Milan have developed a series of correlations which include surface tension. An example of these correlations is that of Premoli et al. (1971) for slip ratio.

$$\Delta K \;=\; K - 1 = f \; (G, \; D, \; \rho_L, \; \rho_G, \; \mu_L, \; \sigma, \; \beta)$$

where σ is the surface tension and β the ratio of the gas volume flow to the total volume flow

where $$\Delta K \;=\; E_1 \left(\frac{y}{1 + y \, E_2} - y \, E_2 \right)^{\frac{1}{2}}$$

where $$y \;=\; \frac{\beta}{1 - \beta}$$

and $$E_2 \;=\; 0.0273 \; We \; Re^{-0.51} \left(\frac{\rho_L}{\rho_G} \right)^{-0.08}$$

$$E_1 \;=\; 1.578 \; Re^{-0.19} \left(\frac{\rho_L}{\rho_G} \right)^{0.22}$$

where $Re = \dfrac{G\,D}{\mu_L}$

$We = \dfrac{G^2\,D}{\sigma\,\rho_L}$

Also if $E_2\,y > \dfrac{y}{1 + E_2\,y}$ then ΔK is assumed to equal zero.

5. NON–PARALLEL FLOWS

5.1 Cross flow over tube banks

Two phase cross flow over tube banks is of importance in a number of heat exchanger applications. The subject has been considered recently in a report by Grant and Murray (1972). A new correlation is proposed for two phase crossflow pressure drop, Δp_{TP}:

$$\frac{\Delta p_{TP}}{\Delta p_{LO}} = 1 + (\Gamma^2 - 1)(x + 0.15\ x^{\frac{1}{2}} - 0.15\ x^{400})$$

where $\Gamma = \left(\dfrac{\Delta p_{GO}}{\Delta p_{LO}}\right)^{0,5}$

and Δp_{GO} and Δp_{LO} are the pressure drops for the total mass flowing as gas or as liquid respectively. The final term on x is introduced to make the prediction satisfy the all-gas condition. Grant and Sutherland also discuss the prediction of window zone pressure drops for two-phase flow in heat exchangers.

5.2 Localized Pressure Losses

When the section of the pipe is gradually changed, it is possible to calculate the additional pressure change by using the homogeneous or separated flow models. However, for sudden changes in cross section, disequilibrium between the two phases occurs: thus, for vapor-liquid flow, one can assume that quality does not vary rapidly near the change in cross section. Equilibrium is then reached again downstream of the obstacle with an additional pressure change, which is not included in the localized pressure loss indicated in the discussion below. It is also assumed below that critical flow is not reached in the pipe; critical flow is discussed in section 6 below.

For practical purposes the Chisholm (1969a) correlation is useful. The basis of this correlation for parallel flows was discussed in 4.5.2 above.

For instance with a short expansion from flow area A_1 to flow area A_2,

$$\Delta p_{TP} = p_2 - p_1 = \frac{W^2}{\rho_L A_1^2} s (1 - s)(1 - x)^2 \left(1 + \frac{C}{X} + \frac{1}{X^2}\right)$$

with $\quad s = \dfrac{A_1}{A_2}$

To obtain this formula, it is necessary to assume that slip ratio K does not change through the expansion. Chisholm (1969a) discusses alternative expressions for when this assumption is invalid. The total pressure drop is composed of reversible and irreversible parts. The reversible part of Δp_{TP} is

$$\Delta p_r = \frac{W^2}{2\rho_L A_1^2} (1 - s^2) \frac{\dfrac{x^3}{\rho_G^2 a^2} + \dfrac{(1-x)^3}{\rho_L^2 (1-a)^2}}{\dfrac{x}{\rho_G} + \dfrac{1-x}{\rho_L}}$$

and the irreversible part is $(p_2 - p_1) - \Delta p_r$

In the expression of Δp_{TP}, we can observe that $\dfrac{W^2}{\rho_L A_1^2} (1 - s) (1 - x)$ is the pressure change when the liquid is flowing alone. Thus:

$$\frac{\Delta p_{TP}}{\Delta p_L} = 1 + \frac{C}{X} + \frac{1}{X^2}$$

which is similar to the expression given for parallel flows. Chisholm suggests the following expression for C:

$$C = Z + \frac{1}{Z} \quad \text{with} \quad Z = 0.19 + 0.92 \frac{p}{p_c}$$

For sudden contractions the rapid acceleration of the flow increases the mixing of the phases so that the homogeneous theory can be used as indicated by Geiger and Rohrer (1966). Thus, the total pressure drop, Δp_{TP}, is given by:

$$\Delta p_{TP} = \frac{W^2}{2\rho_H A_2^2}\left[1 - s^2 + \left(\frac{1}{C_c} - 1\right)^2\right]$$

with $s = A_2/A_1$ and C_c the classical coefficient of contraction, and the irreversible loss is

$$\Delta p_i = \frac{W^2}{2\rho_L A_2^2}\left(\frac{1}{C_c} - 1\right)^2\left[1 + x\left(\frac{\rho_L}{\rho_G} - 1\right)\right]$$

For orifices, Murdock (1962) proposed a simple correlation:

$$\sqrt{\frac{\Delta p_{TP}}{\Delta p_G}} = 1 + 1.26\ X$$

but the Chisholm (1969b) correlation form can also be used for this case for high quality steam/water mixtures. Chisholm proposes the equation

$$\frac{\Delta p_{TP}}{\Delta p_G} = 1 + CX + X^2$$

(which is similar to the one given above for $\Delta p_{TP}/\Delta p_L$) with:

$$C = Z + 1/Z \qquad \text{where} \quad Z = 0.19 + 0.92\ p/p_c$$

Orifices and venturis can be used for the determination of the quality if the total mass flow is measured in another part of the circuit. With high quality and high flow rate the mixture is practically homogeneous, but at lower quality, the slip ratio increases so that the pressure change is less than that in homogeneous flow. Upstream influence of the pipe is important, so that for quality measurements, calibrations are necessary.

For bends, tees and valves, Chisholm (1969a) proposes the use of the formulation:

$$\frac{\Delta p_{TP}}{\Delta p_L} = 1 + \frac{C}{X} + \frac{1}{X^2}$$

with $\quad C = C^*\left(\sqrt{\rho_L/\rho_G} + \sqrt{\rho_G/\rho_L}\right)$

with C^* defined from the following:

(a) - Bends

$$C^* = 1 + 35 \frac{d}{L}, \quad L \quad \text{length of the bend.}$$

If there is some perturbation upstream at a distance less than 56 diameters:

$$C^* = 1 + 25 \frac{d}{L}$$

(b) - Tees

$$C^* = 1.75$$

(c) - Gate valve

$$C^* = 1.5$$

(d) - Globe valve

$$C^* = 2.3$$

(e) - Control valve: homogeneous flow ($C^* = 1$) can be used. Sheldon and Schuder (1965) propose some correlations for steam-water flows.

5.3 Flow Distribution in Two-Phase Fluid Flow

Many practical applications of two-phase flow involve the distribution of a two-phase stream between a number of subsidiary streams. Typically, the initial stream can be divided into a number of parallel paths at right-angles to the main stream flow. Such a situation would occur, for instance, in a two-pass air-cooled condenser in the back header leading to the second pass. The Heat Transfer and Fluid Flow Service at Harwell has been undertaking a fairly extensive study of this problem.

When a portion of the flow is diverted at right-angles at a T-junction the result is a gain of pressure across the junction due to recovery of the forward momentum of the original stream. Similarly, when a flow joins the main stream in the header from the side tube, there is an additional pressure loss due to the acceleration of the fluid. In two-phase flow systems, the pressure change phenomena still occur, but an additional feature is that the quality of the two-phase flow in the side-stream is different from that in the main stream and varies with the amount of the total fluid flow which is taken into the side-stream. Usually, the liquid flow tends to be preferentially along the header and the gas flow preferentially into the side-stream. Typical results for a variety of cases are illustrated in Figure 7.

Equal quality of the side-stream and main header flow would be indicated by equal fractions of the gas flow and liquid flowing into the side-stream. As can be seen, the separation effect is very large indeed.

In the design of headers for heat exchangers etc., the two main alternatives are asymmetric and symetric headers. In the case of asymmetric headers, the flow in the inlet header is in the same direction as that in the outlet header and vice versa for the case of the symmetric headers. If pressure recoveries and losses at the junctions dominate friction losses in the header, then the asymmetric arrangement can give very poor distribution since the pressure drop across the first tube can be very much less than that across the last tube. The symmetric arrangement, on the other hand, gives very much better results for this case, which may be surprising from an intuitive point of view. Experimental work on flow in multi-tube arrays has been carried out at Harwell over the past two years for both single and two-phase flow systems. The type of apparatus used is sketched in Figure 8. By using pressure tappings on each individual tube as illustrated it is possible to deduce the flow rate in the case of single-phase flow. In the case of two-phase flow, since the two-phase quality may vary from tube to tube, it is necessary to have an additional measurement and, in these experiments, the void fraction was measured in each tube by instantaneously closing the valves on each tube as indicated in Figure 8. From the pressure gradient and the void fraction, it was possible to deduce the flow rates of each phase within each tube. Typical results are illustrated in the following table:

Distribution of two-phase flow between parallel tubes

	\multicolumn{4}{c}{Tube No.}	Ratio of maximum to minimum flow
	1	2	3	4	
$W_G/\overline{W}_G$	2.05	1.39	0.37	0.19	10.8
$W_L/\overline{W}_L$	0.46	0.46	1.15	1.93	4.2

where W_G and W_L are the actual flows of the gas and liquid respectively in each of the tubes and $\overline{W}_G$ and $\overline{W}_L$ are the mean flows per tube. As will be seen, a gross maldistribution occurs and this will have a profound effect on the design. The experiments carried out on multi-tube headers for two-phase systems agree well with what had been expected from the single T-junction tests. It can be concluded that the design of systems not taking account of these effects must be open to considerable questions.

6. CRITICAL TWO PHASE FLOWS

For single phase flow, the velocity of sound u_c can be calculated from:

$$u_c = v \sqrt{\left(\frac{\partial p}{\partial v}\right)_s}$$

where v is the specific volume (reciprocal of the density) and $(\partial p/\partial v)_s$ the change of fluid pressure with specific volume at constant entropy. G_c $(= u_c/v)$ represents the "critical" mass velocity, which is the maximum that can be achieved in a duct of constant cross section.

In two phase flow, the velocity of sound and critical mass flow are not as strongly connected as in single phase flow due to the complex nature of the transmission of sound waves in two-phase flow. However, the above equation is usually employed for calculating the critical flow, though a number of alternative definitions of specific volume can be postulated [see Cruver and Moulton (1967)]. For instance, the homogeneous model specific volume:

$$v_H = x \, v_G + (1 - x) \, v_L$$

can be used in which case the expression for the critical velocity u_{cTP} is given by:

$$\frac{u_{cTP}}{u_{cG}} = \frac{x + (1 - x) \, C}{\sqrt{x + (1 - x)D^2 C^2}}$$

where u_{cG} is the critical velocity for the gas phase alone, C the ratio v_L/v_G and D the ratio of the critical velocity in the gas, u_{cG}, to that in the liquid, u_{cL}.   It will be seen that u_{cTP} tends to u_{cG} at $x = 1$ and u_{cL} at $x = 0$ and that, between these two values, it passes through a minimum.

The use of an arbitrarily defined specific volume is questionable and a more detailed study and modelling of the inter-phase interactions is required. Experimental measurements of the detailed pressure, flow, film thickness and temperature profiles in two-phase flow in an annular reaction are reported by Smith et al. (1968). Figure 9 illustrates some of the results obtained and it will be seen that the critical condition is characterized by no upstream transmission of changed in downstream pressure. Further detailed analysis is reported by Smith (1972).

The propagation of sound in a two phase mixture is dependent on the signal frequency and on the flow pattern. In some recent experiments, Chawla et al. (1972) measured the propagation of small pressure disturbances in a tube containing a static layer of liquid with a supernatant gas. It was found that the propagation occurred in three ways: a rapid pulse through the liquid, a less rapid pulse through the gas and, finally, a pulse corresponding to transport through the two phase mixture in the tube. The energy in the respective pulses was inversely proportional to the square of the transmission velocity.

The field of critical and sonic flow in two phase mixtures is one surrounded by considerable controversy and more basic work in elucidating the phenomena would appear to be desirable (Boure and Reocreux 1972).

7. UNSTEADY FLOWS

7.1 Hydrodynamic instabilities

Two phase flow systems are very susceptible to instabilities and these fall into a number of different types, some of which can be described as follows:

(1) System noise. Fluctuating pressures and momentum fluxes occur which are closely linked with nucleation and flow pattern. The slug flow regime is particularly difficult. The fluctuations in two phase pressure gradient are often of the same order of magnitude as the gradient itself (see for instance Chaudry et al. (1965)].

(2) Excusive instability. Two-phase systems can show maxima in pressure drop/flow characteristics at constant heat input. This implies that there is more than one solution for flow at a given pressure drop and an excursion in flow can take place. This problem is particularly severe at low pressures and with high velocity (nominally liquid) cooling systems.

(3) Chugging instability. The induction of liquid superheat (due to the absence of active nucleation sites for boiling) can lead to a rapid ejection of the liquid from the channel once the first vapor is formed. This is a particularly difficult problem with liquid metal systems.

(4) Oscillatory instability. This is the most important kind of instability and it is worth checking for in most boiling systems particularly with parallel channels or natural circulation flow. The mechanism of this form of instability is well

known. It has been clearly given, in particular cases, by Wallis and Heasley (1961) among others. The general acceptance of this mechanism may be traced back to 1967 when it has been explained at the Eindhoven Symposium by several authors, for instance Neal and Ziui, J. Boure, and Davies and Potter. Figure 10 illustrates their explanation of the mechanism. As was seen above, the pressure drop in two phase flow is composed of a number of components which, in the steady state, are additive giving the total pressure gradient. However, if the inlet velocity is fluctuating with time at a given frequency, perturbations occur at the same frequency in the various components of the two phase pressure drop but there is a phase lag between the oscillations in the various components as illustrated in Figure 10. The net result is that, at a given frequency, the oscillation of the inlet velocity does not give rise to a consequent (and counter balancing) oscillation in pressure drop and the velocity fluctuations trigger the oscillatory instability

(5) Acoustic effects. Acoustic effects can be observed as high frequency oscillations often superimposed on other forms of instability. They occur in systems in which pressure disturbances can be transmitted through pipelines and reflected at ends, reservoirs etc. As was discussed in section 7 above, the velocity of sound in two phase flow can be much lower than that in the pure constituent fluids.

A more detailed classification of two phase flow instabilities is given by Bouré et al. (1971).

7.2 Vibration (transient momentum flux)

In recent years, the avoidance of gross over-design in heat exchangers and related equipment has led to important problems in itself. One of the most significant of these has been the increase in incidence of failure of heat exchangers due to tube vibration. This vibration problem is important in single phase flows, but the fluctuations which occur in two-phase flows make the problem even worse in that case. A body suspended in a two-phase flow is subjected to a variable force since the momentum flux of the flow is changing with time. This is an obvious source of vibration and such vibration will be particularly serious when the natural frequency of the element corresponds to the frequency of the variation of momentum flux. Problems of vibration in two-phase flow are discussed, for instance, Gorman et al. (1970).

As was mentioned above, vibration can be caused by the unsteadiness of momentum flux. This flux can be measured by turning the flow through a right angle (i.e. destroying the forward momentum) by using, say, a tee. The force on the tee can then be measured instantaneously by attaching it to a beam whose deflection can be recorded. Measurements of two-phase momentum flux using this technique are reported by Andeen and Kern (1965) Yih and Griffiths (1968) (1970), Andeen and Griffiths (1968) and Vance and Moulton (1965). The steady state momentum fluxes are generally much higher than those calculated by the separated flow model and Andeen and Griffiths (1968) found that, on the whole, the momentum flux calculated from a homogeneous flow model gave the closest agreement to the observations. Such a model would not, of course, actually describe the flow; what probably happens is that the non-uniformity in velocity approximately compensates for the non-homogeneous nature of the flow. The measurements of transient momentum flux can be analysed to give power spectral density data from which the peak frequency can be observed. [Yih and Griffiths (1970)]. There would appear to be much scope for further work on unsteady momentum flux determination, particularly in more realistic heat exchanger geometries. A particularly important case is that of cross flow across banks of cylinders.

8. EQUIPMENT INVOLVING GAS–LIQUID TWO PHASE FLOW

8.1 Classification of two-phase flow equipment

A survey carried out by one of the present authors indicated that about 60% of all industrial heat transfer equipment involved gas-liquid two-phase flow in one form or another. Adiabatic two phase flow is also important and, in addition, two phase flow occurs in a wide range of mass transfer equipment. A possible classification of two phase flow equipment, with illustrative examples is as follows:

Adiabatic flow
    Single component: Geothermal wells.
    Multi-component: Oil/gas pipelines.

Evaporation
    Indirect contact
        Heat flux controlled surface
            Cross flow
                Single component: Shell Boiler
            Parallel flow

Single component: Boiling water reactor
Multi-component: Refinery furnace
Temperature controlled surface
Cross flow
Single component: Cryogenic coil vaporiser
Multi-component: Kettle reboiler
Parallel flow
Single component: Desalination (LTV evaporator)
Multi-component: Vertical thermosyphon reboiler

Direct contact
Liquid continuous
Single component: Flash evaporator
Multi component: Freeze desalination
Gas continuous
Single component: Spray desuperheater
Multi-component: Distillation tower

Condensation
Indirect contact
Filmwise
Cross flow
Single component: Power station condenser
Multi-component: Refinery condenser
Parallel flow
Single component: Air-cooled steam condenser
Multi-component: Plate-fin cryogenic condenser
Dropwise
Cross flow
Single component: Desalination condenser
Direct contact
Liquid continuous
Single component: Steam injection heating
Multi-component: Flash condensation (Safety)

Gas continuous
  Single component: Turbine
  Multi-component: Spray condenser

8.2 Example of Hydrodynamic Problems: Vertical Thermosyphon Reboiler

Distillation columns with vertical reboilers are of common use in chemical industry and the reboilers are generally oversized. In the bundle of the vertical tubes, heated by condensation of steam, the mixture of circulating liquid and generated vapor flows in natural or forced convection. Two-phase flows exist in the upper part of the reboiler and, in design, it is necessary to calculate the rate of production of vapor, and the pressure drop (or circulation rate).

Computer programs for the calculation of pressure drop and circulation rate have been developed by HTFS, Harwell and other organisations. These programs depend on the stepwise integration of thermodynamic and hydrodynamic equations along the channel. Typical results obtained using the HTFS programs are illustrated in Figure 11 (the reader should note that the effect of outlet nozzle diameter shown is specific for the example chosen and may be different for other cases). It is important to be careful about the choice of pressure drop correlation, especially at low pressure where the classical pressure drop correlations are not accurate. This is illustrated in Figure 12 where the experimental results for total pressure drop (in fact, a driving pressure here) are compared with the Martinelli-Nelson correlation for a simple vertical tube.

D. Q. Kern (1950) gives maximum limits of heat flux on the tubes for safe and stable operations with different aqueous or organic solutions. The physical limits are the onset of oscillations or dry-out of the tubes. It was observed that onset of oscillations can be delayed by throttling the flow with a diaphragm, or a butterfly valve at the inlet of the reboiler; in a natural convection loop the maximum power can be increased by a factor of 2 or 3. This was shown by Shellene et al. (1968) and is illustrated by results obtained at CEN Grenoble on water (Figure 13), and water-alcohol mixtures [Croix and Fabrega (1972)]. A model for prediction of onset of oscillations in reboilers was derived by Blumenkranz and Taborek (1971) using the Bouré model (1966). Similarly good agreement with the Shellene et al. data has been observed by HTFS, Harwell using the Davies and Potter (1967) model as a basis.

In conclusion it can be stated a better prediction of the performance of the reboiler (maximum heat rate, outlet quality, stability, safety) depends on a better knowledge of the thermohydraulics of two-phase flows.

9. CONCLUSION

The above brief review illustrates the wide range of topics of active current research interest in this field, and their relevance to the design of heat exchangers. In our second presentation at this Seminar, we shall be dealing with heat transfer respects but it is important to realise that an understanding of the heat transfer processes is very dependent on being able to give an accurate prediction of the hydrodynamics. In some regions (bubble flow, slug flow, the fundamental relationships are better than the empirical ones as a basis for design. However, in annular flow there is still a long way to go largely due to the unresolved problem of liquid entrainement and the best plan here is to have recourse to an empirical correlation. The study of practical heat exchanger systems produces a whole new class of problems - cross flow, vibration, flow distribution etc. - for which design data is sparse and on which much further work needs to be done.

NOMENCLATURE

Symbol	Definition	Units
A	Cross-sectional area of channel	m^2
C	Drag coefficient	
C	Constant in Chisholm model (section 4.5.2)	
C_c	Coefficient of contraction	
C_0	Constant in Zuber/Finlay model (section 4.3.1)	
d_B	Bubble diameter	m
f_{TP}	Two-phase friction factor	
g	Acceleration due to gravity	m/s^2
G	Mass flux	$kg/m^2 s$
G_c	Critical mass flux	$kg/m^2 s$
G_G	Local gas mass flux	$kg/m^2 s$
G_L	Local liquid mass flux	$kg/m^2 s$
K	Slip ratio	
p	Pressure	N/m^2
p_c	Critical pressure	N/m^2
Δp	Pressure drop	N/m^3
Q_G	Volumetric flow rate of gas	m^3/s
Q_L	Volumetric flow rate of liquid	m^3/s
Re	Reynolds number	
Re_{TP}	Two-phase Reynolds number	
S	Channel perimeter	m
u_0	Drift velocity (section 4.3.1)	m/s
u_c	Critical velocity in critical flow	m/s
u_G	Velocity of gas in the axial direction	m/s
u_L	Velocity of liquid in the axial direction	m/s
u_s	Defined as αu_G ¡ $(1-d)U_L$	m/s
u_∞	Bubble rise velocity	m/s
v	Specific volume	m^3/kg
v_G	Specific volume of the gas	m^3/kg
v_L	Specific volume of the liquid	m^3/kg
W	Mass flow rate	kg/s

We	Weber number	
x	Quality	
X	Martinelli parameter	
z	Axial distance or coordinate	m
dp_F/dz	Frictional pressure gradient	N/m^3
α	Void fraction	N/m^3
β	$Q_G/Q_L + Q_G$	
θ	Angle of inclination of channel to the horizontal	
μ_L	Liquid viscosity	
μ_G	Gas viscosity	
μ_{TP}	Two phase viscosity	kg/ms
ρ	Density	kg/ms
ρ^*	Gas to liquid density ratio $\rho^* = \rho_G/\rho_L$	kg/m^3
ρ_G	Gas density	kg/m^3
ρ_H	Homogeneous density	kg/m^3
ρ_L	Liquid density	kg/m^3
σ	Surface tension	kg/s^2
τ_0	Wall shear stress	$poundals/ft^2$
ϕ_G	Martinelli parameter $\phi_G = [(dp_F/dz)/(dp_F/dz)_G]^{1/2}$	
ϕ_L	Martinelli parameter $\phi_L = [(dp_F/dz)/(dp_F/dz)_L]^{1/2}$	

REFERENCES

BOOKS AND MAJOR REVIEWS

[1]   Collier, J.G. (1972)
           Convective boiling and condensation.
           McGraw Hill, 1972, 376 pages.
[2]   Hewitt, G.F. Hall-Taylor, N.S. (1970)
           Annular two-phase flow.
           Pergamon Press, 1970, 310 pages.
[3]   Kern, D.Q. (1950)
           Process heat transfer.
           McGraw Hill, 1950, 871 pages.
[4]   Tong, L.S. (1965)
           Boiling heat transfer and two-phase flow.
           J. Wiley, 1965, 242 pages.
[5]   Wallis, G.B. (1969)
           One-dimensional two phase flow.
           McGraw Hill, 1969, 408 pages.
[6]   Euratom Symposium of Two Phase Flow Dynamics - Vol. I and II -
           Eindoven 1967(see in particular papers by Bouré, Davies and   Potter,
           Neal and Ziut.)

OTHER REFERENCES

[1]   Andeen, B., Kern, R. (1965)
           (1) Two phase momentum flux. (2) The heat pipe.
           Progress Report, MIT DSR Project No. 4547, (June 1965), 22 pages, 15
           Figs.
[2]   Andeen, G. B., Griffith, P. (1968)
           Momentum flux in two-phase flow.
           J. Heat Trans., Vol. 90, No. 2, pp. 211-222, (May 1968).
[3]   Baker, O. (1958)
           Multiphase flow in pipelines.
           Oil and Gas J., Progress Report, pp. 156-167, (November 10th 1958).

[4]   Bankhoff, S. G., (1960)
A variable-density single-fluid model for two-phase flow with particular
reference to steam-water flow.
Trans. A.S.M.E., J. Heat Transfer, Vol. 82, No. 4, pp. 265-272, (1960)

[5]   Baroczy, C. J. (1966)
A systematic correlation for two-phase pressure drop.
Chem. Engng. Prog. Symp. Series, Vol. 62, No. 64, pp. 232-249,
(1966).

[6]   Bennett, A. W., Hewitt, G. F., Kersey, H. A., Keeys, R. K. F., (1968)
Heat transfer to steam-water mixtures flowing in uniformly heated
tubes in which the critical heat flux has been exceeded.
Inst. Mech. Engrs., Thermodynamics & Fluid Mech. Convention,
Bristol, Paper 27, (March 1968), 10 pages.

[7]   Blumenkrantz, A., Taborek, J. (1972)
Application of stability analysis for design of natural circulation boiling
systems and comparison with experimental data.
A.I.Ch.E. Symp. Series, Vol. 68, No. 118, pp. 136-146, (1972).

[8]   Bouré, J. (1966)
The oscillatory behaviour of heated channels. An analysis of the density
effect
Part 1. The mechanism (non-linear analysis)
Part 2. The oscillations thresholds (linearized analysis)
CEA Report R304g (Sept. 1966) 45 pages, 6 figs.

[9]   Bouré, J. A. Bergles, A. E. Tong, L. S. (1971)
Review of two-phase flow instability.
A.S.M.E. Preprint No. 71-HT-42, (1971), 20 pages.
Bouré, J.
Two-phase flow instabilities — Short course in heat transfer in
two-phase flows
Von Karman Institute (1971)
Bouré, J. and Reocreux, M.
General equations of two-phase flows — Application to critical flows
and to non-steady flows
Foure All—Union Heat and Mass Transfer Conf. Minsk (1972)

[10]   Chaudry, A. B. Emerton, A. C. Jackson, R. (1965)
Flow regimes in the co-current upwards flow of water and air.

Symp on Two-Phase Flow, Exter, 1965, paper B2, 7 pages.

[11]   Chawla, J. M., Bae, K. W., Bockh, P. V. (1972)
Sonic and critical velocity in fluid-fluid and fluid-solid mixtures.
European Two Phase Flow Group Meeting, C.N.E.N., Rome, (June 1972), Paper No. C4.

[12]   Chisholm, D., Sutherland, L. A. (1969)
Prediction of pressure gradients in pipeline systems during two-phase flow.
Proc. Inst. Mech. Engrs., Vol. 184, Pt. 3C, pp. 24-32, (1969-70)

[13]   Chisholm, D., (1969a)
Designing for two-phase flow: report of a meeting at NEL, 17th January, 1968. Part IV-Prediction of pressure losses at changes of section, bends and throttling devices.
N.E.L. Report No. 388, (Jan. 1969), 13 pages, 7 Figs.

[14]   Chisholm, D., (1969b)
Theoretical aspects of pressure changes at changes of section during steam-water flow.
Nat. Engng. Lab. Report No. 418, (June 1969), 14 pages, 5 Figs.

[15]   Cousins, L. B., Hewitt, G. F., (1968)
Liquid phase mass transfer in annular two-phase flow: droplet deposition and liquid entrainement.
AERE—R 5657, (March 1968).

[16]   Croix, J., Fabrega, S. (1972)
Determination of the hydrodynamic and thermal conditions for a boiling mixture in an industrial circuit. (In French, English Abstract).
European Two Phase Flow Group Meeting, C.N.E.N., Rome, (June 1972), Paper No. C6, 9 pages, 10 Figs.

[17]   Cruver, J. E., Moulton, R. W., (1967)
Critical flow of liquid-vapor mixtures.
A.I.Ch.E.J., Vol. 13, No. 1, pp. 52-60, (January 1967).

[18]   Davies, A. L., Potter, R. (1967)
Hydraulic stability. An analysis of the causes of unstable flow in parallel channels.
Symp. on Two Phase Flow Dynamics, Eindhoven, Session IX, Paper No. 9.3, (Sept. 1967), 34 pages, 9 Figs.

[19]   Delhaye, J., (1969)
        Application of non-equilibrium thermodynamics to two-phase flows
        with a change of phase.
        C.E.N. Grenoble, Report No. CEA—R—3903, (1969), 26 pages.

[20]   Dukler, A. E., Wicks, M., Cleveland, R. G. (1964)
        Frictional pressure drops in two phase flow. An approach through
        similarity analysis.
        A.I.Ch.E.J., Vol. 10, pp. 44, (1964).

[21]   Geiger, G. E., Rohrer, W. M., (1966)
        Sudden contraction losses in two-phase flow.
        J.Heat Transfer, Vol. 88, pp. 1-9, (Feb. 1966).

[22]   Gill, L. E., Hewitt, G. F., Lacey, P. M. C., (1963)
        Sampling probe studies of the gas core in annular two-phase flow. Part
        2. Studies of the effect of phase flow rates on phase and velocity distribu-
        tion.
        U.K.A.E.A., Report No. AERE—R 3955, (June 1963), 17 pages, 25
        Figures, 9 tables.

[23]   Gorman, D. J., Cali, G. P., Grillo, P., Testa, G. et al., (1970)
        Flow induced vibrations.
        Trans. Amer. Nucl. Soc., Vol. 13 No. 1, pp. 333-338, (1970).

[24]   Grant, I. D. R., Murray, I., (1972)
        Pressure drop on the shell-side of a segmentally baffled shell-and-tube
        heat exchanger with vertical two-phase flow.
        Nat. Engng. Lab., U.K. X Report No. NEL—500, (Feb. 1972), 23 pages,
        13 Figs.

[25]   Greskovich, E. J., Shrier, A. L., (1971)
        Pressure drop and holdup in horizontal slug flow.
        A.I.Ch.E. Journal, Vol. 17, No. 5, pp. 1214-1219, (Sept. 1971).

[26]   Griffith, P., Wallis, G. B., (1959)
        Slug flow.
        Mass. Inst. Technol. , Tech, Report No. 15, (May 1959), 20 pages, 15
        Figs.

[27]   Hewitt, G. F., Roberts, D. N., (1969)
        Studies of two-phase flow patterns by simultaneous x-ray and flash
        photography
        U.K.A.E.A., Report No. AERE—M 2159, (February 1969), 10 pages,
        16 Figs.

[28]   Hewitt, G. F., (1971)
Annular two-phase flow in tubes.
Heat Exchangers Conf., Paris, (June 1971), Paper 2, 14 pages.

[29]   Hewitt, G. F., (1972)
Entrainment mechanisms in annular flow.
European Two Phase Flow Group Meeting, C.N.E.N., Rome, (June 1972), Invited Lecture No. 1, 20 pages, 10 Figs.

[30]   Hewitt, G. F., (1970)
Disturbance waves in annular two-phase flow.
Proc. Inst. Mech. Engrs., Vol. 184, Pt. 3C, pp. 142-150, (1969-70)

[31]   Hutchinson, P., Hewitt, G. F., Dukler, A. E., (1970)
Deposition of liquid or solid disperisions from turbulent gas streams: A stochastic model.
U.K.A.E.A., Report No. AERE–R 6637, (Dec. 1970), 31 pages, 14 Figs.

[32]   Hutchinson, P., Whalley, P. B., (1972)
A possible characterisation of entrainment in annular flow.
U.K.A.E.A., Report No. AERE–R 7126, (May 1972), 7 pages, 1 Fig.

[33]   Lockhart, R. W., Martinelli, R. C., (1949)
Proposed correlation of data for isothermal two-phase, two-component flow in pipes.
Chem. Engng. Progr., Vol. 45, No. 1, pp. 39-48, (1949).

[34]   Martinelli, R. C., Nelson, D. B., (1948)
Prediction of pressure drop during forced-circulation boiling of water.
Trans. A.S.M.E., Vol. 70, pp. 695-702, (August 1948).

[35]   Murdock, J. W., (1962)
Two-phase flow measurement with orifices.
J. Basic Engng. Vol. 84, pp. 419-433, (Dec. 1962).

[36]   Premoli, A., Di Francesco, D., Prina, A., (1971)
A dimensionless correlation for determining the density of two-phase mixtures. (In Italian).
La Termotecnica, Vol. 25, No. 1, pp. 17-26, (1971).

[37]   Scott, D. S., (1964)
Properties of cocurrent gas-liquid flow.
Advances in Chem. Engng., Vol. 4, (1964), 98 pages, 14 Figs., 116 refs.

[38] Sheldon, C. W., Schuder, C. B., (1965)
Sizing control valves for liquid-gas mixtures.
Instrum. and Control Systems, Vol. 38, No. 1, pp. 134-137, (1965).

[39] Shellene, K. R., Sternling, C. V., Church, D. M. et al. (1968)
Experimental study of a vertical thermosyphon reboiler.
Heat Transfer-Seattle, Chem. Engng. Progr. Symp. Series, Vol. 64, No. 82, pp. 102-113, (1968).

[40] Smith, R.V., Cousins, L. B., Hewitt, G. F., (1968)
Two-phase two-component critical flow in a venturi
U.K.A.E.A. Report No. AERE—R 5736, (December 1968), 27 pages, 24 Figs., available from H.M.S.O. Price 8/-net.

[41] Smith, R. V. (1971)
Two-phase, two-component critical flow in a venturi.
Trans. A.S.M.E., J. Basic Engng., Vol. 94, pp. 147-155, (1972).

[42] Stenning, A. H., Martin, C. B., (1967)
Analytical and experimental study of air-liftpump performance.
A.S.M.E., Paper No. 67-WA/FE—1.

[43] Vance, W. H., Moulton, R. W., (1965)
A study of slip ratios for the flow of steam-water mixtures at high void fractions.
A.I.Ch.E.J., Vol. 11, No. 6, pp. 1114-1124, (Nov. 1965)

[44] Wallis, G. B., (1970)
Annular two-phase flow. Part 1: A simple theory. Part 2: Additional effects.
J. Basic Engng., Vol. 92, No. 1, pp. 59-82, (March 1970)
Wallis, G. B., and Heasley, J. H.
Oscillations in two phase flow options
Journal of Heat Transfer 83 - C3 - 1961

[45] Yih, T. S., Griffith, P. (1968)
Unsteady momentum fluxes in two-phase flow and the vibration of nuclear reactor components.
Mass Inst. Technol. Report No. DSR 70318-58, (Nov. 1968), 130 pages.

[46] Yih, T. S., Griffith, P., (1970)
Unsteady momentum fluxes in two-phase flow and the vibration of nuclear system components.

Argonne National Lab., Report No. Anl-7685, (1970), pp. 91-111.

[47]   Zuber, N., Findlay, J. A., (1964)
Average volumetric concentration in two-phase flow systems.
A.S.M.E., Annual Meeting, (dec. 1964), 48 pages, 14 Figs.

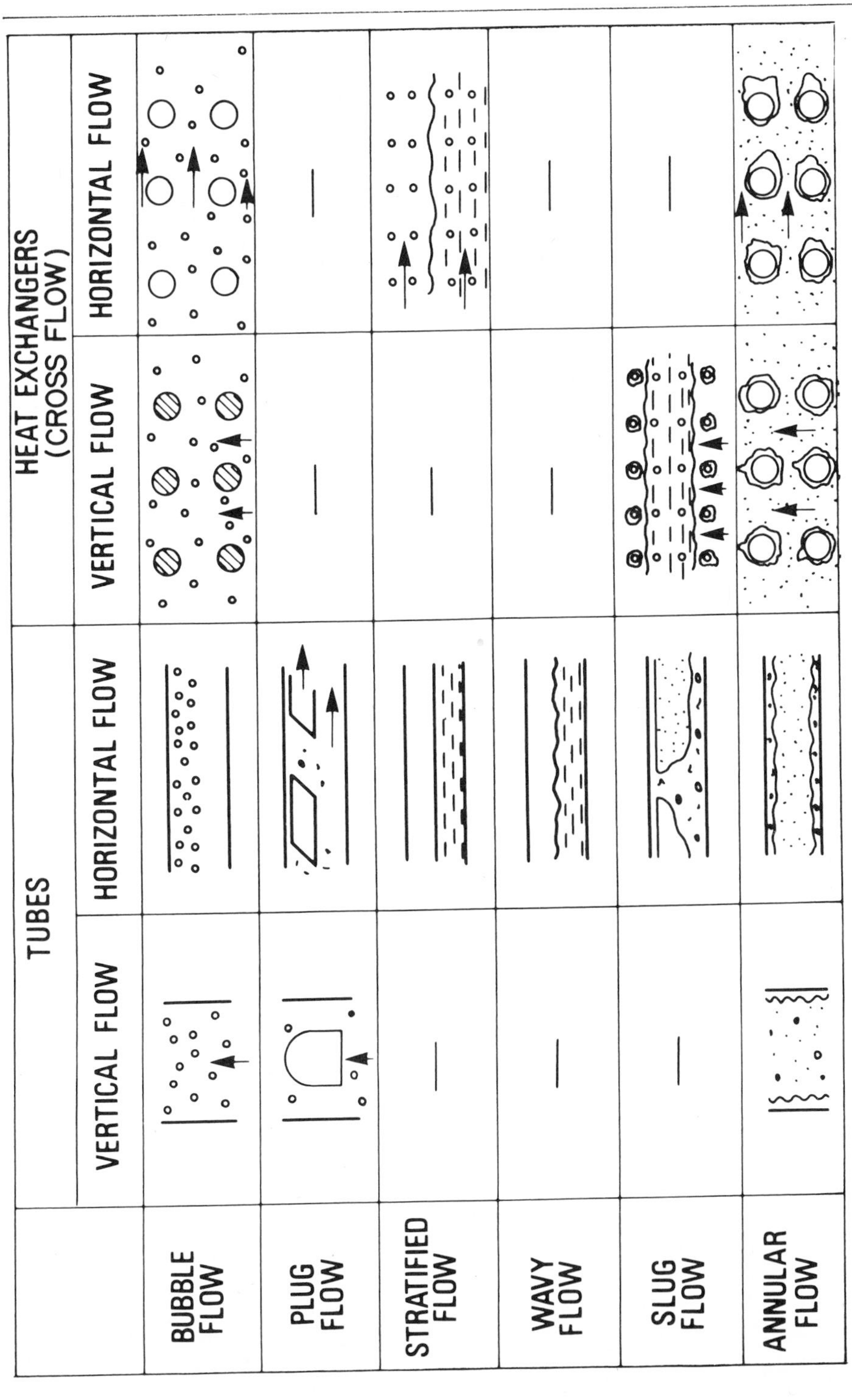

Fig. 1. Flow Regimes.

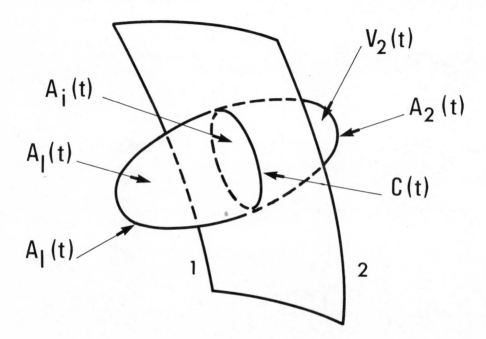

Fig. 2. The Fixed Non-Material Control Volume.

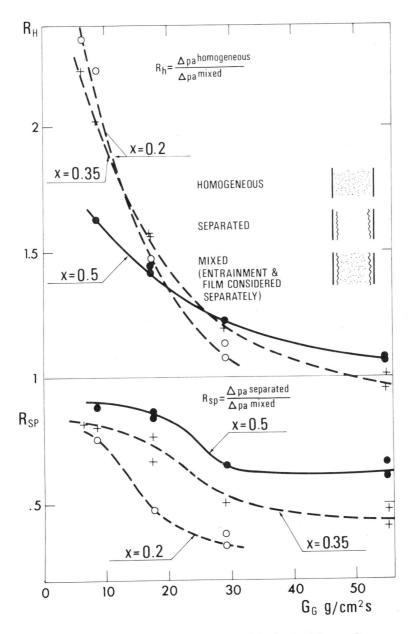

Fig. 3. Comparison of Models for Calculation of Accelerational Pressure Drop.

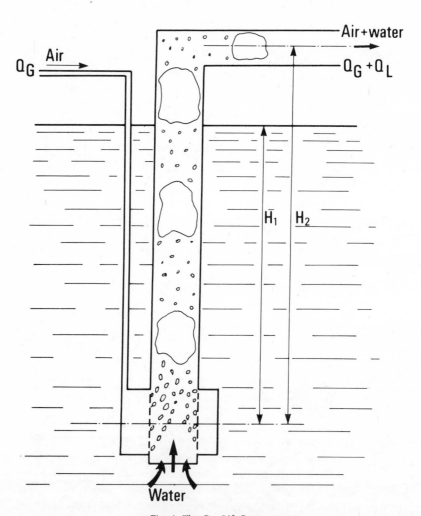

Fig. 4. The Gas Lift Pump.

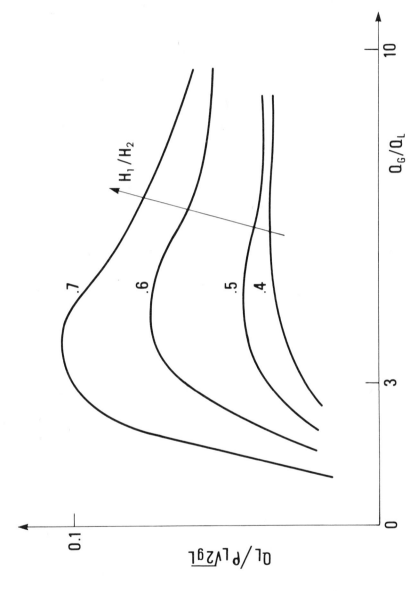

Fig. 5. Performance of a Gas Lift Pump. Results of Stenning and Martin (1968) for Water and Air.

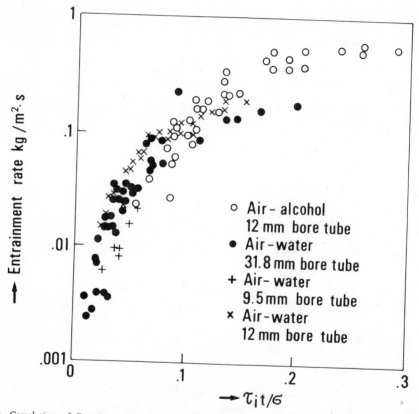

Fig. 6. Correlation of Entrainment Rate in Annular Flow (Hutchinson and Whalley, A.E.R.E. R 7126, 1972).

Symbol	Ratio of liquid to gas density	Orientation of Header	Orientation of Branch	Ratio of gas to liquid volume flow	Mass Velocity
●	690	Horizontal	Horizontal	13	Medium
△	690	Horizontal	Horizontal	460	Low
○	390	Vertical	Horizontal	1.9	High

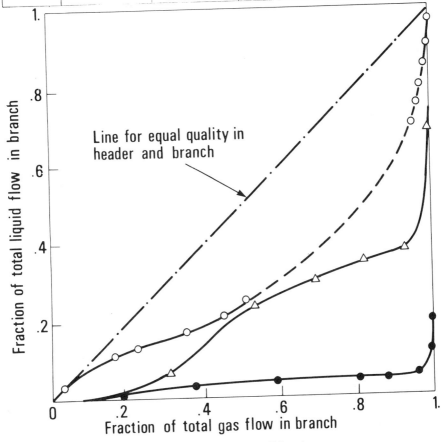

Fig. 7. Two-Phase Flow Split at a T-Junction.

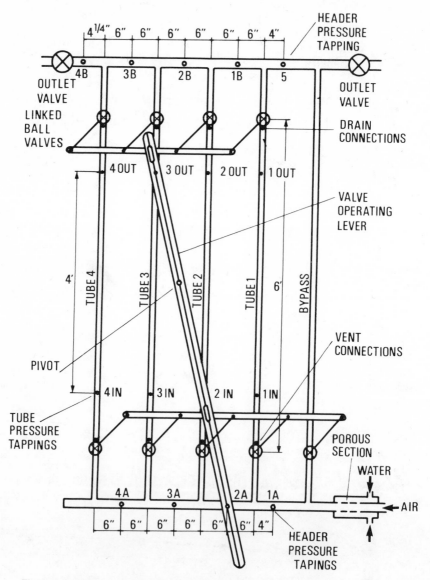

Fig. 8. Apparatus for Studying Two Phase Flow Distribution in Four Parallel Tubes.

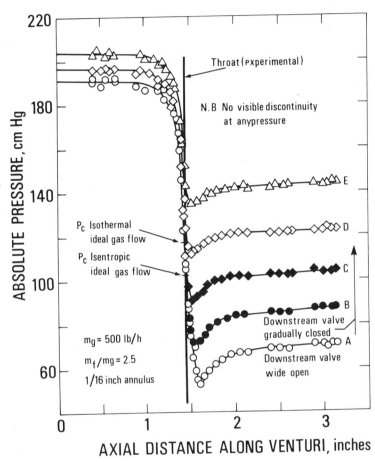

Fig. 9. Critical Flow Pressure Profiles for Annular Venturi Showing Influence of Downstream Pressure on Upstream Profile Smith et.al. (1967).

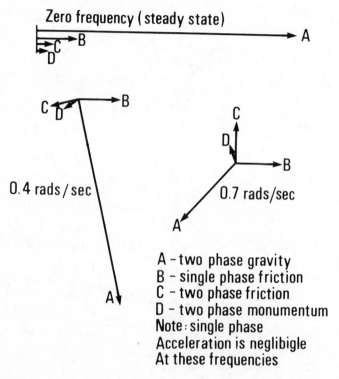

Fig. 10. Components of Pressure Drop in Steady State and Oscillatroy State Once-Through Boiler [Example Described by Davies & Potter (1966)].

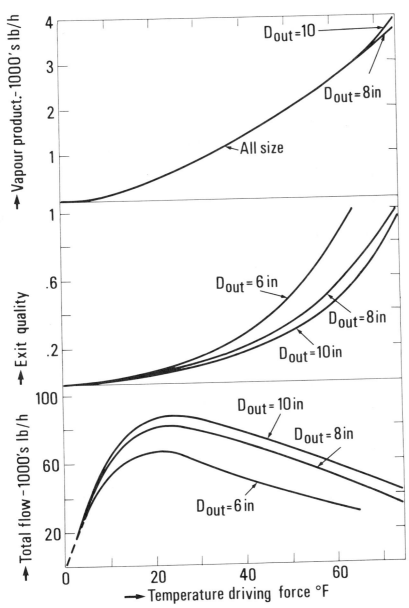

Fig. 11. Calculations Using the HTFS Computer Code on the Effect of Varying the Temperature Driving Force and the Size of the Outlet Nozzle in a Thermosyphon Reboiler.

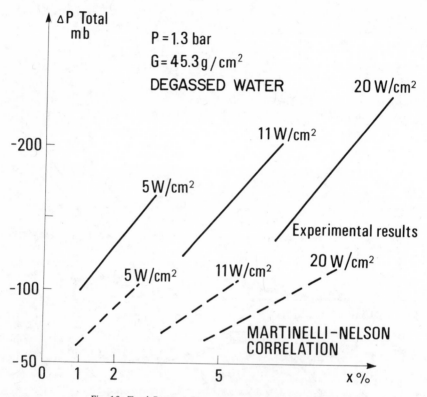

Fig. 12. Total Pressure Drop Along the Boiling Length.

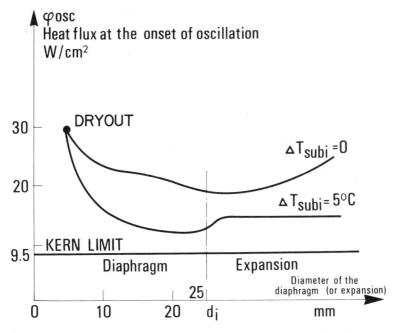

VERTICAL TUBE
$d_i = 24.85$mm
$L_{heated} = 2.9$m
$P_i = 1.3$ bar

Degassed water

φosc
Heat flux at the onset of oscillation
W/cm^2

DRYOUT

$\Delta T_{subi} = 0$

$\Delta T_{subi} = 5°C$

KERN LIMIT

Diaphragm   Expansion

Diameter of the
diaphragm (or expansion)

Fig. 13. Influence of Inlet Restriction (or Expansion) on Maximum Heat Rate.

Chapter 13

ASPECTS OF HEAT TRANSFER IN TWO-PHASE GAS-LIQUID FLOW

R. Semeria(*) and G.F. Hewitt (**)

1. BOILING

1.1. Fundamental aspects

1.1.1. Thermodynamics of phase change

In condensation and in evaporative heat transfer, the most important examples of heat transfer in two-phase flow, we are concerned with a change of phase. Before considering the heat transfer mechanisms associated with this phase change, it is useful to examine briefly the thermodynamic equilibrium situation between two phases and the occurence of departures from this thermodynamic equilibrium in the case of real heat transfer situations.

The situation at thermodynamic equilibrium is illustrated in Figure 1. In the case of the curved interface, represented as the case of a bubble in equilibrium with a liquid, the pressure inside the bubble is higher than that of the surrounding liquid. Furthermore, the vapor pressure over the curved surface is less than the vapor pressure over a planar interface (though the difference is often small). The equilibrium temperature for the curved interface is higher in the case of bubbles and lower in the case of droplets, than the saturation temperature for a planar interface at the pressure of the continuous fluid. Since the bubble of a given size will not grow until the temperature of its surroundings exceeds that for equilibrium, and since most practical situations involve curved interfaces, there is a departure from thermodynamic equilibrium as defined for the planar interface.

In addition to the curvature considerations listed above, there are other

(*) Commisariat à l'Energie Atomique, Grenoble, France

(**) United Kingdom Atomic Energy Authority, Harwell, United Kingdom

N.B. ALL FIGURES QUOTED IN THE TEXT ARE AT THE END OF THE CHAPTER

reasons why a system can depart from thermodynamic equilibrium. These can be divided into two categories as follows:

"Continuum" effects: These are associated with the existence of a fluid in a single phase form at a temperature at which it would be expected to be in a two-phase form. The reason for the existence of such cases is associated with the failure of the second phase to nucleate, and the problems of nucleation will be discussed further below. Cases of such continuum non-equilibrium effects are as follows:

(1) Sub-cooled vapor without liquid droplets.
(2) Superheated liquid without bubbles.

Transient and local effects: Departures from thermodynamic equilibrium under this category include those effects where there are variations across the fluid in the channel giving local phase changes which would not be expected on the basis of the bulk fluid temperature. They also include cases where there is a tendency towards a particular state but where the finite rate at which that change can occur is limiting. Under this category one can define the following cases:

(1) Superheated vapor carrying liquid droplets.
(2) Sub-cooled liquid carrying suspended bubbles.
(3) Sub-cooled boiling in which the wall temperature is higher than the bulk temperature and boiling can occur locally adjacent to the heated surface.
(4) Condensation from superheated vapor, which is the opposite of case (3).

The general conclusion is that careful account should be taken of non-equilibrium effects and that these can often play a dominant role.

1.1.2. Nucleation

1.1.2.1. Homogeneous nucleation

Saturation conditions are usually given for a plane interface, assuming equilibrium between the liquid and the vapor phase. If no interface preexist, liquid (or vapor) does not "know" that vapor (or liquid) may exist and high disequilibrium, that is high superheat for boiling, or high subcooling for condensation, can occur. The extreme limit is the spontaneous formation of nuclei, holes in the liquid, due to the fluctuation of local specific mass by thermal agitation or others energy sources such as radiations (bubble chambers). The maximum temperature T_{NH} is a little less than the critical temperature T_c far from the critical pressure, Spiegler [1963] gives

$$T_{NH} \simeq 27/32 \ T_c$$

Using statistical mechanics, one obtains the rate of formation of critical nuclei, that is nuclei which radius is high enough for a spontaneous growth

$$\frac{dn_L}{dt} = N \ \frac{k_B T}{h} \ \exp - \frac{4}{3} \ \pi \ \frac{r_b^2}{kT} \quad \text{with } r_b = \frac{2\sigma}{P_V - P_{L\infty}}$$

giving $dn_L/dt \sim 10^3$ for instance, a maximum superheat is obtained. Such theoretical results are confirmed in some organic liquids, but not with water. In most cases, the observed initial superheat is much lower and is explained by the presence of microbubbles or non-wettable areas called sites.

1.1.2.2. Heterogeneous nucleation

Similar sites on the walls or in the fluid are involved in the cavitation inception, condensation and generally in all changes of state. The nature of the sites is not completely clear; Westwater has shown that they exist preferentially at the bottom of scratches, pits, etc...

The main parameter of a site is its critical radius r_c calculated from the initial superheat ΔT_c giving the first bubble on the site. For small superheats, using the Clapeyron relation:

$$r_c = = \frac{2\sigma}{J\lambda} \ T_{sat} \ (v_V - v_L) \ \frac{1}{\Delta T_c}$$

On commercial surfaces, the distribution of sites is very sensitive to the roughness, degassing, oxidation of the wall, deposits, and impurities in the liquid so that quantitative prediction of the nucleation properties has been impossible up to now. On carefully polished and cleaned surfaces (smooth glass for instance) high superheats can be observed.

In a temperature gradient (heated wall), initial superheat of the wall at the onset of boiling are higher than in a uniformly heated liquid. The physical explanation is not clear, but models proposed by Hsu [1961] and for forced convection by Bergles and Rohsenow [1964] (fig. 2) may give some explanation of the effect. They assume that the critical bubble is near hemispheric on the wall and that the critical superheat must be reached at a distance near the top. Such models can explain nucleation suppression in thin liquid films in evaporators or in a highly subcooled flow: assuming nuclei of any dimension, the initial wall superheat for the first bubble will be

$$\Delta T_{cw} = \left(\frac{8\sigma \, T_{sat} \, (v_V - v_L) \, \phi}{\lambda \, k_L} \right)^{1/2}$$

In forced convection with a heat transfer coefficient h, the nucleation of bubbles will start when

$$\Delta T_{cw} = \frac{\phi}{h} - \Delta T_{sub}$$

If $\phi/h > \Delta T_{cw}$, nucleation will start before saturation conditions and subcooled boiling occurs.

If $\phi/h < \Delta T_{cw}$, nucleation starts in superheated liquid: flash boiling is observed and may lead to chugging (pulsating boiling) especially at low pressure.

In fact effective boiling is associated with continuous formation of bubbles from the sites. Transient temperature fields in the liquid must be considered by using a more complete description of the phenomena.

The postulation of sites increases rapidly on commercial surfaces with heat flux (or wall temperature) and pressure. An order of magnitude is given by the following formula in water:

$$\frac{N}{A} = 1,20 \; 10^{-2} \phi^2 p \qquad\qquad \phi \quad \text{in w/cm}^2$$

$$p \quad \text{in bars}$$

$$N/A \text{ in number of sites/cm}^2$$

1.1.3. Nucleate boiling

The descriptive analysis of this phenomenon is now satisfying though not very deep for each mechanism involved. Each bubble growing on the wall in leaving it near the end of growth (or condensing), pushes in front of it a part of the superheated liquid film and carries away another part of this film by the "drift" effect in its wake. After departure of each bubble, the superheated film growths again up to the critical nucleation superheat (Semeria 1963).

Due to the statistical distribution of both positions and characteristics (geometry, critical radius), of the nucleation centers on the surface, the frequency, growth velocity and departure diameter of the bubbles are very dispersed and only a qualitative or statistical description is of value. Visualization and statistical analysis of the signal delivered by a microthermocouple (Delhaye 1972) have confirmed the importance of the above-mentioned mechanisms (fig. 4).

Forster and Zuber [1955] and Rohsenow [1952] have proposed heat transfer correlations which can be used as a first approach for the prediction of the wall superheat:

Forster and Zuber correlation:

$$Nu = 0.0015 \ Re^{0.62} \ Pr^{1/3}$$

with:

$$Nu = \frac{\phi r}{k_L \Delta T_{sat}} \ , \ r = \frac{\rho_L c_L \Delta T_{sat} \sqrt{\pi a_L}}{\lambda \rho_v} \left(\frac{2\sigma}{p_v - p_\infty} \right)^{1/2} \left(\frac{\rho_L}{p_v - p_\infty} \right)^{1/4}$$

$$Re = \frac{\rho_L}{\mu_L} \left(\frac{c_L \rho_L \ \Delta T_{sat} \sqrt{\pi a_L}}{\lambda \rho_v} \right)^2 \ , \ Pr = \frac{c_L \mu_L}{\lambda_L}$$

Rohsenow correlation:

$$\frac{c_L \Delta T_{sat}}{\lambda} = C \left[\frac{\phi}{\mu_L \lambda} \sqrt{\frac{\sigma}{g(L-v)}} \right]^{0.33} Pr^n \text{ with } \begin{array}{l} n = 1 \text{ for water} \\ n = 1.7 \text{ for other fluids} \end{array}$$

C takes different values according to the fluid and wall properties: for water C varies from 0.006 to 0.020; 0.013 is the most probable value.

More recently Rohsenow and Mikic [1968] proposed a semi-empirical correlation, where wall influence is separated from the fluid properties.

$$\phi = C_W \cdot C_{1L} \cdot (C_{2L} \Delta T_{sat})^{m+4}$$

where C_W is a dimensional constant which depends on boiling surface properties and gravity, C_{1L} and C_{2L} are functions of fluid properties, and m depends on the nucleation properties of the wall.

For boiling of water at high pressure (50 - 140 bars) in pool boiling or in low quality forced convection boiling, a simple empirical formula was proposed by Jens and Lottes [1951] and is used in nuclear reactors:

$$\Delta T_{sat} = 8 \phi^{0.25} \exp - \frac{p}{62} \qquad \begin{array}{l} \phi \text{ in W/cm2} \\ p \text{ in bars} \end{array}$$

In the case of steam generating units, Thom [1965] proposed a formula giving higher superheats:

$$\Delta T_{sat} = 4.26 \phi^{0.5} \exp - \frac{p}{87}$$

Russian authors have proposed many different formulas obtained by dimensional analysis: some of them are listed in table 1 which illustrates the uncertainty in the influence of different dimensionless numbers. The same result is confirmed by the table 2 where the influence of the physical properties predicted by the various correlations, is indicated; there is a great divergence of results on the various parametric effects.

1.1.4. Nucleate boiling of multicomponent mixtures

One of the most important studies in this field is that of Van Stralen [1970] . For instance, small concentrations of a miscible liquid in water can change the nucleation properties of the wall, principally by decreasing the growth rate and departure diameter of the bubble, but also by increasing their frequency of formation. Heat transfer coefficient and maximum heat flux are affected. Feldkamp [1970] compares different formulas to experimental results for some aqueous mixtures. The main interest in adding small amounts of a second component is that the maximum heat flux can be increased up to a factor of 3 compared to water. However, the result can be very sensitive to the concentration, fouling, etc... Petukhov [1969] has shown that boiling of dissociative liquid $(N_2 O_4)$ is similar to that of binary mixtures.

For pool boiling of immiscible or partly miscible liquid-liquid systems Westwater [1970] gives eleven possible combinations of heat transfer regimes and has studied five of them.

$$Nu_* = A \, Pr^{n_1} \, Pe_*^{n_2} \, K_p^{n_3} \, K_t^{n_4} \, Ar_*^{n_5}$$

Authors	A	n_1	n_2	n_3	n_4	n_5
M.A. Kichigan and Yu. Tobilevich	$1.04.10^{-4}$	0	0.7	0.7	0	0.125
S.S. Kutateladze	$7.0.10^{-4}$	-0.35	0.7	0.7	0	0
V.M. Borishanskiy and F.P. Michenko	$8.7.10^{-1}$	0	0.7	0.7	0	0
G.N. Kruzhilin and Ye. K. Averin	0.082	-0.5	0.7	0	0.377	0
D.A. Labuntsov	0.125	-0.32	0.65	0	0.35	0

$$Nu_* = \frac{h}{k} \left[\frac{\sigma}{g \, (\rho_L - \rho_v)} \right]^{1/2}$$

$$Pe_* = \frac{\phi}{a_L \rho_v \lambda} \left[\frac{\sigma}{g \, (\rho_L - \rho_v)} \right]^{1/2}, \quad a_L = \frac{k_L}{\rho_L c_L}$$

$$Kp = \frac{p}{[g \, \sigma \, (\rho_L - \rho_v)]^{1/2}}$$

$$K_t = \frac{(\rho_v \lambda)^2}{J \, c_L T_{sat_1} [g \, \sigma (\rho_L - \rho_v)]^{1/2}}$$

$$Ar_* = \frac{g}{v^2} \left[\frac{\sigma}{g \, (\rho_L - \rho_v)} \right]^{3/2} \left(1 - \frac{\rho_v}{\rho_L} \right)$$

Table 1 from Kutateladze [1963]

TABLE 2

$$\phi = C\, x_i^{n_i}$$

X_i	ΔT_{sat}	ΔP_{sat}	λ	σ	β	ρ_L	ρ_v	$\rho_L - \rho_v$	μ_L	k_L	c_L	g
Cryder 1932	3.39	0	0	-1.65	0	3.1	0	0	3.45	2.97	0.43	
Bonilla 1941	3.7	0	-2.7	-1.85	0	1.85	2.7	0	1.85	1.83	1.85	
Jakob Rohsenow n 1,7 1952	2.1 / 3	0 / 0	-0.85 / -2	-1.6 / -0.5	0 / -1	2.6 / 0	-1.6 / 0	0 / 0.5	0 / -4.1	1.6 / 5.1	1.6 / -2.1	
Forster-Zuber 1954	1.24	0.75	-0.24	-0.5	0	0.49	-0.24	0	-0.29	0.79	0.45	
Forster Greif 1959	1	1.4	0.8	-0.8	-1	0.4	0	2	0.13	0.87	0.55	
Rohsenow Mikich 1968	m 1 3.5+ ~	0	m-1.7	-(m-1.7)	0	1.96	m-1.7	-0.575	0	0.5	2.2	0.575

In conclusion the physical mechanisms involved in boiling of mixtures are the same as in boiling of pure liquids but it is necessary to take into account the variation of physical parameters such as surface tension, contact angle and increase of the concentration in the less volatil liquid near the interfaces.

Most results are needed in forced convection before any definitive conclusion can be reached.

1.1.5. Interface heat transfer

In addition to nucleate boiling as discussed above, evaporation of a liquid can also take place without nucleation by a process of conductive and convective transport of heat through the liquid phase to an existing interface leading to evaporation of that interface. An example of this kind of process would be evaporation from the falling liquid film and the heat transfer mechanisms are closely analogous to those for condensation heat transfer under the same hydrodynamic conditions. Often, this type of heat transfer is associated with the existence of liquid films on the heat transfer surfaces and is sometimes referred to as "filmwise" heat transfer. However, similar types of heat transfer can occur for other interfacial distributions (flow patterns) and we prefer the name "interface heat transfer" as indicating the generic type of process in which evaporation occurs directly at a pre-existing interface.

A fairly large amount of work has been done on prediction of interface (filmwise) heat transfer and typical examples of solutions are as follows:

Condensation and evaporation in a falling liquid film (Nusselt laminar flow solution):

$$ h = \frac{K_L}{m} = \left(\frac{\rho_L g K_L^3}{3 \mu_L Q} \right)^{1/3} $$

Annular flow with a laminar liquid film and an effectively constant shear stress across the film:

$$ h = \frac{K_L}{m} = \left(\frac{K_L^2 \tau_i}{2 \mu_L Q} \right)^{1/2} $$

Annular flow with a turbulent film. In this case, it is necessary to integrate temperature profile through the film using an effective thermal conductivity which varies with distance from the wall. Semi-empirical equations can be used for evaluation of the effective conductivity and these are discussed in detail by Hewitt and Hall Taylor (1970). The results can be expressed in the form:

$$\text{Nu}_f = \frac{hm}{k_L} = fu\ (\text{Pr},\ \text{W}^+)$$

and data for the case of effectively constant shear stress in the film are illustrated in Figure 5. It will be seen that the value of Nu_f tends to unity at low values of W^+ as might be expected due to the approach to laminar flow.

For evaporative systems, the theoretical models tend to overpredict the heat transfer coefficient, possibly due to averaging effects associated with the occurrence of interfacial waves. For falling film condensation, the results appear to be closer to the theoretical models. Alternative empirical expressions for evaporation are discussed in sections 2.2.2. below.

1.1.6. Inter-related heat and mass transfer mechanisms

In two-phase heat transfer, one is normally concerned with inter-phase mass transfer simultaneous with the heat transfer. The subject area is a very wide one and it is obviously beyond the scope of this present paper to deal adequately with it. We shall merely illustrate some of the problems by discussing briefly the following topics: Interface temperature drop, Effect of incondensible gases, The influence of multi-employment mixtures and, finally, Condensation with imiiscible liquids.

When a substance evaporates or condenses, a driving force is necessary to cause the net deposition or evaporation of molecules at the interface and this takes the form of a small temperature difference between the vapor and the liquid at the interface. This problem is discussed in detail by Schrage [1953] and the temperature difference at the interface between the vapor and the liquid can be calculated to a first approximation from the expression:

$$\frac{\Delta T_i}{\phi} = \frac{2 - \sigma}{2\ \sigma}\ \sqrt{\frac{2\pi\ \text{RT}v}{M}}\ \frac{T_V\ (v_G - v_L)}{\lambda\ 2}$$

where T_V is the vapor temperature, R the gas constant M the molecular weight, ϕ, the heat flux, σ the condensation coefficient and v_L and v_G the specific volumes of the liquid and vapor phase respectively. The condensation coefficient σ represents the ratio of the number of molecules which impinge on the interface per unit area per unit time and "stick," to the total impingement rate. It will be seen that a low value of σ implies a high value of ΔT_i. There has been much discussion in the literature of the magnitude of σ; earlier measurements on water indicated a value of about 0.04 and ΔT_i would, in this case, become very significant, particularly at low pressures where the term $(v_G - v_L)$ is high. However, more detailed measurements (e.g. Mills and Seban [1967] gave indications that the condensation coefficient was much higher than had been indicated in the earlier work). A detailed study of condensation coefficients was made by Meyrial et al. [1970] . They plotted the various measurements of condensation coefficients for liquid metals as indicated in Figure 6. The coefficient shows a tendency to a value of unity as the pressure is reduced. Analysing the results statistically, Meyrial et al. suggested that the coefficient was likely to be always at unity and that the decrease at higher pressures could be ascribed to inaccuracies in the temperature measurement. They argued that there would be a rejection of all data for which the condensation coefficient was greater than unity on the grounds of physical impossibility. This would tend to lead people to reconstruct their apparatus and to swing the results in the direction of values less than unity. They present a rather convincing argument for this point of view. On the whole, it would seem from these measurements that the interfacial resistance is never likely to be very significant, though it is worth calculating from the above expression in the case of operation at low pressure, assuming a value of σ of unity.

The presence of an incondensible gas has the effect of reducing the interface temperature and thus of reducing the driving force for condensation. Due to the build up of a mass-transfer boundary layer, the concentration of incondensibles at the interface is greater than in the bulk vapor and this effect is aggravated by the effective "sweeping" of incondensibles towards the interface by the condensing vapor flow. The importance of this problem is reflected in the very large amount of literature which has been published on it. Reviews of the subject are given by Proven [1963] and by Bell and Müller [1970]. Investigations of the effect of incondensibles in essentially static vapors are reported by, for instance, Sadek [1968] and Minkowycz and Sparrow [1968] an example of investigations in a flowing system in a channel are those of Stewart and Hurd [1963]. The classical treatment of condensation in the presence of incondensible gases is that of Colburn and Hougen [1934]. Their method assumes the existence of a smooth interface in calculating the mass transfer coefficient. In fact, the interface between the liquid film and the vapor core is wavy and this gives rise to an enhancement of the mass transfer coefficient making the Colburn Hougen method conservative. However recent work by the Heat Transfer and Fluid Flow Service at Harwell has indicated that, for horizontal tubes, the Colburn Hougen method may not be conservative due to averaging effects around the periphery of the tube. A further discussion of the effects of incondensible gases in shell-and-tube exchangers is given below.

In many practical cases, condensation occurs simultaneously of several components within a mixture. This problem is reviewed extensively by Bell and Müller [1970]. For binary mixtures, the classical solution for this case is that due to Colburn and Drew [1937]. Again, this method may be conservative due to interfacial roughness effects. For more than two components, other effects such as irreversible thermodynamic phenomena may be important.

In the boiling of multi-component mixtures, similar mass transfer effects occur giving rise to fractionation between the liquid and vapor phases. In nucleate boiling of multi-component fluids, the interface of the growing bubble tends to become denuded of the lighter phase and the boiling coefficient is often much less than would have been expected from the mean physical properties. In both evaporation and condensation, the existence of temperature differences along the interface in multi-component mixtures can cause circulation and redeposition giving rise to sometimes deleterious effects. HTFS is currently working on multi-component evaporation and is conducting both theoretical modeling and experimental measurements.

In a number of practical situations, condensation occurs from a multicomponent vapor into two or more immisicible liquids. Sometimes, the first vapor to condense forms a liquid film on the channel wall and the second vapor to condense cannot be nucleated on the liquid surface. This can lead to extensive vapor sub-cooling.

The area of multi-component evaporation and condensation is one of the most difficult and most uncertain fields in two-phase heat transfer at the present time. There would appear to be scope for much further work.

1.2. Heat transfer in equipment

1.2.1 Pool boiling
The Nukiyama curve:

Heat tranfer in equipments, where pool boiling occurs, is a function of the boiling regime: nucleate, transition or film. This is shown on the Nukiyama curve (fig. 7) where the heat flux (or the heat transfer coefficient) is plotted versus the wall superheat. Nucleate boiling was examined in a preceding chapter; in the film boiling regime, the wall is separated from the liquid by a continuous vapor layer so that the heat transfer coefficient is low, except at very high subcooling.

Correlation of film boiling in saturated liquids are numerous and mostly of the form:

$$Nu = C \cdot (Gr_v \cdot Pr_v)^{1/4} \cdot \left(\frac{\Delta h}{C_L \Delta T_{sat}} \right)^{1/4}$$

$$Nu = \frac{h \, convection \cdot D}{k_L} \; , \quad Gr_v = \frac{g \, (\rho_L - \rho_v) \, D^3}{\rho_v \, \upsilon_v^2} \; , \quad Pr_v = \frac{\upsilon_v}{a_v}$$

$\Delta h - 1/2C_v \Delta T_{sat} + \lambda^1$, the vapor properties are taken at the vapor film temperature that is:

$$\frac{1}{2} (T_w + T_{sat})$$

For horizontal tubes, Bromley [1953] gives:

C = 0.62, D = diameter of the tube

For horizontal plates facing upward, Berenson [1960] gives:

$$C = 0.425, \quad D = \sqrt{\frac{\sigma}{g \, (\rho_L - \rho_v)}}$$

For vertical tubes and plates, HSU and Westwater [1960] and also Borishankii and Fokin [1965] obtain the thickness of the vapor layer and the heat transfer coefficient by methods similar to those used for liquid films during condensation on vertical surfaces.

On the Nukiyama curve, the positions of M and m are of importance for equipments operating at constant or near constant heat flux. The values of ϕ_M and ϕ_m are well predicted for large thick walls in pool boiling with isothermal surface by hydrodynamical models (Kutateladze 1948, Zuber 1959).

$$\phi_{max} = K.\lambda \left(\rho_v \right)^{1/2} \left[\sigma g \left(\rho_L - \rho_v \right) \right]^{1/4}$$

$K \simeq 0.14$ and in the range 0.13 - 0.19 depending of the geometry and surfaces.

$$\phi_{min} \# \frac{\pi}{24} \frac{\lambda \rho_v}{\rho_L + \rho_v} \left[g \, \sigma \left(\rho_L - \rho_v \right) \right]^{1/4}$$

In subcooled liquids

$$\phi_{max} \Delta T_{sub} = \phi_{max_o} \left[1 + a \left(\frac{\rho_L}{\rho_v} \right)^m \frac{C_L \, \Delta T}{\lambda} \right]$$

$$\Delta T = \Delta T_{sub}, \quad a = 0.065, \quad m = 0.8 \quad \text{(Kutateladze)}$$

$$\Delta T = \Delta T_{sub} + \Delta T_{sat}, \quad a = 0.1, \quad m = 0.75 \quad \text{(Yvey and Morris)}$$

If the wall is very thin, (less 0.25mm for stainless steel) ϕ_M can be much lower: this is due to local dry out under the vapor patches which, for thin walls, cannot be cooled by lateral conduction. The same effect occurs in equipment in certain conditions (intermittent gas pocket for instance): looking at the Nukiyama curve, we can see for a constant heat flux heating that no dangerous stable dry out is possible if $\phi < \phi_m$; when $\phi_M < \phi < \phi_m$, an occasional dry out may lead to local overheating. With constant heat flux, transition boiling (intermittent wetting) is not stable so that an excursion to nucleate or film boiling occurs and the prediction of the wall overheating needs a more complete analysis. In fact, a constant heat flux is not possible in these transient conditions so that Nukiyama curve obtained in steady conditions is not valid. This is shown in an analysis of quenching and of boiling belong fins (Beurtheret 1970), (Kovalev and Zhukov 1972, paper H_3 of this seminar).

The hydrodynamic limit for ϕ_M can be eliminated by using liquid film on the wall instead of high depth of liquid: Kopchikov [1969] has shown that the limit was practically the Leidenfrost temperature (fig. 8) (i.e. the temperature for spontaneous dewetting of the wall by spontaneous nucleation in the liquid). Vaporization of liquid films are also of interest in heat pipes and in boilers.

At present the basic knowledge of pool boiling is near complete but more analysis is needed for transient or (and) high anisothermal conditions.

1.2.2. Channel boiling

1.2.2.1. Regimes of boiling in a channel.

In boiling a channel, there is a strong interaction between the hydrodynamics of the two-phase flow in the channel and the heat transfer behavior. Figure 9 shows a conceptual diagram of what happens. Liquid at constant velocity and temperature enters a tube and the state of flow and heat transfer in the tube for equal successive steps in input is illustrated in Figure 9A - 9U. The increment of heat flux is equal to that required to heat the input liquid to the saturation temperature at the end of the tube (X = 0). The flow patterns developed from single-phase liquid flow, through bubbly flow, slug flow, churn flow and into annular flow. The first generation of vapor takes place by nucleation at the wall and the locus of the onset of nucleation is line XX. If the heat flux is low (Figure 9B) then nucleation may be delayed beyond the point at which the thermodynamic quality, x, is zero. In this case, the wall temperature is insufficient to cause nucleation and the liquid can have a considerable bulk superheat before any vapor is formed, liquid metal systems are particularly prone to this kind of behavior which can lead to violent vapor growth and characteristic excursions when nucleation finally does occur. At higher heat fluxes, however, nucleation occurs upstream of a locus of zero thermodynamic quality due to sub-cooled boiling. Initially in the sub-cooled boiling region, the vapor bubbles tend to remain near the wall. However, as the heat flux is further increased, the bubbles detach and undergo condensation to an extent depending on the liquid bulk temperature in the core of the flow. As the fluid proceeds up the tube, further generation of vapor takes place at the nucleation centers and by direct evaporation from the interfaces. The temperature driving force necessary to transfer the heat through the liquid phase to the interface by conduction and convection, decreases as the quality increases. This means that interface heat transfer increases in importance with respect to nucleate boiling heat transfer as quality increases and nucleate boiling is pressed since the wall temperature is no longer high enough to maintain active nucleation sites. Complete suppression is indicated on Figure 9 by locus YY. In the annular flow regime, liquid is lost from the liquid film by evaporation and by entrainment. Eventually, the liquid film flowrate at the end of the channel is reduced to zero (Fig. 9L) and the wall becomes dry. For constant heat input, the temperature in the dryer region is considerably above that in the region below the burnout point. On further increase of the heat flux, the dryout point is propagated downstream as shown by the locus ZZ. In most experiments on burnout, the test is terminated by cutting off the heat flux as soon as the first temperature excursion at the end of the channel occurs. If the heat flux is high enough, the temperature rise at burnout can cause melting of the channel and a possible locus for this is illustrated by line VV on Figure 9. It should be emphasized that the position of VV with respect to Z can vary considerably with mass velocity and, in common with the other locii given in Figure 9, the actual position shown should be regarded as for illustration only. The situations illustrated in Figure 9R - 9U are physically impossible due to the melting of the tube and, to achieve burnout under the heat flux inlet flow conditions, shorter tubes must be used. It will be noted that the line ZZ crosses the lines of constant quality and the regime of flow in which burnout occurs changes from annular through to sub-cooled boiling region, the mechanism is most probably connected with the onset of film boiling rather than with the disappearance of a liquid film as in the annular regime.

1.2.2. Heat transfer in boiling channels

As different types of boiling exist in a boiling channel (see preceding chapter), we have to review the different mechanisms involved along the channel. L. S. Tong [1972] has published recently the recommended various boiling heat transfer equations for use in design.

— In subcooled boiling or bulk boiling at low quality (bubble or plug flow), convective boiling is very similar to pool boiling in the same regime and the main mechanism of heat exchange is the intermittent destruction of the thermal layer by the growth of bubbles on the wall. Another mechanism is the local thinning of the layer when bubbles are sliding along the wall during a part of their trajectory. Due to some compensation effects, for instance influence of the velocity on nucleation centers population, bubble departure diameter and frequency, etc..., heat transfer in vigorous boiling is near that in pool boiling. Much more complicated is the transition between singlephase flow and "established" boiling as is shown for instance by Ricque [1972] for highly subcooled liquids (Fig. 10). Similar difficulties are encountered in flash boiling. In both cases the origin of the unpredictable behaviour is the starting condition for the sites on the wall or in the bulk; therefore gas content and surface effects are of importance.

— In annular regime, heat is transferred from the wall to the liquid film, the flow of which is laminar or turbulent; eventually boiling occurs and bubbles exist in the film. Vaporisation takes place mainly the interface which is saturation temperature. The gas core (with or without droplets) is close to the saturation conditions except in some cases where the axial gradient of pressure is high.

— The simplest case is the heat transfer through the liquid film without nucleation (similar calculation are presented for condensation) which is reviewed briefly in 1.1.1. above and in more detail by Hewitt [1970 chapter 10]; principle of iterative analytical methods are available and can be used in computer codes for the prediction for instance of the performance of vertical film evaporators.

More direct empirical models are of interest for applications in a given range of the parameters: the local heat transfer coefficient is often related to the local quality or the Martinelli parameter:

where H_L is the heat transfer coefficient for the liquid flow flowing alone in the pipe. Collier and Pulling [1962] give:

$$a \; = \; 2.5 \;\; \text{and} \;\; b \; = \; 0.7$$

The correlation proposed by Davis and David 1964 is as follows:

$$Nu_L \; = \; 0.6 \; \rho^{*\,-0.25} \; . \; Re_L^{\,0.87} \; . \; Pr_L^{\,0.4}$$

with $\quad Re_L \; = \; \dfrac{d \; Gx}{\mu_L}$

— When nucleation occurs inside the film, a single phase flow model for the liquid film is no longer valid and the most widely used correlation is that developed by Chen [1966] for forced convective flow: the two phase heat transfer coefficient h_{DP} is assumed to be the sum of a single phase forced convection coefficient h_{FC} and a nucleate boiling coefficient h_{NB}.

$$h_{DP} \; = \; h_{FC} \; + \; h_{NB}$$

h_{FC} is graphically correlated :

$$\frac{h_{FC}}{h_L} \; = \; \text{function of} \; \frac{1}{X_{tt}} \; = \; F$$

h_{NB} is related to coefficient obtained with the Forster - Zuber correlation by a suppression factor S:

$$h_{NB} = S \cdot h_{FZ}$$

S is varying from zero to one when the product $F.Re_L$ decreases.

However the spread of data using such a representation indicates that we are far from a good physical description of the phenomena.

– Mist heat transfer in dispersed flow: No liquid film exists on the wall and the droplets are uniformly distributed in the core [Cumo - 1972]. A simple calculation procedure is to use a vapor heat transfer correlation [Tong - 1972], eventually with a corrective term taking into account the influence of the droplets [Miropolski - 1963]. A more complete analysis, proposed by Forslund and Rohsenow [1968], is more difficult to use for practical applications.

1.2.3. Burnout

The burnout phenomenon was discussed briefly in the context of the discussion of the regimes of boiling in a channel (see fig. 9). The onset of burnout can be defined to correspond to:

(a) for a temperature controlled situation (e.g. a steam-heated evaporator), an inordinate decrease in heat flux or heat transfer coefficient with a small increase in heat flux;

(b) for a heat flux controlled situation (e.g. an electrically-heated channel or a nuclear reactor), an inordinate increase in wall temperature for a small increase in heat flux.

Burnout is taken here to be synonymous with other definitions commonly used in the literature, examples of which are "critical heat flux," "boiling crisis" and "DNB."

Because of the resulting reduction in heat transfer efficiency, and because of the potentially dangerous temperature excursions which can occur, burnout is of great technological importance and has been studied over a wide range of conditions and geometries. In spite of all this work, there is still considerable uncertainty about the mechanism of burnout, particularly in the regime of negative quality (i.e. sub-cooled boiling) and low quality. More work has been done on the mechanisms in the high quality (annular flow) regions, though even there the picture is not yet completely clear and ambiguous.

The mechanisms of burnout are reviewed in detail by Tong and Hewitt.

For the sub-cooled and low quality region, they list the following possible mechanisms:

(1) over-heating due to the spurt of the dry area occurring at a nucleation site during bubble growth;

(2) crowding of bubbles at the heated surface leading to vapor blanketing. This implies the formation of a bubble boundary layer which causes inhibition of heat transfer from the wall to the mainly liquid core leading to bubble coalescence at the wall and a vapor blanketing;

(3) dryout of microfilm under a vapor clot. In this mechanism, a large bubble passes along the channel (e.g. a slug flow bubble) and there is a thin liquid film between the bubble and the channel wall. This liquid film evaporates and overheating occurs.

It seems probable that all three of the above mechanisms can occur depending on the conditions. The first mechanism is likely to happen at very high sub-cooling at the burnout condition and the second mechanism at lower sub-cooling and in the low quality region. The third mechanism is likely to be restricted to relatively low mass velocities and to low pressures where the existence of large vapor bubbles is favored.

In the annular flow regime, the mechanisms listed by Tong and Hewitt [1972] are as follows:

(1) A film-boiling transition under the liquid film of the annular flow.

(2) Instantaneous film disruption by hydrodynamic instability.

(3) Spontaneous breakdown of the liquid film due to the existence of a "minimum working rate" for the surface.

(4) Breakdown due to thermal capillarity effects. In this mechanism, variations in the interface temperature cause bulk movements of the liquid giving rise to dryout of certain areas of the surface.

(5) Equality between droplet deposition and liquid evaporation. Droplets entrained in the gas core are depositing on the surface and when the rate of evaporation exceeds that of deposition, then burnout occurs.

(6) Dryout between waves. Annular flow is characterised by the existence of large disturbance waves, and it is postulated that dryout could occur by a complete evaporation of the thin film between the waves.

(7) Liquid entrainment from the film with no subsequent deposition. This implies that the deposition process is so slow, or is inhibited, as to provide no compensation for the liquid entrainment. Burnout occurs when the film is

therefore entrained and evaporated away.

(8) Film disruption due to nuclear boiling within the film. There are two ways in which nucleate boiling bubbles could affect burnout. In the first, the bubbles could grow due to microlayer evaporation and burst through the film thus producing an incipient dry spot which has to be rewetted. Secondly, nucleate boiling bubbles leaving the film and joining the gas core tend to burst at the interface thus producing a cloud of droplets to add to those already existing from wave entrainment.

(9) Integral entrainment-deposition effects leading to zero film flow rate at the point of onset of burnout. This mechanism implies the existence of both entrainment and deposition of liquid droplets at all positions along the channel, in addition to removal of liquid from the film by evaporation, the net result being a zero film flow at the dryout point.

Of the above mechanism, a film boiling transition underneath the liquid film is contrary to evidence, complete, sudden disruption of the film does not actually take place, spontaneous breakdown does not occur without the preexistence of a dry area and breakdown due to thermal capillarity effects is unlikely to be significant in the evaporation of pure components and unlikely to be significant at anything above moderate mass velocities for multi-component systems. The mechanism involving equality between droplet deposition and evaporation may sometimes occur, though droplet deposition does not usually account for all the evaporation and this implies that there must be some feed of the liquid along the liquid film. Dryout between waves is unlikely since the large waves are suppressed near the onset of burnout. Liquid entrainment with no subsequent deposition is unlikely to occur and film disruption due to nucleate boiling giving rise to dryer patches which spread is also unlikely to be a significant mechanism. However, the influence of nucleate boiling bubbles on promoting entrainment can sometimes be important, though for tubes of lengths used in most practical applications (e.g. 2-3 meters) this mechanism is also thought to be of small importance. The net conclusion from all the evidence examined by Tong and Hewitt [1972] was that, in the annular flow regime, the mechanism was likely to be one of an integral approach to burnout through the cumulative effects of entrainment and deposition. This mechanism is illustrated in figure 11 in which data for liquid entrainment flow against local quality are shown for both the equilibrium case (adiabatic flow) and for the case of the heated tube which has burnout at the end of the tube. It will be seen that the actual entrainment curves for the heated case are displaced to the right of the equilibrium curve, indicating the lag in reaching the equilibrium. However, the entrainment is always trying to proceed to equilibrium, although it never actually does so (except by accident when the two curves cross) due to the changing quality along the tube. It is stressed that these entrainment curves result from local measurements of liquid film flow in a channel which has burnout at its end. Thus, the film flow becomes zero at the end of the channel and the entrained liquid flow equals the total flow as indicated.

Also shown in Figure 11 are data for cases when part of the tube is unheated. The unheated zone (constant quality) results in a shift towards the hydrodynamic equilibrium, giving a decrease or increase in burnout quality depending upon where the unheated zone (cold parch) is situated. This leads to the paradox that an increase in burnout power can be achieved by not heating part of the channel in the region of net deposition.

A very tentative conceptual diagram of the regions in which the various burnout mechanisms occur is given in Figure 12 as a function of mass velocity and quality. The respective regions shown there are as follows:

(I)   Deposition control. Here, the evaporation rate and droplet deposition rate are equal and liquid supply along the channel surface in the film makes a negligible contribution at the burnout point.

(II)   In this zone, the burnout quality changes only slowly with increasing heat flux. In other words, the quality at burnout is independent of tube length for uniform heating at given flow conditions. (A typical heat flux versus quality curve is shown in Figure 12A.)

(III) In this region, the burnout quality decreases more rapidly with increasing heat flux due to two probable effects:
       (a) bubble release from the liquid film causing additional entrainment;
       (b) the high liquid entrainment at the onset of annular flow is retained since there is insufficient length of tube in which to eliminate the "memory" of this early entrainment.

(IV)   In this region, burnout is postulated to occur by the boundary layer crowding mechanism mentioned above. A smooth transition between region IV and III would occur because a bubble layer would be present within the liquid film once annular flow is established (dotted line of Figure 12).

(V)   Here, the sub-cooling is high and the growth of individual bubbles within the sub-cooled boundary layer can lead to the creation of dry patches and the onset of critical heat flux by the mechanism discussed above.

(VI) The slug growth mechanism can occur in limited situations, and the region sketched in Figure 12 for mechanism VI is highly tentative.

The areas occupied by the various regimes on Figure 12 should not be taken as indicative of their relative importance. If the diagram had been plotted in terms of tube length rather than quality, then regions II and III would appear to be dominant ones for practical applications. The relative positions of the lines will also change with reduced pressure and with changes in physical properties.

The phenomena associated with burnout are so complex that it has not yet proved possible to predict the onset of burnout from theoretical or semitheoretical analyses. One has to rely, therefore, on empirical determination of data with the development of correlations for interpolation and (more dangerously) extrapolation. There are many problems associated with interpretation of data and these are reviewed in detail by Macbeth [1968]. There are now a large number of correlations available in the literature and it is difficult for the user to select one for his particular application. This difficulty is discussed by Becker [1972] who carries out selected comparisons between various correlations. A typical comparison with some "standard" data produced as a joint exercise by a number of European laboratories is shown in the following Table.

	No of Runs	RMS Error in %				
		Tong	Cise	Macbeth	Hewitt	Becker
$G < 4000$ kg/m^{2}s	340	13.45	11.83	6.99	8.41	5.08
$G > 4000$ kg/m^{2}s	24	41.51	13.58	14.75	11.85	13.75
All data	364	16.82	11.94	7.75	8.64	6.05

Becker's general conclusion is that the selection of a correlation will depend entirely on the data which are used for comparison. It will also depend on the mass velocity and pressure ranges.

It will be seen, therefore, that the state of prediction methods for dryout even for a single fluid (water) in the simplest form of channel is not completely satisfactory. There are no means at the moment for predicting the behavior of other fluids, though a great deal of work has been done on the use for Freons for scaling water systems (see for instance Stevens et al. [1964], Bouré [1970]. Another situation with great importance within the nuclear industry is that of burnout in bundles of rods. There are two approaches to this problem: overall correlations have not yet proved sufficiently reliable to avoid the need for large scale (and expensive) experimentation and the use of sub-channel analysis would appear to be more promising in the longer term. In this type of analysis, the bundle flow area is divided into a number of "sub-channels" which are considered to exchange mass with adjacent sub-channels so as to maintain an overall pressure balance. The burnout prediction is then based on correlations for burnout in sub-channels at the calculated local quality. The assumptions in the various sub-channel models are discussed by Bowring [1972]. It is clear that a much truer physical picture of what is happening within the sub-channels is urgently required and studies of the mechanisms of flow within sub-channels, the mechanisms of mass transfer between adjacent sub-channels and the mechanisms of film flow in complex geometries would appear to be very desirable.

1.2.2.4. Applications of channel boiling

Description of the evolution of a liquid along a heated channel is given in a preceding section. A classical description of the variation of the different parameters is given in Figure 13 where we can recognize the different difficulties:
— Prediction of onset of boiling; ΔT_{WC}
— Hydrodynamics of subcooled boiling: Prediction of void fraction, pressure drops and wall superheat
— Hydrodynamics of saturated boiling : idem
— Critical heating conditions where the wall becomes dry and overheated; wall temperature fluctuations
— Hydrodynamics of dispersed flow: prediction of the heat transfer coefficient and the true quality.

Depending of the heat exchanger, some of the problems are cancelled or of less importance. This is summarized in the table.

As it was already said, high coupling exists between thermal and hydrodynamical aspects so that iterative methods are necessary, in particular when rapid variations of pressure, fluid properties are occurring along the channel. Very heavy computer codes were built for nuclear applications for simple or complex channels (rod bundle: subchannel analysis) in permanent or transient conditions and are now available for other industrial equipments.

A great effort is now devoted to a better prediction of the influence of obstacles, bends, orientation: for instance the critical heat flux and the dryout limit are very sensitive to the orientation or the shape of the tube (coil, bended tube) as it is shown on figure (14).

1.2.2.5. Cross-flow boiling

There are a number of industrially important applications in which boiling occurs in cross-flow over a heated tube or bundle of tubes. This type of system is important in some kinds of steam generators, and in horizontal and kettle-type reboilers where the cross-flow is induced by circulation patterns within the shell.

Studies of boiling in cross-flow from a single cylinder by McKee and Bell [1969] show that the boiling curve for nucleate boiling is not strongly displaced from the corresponding pool boiling curve, except at low fluxes where the forced convection single-phase coefficient is greater than the pool boiling coefficient.

McKee and Bell found that the critical heat flux was strongly influenced by the existence of a cross-flow and this is in line with other data on cross-flow heat flux for single tubes and tubes within matrixes of other tubes by Kezios and Lo [1958], Kezios et al. [1961], Vliet and Leppert [1962] and Coffield et al. [1967]. Indeed, Coffield et al. showed that the critical heat flux for cross-flow in sub-cooled boiling is higher than that for the same velocity in parallel flow in a tube.

The case of cross-flow in horizontal and kettle-type reboilers is considered by Palen and Taborek [1971] and Palen and Small [1964]. Figure 15 shows the results obtained by Palen and Taborek for heat flux as a function of temperature driving force. It will be seen that the maximum heat flux is considerably below that for the single cylinder. This result is obtained since there is a restriction in circulation giving limitation in the access of the boiling fluid to the tube. Three limiting mechanisms for boiling within a bundle can thus be discerned:
(1) Pool boiling burnout. Palen and Small suggest that the system will approach the pool boiling situation when the ratio $D_s L/A$ is greater than $1/\pi$, , where D_s is the shell diameter, L the length of the bundle and A the area of the heat transfer surface within the bundle.
(2) Circulation around the outside of the bundle and then through the bundle may occur giving an increase in the critical heat flux above that for pool boiling, but then there will be a limitation also for this case and the type of mechanism involved will depend on the flow regime.

TABLE I X important
— less important

	Onset	Subcooled			Saturated		Critical		Dispersed	
	ΔT_{WC}	α	ΔP	ΔT_{sat}	α	ΔP	X_c	Fluctuat of T_W	h	ΔP
Nuclear reactors										
Nominal conditions										
Pressurized water		X	X		X	X	X			
Boiling water		X	X		X	X	X	X	X	—
Safety considerations										
P.W.R.									X	X
B.W.R.									X	X
Fast reactors (sodium)	X				X	X	X			
Research reactors (low pressure water)	X	X	X	X	—	—	X			
Steam generators										
Once through		—	—		—	X	—	X	X	X
Boilers (low pressure)	X			X	—	X	—			
Evaporators	X			X	—	X	—			X

(3) Where circulation through the bundle is small, liquid has to enter the bundle in counter-current flow against the exit vapor stream. This situation will be limited by flooding and it can be shown that a simple flooding criteria will fit the data of Palen and Taborek reasonably well.

Currently, work is being carried out at HTFS Harwell in which the various mechanisms of critical heat flux are being investigated. A two-dimensional model boiler is being used in whlich the circulation pattern and heat flux can be adjusted.

2. CONDENSATION

2.1. Filmwise condensation

The expressions for interface heat transfer, of which filmwise condensation is a case, were described above. We shall confine our discussion here to two specific aspects of the problem: the influence of condensation on interfacial shear and the regimes of heat transfer in a horizontal tube.

When vapor condenses from a fast moving vapor to a slow moving liquid film, there is a transfer of momentum to the film giving rise to an enhancement of the interfacial shear stress τ_i, with a consequent increase in the heat transfer coefficient. A treatment of this problem based on numerical integration of the boundary layer equations is given by Kinney and Sparrow [1970]. An alternative treatment in terms of the Reynolds Flux concept is described by Wallis [1969]. In unpublished work at Harwell, Butterworth has developed an equivalent to the predictions of Kinney and Sparrow and consistent with the semi-empirical method of Wallis. The ratio of the interfacial shear stress to that without mass transfer is given by the expression:

$$\frac{\tau_{iM}}{\tau_i} = \frac{\frac{2}{f_i}\frac{m}{\rho_V U_\delta}}{\exp\left(\frac{2}{f_i}\frac{m}{\rho_V U_\delta}\right) - 1}$$

where τ_{iM} is the interfacial shear stress in the presence of mass transfer, m is the rate of mass transfer (mass per unit area), f_i, τ_i the friction factor and interfacial shear stress in the absence of mass transfer, ρ_V the vapor density and U_δ the velocity of the vapor at the edge of the equivalent laminar layer.

This expression has been developed previously by a number of other authors including Mickeley et al. [1954].

In the case of horizontal flow, two extreme conditions can occur:

(1) At high vapor velocities the flow is annular and can be treated by the methods described above.

(2) At low vapor velocities, the condensate film runs down the inside wall of the tube as illustrated in Figure 16. The classical expression for the heat transfer in this situation is as follows:

$$h_f = 0.725 \ \Omega_1 \left[\frac{K_L^3 \ \rho_L \ (\rho_L - \rho_V) \, g\lambda\Omega_2}{\mu_L \ d_o \ (T_S - T_W)} \right]^{1/4}$$

where k_L is the thermal conductivity of the liquid, T_S is the saturation temperature, T_W is the wall temperature and the parameters Ω_1 is related to the angle ϕ (see Figure 16) and reflects the poor heat transfer through the pool at the bottom of the tube and Ω_2 is a factor including to account for the effect of subcooling of the liquid. The question obviously arises as to where the transition between the two expressions should be considered to occur. Bell et al. [1969] have related the transition to the flow regime map for the horizontal two-phase flow. A position is defined at which there is a transition to the annular regime. A somewhat similar approach has been adopted by HTFS in developing its computer models for condensation in tubes, though more recent work at Harwell is tending to favor a new model giving a smoother transition between the two types of condensation.
There is a need for more data on the distribution of condensation coefficient around the tube in horizontal flow.

Condensation of immiscible mixtures is more complicated and different models are proposed: film-film, film-drop, discontinuous films, etc. (Bernhardt, Sheridan and Westwater - 1972).

2.2. Dropwise condensation

Compared to the film-wise condensation, heat transfer coefficients in drop-wise condensation are 2 to 20 times higher. Several mechanisms had been proposed and now it is generally recognized that dropwise condensation is a nucleation phenomenon like boiling. Analysis of the heat transfer mechanisms shows that the heat is transferred by conduction through the growing drop before their coalescence. Analytical models have been proposed for the prediction of the local heat transfer coefficient around a growing drop, Nijaguna [1971], but for obtaining the mean value, we face to the same difficulty as for boiling: prediction of the number of nucleation centers and of the coalescence time.

To obtain dropwise condensation it is necessary to have a bad wettability of the surface by the liquid. Gold plating (0.25 micron + 6μ silver + 12μ Ni) is very efficient (fig. 17) and durable but expensive (price x 2 for an industrial application). Other promoters are proposed: some of them are thin layers of teflon or organic products (Parylene N from Union Carbide) and others are injected in the steam (octadecylamin, mercaptan, silicon oils, etc...). At the present time, the perfect promoter (non-corrosive, nontoxic, long life, high heat transfer coefficient, low cost) is not known, though many laboratories are testing new products. In the CEN Grenoble research program on desalination, encouraging results have been obtained. (J. Huyghe and F. Lauro 1972): an infinitesimal dose of a new promoter injected in the steam induces droplet condensation during 3000 h with a heat transfer coefficient comparable to the gold plating case. Such improvement is economically interesting if other thermal resistances are reduced; the heat coefficient on the water side in particular must be increased.

2.3. In tube condensation

Some aspects of in-tube condensation were examined in section 2.1. above. We shall now review the main features of the heat transfer during filmwise condensation inside tubes in the context of application in equipments. (Shell-and-tube heat exchanger, air-cooled heat exchanger, et. ... Since all combinations are possible a large number of different cases exist. Different cases are summarized in the table: The way in which condensation occurs in a cross flow system is shown in figure 18.

As in the case of steam generators, high coupling exists between hydrodynamical and thermal aspects so that the prediction of the performance needs a step by step calculation procedure along the tube from the onset of condensation to the completely flooded section (or the outlet). Such computer programs are developed, for instance by H.T.F.S. Harwell. The following features are important:

— the choice of a local heat transfer correlation must be consistent with the choices made for correlation of other parameters, pressure drop, void fraction, transition from laminar to turbulent flow.

— for low velocities, the pressure drop along the tube is negligible and the classical

Geometry (Length-shape of the section	Orientation	Flow pattern	Liquid film regime	Non-conden-sibles	Vapor Super-heat
Circular section	Vertical	Stratified or wavy (horizontal or inclined)	Laminar	YES	YES
Fluted tube (verticle)	Horizontal		Turbulent	NO	NO
	Inclined	Annular (dispersed)			
		Slug			

formula was given above for stratified flow. In vertical tube, Nusselt, formula can be used in the laminar zone. These correlations correspond to a heat transfer resistance concentrated in the liquid film of condensate.

— For high velocities the first problem is the choice of the nondimensional parameters depending on the relative influence of the gravity force compared to the forced convective effects; this is discussed for laminar flow on a vertical surface by Fujii and Uehara [1972]. With thin liquid films (inlet of the tube or (and) very high velocities) the gravity effects are low and an annular model is valuable (Traviss, Baron, Rohsenow - 1971).

A rather simple correlation was proposed by Cavallini and Zecchin [1971] for refrigerants R–11, R–21, R–114:

$$\overline{Nu} = \frac{\overline{h}\, d_o}{k_L} \, , \quad \overline{Re} = \overline{Re}_v \cdot \frac{\mu_v}{\mu_L} \cdot \left(\frac{\rho_L}{\rho_v}\right)^{0.5} + \overline{Re}_L$$

Re_v and Re_I are estimated with reference to the inside diameter of the tube d_o and as if each of the two flows are flowing alone in the whole cross section. The Nusselt number is a mean value along the test section, and is correlated with arithmetic mean between the inlet and outlet of the Reynolds number.

$$\overline{Nu} = 0.05 \, \overline{Re}^{0.8} \, \overline{Pr}_L^{0.33} \quad (15.10^3 < Re < 200.10^3)$$

In conclusion, it may be said that condensation inside tube is as boiling in steam generators, though there is no burnout phenomenon. Hydrodynamic instabilities can also occur in condensation systems. A great effort must be devoted in the future for a better prediction of the pressure drop and heat transfer in condensation equipments.

2.4. Shell side condensation

The most common method of condensing vapors is to use a shell-and-tube condenser with condensation taking place on the shell side. The two main types of condenser are the baffled shell-and-tube condenser and the unbaffled shell-and-tube condenser respectively. The baffled type is most commonly used in the process industry whereas the unbaffled type is most commonly used in Power Stations, etc.

The principal problems in the design of shell-and-tube condensers are as follows:

(1) To reduce the pressure drop. This is particularly important in Power station condensers since it can effect seriously the thermal efficiency.

(2) To avoid poor performance of the condenser due to the build-up of incondensible gases.

(3) Estimation of the filmwise heat transfer coefficient, particularly in the context of large amounts of inundation from tubes above the tube in question and the effects of vapor cross-flow on the coefficient.

Considering the case of the unbaffled condenser, the problems of designing such units are principally those of providing suitable "lanes" through which the vapor accesses the various regions of the bundle and, secondly, the removal of accumulated incondensible gases to prevent the effective poisoning of incondensibles in the inlet vapor can give rise to large deterioration in heat transfer in unbaffled condensers and the positioning of the vent lines is of paramount importance.

Work on flow and heat transfer in unbaffled condensers is proceeding at the National Engineering Laboratory in England. This work is reviewed in a recent paper by Wilson [1972] and is proceeding on two main fronts:

(1) The use of an experimental unbaffled condenser in which the performance of

various vent arrangements can be assessed in terms of local concentration measurements of the air. Typical results from this condenser are shown in Figure 19 for various air leakage rates into the inlet. The very high concentrations of air obtained in the center zones of the condenser will be observed.

(2) The development of two-dimensional computer programs for the prediction of flow and heat transfer in the presence of inert gases. Wilson [1972] gives results which illustrate that these computer models are now capable of predicting many of the concentration effects which are observed in unbaffled condensers but, at the present time, the computer times involved in these calculations are so large as to make their use difficult in practice for design work.

The effects of inundation and vapor cross-flow on condensation are likely to be rather complex. The effect of inundation of condensate is to reduce the condensing coefficient, but the reduction is very much less than would have been expected theoretically from the expected accumulation of condensate. This is partly due to the fact that the "dripping" action generates turbulence and also the fact that splashing occurs such that the condensate does not all fall onto the next tube down. An empirical correlation for inundation effects is given by Grant and Osment [1968]. The effect of vapor velocity on condensation of the outside of a tube has been investigated, for instance, by Berman and Tumanov [1962] and it is generally accepted that increasing the vapor velocity will increase the condensation heat transfer coefficient. Detailed work carried for HTFS (*) at A.E.E. Winfrith, (England) has shown that the coefficient does increase with vapor velocity under some circumstances, but that there may be conditions under which it actually decreases.

For the past three years, HTFS has had a program of work on baffled shell and tube condensers. The types of arrangement commonly found in baffled condensers are illustrated in Figure 20. The baffled arrangement has the advantage of promoting a one-dimensional flow and reducing the possibility of large areas of the bundle being poisoned by incondensibles. However, the effect of incondensible gases is still significant. Experimental work on a baffled shell and tube condenser carried out at Harwell has shown, for instance, that a concentration of 7% by volume of nitrogen in the input vapor stream can reduce the overall heat transfer coefficient by the order of a factor of two (this implies a very much greater

(*) Heat Transfer and Fluid Flow Service, Harwell, England.

reduction in the shell coefficient).

Prediction methods for shell and tube condensers have been developed for HTFS by the National Engineering Laboratory at East Kilbride. The programs can handle up to four levels of condensers, each level feeding vapor to the following level. Each level consists of any number of parallel condensers with various shell side arrangements and up to 16 passes on the tube side. Any vapor non-condensible gas and coolant combination whose properties are known can be calculated. The program determines the condensation rate for various combinations of inlet and outlet conditions and flowrates. The condenser is divided into sections separated by the baffles and an average heat transfer coefficient is assigned to each section. This heat transfer coefficient takes account of the mass transfer of the vapor through an incondensible gas layer and the effects of inundation, vapor share and leakage and by-pass are taken into account. The temperature profiles on the coolant side are estimated by an iterative process such that the outlet temperature of pass n is equal to the inlet temperature of pass (n + 1).

The powerful tool that this program represents is best illustrated by an example. It is assumed that a condenser is to be designed to condense 4.0kg/sec of butane which contains 0.5% by weight of methane. It is assumed that none of this methane condenses. Two alternative arrangements of shells are compared:
(1) Four shells in parallel, each 3.0 meters long having 10 equal sections divided by baffles; these shells feed to parallel similar condensers.
(2) In the second arrangement, the velocity of the vapor has been maintained by the use of a non-uniform baffle spacing. The first level condensers are 2.5 meters long and feed a single second level condenser 3.0 meters long containing 21 non-uniform sections.

The computer calculation indicates that the performance of the two arrangements is virtually identical and the pressure drops small. There has, therefore, been a saving of 36% of the required surface area by using a non-uniform baffle spacing, resulting in five rather than six shells being required, four of them being shorter than the originals.

The computer prediction methods described above are relatively complex and there is still a place for rapid estimation methods in the calculation of a "first guess" of the design required. Such methods are discussed, for instance, by Gloyer [1970] and by Bell and Muller [1970].

An interesting side-light on the effect of non-condensible gases is described by Wild [1969]. In the absence of incondensibles, and in the presence of a layer of sub-cooled liquid at the bottom of the shell, Wild found that there were vibrations within the shell which caused mechanical damage. He ascribed these to oscillations in the sub-cooled liquid level caused by the following mechanism: a small surge in the vapor causes a displacement of the liquid level downwards which frees surface for condensing causing a rapid depressurisation and a rise of the liquid level. This overshoots, reduces the amount of surface available for condensation and causes the vapour to push the liquid down again. The addition of a small amount of incondensible gas completely eliminated this effect since, during very rapid condensation, the incondensibles have a very severe restrictive influence, though in the average condensation rate of the condenser, their influence was small.

It would seem that the best approach to predicting the performance of shell and tube condensers is to use programs which integrate the appropriate local effects. However, much further work is required in confirming the validity of this integration and in throwing clearer light on the detailed local parameters.

3. CONCLUSION

General conclusions on two-phase flows are given in the preceding chapter so that the following ones are confined to the heat transfer aspects.

The present knowledge of the phenomena of boiling and condensation is qualitatively satisfactory in spite of some irritating questions such as the nature of sites, the role of interfacial parameters. Therefore models can be proposed for the description of a simple (or simplified) heat transfer equipment. The main difficulties come from the complexity of the computational procedures due to the coupling between hydrodynamics and heat transfer and from the great number of experimental results needed for the adjustment of the empirical correlations used in the models. This is the same reason which limits the local analysis of two-phase flow in complex geometries or (and) in rapid transient conditions.

New devices or running conditions are of interest (Semeria [1972]): heat pipes, that is two phase flows in porous media, anisothermal boiling on complex walls (multilayers, corrugated, etc...), high speed two-phase flows,... We are far from the end!

NOMENCLATURE

A	Cross-sectional area of the channel
c_p	Specific heat
d_o	Tube diameter
D	Diffusivity
g	Acceleration due to gravity
G	Mass flux
h	Planck constant
h	Heat transfer coefficient
k	Conductivity
k_B	Boltzmann constant
l	length
m	Film thickness
M	Molecular weight
dn/dt	Number of nuclei per unit volume and per unit time
N	Number of molecules per unit volume of liquid
P	pressure
Pr	Prandtl number $Pr = C_p \mu / k$
Q	Volumetric flowrate per unit perimeter
r	Radius
r_b	Equilibrium radius of a vapor bubble
r_c	Critical radius
R	Gas constant
Re	Reynolds number
t	Time
T_c	Critical temperature
T_{NH}	Temperature of homogeneous nucleation
T_{sat}	Saturation temperature
ΔT	Temperature difference
ΔT_{sat}	Degree of superheat
ΔT_{sub}	Subcooling $\Delta T_{sub} = T_{sat} - T_L$
u	Velocity
v	Specific volume
W	Mass flow rate

W^+	Dimensionless mass flow rate $W^+ = W_{LF}/2\pi r_0 \mu_L$
x	Quality
X	Martinelli parameter $X^2 = [(dp_F/dz)_L/(dp_F/dz)_G]$
X_{tt}	Martinelli parameter for turbulent liquid and gas phases
α	Void fraction
λ	Latent heat of vaporisation
υ	Kinematic viscosity
ρ	Density
ρ^*	Gas to liquid density ratio $\rho^* = \rho_G/\rho_L$
σ	Condensation coefficient
σ	Surface tension
τ	Shear stress
τ_i	Interfacial shear stress
τ_o	Wall shear stress
ϕ	Heat flux
μ	Viscosity

Indices

L	Liquid
v	Vapor
Bo	Burnout

BIBLIOGRAPHY

Books, Proceedings etc...

Borishanskii V.M. and Paleev I.I.
Convective heat transfer in two-phase and one phase flows.
(English translation), Israël. Program for scientific translations. Jérusalem 1969.
(Isdatel'stvo Energiya - Moskva 1964)

Collier J.G.,
Convective Boiling and Condensation
McGraw Hill 1972.

Hewitt G.F. and Hall Taylor N.S.
Annular two phase flow
Pergamon Press, 1970.

Kutateladze S.S.
Fundamentals of heat transfer
(English translation). Edward Arnold 1963.

Rhodes E. and Scott D.S.
Cocurrent gas-liquid flow
Plenum Press, New York, 1969.

Rohsenow W.M.
Developments in Heat transfer,
M.I.T. Press 1964.

Tong L.S.
Boiling heat transfer and two-phase flow, John Wiley, 1965.

Advances in Heat Transfer, Contribution on boiling in volume 1 and 6, 1970, Academic Press

Advances in Chemical Engineering, Academic Press, 1956 and 1958 (Contributions of J.W. Westwater).

Fourth International Heat Transfer Conference, Paris – Versailles 1970, Volumes V and VI.

Chem Ing. Technik 11, Nov. 1963: papers from W. Fritz, H. Brauer, K. Stephan, W. Kast (in German).

Journées Internationales sur les échangeurs de chaleur.
Institut français des combustibles et de l'énergie and institute of fuel. Paris, Juin, 1971 (in French and English).

REFERENCES

Becker K.M., Badger J., and Djursin D.
"Burnout correlations in simple geometries; more recent assessments".
Paper presented at the Seminar on Two-Phase Flow Thermal Hydraulics organised by the Italian National Commettee for Nuclear Energy, Rome, June 1972.

Bell K.J., and Müller A.C.
"Condensation heat transfer and condenser design".
A.I.Ch.E. Today Series, American Institute of Chemical Engineers, New York, 1970.

Bell K.J., Taborek J. and Fenoglio F.
"Interpretation of horizontal in-tube condensation heat transfer correlation using a two-phase flow regime map".
11the National Heat Transfer Conference, A.I.Ch.E., A.S.M.E., Minneapolis, (A.I.Ch.E. preprint no. 18), 1969.

Berenson P.J.
M.I.T. Heat Transfer Laboratory, Technical Report no. 17, 1960.

Bergles A.E., and Rohsenow W.M.
Journal of Heat Transfer ASME, 86 C, page 365, 1964.

Berman L.D. and Tumanov V.A.
"Investigations of heat of heat transfer during the condensation of flowing steam on a horizontal tube bundle".
Teplo Energetika, 9, 77, 1962.

Bernhardt S.H., Sheridan J.J. and Westwater S.W.
"Condensation of immiscible mixtures".
A.I.Ch.E. Symposium Series 68, page 21, 1972.

Beurteret C.A.
Com. B4.2 — H.T. International Conference, Versailles, 1970.

Borishanskii V.M. and Fokin B.S.
In2. Fizicheski Zhurnal 8, page 290, 1965.

Bouré J.
Paper B5.6. Fourth International Heat Transfer Conference.
Paris — Versailles — 1970.

Bowring R.W. (1972)
"Heat transfer and fluid flow in complex geometries — assumptions and deductions of the main sub-channel codes".
Paper presented at the Seminar on Two-Phase Flow Thermal Hydraulics organised by the Italian National Committee for Nuclear Energy. Rome, June, 1972.

Bromley L.A., Leroy N. and Robbers J.A.
"Heat transfer in forced convection film boiling."
Ind. Eng. Chem., 45, 1953, page 2639.

Cavallini A. and Zecchin R.
"High velocity condensation of organic refrigerants inside tubes".
Symposium, International congress of refrigeration, Washington, 1971.

Chen J.C.
"Correlation for boiling heat transfer to saturated fluids in convective flow."
Ind. Eng. Chem. Proc., Des. Dev., 1966, page 322.

Coffield R.J., Rohrer W.M. and Tong L.S.
"An investigation of departure from nuclear boiling in a crossed-rod matrix with normal flow of Freon 113 coolant".
Nuclear Engineering and Design, 6, 1967.

Colburn A.P. and Hougen O.A.
"Design of cooler condenser for mixtures of vapours with non-condensing gases."

Ind. Eng. Chem. 26, 1178, 1934.

Colburn A.P. and Drew T.B.
"The condensation of mixed vapors,"
A.I.Ch.E. transactions, 42, 197, 1937.

Collier J.G. and Pulling D.J.
"Heat transfer to two-phase gas-liquid systems."
Trans. Ind. Chem. Ing., 33, 1962, page 127.

Cumo M. et al.
"On two-phase highty dispersed flows."
European Two-phase flow group Meeting – C.S.N. Casaccion, Rome, June, 1972.

Davis C.J. and David M.M.
"Two-phase gas-liquid convection heat transfer."
Ind. Engng. Fundamentals, 3, 1964, page 111.

Delhaye J.M., Semeria R. and Flamand J.C.
"Void fraction, vapor and liquid temperatures local measurements in two-phase flow using a microthermocouple."
Paper 72-HT-13, ASME–AIChE, Denver, Colorado, 1972.

Fabrega S. – Huyghe J. and Rique R.
"Eléments de calcul des échangeurs pour fluides en vaporisation."
Journées internationales des échangeurs de chaleur. Paper 3 – I.F.C.E. – Insitute of fuel, Paris, 1971.

Feldkamp K.
"Der Wärmeübergang beim Sieden von Wässrigen Lösungen."
4th International Heat Transfer Conference. Paris – Versailles, paper B7, 2, 1970.

Forster H.K. and Zuber N.
"Dynamics of vapour bubbles and boiling heat transfer."
A.I.Ch.E. Journal, 1, 1955, page 531.

Fujii T. and Uehara H.
"Laminar filmwise condensation on a vertical surface."
Int. J. Heat Mass Transfer, **15**, 1972, page 217.

Gloyer W.
"Thermal design of mixed vapor condensers".
Hydrocarbon Processing, **49**, (6), 103, 1970.

Grant I.D.R. and Purser B.G.
"Measurement of air distribution in a condensers."
NEL Report no. 477, 1971.

Hsu Y.Y.
"On the size range of acitve nucleation cavities on a heating surface." Paper
61–WA–177, A.S.M.E., New York, 1961.

Huyghe J. and Lauro F.
"Amélioration des performances des condensers."
To be published in Bulletin des Informations Aérauliques et Thermiques,
CETIAT–Paris, 1972.

Jens W.H. and Lottes P.A.
ANL 4627, 1951.

Kezios S.P., Kim T.S. and Rafchiek F.M.
"Burnout in crossed-rod matrices and forced convection flow of water."
Internation developments in Heat Transfer, Part II, paper no. 31, page 262, 1961.

Kezios S.P. and Lo R.K.
"Heat Transfer for rods normal to sub-cooled water flow non-boiling conditions
up to and including burnout."
ANL – 5822, 1958.

Kinney R.B. and Sparrow E.M. (1970)
"Turbulent flow, heat transfer and mass transfer in a tube with surface suction."
J. Heat Transfer, **92**, 117.

Kopchikov I.A. et al.
"Liquid boiling in a thin film."
International Journal Heat Mass Transfer, **12**, 1969, page 791.

Kutateladze S.S.
"Heat transfer in condensation and boiling."
U.S.A.E.C. Report AEC — tr — 3770, (1952), 1948.

Macbeth R.V.
"The burnout phenomenon in forced convection boiling."
Advances in Engineering, **7**, 208, Academic Press, New York and London, 1968.

Mac Kee, H.R. and Bell, K.J.
"Forced convection boiling from a cylinder normal to the flow."
Chem. Eng. Prog. Symp. Series, volume 65, no. 92, page 222, 1969.

Meyrial P.M., Morin M.M., Wilcox S.J. and Rohsenow W.M.
"Effect of precision of measurement on reported condensation coefficients for liquid metals — including condensation data on a horizontal surface."
Fourth Int. Heat Conference, Paris, 1970, Paper Cs1.1, 1970.

Mickley H.S., Ross R.C., Squyers A.L. and Stewart W.E.
"Heat, mass and mementum transfer in flow over a flat plate with blowing or suction."
NACA — IN — 3208, 1954.

Mills A.F. and Seban R.A.
"The condensation coefficient of water"
Int. J. Heat Mass Transfer, **10**, 1815, 1967.

Minkowycz W.J. and Sparrow E.M.
"Condensation heat transfer in the presence of non-condensibles." Int. J. Heat Mass Transfer, **9**, 1125, 1966.

Miropolski Z.L.
Teploenergetika **10**, no. 5, 49, 1963.

Nijaguna B.T. and Abdelmessih A.H.
"Precoalescence drop growth model for dropwise condensation."
Paper 71 — WA/HT — 47 ASME, Washington, 1971.

Palen J.W. and Small W.M.
"A new way to design kettle and internal reboilers."
Hydrocarbon processing, **43**, (11), 119, 1964.

Palen J.W. — Yarden A. and Taborek J.J.
"Characteristics of boiling outside large scale horizontal mulitube bundle."
A.I.Ch.E., Symp., Series 68, nb 118, 50, Tulsa — 1971.

Petukhov B.S., Kovalev S.A. and Kolodtcev I.K. "Experimental investigation of heat transfer in boiling dissociative liquid."
Paper 69 — HT — 58, ASME–AIChE, Minneapolis, 1969.

Provan T.F.
"Condensation of pure vapours in the presence of non-condensible gases: a survey to 1962."
NEL Report no. 114, 1963.

Rique R. and Siboul R.
"Etude expérimentale de l'échange thermique à flux élevé avec l'eau en convection forcée à grande vitesse dans des tubes de petit diamètre avec et sans ébullition."
Int. J. Heat Mass Transfer, **15**, 1972, page 579.

Rohsenow W.M.
"A method of correlating heat transfer data for surface boiling of liquids."
Trans. ASME, 74, 1952, page 969.

Rohsenow W.M. and Mikic
"A new correlation of pool boiling data including the effect of heating surface characteristics."
Paper 68 — WA/HT — 22, ASME — New York, 1968.

Schrage R.W.
"A theoretical study of inter-phase mass transfer."
Columbia University Press, New York, 1953.

Sedek S.E.
"Condensation of steam in the presence of air."
I and EC fundaments, 7, 321, 1968.

Semeria R.
"Les échanges thermiques en ébullition nuclée."
Publ. Scient. et Techn. du Ministère de l'Air, no. 417, 1963, page 59.

Semeria R.
"Thermique des fluides diphasiques bouillants."
Revue générale de thermique to be published, 1972.

Spiegler P. et al.
Int. J. Heat Mass Transfer, 6, 1963, page 987.

Stevens G.F., Elliot D.F. and Wood R.W.
"An experimental investigation into forced convection burnout in Freon with reference to burnout in water: uniformly heated tubes with vertical up-flow."
AEEW – R321, 1964

Stewart P.B. and Hurd S.E.
"Condensation heat transfer in steam-air mixtures in turbulent flow"
Sea Water Conversion Laboratory, The University of California, Report no. 63-5, June 1966.

Thom, J.R.J. et al.
"Boling in subcooled water during flow up heated tubes or annuli."
Proc. Int. Mech. Eng., 180, paper 6, 1965, page 226.

Tong L.S.
"Heat transfer mechanisms in nucleate and film boiling."
Nucl. Eng. Des. 21, 1, 1972.

Tong L.S. and Hewitt G.F.
"Overall viewpoint of flow boiling mechanisms"
ASME/AIChE 13th National Heat Transfer Conference, Colorado.
August 1972, ASME Paper 72 − HT − 54.

Traviss D.P., Baron A.B. and Rohsenow W.M.
"Forced convection condensation inside tubes."
Tech. Report no. DSR 72 591 − 74, Heat Transfer Laboratory M.I.T., 1971.

Van Stralen J.J.D.
"The mechanism of nucleate boiling in pure liquids and in binary mixtures."
Int. J. Heat Mass Transfer, **9**, 995 and 1021, **10**, 1469, 1485, 1905, 1970.

Vliet and Leppert
"Critical heat flux for sub-cooled water flowing normal to a cylinder."
ASME Paper no. 62−WA−174, 1962

Wallis G.B.
"High velocity condensation in straight pipes."
Dartmouth College, Hanover, New Hampshire.
Report on NSF grant GK 1841, 1969.

Westwater J.W. and Bragg J.R.
"Filmboiling of immiscible liquid mixtures on a horizontal plate."
4th International Heat Transfer Conference, Paris − Versailles, paper B7.1.,
1970.

Wild N.H.
"Non-condensible gas eliminates hammering in heat exchanger."
Chemical Engineering, **76**, (9), 132, 1969.

Wilson J.L.
"The design of condensers by digital computers"
Paper presented at the Symposium on design, decision and the computer,
Institution of Chemical Engineers, England, 1972.

Zuber N.
"On stability of boiling heat transfer."
Trans. ASME 80, 1958, page 711.

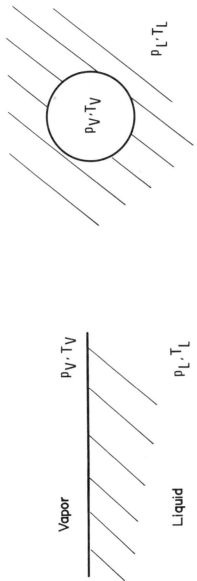

Fig. 1. Equilibrium at planar and curved interfaces between a liquid and its vapour

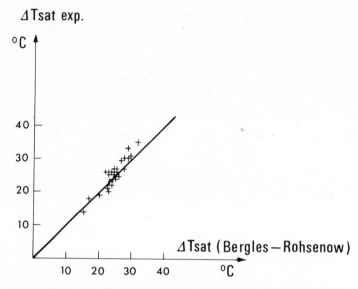

Fig. 2. Wall superheat at the onset of boiling (Ricque 1972)

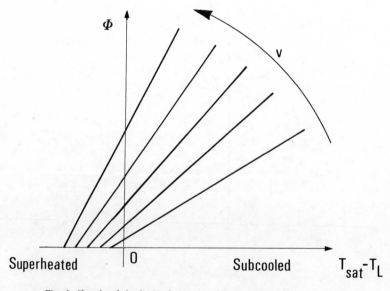

Fig. 3. Sketch of the limits for onset of boiling in forced convection

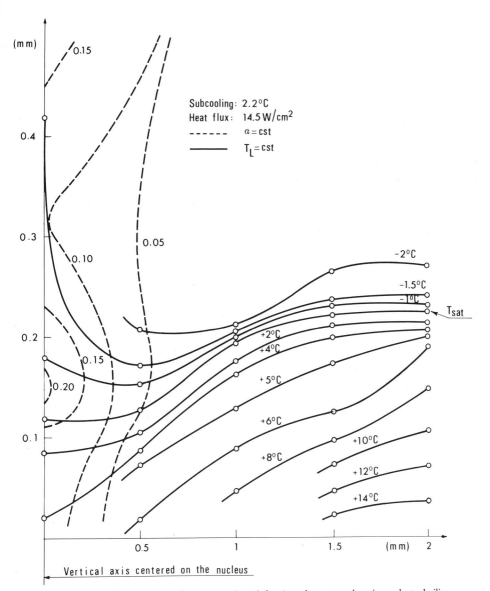

Fig. 4. Most probable liquid temperature and void fraction above a nucleus in nucleate boiling

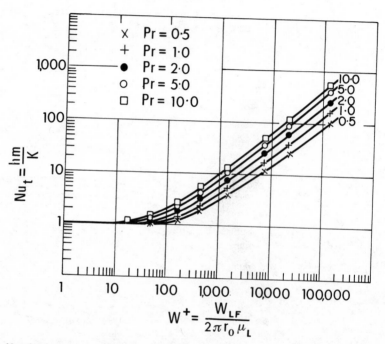

Fig. 5. Nusselt number in annular flow with effectively constant shear stress across the liquid flow

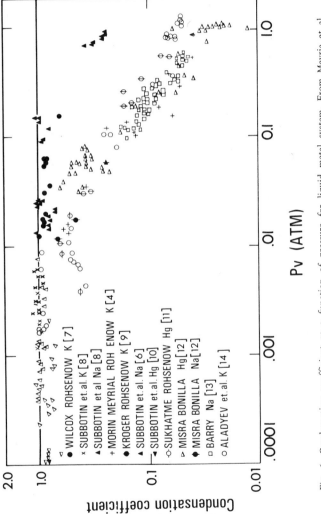

Fig. 6. Condensation coefficient as a function of pressure for liquid metal system. From Meyria et al (1970)

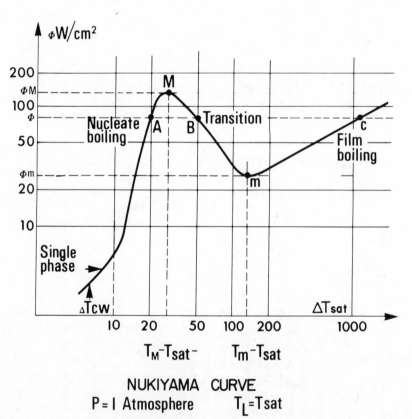

Fig. 7. Nukiyama curve; P = 1 Atmosphere; T_L = Tsat.

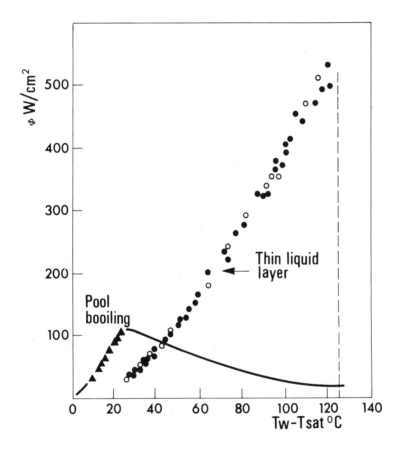

Fig. 8. From Kopchikov

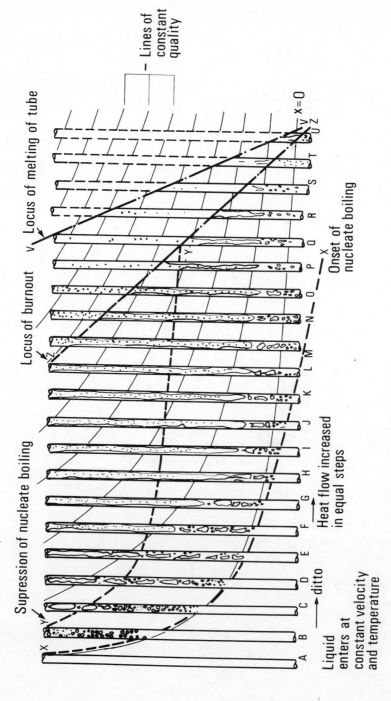

Fig. 9. Regimes of fluid flow and heat transfer in the evaporation of a liquid in a tube

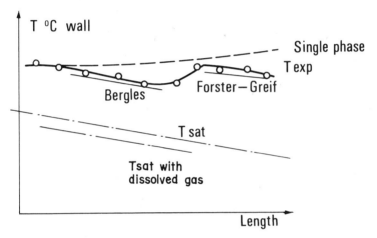

Fig. 10. Subcooled boiling (from Ricque 1972)

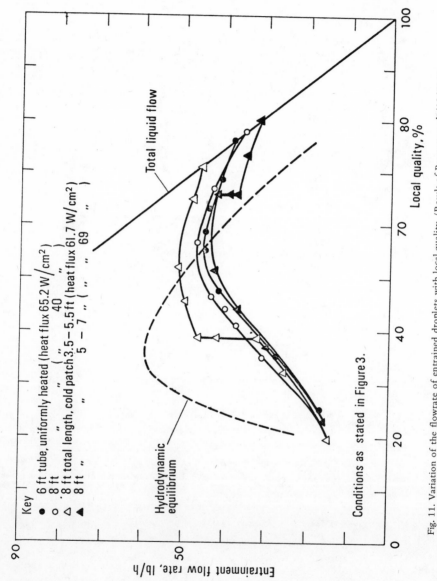

Fig. 11. Variation of the flowrate of entrained droplets with local quality. (Result of Bennett et al. (1967) Mass velocity 2.19×10^5 lb/h ft². Tube diameter 0.366 in. Inlet pressure 40 psia.

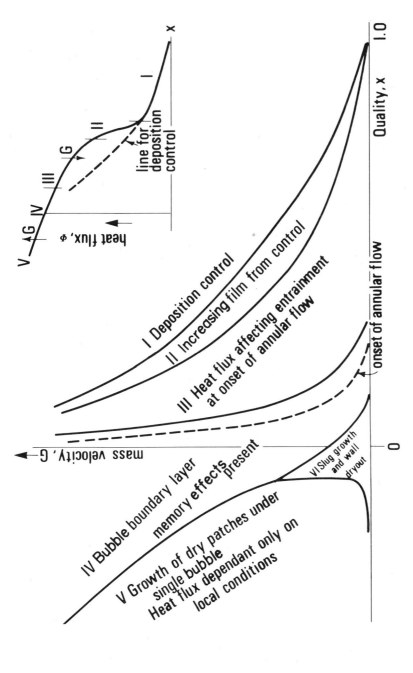

Fig. 12. Tentative qualitative representation of the regions of Burnout

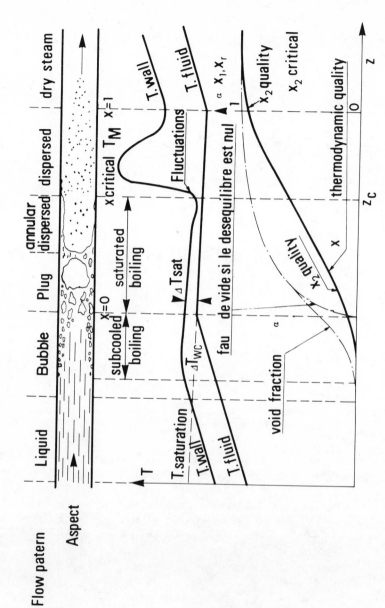

Fig. 13. Example of two phase flow evolution in constant heat flux channel.

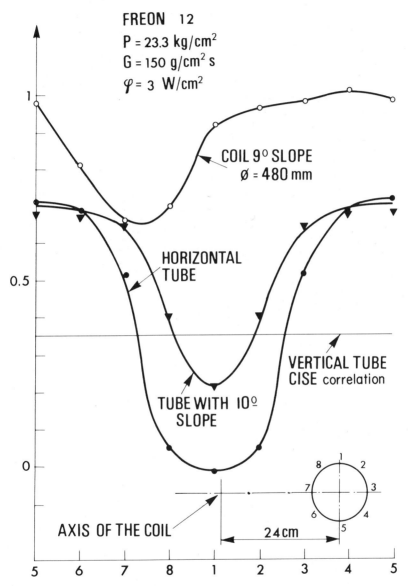

Fig. 14. Critical qualities for different goemetries from Fabrega et alt. (1971)

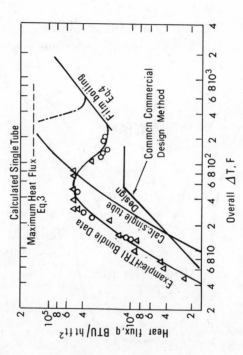

Fig. 15. Typical tube bundle boiling data compared to single tube. From Palen, Yarden and Taborek (1971).

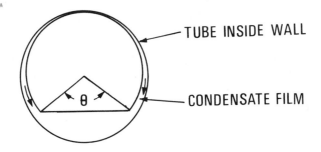

LOW VAPOUR VELOCITY

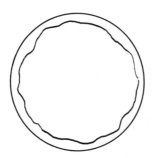

HIGH VAPOUR VELOCITY (ANNULAR FLOW)

Fig. 16. Modes of heat transfer in condensation in a horizontal tube

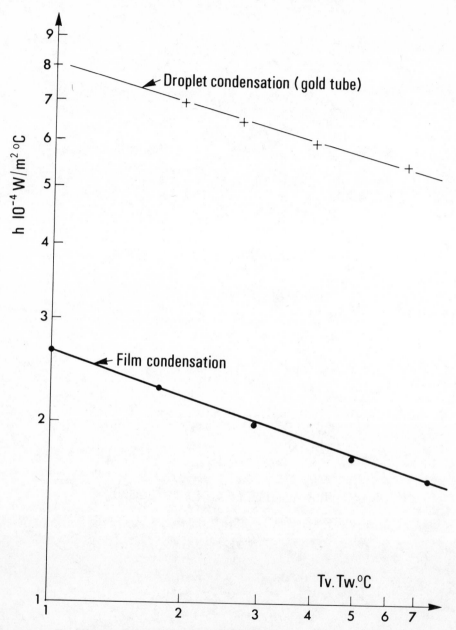

Fig. 17. Condensation on a single horizontal tube (from Huyghe et alt. 1972)

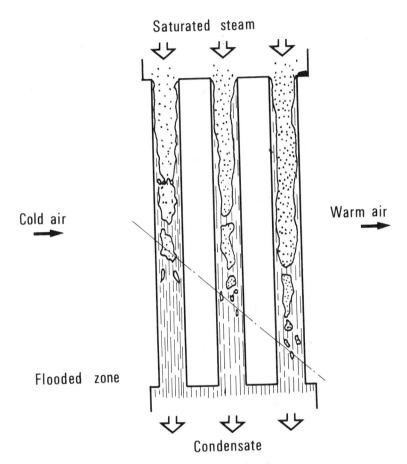

Fig. 18. Condensation in tube

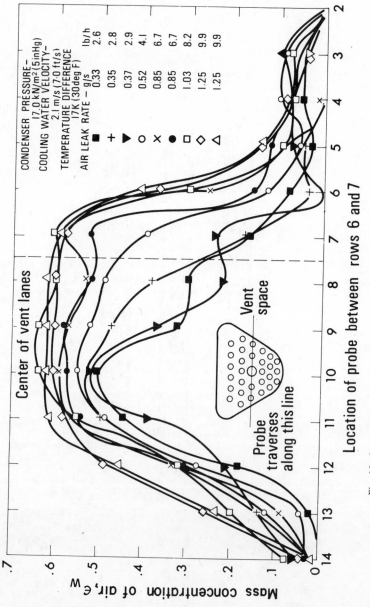

Fig. 19. Concentration profile in an experimental condenser (Grant and Purser (1971))

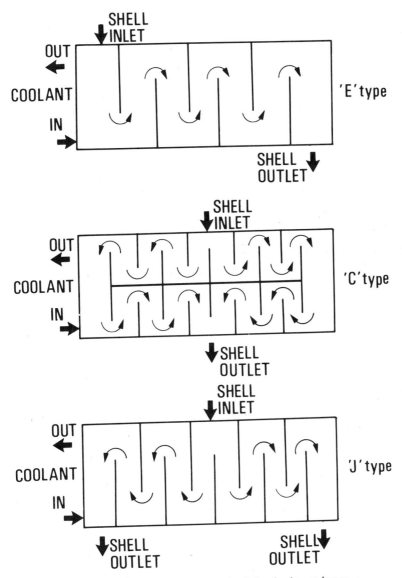

Fig. 20. Types of baffle arrangement in shell-and tube condensers -

Chapter 14

EVAPORATION AND CONDENSATION PHENOMENA
IN PROCESS HEAT EXCHANGERS

H. Baldus and E. Abadzić (*)

Introduction

Processing fluids in the chemical and process industry are frequently complex mixtures of various chemical compounds. With heat transfer to a single phase, the composition of the fluid does not cause major difficulties. But where there is a change of phase, i.e., evaporation or condensation of the mixture, the situation may be so complicated that even a definition of the problem may be difficult. This whole subject is termed multicomponent multiphase flow in heat exchangers. The behavior of such a system is determined by the interaction of:

a) The specific thermodynamic behavior of the mixture during change of phase. and

b) the hydrodynamics of the two phase flow, with

c) the heat and mass transfer process.

In addition the performance of the heat exchanger can be greatly affected by hydromechanical, chemical or electrochemical effects, briefly described as fouling of the heat transfer surface.

Even if we recognize that our present knowledge of this complex phenomena is far from complete, the process engineer is still faced with the problem of designing actual equipment to operate under such conditions. Usually the equipment is overdesigned, resulting in unnecessarily high capital costs. However, the underdesign of only one exchanger in a process plant can be far more costly due to production difficulties, excessive downtime, loss of capacity and sometimes serious equipment failure.

For the majority of process heat transfer applications, the baffled

(*) Linde Aktiengesellschaft, Munich

N.B. ALL FIGURES QUITED IN THE TEXT ARE AT THE END OF THE CHAPTER shell-and-tube heat exchanger is used. For special service, for example low temperature processes, the coiled-tubular and plate-fin exchanger are important. In addition, there are many other types in use. The selection of the best type of heat exchanger for a given duty is often more influenced by factors not directly related to the heat transfer considerations, such as ease of maintenance, cleaning, etc.

A. Process Requirements

The first step in solving a process heat transfer problem is usually to define the process requirements. These are conventionally defined on the left hand side of the basic differential equation

$$\frac{dQ}{\Delta T} = UdF$$

In contrast to single-phase flow, the heat load dQ and the temperature of the mixture undergoing a phase change and thus ΔT, are generally functions of

a) the specific thermodynamic behavior of the mixture in the two-phase region, and of

b) a local equilibrium approach between the vapor and liquid phases during the process.

The first function is usually given by the concept of equilibrium phase change of the mixture. Gibbs Phase Rule gives qualitative information on the expected behavior of the mixture during the phase change. It may be shown that condensation or evaporation of a binary mixture (C = 2) of two immiscible compounds (P = 3) at constant pressure will be isothermal (resulting degrees of freedom $F = C - P + 2 = 1$).

A similar calculation for a miscible binary mixture indicates the nonisothermal nature of this process.

The quantitative estimation of the heat load - temperature relationship for mixtures during phase changes may be obtained by measurement or from known empirical temperature-enthalpy relations [1,2]. The temperature-enthalpy relation for a given mixture with sufficient accuracy for engineering work can be predicted by means of either an Equation of State [1,3] or by the Principle of Corresponding States [2,4].

The calculation can be performed relatively easily nowadays on a digital computer, and can take account also of the expected pressure drop gradient in the heat exchanger. However, the results are valid only for the assumed local equilibrium between the vapor and liquid phases during the process, which strictly speaking never occur in a heat exchanger. In practice, the approach to the desired equilibrium depends on the flow conditions and a mixture can be made to evaporate or condense in a number of basically different ways. This will be illustrated for two somewhat idealized cases of condensation of a binary mixture with reference to the temperature-composition diagrams shown in figures 1 and 2.

For the first case of condensation of a saturated vapor mixture within a tube as shown in figs 1a and b, the first quantity of equilibrium liquid will have a concentration x_1 corresponding to its dew point temperature t_1. The necessary temperature of the coolant medium T_{c1} is slightly below t_1. Assuming adequate turbulence within the vapor and concurrent condensate flow, the process of condensation will be performed at near equilibrium conditions. The temperature of the fluid near the end of the condensation process will be close to the bubble point temperature of the mixture t_B. The wall temperature along the tube t_w and the required coolant medium temperature are indicated by the broken line in diagram 1b. This case is called integral condensation and the heat load-temperature profile (dQ/T) of the process can be closely predicted by the enthalpy-temperature calculations.

For the second case of condensation of a saturated vapor mixture within a tube, a turbulent flow of the vapour phase, but an idealized laminar flow of the cocurrent liquid phase as shown in figs. 2a and b will be assumed. The first condensate of the mixture will again occur at the dew point temperature t_1. This will form the first layer of liquid adjacent to the tube walls.

The vapor, with a rather greater concentration of the heavy component, then forms a second layer of liquid. Assuming that there is no mixing between these two and all successive layers, the final vapour at the end of the process will have a concentration close to that of the pure high boiling component. Therefore the coolant temperature for the end of the process must be much lower than for the previous case. This process is called differential condensation.

The essential differences between the integral and differential condensation modes can be demonstrated by a more practical example similar to that given by K. Bell [5]. Assuming that a total flow rate $\dot{M}$ of 100 kmol of the equimolal mixture of normal pentane (n-C5H12) and normal butane (n-C4H10) is to be condensed, the calculated heat load (dQ) versus temperature (T) dependency during the process together with the amounts of liquid condensed $(\dot{M}_L)$ for equilibrium conditions (integral condensation) at a pressure of 1.5 bar is shown in figure 3. The behavior of the chosen mixture at the observed pressure is near ideal and the equilibrium constants may be calculated from the partial pressures. The calculated dew point is about 37°C and the bubble point is near 23°C.

Assuming a countercurrent flow in a suitably sized exchanger and cooling water with an inlet temperature of 18°C, total condensation of the mixture can be easily accomplished.

But if differential condensation is expected, the situation will change drastically. This condition may be simulated by a stepwise calculation of condensation of the same mixture in a tube bank as shown in figure 4, divided into 10 sections. Assuming that the condensation in the first section gives 10% condensate of the incoming vapor and the second section 20% from the remaining vapor and so on, condensation will end in the tenth section. For this situation it is assumed that the liquid condensate from each section is not in contact with the remaining vapor. This can be the case with a poorly designed exchanger. The resulting Q-T and M_L-T curves represented by broken lines in figure 3 indicate the same dew point of the mixture at the condenser inlet as for the previous case. However, the condensation will end at a temperature of about 8.6°C, because the vapor to the tenth section is almost pure n-C4. It is obvious that for this condensation mode, the heat to be removed is not much different from the previous case, but the cooling water temperature of 18° C will be too high for the total condensation of the mixture.

Similar calculations and conclusions can also be made for evaporation of the mixture. Generally it may be assumed that a tube side flow of the mixture in the heat exchanger will give a closer approach to the ideal integral phase change mode than by the shell-side flow.

After obtaining a realistic Q-T plot for the process fluids, the mean temperature difference T_m and the process requirements $\Delta Q/\Delta T_m$ may be calculated. For the case of a multistream heat exchanger duty, such a coiled-tubular or plate-fin exchanger, the temperature-enthalpy relationship becomes more complex [6] and especially careful consideration has to be given to possible maldistribution [7].

Besides the heat transfer requirements for the process a certain optimum pressure drop in the exchanger is required. This requirement can only be satisfied by iteration of the calculation for sizing the heat exchanger.

Heat exchanger performance, UdF

According to the basic equation $dQ/\Delta T = UdF$, the prediction of the performance of a heat exchanger for given process requirements reduces in practice to the prediction of an overall heat transfer coefficient U, which results from the individual film heat transfer coefficients. The heat transfer coefficient of a multicomponent multiphase flow is determined by a large number of factors including the flow conditions for all phases, the flow orientation and particular geometry of the exchanger, and on the various physical properties. Usually special consideration of the expected heat transfer mechanism in the liquid phase such as nucleate or film boiling or condensation process together with the mass transfer resistances mainly in the vapor phase should be made. In contrast to the single phase turbulent flow in an exchanger, the performance with changer phase of a fluid will to a considerable extent depend on the driving force ΔT.

The present state of knowledge and a large number of investigations are reviewed elsewhere. It suffices to point out here that all heat transfer correlations used for related problems are based largely on empirical data. At present there is no general theoretical treatment of them, and their validity is generally restricted to the conditions used in the experiments. Besides published data a large number of proprietary methods and correlations are used in industry.

In order to illustrate the complexity of the problems, some industrial approaches will now be considered. Pool-boiling of the liquid may be the controling heat transfer mechanism for some types of heat exchanger (kettle-type reboiler). In the case of pool-boiling of miscible mixtures, poorer heat transfer coefficients than for the pure components can usually be expected. For a given temperature difference, geometry, surface conditions etc., heat transfer coefficients for the nucleate boiling of compounds can be predicted. The values for a mixture may then be estimated using any of several published empirical correlations. To account for the normal decrease in the heat transfer coefficient for mixtures Pallen and Small [8] developed an equation based on the temperature difference between the dew- and bubble-points. Stephan and Körner [9] and Afgan [10] used the concentration difference between the vapor and liquid in a similar manner. However, for some important technical mixtures, such as refrigerants with low hydrocarbon content, a remarkable increase in the heat transfer coefficient by pool-boiling was observed.

Boiling heat transfer by forced convection in vertical channels is very important for a variety of industrial vaporisers (thermosyphon reboilers, evaporators etc.). For such exchanger geometry, the effects of the different two-phase fluid-flow patterns (bubbles, slug, annular, dispersed etc.) are dominant for the heat transfer mechanism. Whereas under similar conditions in a kettle-type reboiler a nucleate heat transfer process occurs, with this type of exchanger a surpression of the boiling mechanism by forced flow is observed. The related heat transfer correlations are usually based solely on a liquid convection term and a two-phase factor [11,12]. This approach often gives reasonable agreement with measured results. Another very important problem in industrial heat exchangers is the condensation of a multicomponent mixture. The non-isothermal nature of the process by miscible fluids is often handled by the two terms for so-called latent and sensible heat transfer. In fact the treatment of simultaneous heat and mass transfer goes back many years in the literature, and most theoretical works in this field are based on the classical approach of Colburn, Hougen and Chilton [13,14]. However, for complex mixtures this approach becomes very complicated even if all the necessary physical thermal and diffusional properties can be predicted, which is seldom possible. For these reasons, industry must have recourse to a large number of somewhat crude methods.

One of these recently proposed by K. Bell [15] ,is based on several assumptions which may be briefly described as the equilibrium concept for the temperature and enthalpy of the vapor and liquid phases, with idealized separated flow conditions, as shown in Figure 5. A further assumption is that the total heat of condensation and sensible heat of the cooling liquid are transferred through the entire thickness of the liquid film. Assuming that the sensible heat transfer of the vapor h_v may be considered by an adequate single-phase correlation based on a superficial velocity, the final equation is obtained as

$$A_o = \int_o^T \frac{1 + \dfrac{U_o' Z}{h_v}}{U_o' (T_v - T_i)} \, dQ_T$$

REFERENCES

[1] Lewis C. Yen, R.E. Alexsander; AIChE Journal, March (1965)

[2] R.F. Curl, Jr., K.S. Pitzer; Industr. and Eng. Chem. Vol. 50, No. 2 (1958)

[3] E. Bender Habilitationsschrift Ruhr-Univ. Bochum (1971)

[4] G.D. Fischer, Th. Leland Jr.; Ind. Eng. Chem. Fundam. Vol. 9, No. 4 (1970)

[5] K.J. Bell, A.C. Mueller; AIChE Today Series (1971)

[6] E. Abadzic, H.W. Scholz; Adv. in Chem. Eng., Vol 18 (1972)

[7] R.F. Weimer, D.G. Hartzog; 13th Nat. Heat Transfer Conf. Denver (1972)

[8] J.W. Palen, W.M. Small; Petroleum Refiner 43, No. 11 (1964)

[9] K. Stephan, M. Körner; Chemie Ing. Techn. 41, No. 7 (1969)

[10] N. Afgan; Int Heat Transf. Conference, Chicago (1966)

[11] J.R. Fair; Chem. Eng., July (1963)

[12] J.C. Chen; Paper 63–HT–34, Heat Transfer Conference (1963)

[13] A.P. Colburn, O.A. Hougen; Ind. Eng. Chem., Vol. 26 (1934)

[14] T.H. Chilton, A.P. Colburn; Ind. Eng. Chem., Vol. 26 (1934)

[15] K.J. Bell and M.A. Chaly; 13the Nat. Heat Transf. Conf. Denver (1972)

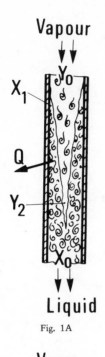

Vapour

X_1

Q

Y_2

X_0

Liquid

Fig. 1A

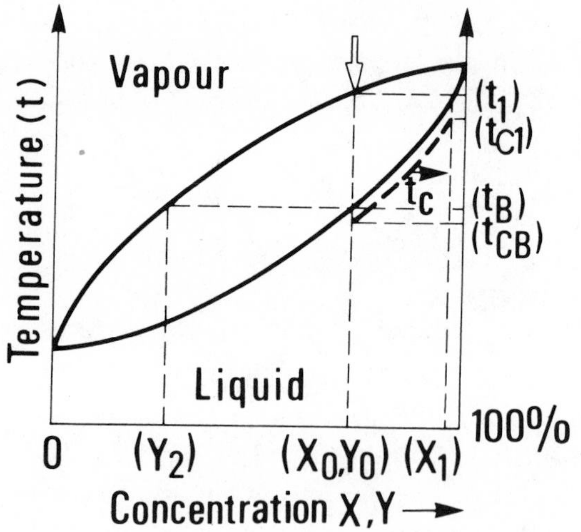

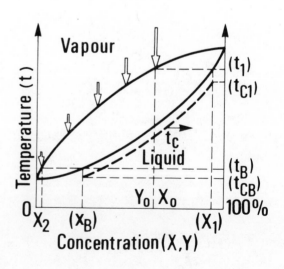

Fig. 1B

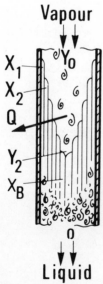

Vapour

X_1
X_2

Q

Y_2
X_B

0

Liquid

Fig. 2A

Fig. 2B

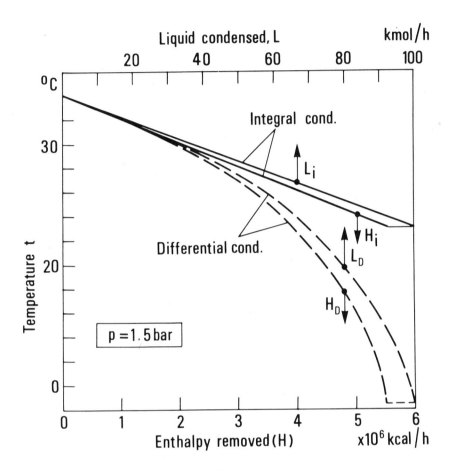

Fig. 3

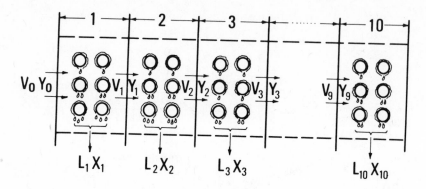

Fig. 4

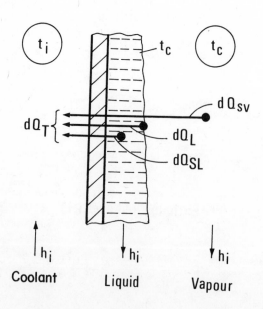

Fig. 5.

Chapter 15

FLUIDIZED BED APPLICATIONS IN HEAT TRANSFER—
CONTROLLED CHEMICAL PROCESSES

L. Massimilla and G. Donsi (*)

1. Introduction

Fluidization is the technique by which beds of particulate solid materials come into contact with upward moving fluid under such conditions that pressure drop through the bed remains substantially unchanged, whatever the fluid flow rate.

In theory, at the minimum velocity for fluidization, solid weight is just balanced by upward drag forces. Expansion is the bed reaction to the increase of velocity above the minimum. With gas fluidization, to which this presentation is limited, expansion is mainly related to the appearance of "bubbles".

According to the two-phase fluidization theory, there are: 1) a particulate (or dense) phase, containing by far the largest part of the solids, and whose properties are essentially those of the bed at incipient fluidization, and 2) a bubble (or lean) phase with a very low solids concentration, through which the fraction of flow rate exceeding that required for the fluidization of the material is conveyed.

Solids motion is the main feature of a fluidized bed, to which some of the peculiar heat transfer properties of such systems are related. Bubbles rising up the bed are the main, often the only, cause of solids motion. They also characterize gas/solids contact.

Gas fluidization has been used, or suggested, for a number of

(*) Istituto di Chimica Industriale e Impianti Chimici, Università Laboratorio di Ricerche sulla Combustione, C. N.R. - Napoli.

N.B. ALL FIGURES QUOTED IN TEXT ARE AT THE END OF THE CHAPTER

applications: drying, mixing, adsorption, reactor engineering, but examplifications reported here are only related to the latter. The purpose of this paper is to outline how the performance of different classes of fluid bed chemical processes rely on overcoming resistances related to specific heat transfer steps. To this end theories on heat transfer properties of fluidized bed will be briefly reported. For more detailed discussion reference should be made to the pertinent literature, particularly to books which since 1956 have marked the fast growing interest in fluidization (1–7).

2. Heat Transfer Steps in Fluidized Beds

These might be classified as follows:
a) wall to bed heat transfer
b) gas to solids heat transfer   1 - between gas and particles in the particulate phase
   2 - between bubble phase and particulate phase
c) heat transfer inside the particle

a) Wall to bed heat transfer

Conductive wall to bed heat transfer mechanism is based on the unsteady state transfer developing when particle aggregates at the bulk bed temperature reach the bed/wall interface, which is at a different temperature (8).

One approach is to consider the actual structure of the aggregate as being made up of particles touching, or in neighborhood of, the surface and of the gas in which they are immersed (9). Then, unsteady state heat transfer calculations are made for conductive transfer between the surface and the gas, and between the gas and the particle. Because of the very large difference between gas and solids specific heats, the rate of heat transfer may be estimated from the change in the cumulative heat content of the particles, as a function of the particle residence time at the heat transfer surface. This means very high rates of heat transfer to a dense bed of small particles touching the surface if they are replaced periodically, so that particle residence times at the transfer surface are very short (10). Figure 1 compares calculated and experimental heat transfer coefficients as obtained for a stirred, air fluidized bed in which the particle residence time could be controlled (11).

Another approach is to consider the aggregate as a whole (12). The instantaneous heat transfer coefficient is evaluated by assuming monodirectional unsteady state heat conduction and accounting for the time elapsed since aggregate renewal at the wall. It is:

$$h_i = \frac{q_i}{T_e - T_B} = \sqrt{\frac{k_{ea} c_{ps} \rho_B}{\pi t}} \tag{1}$$

The effective mean value of the heat transfer coefficient, referred to the entire heat transfer surface is then:

$$h_e = 2(1 - fo) \sqrt{\frac{k_{ea} c_{ps} \rho_B}{\pi t_m}} \tag{2}$$

after allowance is made for the heat transfer surface occupied by gas bubbles.

The trend of h_e as a function of U is evident in equation (2). Increasing U, a reduction of t_m is expected because of the faster solid mixing promoted by the more vigorous bubbling (13) but, on the other hand, $(1-fo)$ should also decrease because of the increasing fraction of heat transfer surface occupied by bubbles. The different weight of these opposite effects in different ranges of gas velocities should explain the maximum found in experimental curves of h_e as a function of U (Figure 2).

From equation (2), the correlation:

$$h_e = \frac{(1 - \epsilon)/(1 - \epsilon_{mf})}{R_w + 0.5 \sqrt{\frac{\pi}{k_{ea} c_{ps} \rho_B} \left[\frac{1-\epsilon}{1-\epsilon_{mf}} (U - U_{mf}) \right]^{1/2} \frac{1-\epsilon}{\epsilon - \epsilon_{mf}} M}} \tag{3}$$

has been derived. A wall resistance R_w is required in equation (3) for correlating experimental results, and is related to the special packing conditions there. Other terms are the same as equation (2), after expressing $(1-fo)/\sqrt{t_m}$ in function of fluidization parameters. The maximum in h_e is determined by decreasing of both $(1 - \epsilon)/(1 - \epsilon_{mf})$ and $(1 - \epsilon)/(\epsilon - \epsilon_{mf})$ with U. M is an empirical parameter, which should remain constant throughout the range of U and take into account effects, including that of particle size, not considered by the model. Figure 2 shows data fitting with equation (3) by using M calculated at U for which an optimum of h_e is obtained.

Data for coarser materials in Figure 2 indicate that there is a reversal of the trend for h_e at increasing U. This should account for the fact that in these conditions a turbulent component arises, increasing the effective thermal conductivity of the aggregate.

Figure 3 shows the influence of gas velocity and solid particle size on heat transfer. The effect of decreasing particles size is that expected from the theoretical model based on conductive heat transfer between single particles immersed in a fluid and a heat exchange surface (10). It will be noted that the finer the particles, the greater is the ratio U/U_{mf} at which the largest values of h_e are obtained.

As regards heat transfer at surfaces of bundles of tubes inside the bed, it should be noted that differences between values of h_e for horizontal, sloping and vertical tubes are small, so that their arrangement in a fluidized bed may be decided on the basis of other requirements. There is the tendency to take advantage of such internals to control bubble growth and improve exchange between bubble-and particulate-phase (14,15). However, care should be taken to check whether their baffling effect on the solids mixing does not lower heat transfer coefficients to unacceptable values (16). Correlations are available for this prupose in terms of the ratio of the transverse (or longitudinal) pitch of the tube bundle to the tube diameter (12).

Temperature effect on heat transfer coefficients is limited to the increase of k_{ea} with T up to about 800°C. Above this, a radiative component should be added to the heat transfer coefficient. Its value rises considerably (up to hundreds kcal/m² h°C), the higher the bed temperature, and the lower the difference between the temperatures of the bed and of heat transfer surface (17).

b) Gas to solids heat transfer
b.1. Between gas and particles in the particulate phase

According to the two-phase theory of fluidization, the particulate phase might be at first regarded as a "packed bed" of the solids permeated by gas velocity U_{mf}.

For conduction transfer, heat transfer coefficients for particles in array are available in terms of Nusselt number as a function of Reynolds and Prandtl numbers. In comparison to values of Nu from equations for an individual (spherical) particle in an indefinite medium, for instance (18):

$$Nu = \frac{h_p D_p}{k_g} = 2 + 0.74 \, Re^{1/2} \, Pr^{1/3} \qquad (4)$$

Nusselt numbers for packed beds are greater at Reynolds number above 10^3 (Figure 4). Below, they can become even less than the minimum value 2 for an individual particle in stagnant medium, which raises the question of whether experimental values of heat transfer coefficients involved in such correlations are "true", or rather "effective", coefficients, depending on spurious phenomena (19-22). The answer should be found in the incomplete accessibility of the transfer surface to the fluid at low Re. This causes a part of the exchange surface to operate under negligible temperature driving force, resulting in exceedingly low values of h_p, when it is calculated extending to all the solids surface temperature differences between solids and gas averaged over the entire bed cross section.

Conclusions drawn for fixed, packed beds of particles might be extended to the particulate dense phase of a fluidized bed, with two considerations, both related to particle mobility in such phase, but leading to opposite results about values of Nu to be assumed at low Re: 1) particle migration from one aggregate to the other should occasionally remove solids gas contact stagnations and increase effective coefficient toward the expected true values, whereas 2) particle aggregate adjustment in response to local increase of gas flow should favor non-uniformities in gas distribution with consequent reduction in effective gas to solids heat transfer coefficients. Cavities and little channels observed in the structure of the fluidized particulate phase of fine materials (23) might act for the fluid flow as those "microbreak-throughs" which have been considered responsible for low Nu (19).

In establishing which of the effects above might prevail in actual operation the results of measurements of effective gas to solids heat transfer coefficient in fluidized beds cannot be of much help. As shown in Figure 4, these decrease as Re decreases, but 1) dispersion of data hampers any comparison with results of packed bed experiments and 2) they also include the effect of heat exchange between the bubble and particulate phase, which also tends to reduce values of Nu.

Only with these limitations might heat transfer coefficients between gas and solids in the particulate phase be derived from correlations for packed beds. According to one of these (21), it is:

$$Nu = 0.42 + 0.35\,Re^{0.8} \tag{5}$$

with an average voidage of packed bed of 0.394. The asymptotic value of 0.46 for Re = 0 has been confirmed by another recent work (22).

The radiant component of heat transfer coefficient between particles and gas is usually negligible because of the considerably high thermal transparency of gases and the thinness of gas layers around the particles (17). Provided all particles are equally acting either as heat sources or sinks, radiative transfer within the particulate phase can be disregarded, whatever the solids temperature.

b.2. Between bubble phase and particulate phase

Gas to solid heat transfer under these conditions occurs in different forms according to whether the bubble phase is actually made of bubbles or concentrated in high velocity jets at the distributor.

Bubbles are found throughout the bed in the case of experimental apparatus with porous plate distributors. Bubble diameters tend to increase from their original small values as bubbles coalesce at upper bed level. With fine particles, usually less than 100 μ, U_{mf} = 0.1-0.2 cm/sec and fluidization ratios U/U_{mf} as high as 300-400 are reached. At these ratios bubbling regime switches into a turbulent regime, considerably enhancing gas and heat exchange rates with the surrounding fluidized dense phase (24-25).

In industrial apparatus, the grid distributor at the bottom produces gas jets, which in turn are dispersed in the bed to form bubbles. Jet and bubbling regimes in fact coexist at different bed heights in industrial fluidized bed reactors.

Extending concepts and formulae developed for gas exchange (26) to heat transfer, the phenomena dominating heat exchange between the two phases in the bubble regime are the convective component, due to the gas flowing in and out of the bubble (q_1), and the diffusional component, due to the conductive transfer through the wall (q_2). We get:

$$q_1 = 3/4 \ \pi \ D_e^2 \ U_{mf} \rho_g \ c_{pg} \tag{6}$$

$$q_2 = 0.975 \ \pi \ (\rho_g c_{pg} k_g)^{1/2} \ g^{1/4} \ D_e^{7/4} \tag{7}$$

and

$$h_b = 3/4 \ \rho_g c_{pg} \left[U_{mf} + 1.23 \left(\frac{k_g}{\rho_g c_{pg}} \right)^{1/2} g^{1/4} \ D_e^{-1/4} \right] \tag{8}$$

in which D_e is the diameter of the sphere having the bubble volume and h_b an overall heat transfer coefficient between the bubble phase and the particulate phase.

The convective component of h_b reduces rapidly with decrease in D_p, being $U_{mf} \propto D_p^2$. With fine particles the diffusional transfer is practically coincident with the overall exchange between the bubble and the particulate phase. In this case values of h_b = 30 − 40 kcal/m^{2}h°C would be predicted by equation (8) in the range of D_e = 5 ÷ 10 cm, assuming air thermal properties at room temperature. Bubble interaction at the onset of turbulence promotes exchange, which might occur at rates up to 5-10 times greater. It should be noted that D_e is an average bubble diameter at a given level of the bubbling bed. Limiting bubble size increases heat exchange rate per unit volume of bubble phase because of the favorable effects on the diffusional component of the heat transfer coefficient and, especially, because of the increase of the exchange surface.

Heat transfer at gas injection into the bed is likely to occur through two mechanisms 1) an unsteady state conduction at the jet/bed interface, through which particles exchange heat with gases at a rate determined by the average time the particles remain exposed to the gases and 2) a heat exchange between gases and particles entrained in the jet. The first mechanism prevails in downward jets at low velocity (27); the second one should occur with turbulent jets at grid distributors (28-30). Rates of heat transfer from such jets have been expressed by means of effective heat transfer coefficient h_j referred to a gas column of diameter equal to the nozzle diameter. With fine particles, they can be as high as 1500-6000 kcal/m^{2}h°C, depending on inlet velocity and nozzle diameter (30). Experimental results (28) indicate that $h_j \propto D_p^{-2.1}$, contrary to the tendency of the convective component of h_b to increase with particle size.

Comparison between the two modes of heat transfer between the bubble and the particulate-phase indicates that exchange through jet/bed interfaces might be more effective than that through bubble/bed interfaces, thus featuring a similar favorable entrance effect noted in fluid bed catalytic reactors (25).

Gas to solids heat transfer conditions for particles dispersed inside bubbles or entrained in jets are similar to those for diluted solid gas systems, and relative heat transfer coefficients can be evaluated from Figure 4. There might be, however, differences in behaviour between these particles and those forming the particulate phase because of the different reactive conditions they might find in relation to actual concentration of reactants in the surrounding gas (31-32).

Due to the large size that bubbles and jets can reach, a check should be made for any radiative transfer between the bubble and the particulate phases, especially in the case of optically thick process gases.

c) Heat transfer inside the particles

For steady state operation, a check for the relevance of this step is made on the basis of the Biot number $h_p D_p/2/k_s$. Sizes of fluidized particles are commonly in such range $Bi < 0.25$, i.e. less than the minimum for which internal thermal resistance limits heat transfer rate (12).

3. Exchange Equations

Heat transfer steps involved in fluidized bed reactions are analyzed with the assumption that solids and gas are perfectly mixed in the particulate phase, whereas gas in bubble phase passes through the reactor with plug flow. An analogous treatment is easily obtained for the case of plug flow of gas in both phases, by re-writing, for the heat conservation problem, the conversion equations derived for fluid bed reactors operating under these conditions (26). For the sake of simplicity only exchange through bubble surface is considered. Entrance effects due to distributors are disregarded. According to notations in Figure 5, heat conservation equations are:

for each **bubble in the bubble phase**:

$$(q_1 + q_2)\ (T_p - T_b) - U_b V_b c_{pg} \rho_g \frac{dT_b}{dy} = 0 \qquad (9)$$

in the **particulate phase**:

$$N\ (q_1 + q_2) \int_0^H (T_b - T_p)\ dy + U_{mf}\ (T_o - T_p)\ c_{pg} \rho_g +$$

$$+ h_e a_e H\ (1 - NV_b)\ (T_e - T_p) + QH\ (1 - NV_b) = 0 \qquad (10)$$

being:

$$h_p a_v (T_p - T_s) = Q \qquad (11)$$

the heat continuity equation at the interface between the solid and the gas in the particulate phase, with Q the amount of heat evolved or absorbed per unit time and unit volume of particulate phase.

Equations (9-11) do not take into account possible thermal effects of solids transfer through the reactor. Should such a transfer occur, another term should be added to equation (10), unless T_s is the inlet and outlet solids temperature. A full verification of the two-phase theory of fluidization is also assumed, with no solids present in the bubble phase. A further supposition is that the particle internal heat transfer resistance is negligible.

Integration of (9) and (10) with the boundary conditions:

$$T_p = T_b = T_o \qquad \text{at} \qquad y = 0 \qquad (12)$$

gives

$$T_{bH} = T_p + (T_o - T_p) e^{-X_h} \qquad (13)$$

$$T_p = \frac{T_o(1 - \beta e^{-X_h})}{1 - \beta e^{-X_h} + A} + \frac{QH_o}{Uc_{pg}\rho_g(1 - \beta e^{-X_h} + A)} + \frac{AT_e}{1 - \beta e^{-X_h} + A} \qquad (14)$$

$$T_H = \beta T_{bH} + (1 - \beta) T_p \qquad (15)$$

$$T_s = T_p - \frac{Q}{h_p a_v} \qquad (16)$$

being

$$\beta = \frac{U - U_{mf}}{U} \; ; \; X_h = \frac{(q_1 + q_2)H}{U_b V_b c_{pg}\rho_g} \; ; \; A = \frac{h_e a_e H_o}{Uc_{pg}\rho_g} \qquad (17)$$

respectively, the fraction of gas passing as bubble phase, a heat transfer factor between the particulate and the bubble-phases and heat transfer factor between the particulate phase and the external heat transfer surface.

Sometimes Q can be expressed as a function of solids temperature and concentration in the bed. In other applications Q derives from heat absorbed or generated as the fluid reacts at the solid surface, the reaction velocity being a function of the reactant concentration on the solid surface for a given solids concentration in the bed. In the latter, assuming a first order kinetics in respect to the reacting gas, it is:

$$r = kc_s \qquad (18)$$

and

$$Q = kc_s (- \Delta H) \qquad (19)$$

Equation (19) links the heat to the reactant-conservation equation. Consistent with the model of gas perfect mixing in the particulate phase and plug flow in the bubble phase, it is in fact:

$$c_p = c_o \frac{1 - \beta e^{-X_G}}{1 - \beta e^{-X_G} + k'} \qquad (20)$$

being:

$$X_G = \frac{(3N \, U_{mf} D_e^2 / 4 + k_{Gb} S) \, H}{U_b V_b} \qquad k' = - \frac{kH_o}{\left(1 + \dfrac{k}{k_{Gp} a_v}\right) U} \qquad (21)$$

respectively, a gas transfer factor between phases and a dimensionless first order reaction velocity constant including diffusional resistance between gas and solids. Parallel to equations (13), (15) and (16) the following equations hold for reactant exchange:

$$c_{bH} = c_p + (c_o - c_p) \, e^{-X_G} \qquad (22)$$

$$c_H = \beta c_{bH} + (1 - \beta) \, c_p \qquad (23)$$

$$c_s = c_p - \frac{r}{k_{Gp} a_v} \qquad (24)$$

4. Process Applications

a) Adiabatic endothermic processes without gas/solid chemical interaction.

Calcination, dehydration and carbonization processes fall in this class, and are usually carried out in continuous multistage reactors.

Relevant process parameters reduce to c_s and T_s, being $Q = Q(c_s, T_s)$. At steady state c_s is constant throughout the bed, its value depending on the loss of unreacted solids that can be afforded at the outlet. T_s, somewhere in the range

500°–1000°C, will be the highest temperature at which the material can be raised without overburning. According to equation (16), from Q temperature differences $(T_p - T_s)$ are evaluated. These are generally in the range of few degrees due to the low solids reactivity at working temperatures.

Equation (14), simplified for adiabaticity with A = O, shows that, once T_p is fixed, heat requirements can be satisfied by suitably choosing between inlet temperature T_o and velocity of hot gases U. Increasing T_o and lowering U might prove to be the best solution. Among other advantages, it results in an increase of heat utilization efficiency in the bed because, 1) for a given T_p, the heat in exhausted gas downstream the bed is smaller with lower gas flow rate, and 2) a reduction of U might increase the transfer parameter X_h through a decrease of D_e. In turn, as shown by equations (13) and (15), an increase of X_h causes T_H to approach T_p better with bubbling shallow beds.

A lower limit of the fluidization velocity is fixed by the fact that U should be large enough to ensure solid mixing and bed temperature uniformity with the rather coarse, widely sized raw materials being used. With upper particle size of 5–8 mm, velocities up to 5 m/sec (at the bed temperature) might be required.

Inlet gas temperatures T_o are the highest permitted in relation to distributor life and need to prevent bed material sintering in dead spots between active gas jets. Operation might involve the combustion of fuel directly in the bed. A case of submerged combustion is reported in which undiluted combustion gases (theoretical temperature above 2000°C) supply heat to a bed of phosphatic ores at 550°C (33). Hot gases jet into the bed horizontally at the side of slotted downward combustor. Solids temperature control is achieved because particles entrained in the high turbulent jets or forming the jet/bed interfaces are exposed only briefly to the flames.

b) Adiabatic endothermic processes with gas/solids chemical interaction.

Pertinent examples are the processes for the reduction of iron ores with hot reducing gases (34). Catalytic fluid cracking of hydrocarbons and coking should also be considered in this class, with the peculiarity that in such processes reaction heat is supplied with proper temperature and flow rate of a stream of particulate material.

Contrary to processes under a) there is a relationship between heat and mass transfer, and reaction which involves solid and gaseous reactants. As shown by equations (19), (20) and (24), the amount of heat absorbed per unit time and unit bed volume now depends on the rate of gas being exchanged between the bubble and the particulate phases, and reacting on the solids at temperature T_s, for given values of X_G and k'. Inlet T_o is then calculated by equation (14), once the gas temperature T_p in the particulate phase is obtained through equation (16). In this calculation U is already fixed with k'.

According to equations (13) and (22), temperature and reactant concentration profiles up the bed: $T_b = T_b(y)$ and $c_b = c_b(y)$ can be calculated as a function of $X_h y/H$ and $X_G y/H$, respectively. It appears that trends of T_b and c_b towards the asymptotic values of T_p and c_p are the same only as far as the convective components of transfer factors are relevant. In this case it is $X_h/X_G = 1$.

Temperature driving forces $(T_p - T_s)$ at the particle interface might be relatively high. For instance, in H-iron process, values of $(T_p - T_s)$ about 20-30°C are estimated for particles of 0.8 mm upper size (35).

c) Exothermal solids consuming processes

Fluid bed applications under this heading include processes such as pyrite roasting and fossil fuel combustion.

As in processes under b), U is fixed with k'. This, however, does not result in a limitation of T_o. According to equation (14), T_o can be suitably chosen for any value of T_s by properly selecting A.

Upper particle sizes of solids are in the range of $3 \div 7$ mm, it being the result of an optimization of a number of factors including particle elutriation, external heat transfer coefficients, reaction rates referred to unit mass of reactive solids in the bed, permitted loss of consumable solids in the unburnt particle core at bed outlet. Besides corrosion problems, upper solids temperatures, generally ranging between 800° and 1000°C, are limited by process requirements. These might be, for instance, the complete oxidation of carbon to carbon dioxide in coal combustion or the minimization of loss of valuable metal as insoluble solid product in the ashes from rosting.

Even with limits on solids temperature, reaction rate constants are very high in processes of this class. As a result, it is from equation (20) $c_p \simeq 0$, and from equations (22) and (23) $C_H/C_o \simeq \beta e^{-X}G$. This expresses the determining effect of gas exchange between the bubble phase and the particulate phase on the fraction of reactant flow rate converted on the solids, and thus on the overall process. In other conditions, mass diffusional resistance might rise according to equation (24) around particles (36).

Temperature driving forces $(T_s - T_p)$ are usually relevant. Values about 200°C and 80°C can be estimated, respectively, in the coal combustion and in blendas roasting.

External heat transfer might become a relevant problem in equipment design, especially where heat recovery from outlet gases is imposed by more general cycle optimization and the inlet temperature T_o is raised. With T_p fixed at the highest level possible in respect to T_e, and Q also high, heat balance according to equation (14) can only be obtained by increasing A. As also h_e and a_e are substantially fixed for a given U and average particle size, an exchange surface increase is only obtained with a proper bed depth, which might remain defined by this need rather than by solid residence time or other reaction requirements. This is the case with coal combustion, in which the need for sufficiently high H_o to locate all the necessary heat surface, combines with the opportuneness of keeping down the available carbon surface in the bed for complete oxidation. Operationally, this results in a very low concentration of carbon in the bed, as low as a few percent (37).

d) Exothermic catalytic processes

This class of fluid bed applications includes the processes of partial oxidation, ammonoxidation, cloruration and oxicloruration of hydrocarbons. They are characterized by the occurrence of degradative reactions and the advantage of using fluid in respect to fixed bed reactors lies in better control with regard to reaction run-away. This is partly made more easy by the possibility of avoiding reactant preheating. Equation (14) shows that the lower T_o, the smaller may be the value of A required to obtain the desired temperature in the particulate phase.

Fluid bed application to catalytic exothermal reactions also shows the inherent relationship discussed for processes under c) concerning gas exchange between the bubble and the particulate-phase and heat being developed in such a phase and to be transferred to the internal cooling system. However, requirements for catalytic exothermal reactions are usually more constraining, due to the narrowness of useful temperature range. Temperature levels are also lower, usually in the range of 300°-600°C. Thus, lower temperature differences $(T_p - T_e)$ are available for outer heat exchange.

Such circumstances call for both, a greater bed temperature stability and a high rate of heat extraction from the bed, which are requirements both to be fulfilled with adequately high values of A. As shown in Figure 3, use of fine catalyst particles, fluidized at large U/U_{mf} can assure sufficiently large heat transfer coefficients to reduce the exchange surface to an acceptable value and, at the same time, to assure throughputs required for intensive processes. In some of these, powders less than 100 μ are used, which further enhances thermal bed properties as far as external heat transfer and gas/solid contact between bubble and particulate phases are concerned. As suggested (24), this might be the result of switching flow properties from plain bubbling to turbulent bed behavior. Temperature differences (T_s-T_p) for such small particles are neglible.

e) Quenching processes

Although recognized for a long time (38) the possibility of using fluid beds of inert solids as quenchers has not so far received much attention. One application is related to the production of acetylene from sooty flames of hydrocarbons and oxygen discharged into the beds at temperature of 1300-1500°C.

The interesting feature of this class of processes is that the relevant heat transfer step is all confined to the jet zone.

5. Conclusions

Considerations above show that heat transfer steps in fluidized bed chemical reactors are usually associated with other process steps involving reaction or mass exchange.

Flame quenching is of course an application of fluidization to chemical reaction engineering in which, for a given burner performance, the process is entirely controlled by a heat transfer step. In other applications, factors limiting process productivity are related to reactivity and/or to mass transfer, while, according to the case, one or other of the intervening heat transfer steps might become controlling. This in the sense that reactor design has to take into account the thermal resistances arising in that step.

For instance, choice of bed height and of combination between flow rate and temperature of inlet gas stream in adiabatic endothermal processes is directed to face resistance to heat transfer between the bubble and particulate phases, the only relevant ones in cases of low solid reactivity. Similar reasoning applies to the need for the bed height to locate heat transfer surfaces in exothermal, solid consuming process, whose output depends on how large a gas flow rate can be passed through the bed with efficient mass exchange between the particulate- and the bubble-phase. Also in catalytic processes, which usually operate under conditions for mixed chemical and mass exchange regimes, particle size and size distribution inlet gas temperature, size and distribution of heat transfer surfaces in the bed are designed to achieve the high rates of external heat transfer required to obtain the necessary close control of bed temperature.

It appears that from one process to the other needs can change greatly, the only feature constantly assumed being the uniformity of solids temperature and composition throughout the bed. Then, based on process peculiarities, design features are changed according to whether it is exchange between solids and gas in the particulate phase, or between the bubble and particulate-phase, or between the bed and inside heat transfer surfaces which has relevance, always in the favorable situation of a low pressure drop through the bed, whatever the gas flow rate and the particle size. This flexibility is perhaps among the most interesting characteristics of fluidization as an unit operation.

NOMENCLATURE

a_e	area of heat transfer surface per unit volume of particulate phase m^2/m^3
a_v	surface area of solid-gas interface per unit volume of particulate phase, m^2/m^3
c	reactant concentration, $kgmole/m^3$
c_{pg}	specific heat of the fluid, $kcal/Kg°C$
c_{ps}	specific heat of solids, $kcal/Kg°C$
D_e	diameter of sphere having the bubble volume, m
D_p	particle diameter, m
f_o	fraction of heat transfer surface occupied by bubbles
g	acceleration of gravity, m/hr^2
H	bed height, m
H_o	bed height at the minimum velocity for fluidization, m
h	heat transfer coefficient, $kcal/m^2 hr°C$
h_b	heat transfer coefficient relative to a bubble, $kcal/m^2 hr°C$
h_p	heat transfer coefficient relative to a particle, $kcal/m^2 hr°C$
k	reaction velocity constant, $1/sec$
k_{ea}	effective thermal conductivity of an aggregate of solid particles, $kcal/mm \quad hr°C$
k_g	thermal conductivity of the gas phase, $kcal/m \ hr°C$
k_{Gb}	Mass transfer coefficient between particulate and bubble-phase, m/hr
k_s	thermal conductivity of solids, $kcal/m \ hr°C$
k_{Gp}	mass transfer coefficient between the fluid and the particle, m/hr
M	dimensional parameter, $hr/m^{1/2}$
N	number of bubbles per unit volume of bed, $1/m^3$
Q	heat evolved per unit volume of particulate phase and unit time, $kcal/hr \ m^3$
q_1	instantaneous value of heat flux, $kcal/hr \ m^2$
q_1, q_2	rate of heat exchange from a bubble, $kcal/hr°C$
r	reaction rate, $Kgmole/m^3 h$
R_w	thermal resistance, $m^2 hr°C/kcal$
S	bubble surface, m^2
T	temperature, $°C$

t	time, hr
t_m	mean time, hr
U	gas velocity, m/hr
U_b	bubble rising velocity, m/hr
U_{mf}	minimum gas velocity for fluidization, m/hr
V_b	bubble volume, m^3
y	vertical coordinate, m

Greek symbols

ϵ	voidage fraction of the bed
ϵ_{mf}	voidage fraction at U_{mf}
ρ_B	bulk density of an aggregate of particles, Kg/m^3
ρ_g	density of the gas phase, Kg/m^3

Subscripts

B	relative to the bulk properties of an aggregate
b	within the bubble
e	at the heat transfer surface
i	instantaneous
j	relative to a jet
H	at the bed outlet
p	within the particulate phase
s	at the surface of the solid particle
o	at the inlet of the bed

REFERENCES

[1]   Othmer, D.F., Fluidization, Reinhold, New York (1956)

[2]   Leva Max, Fluidization, McGraw-Hill, New York (1959)

[3]   Zenz, F.A. and Othmer, D.F., Fluidization and Fluid-Particle Systems, Reinhold, New York (1960).

[4]   Davidson, J.F. and Harrison, D., Fluidisation Particles, Cambridge Press (1963)

[5]   Zabrodsky, S.S., Hydrodynamics and Heat Transfer in Fluidized Beds, M.I.T. Press (1966)

[6]   Kunii, D. and Levenspiel, O., Fluidization Engineering, Wiley, New York (1969)

[7]   Davidson, J.F. and Harrison, D., Fluidization, Academic Press, London (1971)

[8]   Mickley, H.S. and Trilling, C.A., Ind. Eng. Chem., 41, 1135 (1949)

[9]   Botterill, J.S.M., Redish, K.A., Ross, D.K. and Williams, J.R., in Rottenburg, Symposium on the Interaction Between Fluids and Particles, Instn. Chem. Engrs, London, 1962, p. 183.

[10] Botterill, J.S.M. and Butt, M.H.D., 60th Annual A.I.Ch.E. Meeting New York, (1967)

[11] Botterill, J.S.M. and Williams, J.R., Trans. Instn. Chem. Engrs. 41, 217 (1963)

[12] Gelperin, N.I. and Einstein, V.G., in Davidson and Harrison (ref. 7) p. 471

[13] Rowe, P.N., Partridge, B.A., Cheney, A.G., Henwood, G.A. and Lyall, E., Trans. Instn. Chem. Engrs, 43, T 271 (1965)

[14] Grace, J.R. and Harrison, D., I. Chem. E. Symposium Series No. 27 Instn. Chem. Engrs., London (1968)

[15] Volk, W., Johnson, C.A. and Stotler, H.H., Chem. Eng. Progr. 58, 44 (1962)

[16] Massimilla, L., Bracale, S. and Cabella, A., Ric. Sci. 27, 1853 (1957)

[17] Zabrodski, S.S., ibidem, p. 235

[18] Rowe, P.N., Claxton K.T. and Lewis, J.B., Trans. Instn. Chem. Eng., 43, T14 (1965)

[19] Zabrodski, S.S., Int. J. Heat and Mass Transfer, 6, 23 (1963)

[20] Rowe, P.N., Int. J. Heat and Mass Transfer, 6, 989 (1963)

[21] Littman, H. and Sliva, D.E., 4th International Heat Transfer Conference, Versailles, Sept. 1970

[22] Wakao, N., paper to be published

[23] Massimilla, L., Donsi', G. and Zucchini, C., to be published in Chem. Eng. Sci.

[24] Kehoe, P.W.K., and Davidson, J.F., Chemical Engineering Conference Chemeca 70, Canberra (Australia 1970)

[25] Crescitelli, S., Ginnasi, A., Massimilla, L. and Maviglia, G., to be submitted to Ing. Chim. Ital.

[26] Davidson, J.F. and Harrison, D., ref. 4, p. 97

[27] Russo, G. and Massimilla, L., Symposium on Chemical Reaction Engineering, Amsterdam (1972)

[28] Baerns, M. and Fetting, F., Chem Eng. Sci., 20, 273 (1964)

[29] Merry, J.M.D., Trans. Instn. Chem. Engrs. 49, 189 (1971)

[30] Behie, L.A., Bergougnou, M.A. and Baker, C.G.J., paper presented at the 21st Canadian Chemical Engineering Conference, Montreal Quebec (1971)

[31] Kunii, D., paper presented at the International Symposium on Fluidization Fundamentals, 64th Annual A.I.Ch.E. Metting, S. Francisco, California (1971)

[32] Gilliland, E.R., Discussion on paper of ref. 31

[33] Goldney, L.H. and Hoare, J.S., Mech. and Chem. Eng. Transactions, The Institution of Engineers, Australia, Nov. (1968)

[34] Rogers, R.R., Iron Ore Reduction, Pergamon, Oxford (1962)

[35] Squire, A.M. and Johnson, C.A., J. of Metals, 10, 586 (1957)

[36] Avedesian, M.M. and Davidson, J.F., paper to be published

[37] Skinner, D.G., The Fluidized Combustion of Coal, National Coal Board, London (1970)

[38] Fetting, F. and Wicke, E., Chemie. Ing. Techn. 28, 88 (1956)

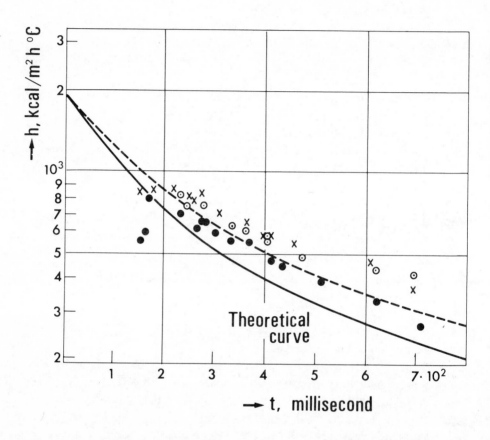

Fig. 1. (Botterill and Williams, ref. 11) ● = Superficial air velocity 0.228 ft/s; ○ = Superficial air velocity 0.315 ft/s; X = Superficial air velocity 0.378 ft/s; comparison of the practical data for the stirred, air fluidized beds with theoretical predictions.

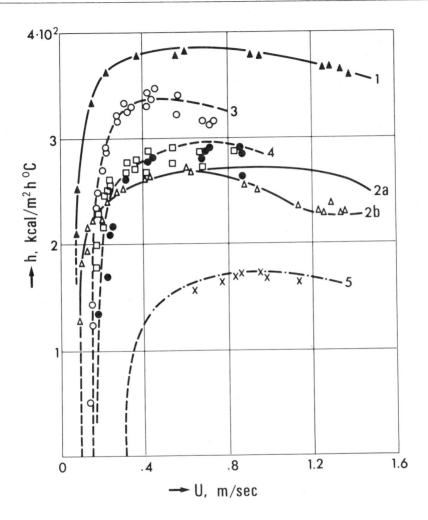

Fig. 2. (Gelperin and Einstein, ref. 12) Comparison of equation (3) with experimental results. Horizontal tubes (Gelperin et al.); 1 ▲ quartz sand (d_e = 0.263 mm) air, single tube, D_R = 20 mm; 2 △ quartz sand, middle tube of horizontal row, s_h = 40 mm, D_R = 20 mm; 2a, 2b, curves calculated for mean and local voidage fractions. Vertical tubes (Gelperin et al.); 3 ○ quartz sand (d_e = 0.352 mm) air, single tube, D_R = 20 mm; 4 quartz sand, D_R = 20 mm, tube bundle, s_h = 40 mm; distance of tube from center line, ● r = 0 mm; ○ r = 80 mm; ○ r = 160 mm. Outer cylindrical wall (Massimilla et al.); 5 Glass spheres (d = 0.7 mm) air; D = 90 mm. D_R = tube diameter; s_h = pitch of tube bundles.

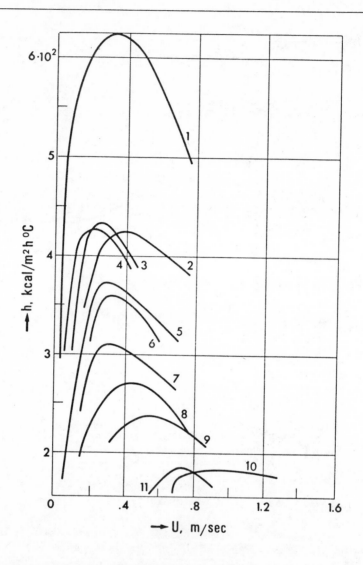

Fig. 3. (Gelperin and Einstein, ref. 12) Influence of air velocity and solid particle dimensions on heat transfer from sphere 20 mm diameter (Variggin 1959). 1) ferrosilicon, d_e = 0.082 mm; 2) hematite, d_e = 0.173 mm; 3) carborundum, d_e = 0.137 mm; 4) quartz sand, d_e = 0.140 mm; 5) quartz sand, d_e = 0.198 mm; 6) quartz sand, d_e = 0.216 mm; 7) quartz sand, d_e = 0.428 mm; 8) quartz sand, d_e = 0.515 mm; 9) quartz sand, d_e = 0.65 mm; 10) quartz sand, d_e = 1.11 mm; 11) glass spheres, d_e = 1.16 mm.

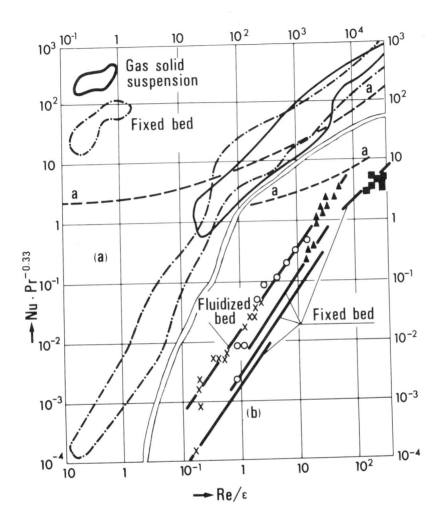

Fig. 4. (Gelperin and Einstein, ref. 12) Interphase heat transfer in various fluid-particle systems a) in gas/solid suspension and fixed bed; b) at transition of fixed bed to fluidised state. X Particles of alumina, $d_e = 0.09$ mm; ○ glass spheres, $d_e = 0.132$ mm; Δ glass spheres, $d_e = 0.444$ mm; ■ glass spheres, $d_e = 1.10$ mm.

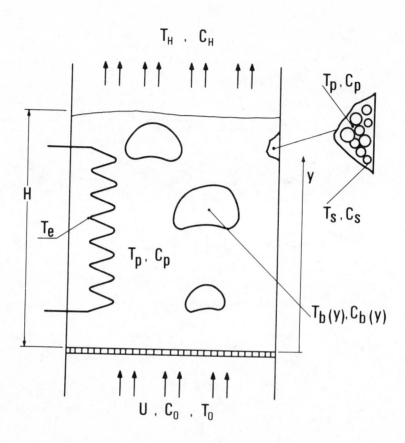

Fig. 5. Fluid bed model showing temperatures and concentrations.

Chapter 16

DESIGN OF COOLER CONDENSERS
AND EVAPORATIVE COOLERS

T. Mizushina (*)

Simultaneous heat and mass transfer complicates the design calculations for cooler condensers and evaporative coolers. In addition, the non-linearity of the relation between saturated vapor pressure and temperature causes further difficulty in the calculation process.

The author and his group have developed design methods for such heat exchangers, which will be described briefly in the following. The details are presented in the original papers.

1. Driving force.

In the case of simultaneous heat and mass transfer in an air-water vapour system, F. Merkel (1925) proposed to use the difference of enthalpy of air as a driving force.

$$- G_B \frac{di}{dA} = \frac{h_g}{C_s} (i_g - i_i) \qquad (1)$$

$$i = C_s t + \lambda_o H \qquad (2)$$

This combination of two different driving forces is due to "Lewis' law" which is valid for dilute mixtures of air-water vapor. However, "Lewis' law" is not applicable to other cases. Mizushina and Kotoo [1] proposed the use of a "modified

---

(*) Dept. of Chemical Engineering, Kyoto University, Kyoto, Japan

N.B. ALL FIGURES QUOTED IN TEXT ARE AT THE END OF THE CHAPTER

enthalpy driving force" in the design of cooler condensers for general gaseous mixtures. The modified enthalpy was defined as

$$i' = aC_s t + \lambda_o H \tag{3}$$

and the transfer equation is written as

$$-G_B \frac{di}{dA} = \frac{h_g}{aC_s} (i'_g - i'_i) \tag{4}$$

In a dilute gas mixture the value of α is calculated by the following equation.

$$a = C_s (Sc/Pr)^{1/2} \tag{5}$$

This relation was obtained from the equation of analogy [2] between momentum transfer and heat or mass transfer,

$$a = C_s \frac{1 + (Sc-1) \ 6.3 \sqrt{f}/Sc^{0.2}}{1 + (Pr-1) \ 6.3 \sqrt{f}/Pr^{0.2}}$$

$$\simeq C_s (Sc/Pr)^{0.5} \tag{6}$$

and confirmed experimentally with a wet and dry bulb hygrometer by Mizushina and Nakajima [3].

2. Design of Cooler Condensers.

The graphical method proposed in the above mentioned paper [1] enables us not only to calculate the surface area of a cooler condenser but also to trace the change of gas condition in the apparatus.

However, this deals only with dilute mixtures, and should be modified for concentrated mixtures. Mizushina et al. [4, 5] revised the above mentioned method and proposed a graphical method which is applicable to practical design.

For concentrated mixtures, the ratio of the heat and mass transfer coefficients is expressed as

$$\frac{h_g}{k'_g} = aC_s \tag{7}$$

where

$$a = \frac{P_{BM}}{P_t - P_i} \left(\frac{Sc}{Pr} \right)^{1/2} \tag{8}$$

$$h_g = j \, C_g \, G' \, / \, (Pr)^{1/2} \qquad (9)$$

The enthalpy balance is

$$- G_B \, di_g = \frac{h_g}{a \, C_s} \, (i'_g - i'_i) \, dA \qquad (10)$$

where i' is the "modified enthalpy" and is defined as

$$i'_g = a \, C_s \, t_g + \lambda_o \, H_g \qquad (3)$$

Therefore

$$A = a \, G_B \int_{i_{g2}}^{i_{g1}} \frac{C_s \, di_g}{h_g \, (i'_g - i'_i)} \qquad (11)$$

To take account of the change of gas velocity due to the condensation of vapor, the author derived the following equation,

$$\frac{h_g}{C_s} = \frac{h_B}{\beta \, C_B} \qquad (12)$$

where

$$\beta = (1 + H_g)^n \qquad (13)$$

and

n = 0.4 for the flow across tube banks
n = 0.2 for the flow inside tubes   (14)

Substituting equation (12) into equation (11), one obtains

$$A = \frac{a C_B \, G_B}{h_B} \int_{i_{g2}}^{i_{g1}} \frac{\beta \, di_g}{i'_g - i'_i} \qquad (15)$$

A heat balance yields the slope of the operating line as

$$\frac{di_g}{dt_\ell} = \frac{L \, C_\ell}{G_B} \qquad (16)$$

The slope of the path line of the gas-vapor mixture is

$$\frac{di_g}{dt_g} = \frac{1}{a} \frac{i_g' - i_i'}{t_g - t_i} \tag{17}$$

and the slope of the tie line is

$$\frac{i_g' - i_i'}{t_\ell - t_i} = - \frac{a \beta h_0 C_B}{h_B} \tag{18}$$

The difference between the enthalpy and the modified enthalpy is given by

$$i_g' - i_g = (a - 1) C_s t \tag{19}$$

Design procedure

(1) Assuming that the mixture is saturated at the exit, calculate the values of β at the inlet and exit.

(2) Taking as a first approximation $\alpha = 1$. i.e. $hg/kg' = C_s$, construct tie lines of slope $-\beta h_0 C_B/h_B$ through the points A and Z on the operating line and locate $A'''(t_{iA}''', i_{iA}''')$ and $Z'''(t_{iZ}''', i_{iZ}''')$ on the saturation curve as shown in fig. 1.

(3) Calculate the values of α_A''' and α_Z''' by equation (8) using values of P_{iA}''' and P_{iZ}''' which correspond to t_{iA}''' and t_{iZ}''' respectively.

(4) Take the arithmetic mean value of α_A''' and α_Z''' for α. Usually, the value of α is less than one.

(5) Using the value of α obtained, follow the usual procedure of graphical calculation.

The error in calculating the heat transfer surface area was within ±10% in comparison with the experimental value. In addition, the calculation was in good agreement with experiment in tracing the gas temperature in the experimental apparatus, as shown in Fig. 2.

3. Calculation for designing cooler condensers with a computer.

In calculation with a computer, the following Colburn-Hougen type equation of heat balance is easier to use than an equation of modified enthalpy driving force such as equation (10).

$$h_g (t_g - t_i) + \lambda_m k_m \ln \frac{P_t - P_i}{P_t - P_g} = h_o (t_i - t_\ell) \qquad (20)$$

where

$$h_g = j . C_g . G' / (Pr)^{1/2} \qquad (9)^*$$

$$k_m = j . G' / M_g (Sc)^{1/2} \qquad (21)^*$$

The total area of the heat transfer surface is calculated by

$$A = \Sigma \Delta A = \Sigma \frac{\Delta Q}{h_o (t_i - t_\ell)} \qquad (22)$$

and the change of condition of the gas-vapor mixture by

$$\frac{\Delta p_g}{\Delta t_g} = \frac{P_t - P_g}{P_{BM}} \left(\frac{Pr}{Sc} \right)^{1/2} \frac{P_g - P_i}{t_g - t_i} \qquad (23)^*$$

The flow charts for the stepwise calculation program are shown in Fig. 3, 4 and 5. The process of calculation is as follows.

(1) Calculate the heat transfer coefficient h_g and mass transfer coefficient k_m at the inlet for the gas-vapor mixture. This process corresponds to "CALCULATION OF INITIAL CONDITION" and "CALL SUB I" in flow chart.

(2) Taking into account that p_i is the saturated vapor pressure as t_i, determine the values of t_i and p_i by substituting the values of h_g and k_m into equation (20). This process corresponds to "CALL SUB II" in flow chart.

(3) Choosing the proper increment of temperature of gas vapor mixture Δt_g, calculate t_g and p_g at the following step equation (23). This calculation corresponds to "CALCULATION OF MIXED GAS CONDITION" in flow chart.

(4) If p_g is larger than the saturated vapor pressure p_s at t_g, put $p_g = p_s$.

(5) Add the value of the heat transfer area ΔA required for an

---

(*) In the case of a gas system (3), the power to the Prandtl and Schmidt moduli is 1/2 rather than 2/3.

increment to A, using equation (22).

(6) Calculate the value of the enthalpy change of the gas-vapor mixture in an increment of temperature, ΔQ, and calculate the cooling water temperature t_ℓ from the heat balance by using the value of ΔQ.

(7) If t_g is larger than t_{g2}, repeat the procedure (1) $\simeq$ (6). If not, proceed as follows.

(8) Calculate $\Sigma\Delta A$ at the exit gas temperature t_{g2}. The required area of heat transfer is given by $\Sigma\Delta A$.

Though the error in calculating the heat transfer surface area was the same as for the graphical method mentioned above, the agreement between calculation and experiment in tracing the gas temperature change in the apparatus was not as good as with the graphical method.

4. Simplified calculation for designing cooler condensers.

Though calculation with a computer makes the design process easier, a quick design method is desirable.

Mizushina et al. proposed a "Three points method" [6], a simplified calculation which takes a short time to perform and gives fairly accurate results. In addition, this method can be applied to gas-multicomponent vapor systems. The procedure used for the calculation is as follows.

(1) The gas rate, the vapor concentration and the gas temperature at the inlet, the mole-percentage of vapor to be condensed and the temperature of the cooling water available may be given.

(2) Assuming that the outlet gas mixture is in equilibrium with liquid of the concentration of the vapor component ratio in the inlet gas mixture, and calculating the total vapor pressure there from the values in (1), the temperature and the partial pressure of each vapor component are determined. Since this is not the actual case, the values thus determined may be erroneous. However, this does not cause any serious error in the final results for the heat transfer surface area.

(3) From the total heat balance, the total heat transferred, the rate and the outlet temperature of the cooling water are determined. In this balance, the temperature of the condensate is assumed to go down to that of cooling water for the sake of conservative calculation.

(4) A fictitious middle point is assumed where the gas temperature, the rate of each component of vapor, and the cooling water temperature are the arithmetic mean of those at the gas inlet and outlet.

(5) The heat balance of Colburn-Hougen type as shown below, are applied to the three points, viz. inlet, middle and outlet, to calculate the temperatures at the interface between condensate film and gas at those points.

$$h_o (t_i - t_\ell) = h_g (t_g - t_i) + \lambda_m k_m \ln \frac{P_t - \Sigma p_i}{P_t - \Sigma p_g} \qquad (24)$$

where Σp means the sum of partial pressures of vapor components.

The average value of the combined conductance h_0 can be calculated from the heat-transfer coefficient of cooling water and the condensation coefficient computed from Nusselt's equation. This average value is used for all three points.

The gas film coefficient of heat transfer, h_g, and that of mass transfer, k_m, are calculated from the following equations:

$$h_g = \frac{j \, C_g \, G'}{(Pr)^{1/2}} \qquad (9)$$

$$k_m = \frac{j G'}{M_g (Sc)^{1/2}} \qquad (21)$$

In these equations the average values of Pr and Sc are used for all the three points.

(6) As shown in Fig. 6 the relation of $\Delta t = t_i - t_\ell$ and Q is assumed to be represented by two straight lines — the dotted lines — between the inlet and the middle point and between the middle and the outlet point respectively, though the actual temperature difference is supposed to follow the solid line curve. Hence, the heat-transfer surface area can be calculated by

$$A = \frac{Q/2}{h_o \left\{ \left[(\Delta t)_1 - (\Delta t)_m \right] / \left[2.3 \log \frac{(\Delta t)_1}{(\Delta t)_m} \right] \right\}} +$$

$$+ \frac{Q/2}{h_o \left\{ \left[(\Delta t)_m - (\Delta t)_2 \right] / \left[2.3 \log \frac{(\Delta t)_m}{(\Delta t)_2} \right] \right\}} \qquad (25)$$

On the other hand, on the assumption that the curve of $1/\Delta t$ vs. Q is parabolic, Simpson's method was applied to calculate the surface area, but the error was larger than that obtained using equation (25).

Since the solid line in Fig. 6 is always convex in the upper direction, the calculated results should be conservative. Comparison of the calculated results for the heat transfer area and the experimental values shows that the error of calculation for air-water, air-benzene, air-toluene, air-methanol and air-chloroform mixtures is within +13% and −6%, that the error of calculation for air-methanol-water, air-methanol-benzene, and air-benzene-water mixtures is within +21% and −4%, and that the error of calculation for air-acetone-methanol-water mixture is within +20% and +5%..

5. Design of multi-pass cooler condensers.

The methods mentioned above are applicable only to single-pass type heat exchangers. Attempts to apply the methods to the design of multi-pass cooler condensers were so complicated as to be useless.

Mizushina et al. [7] presented two simplified methods and one step-wise method for designing multi-pass cooler condensers, and compared these three methods with each other and with the experimental results. The conclusion is the simplest method is useful, and the calculation error is not too large for design purposes. Therefore, only this simplest method will be described here, as follows.

(1) Assume that the temperature of the cooling water is constant at the arithmetic mean of its inlet and outlet temperatures.

(2) Then apply the "Three points method" mentioned above.

The maximum error in the calculation of the heat transfer surface area is −6.1% compared with the step-wise method and +19.0% compared with the actual area used in experiments with Air-H_2O, Air-CCl_4, Air-CH_3OH and Air-C_6H_6 systems.

6. Cooler condensers with finned tubes.

Finned tubes are often used in gas coolers or heaters, and a factor for fin efficiency is introduced in designing them. Though finned tubes are used sometimes in cooler condensers, a design method has not been established for them, because the fin efficiency in this case is different from that for transferring only sensible heat.

Owing to the additional transfer of latent heat, the fin efficiency in a cooler condenser is much smaller than that in an ordinary sensible heat exchanger and varies from the inlet to the exit of the exchanger.

Mizushina et al. [8] defined fin efficiency in cooler condensers and presented a design method for cooler condensers with finned tubes.

The fin efficiency is defined as the ratio between the actual heat in through the fin, and the heat in on the assumption that the whole surface of the fin is at the temperature of the tube surface at the root of fin. Because of the nonlinear relation between saturated vapor pressure and temperature, accurate calculation is quite complicated and should be conducted with a computer. To simplify the calculation, the relationship between vapor pressure and temperature was assumed to be linear in the range between the temperature of the gas mixture and of the root of a fin at a cross-section. Though the values of fin efficiency calculated by the accurate and simplified methods are different from each other, as shown in Fig. 7, the results for the calculation of heat transfer are not different, and were confirmed to be within the accuracy of practical use by comparing them with the experimental value (the calculation error is within ±10%). Accordingly, the simplification mentioned above is valid and practical for design purposes.

Further simplification of the calculation is achieved by adopting the "Three points method". Namely, the temperature of the inside surface of the tube is calculated at the gas inlet and outlet and a fictitious middle point using the following equation, in which the thermal resistance of tube wall is neglected.

$$ h' \, (t_g - t_w) \, \{ A_f \Omega + (A_t - A_f) \} \; = \; h_\ell (t_w - t_\ell) \, A_\ell \qquad (26) $$

where A_f, A_t and A_ℓ are the area of fin, total outside and inside surface respectively, and h' is an apparent heat transfer coefficient on the surface of fin calculated from

$$ h' \; = \; \cfrac{1}{\cfrac{1}{h_g + k_g a} + \cfrac{1}{h_v}} \qquad (27) $$

where a is the inclination of the linear equation for vapor pressure and temperature assumed at these points as shown in Fig. 8, and calculated by

$$ a \; = \; \frac{p_g - p_w}{t_g - t_w} \qquad (28) $$

Ω is calculated making use of the apparent heat transfer coefficient h' and charts, such as those of Gardner, used for calculations on ordinary finned-tube exchangers. Then, one can calculate the heat transfer surface area as described in Section 4, using h_ℓ and $(t_w - t_\ell)$ at the three points. The calculation accuracy of this, the simplest method, is a little worse than that of the more accurate method described above, but is still good enough for practical use. (The error is within ±15%).

7. Design of evaporative coolers

In evaporative coolers, water is circulated and sprayed on the outside surface of the tube bundle while air is blown from the bottom, and the process fluid inside the tube is cooled by utilizing the latent heat of evaporation. Heat is transferred from the process fluid to the water spray from which sensible and latent heats are transferred to the air stream.

Thus, the heat transfer mechanism involves heat transfer between three kinds of fluids and is complicated. In addition the temperature of the circulating water varies in the apparatus as shown in Fig. 9.

Mizushina et al. [9,10,11,12,13] proposed a design method, and presented the correlating equations for three transfer coefficients between the process fluid and the inside wall of the cooling tube, the outside wall of the cooling tube and circulating water, and between circulating water and blowing air, which are measured experimentally. Since the range of operating condition is limited automatically from the heat balance between the three fluids when the condition of one fluid is fixed, it is necessary to determine the condition of each fluid at the inlet and exit of the apparatus as a first step. Then the heat transfer surface area can be calculated.

Mizushina et al. developed a design procedure and compared the results of their calculation with the experimental results. Both analytical solution and numerical calculation of the basic differential equation to obtain the heat transfer area gave results which were accurate enouth for practical design. (The error is mostly within ±10%, the worst case being ±20%.

In addition, a simple equation to calculate cooling surface area was derived from the assumption that the temperature of the circulating water was constant throughout the apparatus.

$$A = \frac{L_f C_f}{h_o} \ln\left(\frac{t_{f_1} - t_c}{t_{f_2} - t_c}\right) = \frac{G_B}{k_{og}} \ln\left(\frac{i_c - i_{g1}}{i_c - i_{g2}}\right) \tag{29}$$

The calculation result is again accurate enough (the error being generally within ±10%), and this simple method may be recommended for practical use.

NOMENCLATURE

A	area of transfer surface	m^2
A_f	area of fin surface	m^2
A_t	total area of finned surface	m^2
C	specific heat	$kcal/kg°C$
C_S	humid heat	$kcal/kg$ of dry gas$°C$
f	friction factor	—
G_B	flow rate of non-condensable gas	kg of dry gas/hr
G'	mass velocity of gas vapor mixture	kg/m^2 hr
H	humidity	kg/kg of dry gas
h	heat transfer coefficient	$kcal/m^2\,hr°C$
h'	apparent coefficient of heat transfer	$kcal/m^2\,hr°C$
h_B	fictitious coefficient of heat transfer when	kg/m^2 hr
	only non-condensable gas flows	$kcal/m^2\,hr°C$
h_0	combined conductance other than the gas film	$kcal/m^2\,hr°C$
h_v	heat transfer coefficient of condensation	$kcal/m^2\,hr°C$
i	enthalpy of gas vapor mixture	$kcal/kg$ of dry gas
i'	modified enthalpy of gas vapor mixture	$kcal/kg$ of dry gas
j	j factor in heat and mass transfer	—
k_g'	mass transfer coefficient	$kg/m^2\,hr(kg/kg$ of dry gas$)$
k_g	mass transfer coefficient	kg/m^2 hr mm H_g
k_m	mass transfer coefficient	kg-mole$/m^2$ hr
k_{og}	overall coefficient of mass transfer	$kg/m^2\,hr(kg/kg$ of dry gas$)$
L	flow rate of cooling water	kg/hr
L_f	flow rate of process fluid	kg/hr
p	partial vapor pressure	mmHg
p_t	total pressure in condenser	mmHg
p_w	saturated vapor pressure at t_w	mmHg
p_{BM}	average partial pressure of non-condensable gas in gas film	mmHg
Q	total heat to be transferred	$kcal/hr$
t	temperature	$°C$
α	constant defined by equations (5) and (8)	—
β	variable defined by equation (13)	—
λ	latent heat of vaporization	$kcal/kg$

λ_m	latent heat of vaporization	kcal/kg-mole
λ_0	latent heat of vaporization at 0°C	kcal/kg
Ω	fin efficiency	
Pr	Prandtl number	
Sc	Schmidt number	

Subscripts

B	non-condensable gas
c	circulating water
f	process fluid
g	gas vapor mixture or gas film
i	interface
ℓ	cooling water
w	tube wall
1	gas inlet
2	gas outlet

REFERENCES

[1] Mizushina T. and T. Kotoo, Calculation of simultaneous heat and mass transfer, Kagaku-Kikai, 13, 75 (1949)

[2] Mizushina T., Analogy between heat and mass transfer, Ibid. 13, 22 (1949)

[3] Mizushina T. and M. Nakajima, Simultaneous heat and mass transfer, Memoirs of the Faculty of Engineering, Kyoto University, 8, 40 (1951)

[4] Mizushina T., N. Hashimoto and M. Nakajima, Design of cooler condensers for gas vapor mixtures, Chem. Engng. Science, 9, (1959)

[5] Mizushina T., M. Nakajima and T. Oshima, Study on the cooler condensers for gas-vapour mixtures, Ibid., 13, 7 (1960)

[6] Mizushina T., H. Ueda, S. Ikeno and K. Ishii, Simplified calculation for cooler condensers for gas-multicomponent vapour mixtures, Int. J. Heat Mass Transfer, 7, 95 (1964)

[7] Mizushina T., R. Ito, H. Kamimura and A. Nakamura, Simplified calculation for multi-tube cooler condensers, Kagaku-Kogaku, 33, 429 (1969)

[8] Mizushina T., S. Iuchi, T. Oshima, I. Ito and T. Hamaura, Calculation for finned tube cooler condensers, Ibid. 34, 292 (1970)

[9] Mizushina T., R.Ito and H. Miyashita, Experimental study of an evaporative cooler, Int. Chem. Engng., 7, 727 (1967), Translated from Kagaku-Kogaku, 31, 469 (1967)

[10] Mizushina T., R. Ito and H. Miyashita, Characteristics and methods of thermal design of evaporative coolers, Ibid., 8, 532 (1968), Translated from Kagaku-Kogaku, 32, 55 (1968)

[11] Mizushina T. and H. Miyashita, Study on froth contact coolers, Kagaku-Kogaku, 32, 987 (1968)

[12] Mizushina T. and H. Miyashita, Study on evaporative coolers, Ibid. 33, 651 (1969)

[13] Mizushina T. and H. Miyashita, Thermal design of cooler condensers, Ibid., 35, 693 (1971)

Microfilms of these papers are deposited in the International Centre for Heat and Mass Transfer.

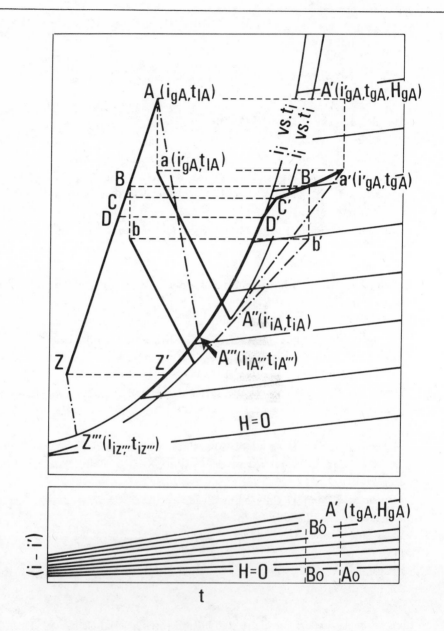

Fig. 1. Enthalpy and modified enthalpy-temperature chart

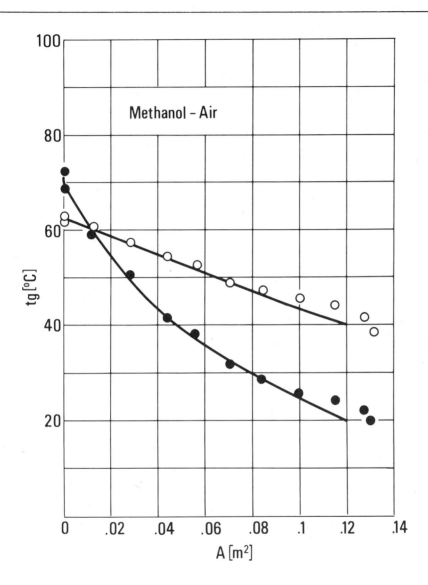

Fig. 2. Comparison of the results of calculation and experiment for a gas temperature change

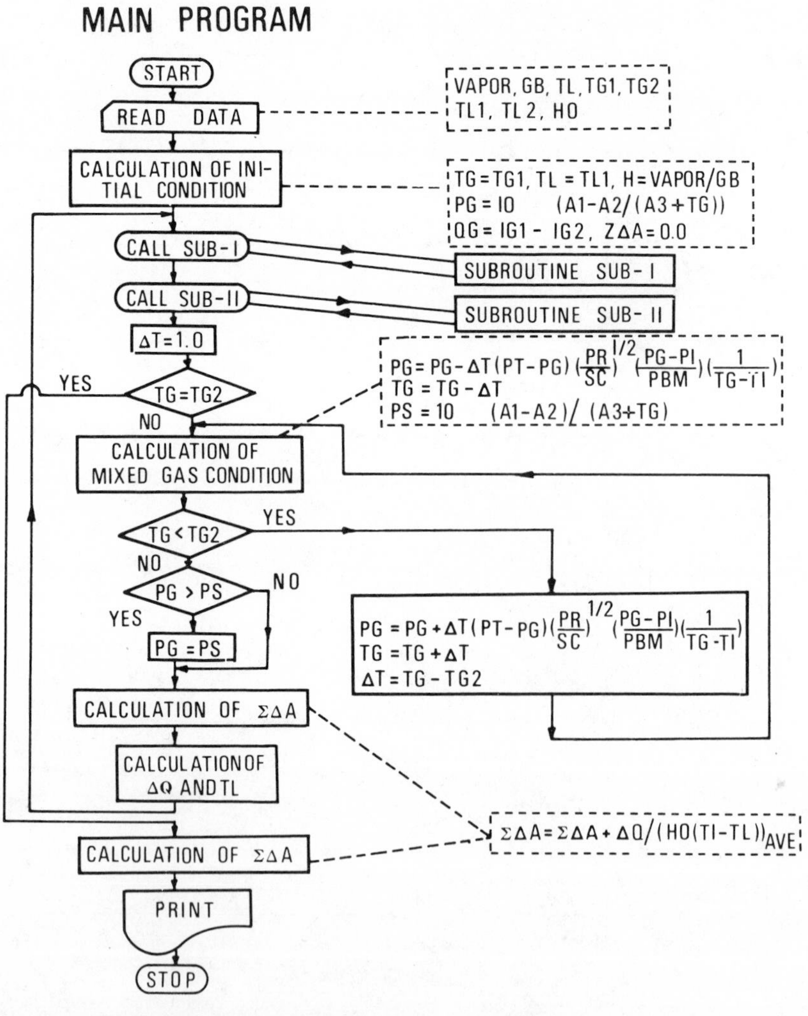

MAIN PROGRAM

START

READ DATA - - - - - - - - VAPOR, GB, TL, TG1, TG2
TL1, TL2, H0

CALCULATION OF INI-
TIAL CONDITION - - - - - - - - TG = TG1, TL = TL1, H = VAPOR/GB
PG = 10   (A1−A2/(A3+TG))
QG = IG1 − IG2, ZΔA = 0.0

CALL SUB-I ⟶ SUBROUTINE SUB-I

CALL SUB-II ⟶ SUBROUTINE SUB-II

ΔT = 1.0

$PG = PG - \Delta T (PT - PG) \left(\frac{PR}{SC}\right)^{1/2} \left(\frac{PG-PI}{PBM}\right) \left(\frac{1}{TG-TI}\right)$
$TG = TG - \Delta T$
$PS = 10 \quad (A1-A2)/(A3+TG)$

TG = TG2   YES

NO

CALCULATION OF
MIXED GAS CONDITION

TG < TG2   YES

NO

PG > PS   NO

YES

PG = PS

$PG = PG + \Delta T (PT - PG)\left(\frac{PR}{SC}\right)^{1/2}\left(\frac{PG-PI}{PBM}\right)\left(\frac{1}{TG-TI}\right)$
$TG = TG + \Delta T$
$\Delta T = TG - TG2$

CALCULATION OF ΣΔA

CALCULATION OF
ΔQ AND TL

CALCULATION OF ΣΔA - - - - - - - - $\Sigma\Delta A = \Sigma\Delta A + \Delta Q/(H0(TI-TL))_{AVE}$

PRINT

STOP

Fig. 3. Flow chart of stepwise calculation program for cooler condenser

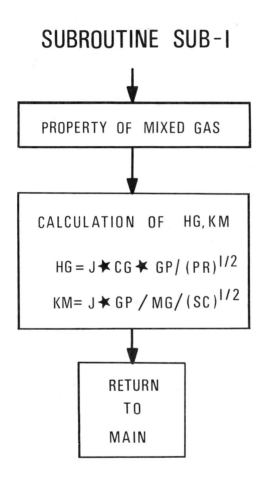

Fig. 4. Subroutine of calculation

SUBROUTINE SUB-II

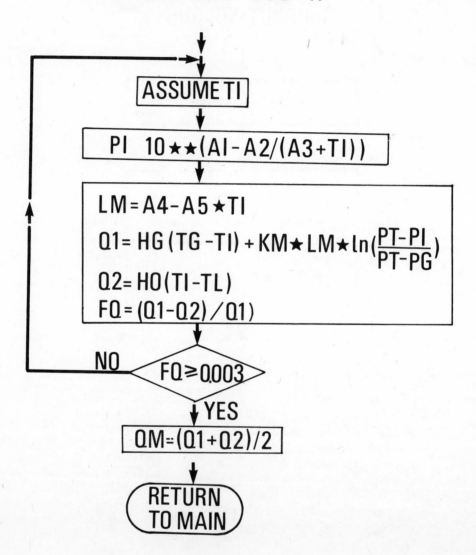

Fig. 5. Subroutine of calculation

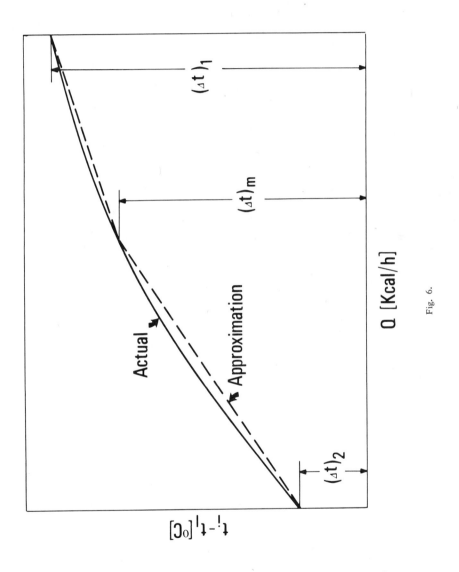

Fig. 6.

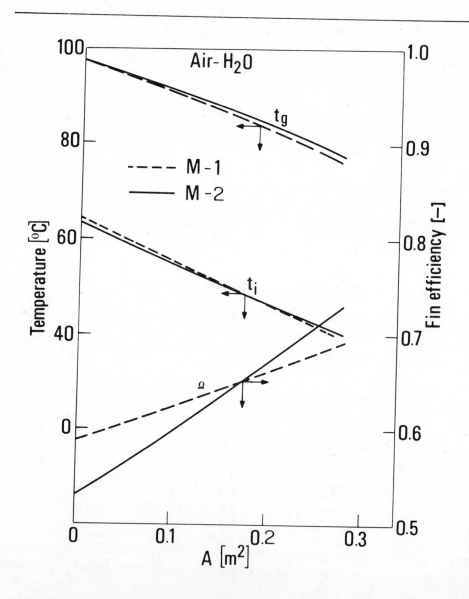

Fig. 7. Comparison of simplified method (M-1) and accurate method (M-2)

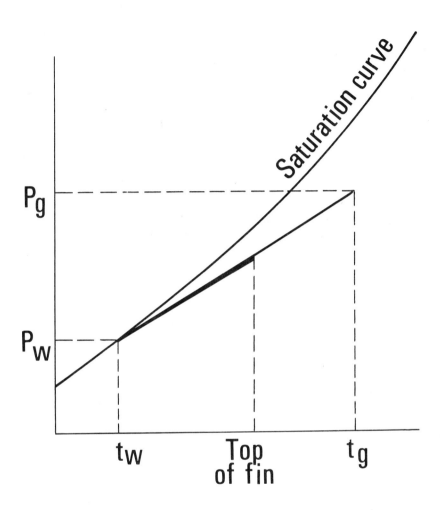

Fig. 8. Assumption of lineality of relation between p and t

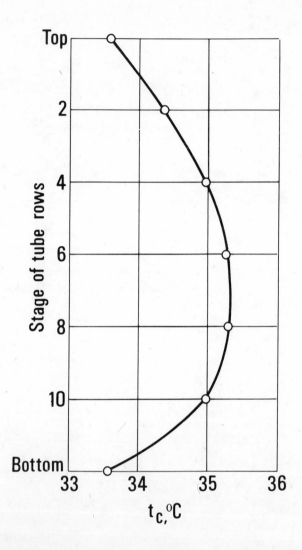

Fig. 9. An example of temperature distribution of circulating water

Chapter 17

SOME RECENT DEVELOPMENTS IN AUGMENTED HEAT EXCHANGE ELEMENTS

T.C. Carnavos (*)

Introduction

In recent years, the need to augment or intensify heat transfer has had engineers urgently searching for new methods and techniques. Motivation has not only come from economic pressures but also from the developing technology requiring more efficient equipment smaller in size and weight.

The heat transfer design engineer has always been receptive to methods which improve heat transfer coefficients. It is only recently that comprehensive experiments have been performed and reported which more clearly defines the conditions an augmentative technique will improve heat transfer. These techniques may be broadly classed as follows:

(1) Augmentive stimulation is continuously supplied.

(2) Turbulence promotion.

(3) Extended heat transfer surfaces.

(4) Enhanced heat transfer surfaces.

Typical examples of (1) are fluid additives, electrostatic fields, and surface or fluid vibrations. Some examples of (2) are coiled wires, inlet vortex generators, twisted tapes, displaced promoters, and spirally grooved tubes. Category (3) houses the large and well established family of finned surfaces. Tubes with externally finned surfaces are the most commonly used. The use of internally finned tubes is not unknown in the art but not commonly used. Interest in this type of surface has increased in recent years. Enhanced heat transfer surfaces are a relatively new class of surfaces that yield heat transfer coefficients that are several orders of magnitude greater than

---

(*) Noranda Metal Industries, Inc. Newton, Connecticut 06470 U.S.A.

N.B. ALL FIGURES QUOTED IN TEXT ARE AT THE END OF THE CHAPTER

conventional surfaces. These gains are usually brought about by permanent surface treatment. Some examples of (4) are applied hydrophobic coatings for condensing, applied porous coatings for boiling, and textured or formed surfaces for both condensing and boiling.

The absolute classification of an augmentation technique is not possible since generally it can be beneficial for several modes of heat transfer. Externally finned tubes are a good example wherein they have been used to good advantage in natural and forced convection, pool and forced convection boiling, and condensing applications. The use of any augmentation technique will heavily depend on the economics of a particular situation with the final decision being made on the basis of total minimum cost comprised of operating and first costs. Many published studies provide the data necessary for the design engineer to establish the impact on operating costs. Such studies cannot provide data to assess the impact on first costs since each situation is unique and such information must come from the equipment manufacturer.

This presentation reports on the performance of four augmented heat exchange elements that are receiving increased attention and have reached the stage of commercial use. They fall into the general classifications of (2), (3) and (4) as defined above. Much of the data are recent and not previously available in the open literature. Some of the data have been reported upon in the open literature at various times but may not be broadly known and are presented here for the sake of completeness.

(A) Fluted Surfaces

The first reported use of fluted enhanced surfaces was by Gregorig [1] who showed by calculation and experiment that fluting the outside surface of a vertical tube in a steam condensing application gave heat transfer coefficients between 10,000 and 20,000 BTU/hr. ft.^{2} °F. Condensation was in a filmwise manner yet coefficients ware produced in the dropwise range. Diedrich and Lotz [2], using the basic ideas of Gregorig, developed a tube fluted on both sides for Vertical Tube Evaporators in desalination applications. A final tube configuration developed is shown in Figure A-1 and differs from Gregorig's tune in the number of flutes per unit of circumference and their shape. Gregorig's tube had a higher flute density and were more sinusoidal in shape.

The principle of operation for the double fluted tube is pictorially shown in Figure A-2. The condensate film is subjected to surface tension forces acting to drain it from the crests into the grooves. This results in the major portion of the flute having a very thin film of condensate, which greatly reduces the resistance to heat flow through the crest area. The condensate in the grooves is channeled off by gravity with the heat flow through this area somewhat less. The falling film of feed on the boiling side also tends to drain into the grooves. Boiling takes place in the thin film on the crests as well as in grooves. The crests are kept constantly wet by the splash from the boiling action in the grooves. The generated vapor is preferably exited out the bottom of the tube. The flutes perform a secondary but very important function in organizing and controlling the falling film to assure a uniform distribution down the tube length.

Condensing and boiling heat transfer performance for double fluted tubes was reported by Carnavos [3]. Typical condensing performance is shown in Figure A-3 left side. Data by Lustenader and Staub [4] are shown on the right hand side for a smaller diameter fluted tube. The filmwise coefficient, based on total outside surface, for fluted tubes are 4-1/2 to 7 times greater than obtained for a comparable smooth tube at a given heat flux and are within the accepted range of dropwise condensation.

The novelty of the double fluted tube is that it provides enhancement both sides. Typical falling film boiling coefficients for sea water feed are shown in Figure A-4 and are compared to smooth surface data of Young and Hummel [5]. At a typical operating heat flux of 40,000 BTU/hr. ft.^{2} a three-fold increase in coefficient was realized.

Chia [6] has recently reported the overall performance for seven configurations of 2″ O.D. X 10ft. long double fluted tubes. Also tested under identical conditions was a smooth tube. Chia reports overall heat transfer coefficients, based on nominal or rubber band outside surface area, that are consistently a factor of 2.4 times higher than smooth tube for the best configuration tested. Double fluted tubes of this type typically have from 1.25 to 1.35 more outside surface than a smooth tube of equal diameter. Taking a 1.35 surface correction, the enhancement factor is about 1.8 which would be consistent with the individual condensing and boiling coefficients previously mentioned factoring in tube wall resistance.

Double fluted tubes have potential use as feed preheaters and condensers in VTE desalting plants as well as vertical tube condensers for other applications. Overall heat transfer coefficients were reported by Carnavos [7] and shown in Figure A-5 for this type tube operated with a spray film tubeside. Overall coefficients of 1200 to 1400 BTU/hr. ft.^{2} °F based on total outside surface were realized dependent upon the tube material and is typical of the effect metal resistance can have on enhanced tubes. More recently Eissenberg [8] reported data for a 1″ O.D. tube operated with a flooded tube side. Figure A-6 shows the tube configuration tested with Figures A-7 and A-8 detailing the heat transfer and pressure loss performance obtained all based on rubber band diameters and areas. The convective constant C_i, defined by the Colburn type equation:

$$Nu = C_i R_e^{0.8} P_r^{0.4}$$

had a value (0.0312) higher than for a smooth tube (0.023-0.027) by slightly more than the amount expected for the developed area produced by the flutes. The estimated developed area ratio is 1.13. The friction factor curve lies about 25% above the conventional smooth tube line, which reflects the loss in flow area due to fluting, since the friction factor was defined based on the nominal inside diameter. The overall coefficient of 1350 BTU/hr. ft.^{2} °F at 5.25 ft./sec. tube side velocity is almost double that for a vertical smooth tube. The experimentor noted the grooves of the flutes were almost completely filled at the higher heat flux (31,000 BTU/hr. ft.^{2}) runs.

(B) Spirally Grooved Surfaces

The heat transfer performance of spirally grooved surfaces are of interest due to the high coefficients that can be obtained, both sides, by means of this enhanced surface configuration. An increase in pressure drop will naturally occur in forced convection sensible heat transfer since the augmentation mechanism is turbulence promotion. The high tubeside pressure drop for these tubes do not negate high utilization in applications where this is not a limiting condition but compactness and weight are primary considerations. A typical spirally grooved tube configuration is shown in Figure B-1.

Palen, Cham, and Taborek [1] have reported on the performance of a spirally grooved tube configuration in a baffled shell and tube condenser. An identical bundle constructed with smooth tubes of the same diameter was also tested for comparison. Figure B-2 shows the spirally grooved tube bundle tested while Table B-1 details tube and bundle geometry as well as the range of experimental conditions. Figures B-3 and B-4 present the overall heat transfer coefficient obtained corrected for tube wall resistance as functions of the tubeside flow rate and condensate rate respectively. Within the range tested, this spirally grooved tube exhibited an overall enhancement factor of somewhat over 2 and was relatively insensitive to the condensate loading or temperature difference at constant tubeside flow rate. The increase in overall heat transfer coefficient with an increase in tubeside flow rate indicates the tubeside heat transfer resistance is a significant portion of the total resistance. Figure B-5 presents condensing film coefficients as a function of total condensate rate. These data are derived from Wilson plot analysis of the test results and cannot be given the high level of confidence of the directly acquired data due to assumptions that must be made. Within this framework, certain trends can be ascertained. The effect on the condensing coefficient of shellside pressure over the range of 55 to 105 psig and condensate loading is small. The condensing coefficient was enhanced by a factor of slightly over 2 as was the overall coefficient for this spirally grooved tube. The experimentors report no statistically significant increase in shell side pressure drop over the plain tubes for this particular baffle arrangement. They concluded the increased condensing coefficient for this spirally grooved tube was probably due to better film drainage and/or promotion of turbulence in the otherwise laminar condensate film. Special surface treatments and tests preclude the possibility that dropwise condensation was occurring. A more general conclusion was reached, within the accuracy of the Wilson plot technique, that this particular configuration increases both tubeside and shellside coefficients by about a factor of 2 as compared to a smooth tube under equal flow conditions.

It was not possible to obtain meaningful tubeside pressure drop measurements since tube length was small and entrance and exit effects obscured frictional losses. Separate measurements were made and the results are presented in Figure B-6. It is quite evident the conversion of pressure drop to heat transfer is inefficient for this spirally grooved tube relative to a smooth tube. However, if pressure loss is not the omitting design condition, much higher heat transfer rates are obtainable from these tubes at a given flow rate.

Blumenkrantz and Taborek [2] have reported on the heat transfer and pressure loss performance of spirally grooved tubes in the turbulent regime. A wide range of configurations were tested and Table B-2 presents the dimensional characteristics of the tubes that will be reported on here. Water or air were used as test fluids circulating inside the spirally grooved tube, while the service fluids, condensing steam or cooling water, were circulated in an external annulus. The system was operated as an open loop when air was used. Tubes were tested ranging from 3/8″ to 1″ outside diameters, including a smooth tube for compariosn, with the tube length kept constant at 8 ft.

The experimenters have reported their results in a most complete manner in the reference. The design engineer is most interested in the comparison of the tube performance in terms of heat transfer rate versus pressure drop for a rational selection of the most suitable configuration. Figure B-7 presents UA as a function of pressure loss. Several observations were made as follows:

(1) UA increases approximately with the 1/3 power of the pressure loss.

(2) The pressure loss required to transfer the same amount of heat increases as the tube diameter decreases.

(3) Tube 21 was basically the outstanding performer and, in its diameter class, had the shallowest groove depth and coarsest pitch.

This suggests the tight pitched and deep grooved tubes obtain their enhancement from the smaller free stream cross-section and increased core velocity at the cost of pressure loss. The fluid in the deep and narrow grooves is almost stagnant reducing considerably the ability of this portion of the tube surface to transfer heat.

In the cases where the service fluid was condensing steam, Wilson plot analysis of the data was performed to separate liquid-side and steam-side contributions to heat transfer. The results of this investigation were consistent with those obtained in Reference [1] this section. The groove valleys contribute in easing the drainage of condensate, thereby decreasing the film resistance at the top and, may, at lower condensate loadings, induce film turbulence. The mechanism is analogous to that set forth in Section A for fluted surfaces operating in the condensing mode.

(C) Porous Surfaces

Much effort has been expended in developing surfaces that will enhance nucleate boiling heat transfer. Generally accepted theory is that nucleate boiling requires the presence of microscopic vapor nuclei in the form of bubbles entrapped on the heat transfer surface. Surface tension at the vapor-liquid interface of the bubbles exerts a pressure above ambient on the vapor. This additional pressure requires the liquid to be superheated in order for the bubbles to exist or grow. On conventional surfaces the starting vapor nuclei contained within surface pits or scratches are quite small and require substantial superheat to grow. Surfaces that are artificially roughened, scored, cross-knurked, etc., have been developed to improve performance over smooth surfaces.

A novel surface has been developed that increases boiling heat transfer coefficients significantly, extends the nucleate boiling range to very low temperature differences and delays the onset of vapor binding. The surface consists of a porous metallic matrix which is bonded to the outer or inner diameter of a tube. Figure C-1 shows such a tube. This surface appears as a roughened finish, is on the order of .01 to.02 inches thick, and has a void fraction of 50 to 65%. Figure C-2 pictorially represents the principle of operation. The porous surface is able to substantially reduce the superheat required to generate vapor by stably entrapping a high density of relatively large vapor nuclei in the sites or cavities. The entrances to the sites are restricted which prevents them from being flooded out but many are interconnected so that active ones are continuously supplied with new liquid. A second temperature difference exists across the liquid film between the wall and vapor. This temperature difference can be made small if the liquid film is thin and the total area of the film per unit of superficial surface is large. The high density of active nuclei fulfills these conditions. Figure C-3 presents comparative boiling performance data for three typical organic fluids. The enhancement provided by the porous surface is quite remarkable. Figure C-4 presents overall performance for a Vertical Thermosiphon Reboiler condensing steam and boiling 15% ethylene glycol at 45 psia. The porous surface tube exhibited a seven-fold increase in heat flux capability over smooth tube at equal overall temperature differences. Figure C-5 presents long term performance in an ethylene-propylene reboiler-condenser. Ethylene is condensed at 250 psig against propylene boiling at 6 psig. The propylene refrigeration system was of the reciprocating type and, at times, oil concentrations in the propylene were as high as 9%. In spite of these conditions, performance was maintained for 2-1/2 years at about six times that obtainable from conventional tubing.

(D) Internally Finned Tubes

Finned surfaces have been extensively investigated for both the sensible and two-phase conditions. The majority of the reported information has been confined to externally finned tubes and plate-fin type surfaces. In recent years new manufacturing techniques have made it possible to produce a wide variety of tubing with integral internal fins that may be straight or spiralled. Figure D-1 shows such a tube. These types of surfaces provide augmentation by the additional heat transfer surface and, in the case of the spiralled fins, the vortex motion produced.

Prior work with internally finned tubes have been reported by Hilding and Coogan [1] testing with air, Lavin & Young [2] evaporating refrigerants, and more recently Bergles et al. [3] with water in the turbulent regime. In the sensible heat transfer tests augmentation was reported over smooth tubes in the range of

110% to 190% at constant pumping power. Nucleate boiling coefficients for refrigerant R-12 were reported to be much higher on a square foot of area basis than those for a plain tube. Other experimenters boiling on external finned tubes have noted this same effect. It was postulated the sharp corners found in each fin tube configuration provided conditions favorable to nucleation. Three tubes with approximately the same corner length per foot were correlated by a single line. A fourth tube with about one-third the corner length per foot fell about half-way between the plain tube and other data which appears to support the explanation.

More data [4, 5] than previously reported in the open literature, are now available for a wider range of internally finned tube configurations with water in turbulent flow. Seventeen internally finned tubes, five with straight fins and twelve with spiral fins, were tested. A plain tube was also tested for comparison. Figure D-2 presents cross-sectional views of these tubes and Table D-1 lists important dimensions. Figures D-3, 4 and 5, present friction factors based on the equivalent diameter of each tube tested. Figure D-6 presents equations correlating this data for straight and spiral finned tubes within acceptable standard deviations. Figures D-7, 8 and 9 present heat transfer performance based on the inside tube diameter and nominal area. In essence, these plots indicate the effect on performance when replacing a smooth tube with a finned tube. Several observations are made as follows:

(1) The performance of either a spiral or straight finned tube approaches that of a smooth tube as the Reynolds Number increases.

(2) In the lower Reynolds Number range (10,000) spiral finned tunes were superior performers.

(3) In the higher Reynolds Number range (100,000) both spiral and straight finned tubes were nearly equal in performance except the low spiral Tube 19 which is higher.

Figure D-10 presents the heat transfer performance on an equivalent diameter and effective area basis. In order to correlate the data it was necessary to use an approach similar to the friction factor correlations. Figure D-11 presents equations that correlated the data within acceptable standard deviations. Tube 19 with 50 very low fins on a tight spiral which almost approaches a thread type roughness gave substantially higher Nusselt numbers than predicted by the spiral fin tube correlating equation and was excluded. The experimenters report their work is essentially in agreement with that of Bergles et al. except for Bergles Tube 5 which also gave higher heat transfer performance. The Bergles Tube 5 configuration also approaches a thread type roughness but with a higher density of very low fins of the same height as Tube 19 and with a slightly tighter pitch. It should be noted the fin shape of Tube 19 is very near triangular as were the fins of Bergles Tube 5 which may have influenced performance favorably. The fin shape of the other tubes tested either in this work or Bergles et al. are more nearly trapezoidal. More work should be done with trapezoidal and triangular shaped fins to better understand their influence on performance.

Figures D-12, 13 and 14 present the heat transfer coefficient ratio at constant pumping power as a function of Reynolds Number which is one of many ways to compare performance. The best performance was exhibited by the spiral fin tubes with Tube 18 in the low and Tube 19 in the high Reynolds Number range. The sudden increases in performance ratio for Tubes 15 and 18 coincide with the decrease in friction factor. Straight fin tubes gave a slightly lower performance than spiral fin tubes as might be expected upon inspecting the correlating equations which reveal the Nusselt Number increases more rapidly with spiralling than does the pressure drop. None of the straight fin tubes exceeded an enhancement ration of 1.6.

Closure

The work to date in the art of augmentation is only the beginning. It is so obviously beneficial that it will only be a question of time before conventional tubing will be passé. Much work has yet to be done and as it gets done augmentation of both sides of the heat transfer surface will be more the common occurrence than the exception as it is today.

Acknowledgement

The preparation of this chapter has been made possible with the cooperation of the following organizations:

(1) Nuclear Desalination Program, Oak Ridge National Laboratory, Oak Ridge, Tenessee, 37830, and the Office of Saline Water, Department of the Interior, Washington, D.C. 20240.

(2) Heat Transfer Research, Inc., Alhambra, California, 91803, and Spiral Tubing Corporation, New Britain, Connecticut, 06051.

(3) Noranda Research Centre, Pointe Claire 730, Quebec, and Noranda Metal Industries, Ltd., Montreal 101, Quebec.

REFERENCES

[A-1]   Gregorig, R., "Film Condensation of Fine Grooved Surfaces with Consideration of Surface Tension," ZAMP, 5, 36-49, 1954.

[A-2]   Diedrich, G.E. and Lodtz, C.W., "Distillation Apparatus Having Corrugated Heat Transfer Surfaces." U.S. Patent 2, 291, 704, 1966.

[A-3]   Carnavos, T.C., "Thin-Film Distillation" First International Symposium on Water Desalination" SWD-17, Washington, D.C. October, 1965.

[A-4]   Lustenader, E.L. and Staub, F.W., "Development Contributions to Compact Condenser Design," Session II, International Nickel Company Power Conference, May, 1964.

[A-5]   Young, R.K. and Hummel, R.L., "Higher Coefficients for Heat Transfer with Nucleate Boiling" 7th National Heat Transfer Conference, AICHE-ASME, August, 1964.

[A-6]   Chia, W.S., "Heat Transfer Testing of Enhanced Tubes". United States Department of Interior, Office of Saline Water Research and Development Progress Report No. 73, December, 1971.

(A-7)   Carnavos, T.C., "Augmenting Heat Transfer in Desalination Equipment with Fluted Surfaces" Office of Saline Water Symposium "Enhanced Tubes for Distillation Plants" Washington, D.C., March, 1969.

(A-8)   Eissenberg, D.M., "Performance of a Double-Fluted Tube as a Vertical Condenser," Task Report No. 6, Office of Saline Water, Oak Ridge National Laboratory, April, 1972.

(B-1)   Palen J., Cham, B., and Taborek, J. "Comparison of Condensation of Steam on Plain and Turbotec Spirally Grooved Tubes in a Baffle Shell-and-Tube Condenser." HTRI Report 2439-300/6, January, 1971.

(B-2)   Blumenkrantz, A. and Taborek, J. "Heat Transfer and Pressure Drop Characteristics of Turbotec Spirally Deep Grooved Tubes in the Turbulent Regime," December, 1970.

(C-1)   Czikk, A.M., Gottzmann, C.F., Ragi, E.G., Withers, J.G., Habdas, E.P. "Performance of Advanced Heat Transfer Tubes in Refrigerant-Flooded Liquid Chillers" American Society of Heat, Refrigerating, and Air Conditioning Engineers, Volume 76, Part 1, No. 2132.

(C-2)   O'Neill, P.S., Gottzmann, C.F., Terbot, J.W., "Heat Exchanger for NGL," Chemical Engineering Progress, Volume 67, No. 7, July, 1971.

(C-3)   Pamphlet F-3453, Union Carbide Corporation, Linde Division, Tonawanda, New York.

(D-1)   Hilding, W.E., and Coogan, C.H., Jr., "Heat Transfer and Pressure Loss Measurements in Internally Finned Tubes," Symposium on Air-Cooled Heat Exchangers, ASME, New York, 1964 pp 57-85.

(D-2)   Lavin, J.G., and Young, E.H., "Heat Transfer to Evaporating Refrigerants in Two-Phase Flow," AICHE Journal Volume 11, No. 6, November 1965, pp 1124-1132.

[D-3]   Bergles, A.E., Brown, G.S., Jr., Snider, W.D., "Heat Transfer Performance of Internally Finned Tubes," Tulsa Heat Transfer Conference, August, 1971, ASME Preprint 71-HT-31.

[D-4]   Watkinson, A.P., and Miletti, D.L., "Heat Transfer and Pressure Drop of Forge-Fin Tubes in Turbulent Water Flow," Noranda Research Centre, Pointe Claire, 730, Quebec, Report No. 255, January, 1972.

[D-5]   Watkinson, A.P., Miletti, D.L., Tarassoff, P., "Turbulent Heat Transfer and Pressure Drop in Internally Finned Tubes," Colorado Heat Transfer Conference, August, 1972, AICHE Preprint No. 10.

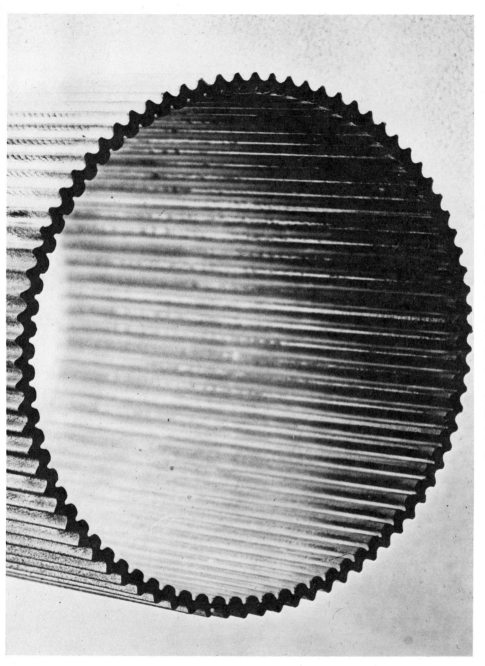

Fig. A-1. 3" Od nominal double-fluted tube configuration.

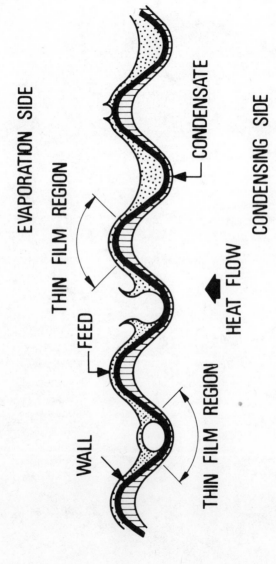

Fig. A-2. Double-fluted tube principle of operation.

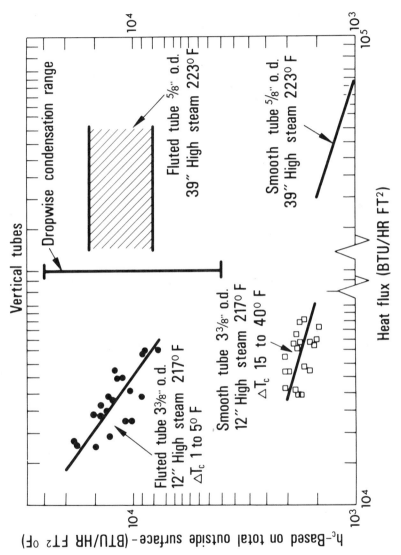

Fig. A-3. Fluted tube condensing heat transfer.

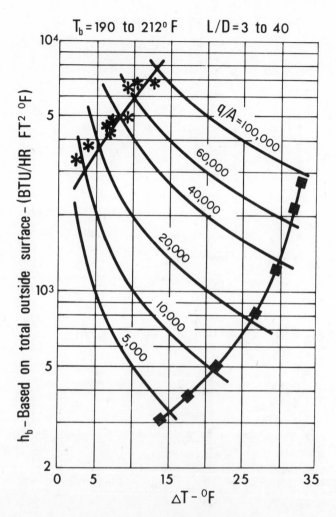

Fig. A-4. Fluted tube boiling heat transfer.

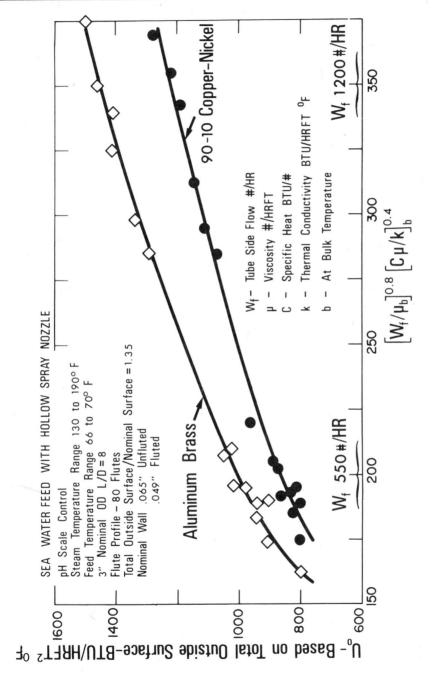

SEA WATER FEED WITH HOLLOW SPRAY NOZZLE

pH Scale Control
Steam Temperature Range 130 to 190° F
Feed Temperature Range 66 to 70° F
3″ Nominal OD L/D = 8
Flute Profile – 80 Flutes
Total Outside Surface/Nominal Surface = 1.35
Nominal Wall .065″ Unfluted
                     .049″ Fluted

Aluminum Brass

90-10 Copper-Nickel

W_f – Tube Side Flow #/HR
μ – Viscosity #/HRFT
C – Specific Heat BTU/#
k – Thermal Conductivity BTU/HRFT °F
b – At Bulk Temperature

W_f 1200 #/HR

W_f 550 #/HR

$\left[W_f / \mu_b \right]^{0.8} \left[C\mu/k \right]_b^{0.4}$

U_o – Based on Total Outside Surface–BTU/HRFT² °F

Fig. A-5. Double-fluted tube spray film condenser.

Fig. A-6. 1" Od nominal double-fluted tube configuration.

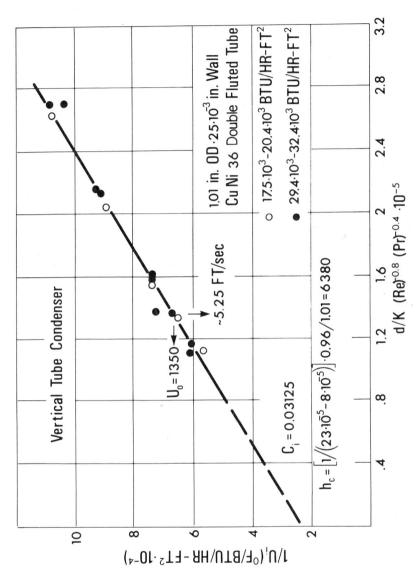

Fig. A-7. Wilson plot of 1" double-fluted condenser tube.

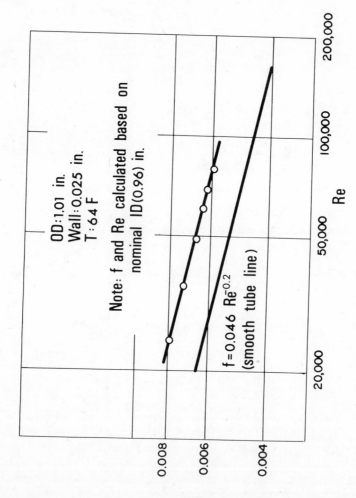

OD: 1.01 in.
Wall: 0.025 in.
T: 64 F

Note: f and Re calculated based on
nominal ID(0.96) in.

$f = 0.046 \, Re^{-0.2}$
(smooth tube line)

Fig. A-8. Friction factor plot of 1" double-fluted condenser tube.

Fig. B-2. Tube bundle with spirally grooved tubes.

TABLE B-1

TUBE GEOMETRY	Specification	Tube Type	
		Plain	Spirally Grooved Tube
	Tube plain end OD, in, DTO	1.0	1.0
	Wall thickness, in, t	0.036	0.038 (plain end)
	Number of Groove Starts, n	0	4
	Groove Depth, in, e	0	0.19
	Pitch Length, in, P	0	2.25
	Tube Material	90−10 cupro-nickel	97.5% copper

Thermal Conductivity Equation:

$$k_w(\text{BTU/hr ft }^\circ\text{F}) = \quad 25.75 + 0.019T \qquad\qquad 150$$
$$T \text{ in } ^\circ\text{F}$$

TUBE GEOMETRY ILLUSTRATION

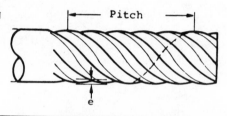

Bundle Geometry.

a) Tube length between tubesheets (L): 3.75 ft.

b) Number of tubes (NTT): 196

c) Shell ID (DS): 24 in.

d) Number of tubes in vertical row at bundle diameter (N_{rv}): 16

e) Number of baffles (NB): 3

Range of Experimental Conditions.

Shellside (Steam)

Steam pressure	55 psig and 105 psig
Condensate temperature, (saturation)	303 F and 341 F
Condensate flow rates	3000−10,000 lb/hr
Fraction condensed	0.75 − 0.93

TABLE B-2

TUBE CODE NO.	NOMINAL O.D. (in)	MINIMUM INSIDE DIA (in)	NUMBER OF STARTS	GROOVE DEPTH (in)	PITCH (in)	AREA* (ft^2)	SYMBOL LINE
3	1.00	0.585	4	0.184	2.227		◁ ·····
5	1.00	0.650	3	0.204	2.326		⊕ ·—··
9	5/8	0.353	4	0.0625	1.344		⬙ –·–·–
10	5/8	0.396	4	0.078	1.391	1.359	▶ ————
11	5/8	0.372	3	0.0938	1.531		◆ ————
14	1.00	0.668	4	0.166	2.594	2.279	■ —·—
17	1.00	0.582	5	0.153	1.922	2.493	▷ ········
19	3/8	0.219	3	0.064	0.781	0.738	◇ ————
21	1.00	0.840	3	0.106	4.250	1.977	□ – – –
Plain	1.00	0.916				1.918	○ ————

*This is the inside heat transfer area of an 8 ft. long tube

Fig. B-1. Spirally grooved tube.

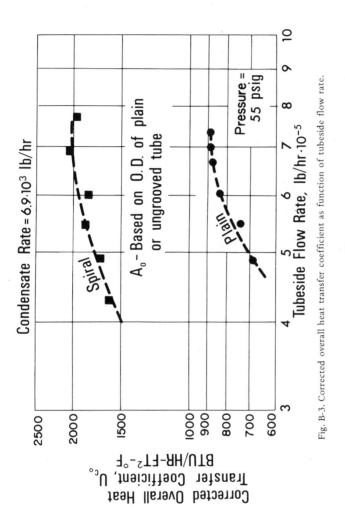

Fig. B-3. Corrected overall heat transfer coefficient as function of tubeside flow rate.

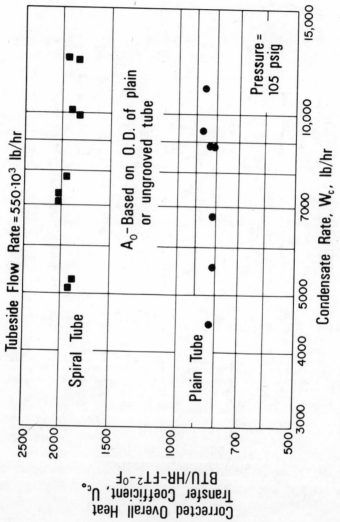

Fig. B-4. Corrected overall heat transfer coefficient as function of condensate rate.

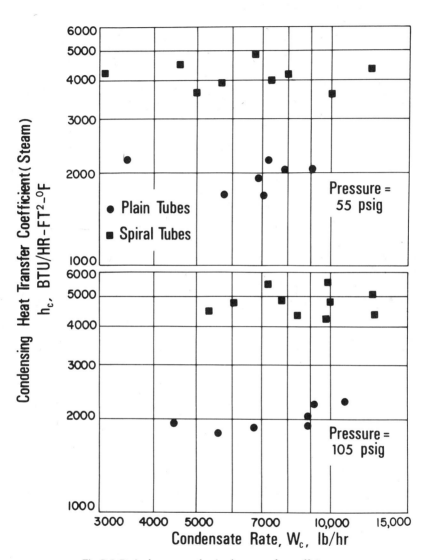

Fig. B-5. Derived steam condensing heat transfer coefficients.

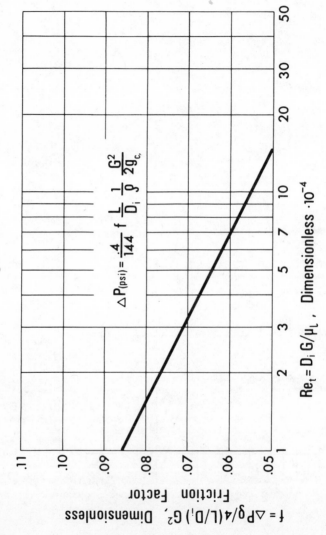

$$\Delta P_{(psi)} = \frac{.4}{144} f \frac{L}{D_i} \frac{1}{\wp} \frac{G^2}{2g_c},$$

$f = \Delta P \wp / 4 (L/D_i) G^2$, Dimensionless
Friction Factor

$Re_t = D_i G/\mu_L$, Dimensionless $\cdot 10^{-4}$

Fig. B-6. Friction factor for spiral tube based on plain end inside diameter.

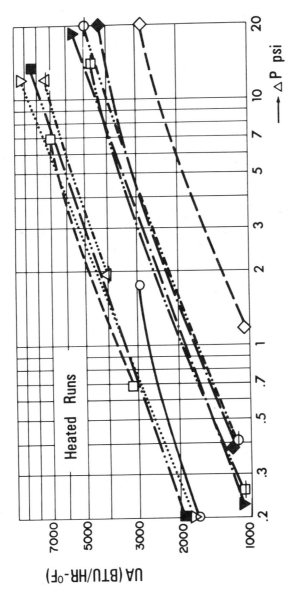

Fig. B-7. Heat transfer (UA) vs. pressure drop.

Fig. C-1. 1″Od nominal high flux tubing.

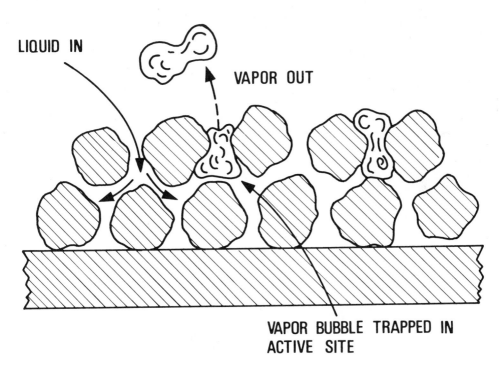

LIQUID IN

VAPOR OUT

VAPOR BUBBLE TRAPPED IN
ACTIVE SITE

Fig. C-2. Porous surface principle of operation.

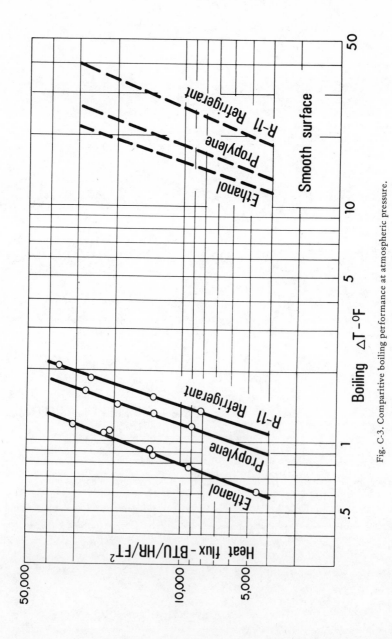

Fig. C-3. Comparitive boiling performance at atmospheric pressure.

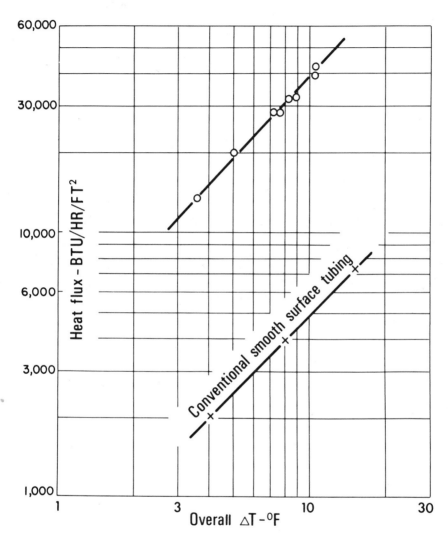

Fig. C-4. Vertical thermosiphon reboiler condensing steam; boiling 15% ethylene glycol @ 45 PSIA.

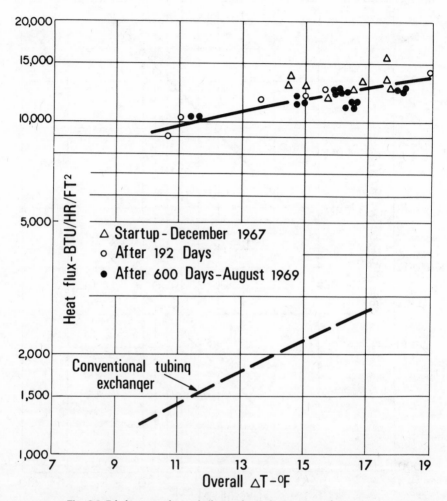

Fig. C-5. Ethylene-propylene reboiler-condenser long term performance.

Fig. D-1. Integral inner finned tube.

TABLE D-1

DIMENSIONS OF TUBES TESTED

Tube No.	1	2	3	4	5	9	10	11	13	14	15	16	17	18	19	20	21	22
Type (1)	High SP	High SP	High SP	High SP	High SP	High SP	Smooth	High ST	High ST	High ST	Low SP	High ST	Low SP	High SP	Low SP	High ST	Low SP	Low SP
Number of Fins	16	16	16	6	16	16	0	14	6	10	30	12	30	8	50	20	32	32
Outside Diameter (in.)	0.8725	0.7485	0.626	0.5035	0.500	1.050	1.125	0.750	0.503	0.624	1.050	0.500	0.837	1.250	1.287	1.04	1.020	0.625
Wall Thickness (in.)	0.031	0.035	0.037	0.018	0.039	0.030	0.050	0.020	0.025	0.033	0.020	0.022	0.025	0.027	0.023	0.017	0.026	0.023
Inside Diameter (in.)	0.8025	0.675	0.550	0.465	0.420	0.999	1.025	0.710	0.454	0.559	1.01	0.457	0.787	1.196	1.240	0.998	0.976	0.579
Fin Height (in.)	0.078	0.071	0.065	0.073	0.060	0.084	0	0.070	0.067	0.062	0.037	0.043	0.029	0.178	0.032	0.073	0.041	0.027
Fin Height/Inside Diameter	0.097	0.105	0.119	0.158	0.144	0.088	0	0.099	0.148	0.111	0.037	0.094	0.037	0.149	0.026	0.073	0.042	0.047
Cross-Sectional Area (ft.)² x 100	0.334	0.235	0.150	0.108	0.0824	0.510	0.569	0.255	0.107	0.154	0.556	0.108	0.322	0.727	0.819	0.514	0.516	0.173
Nominal Area (ft.)²	1.681	1.414	1.152	0.974	0.879	2.092	2.147	1.487	0.951	1.171	2.115	0.957	1.648	2.505	2.597	2.092	2.044	1.213
Unfinned Area (ft.)²	1.043	0.805	0.624	0.736	0.500	1.200	2.147	0.939	0.736	0.694	1.317	0.592	0.855	1.634	0.905	1.203	0.941	0.507
Finned Area (ft.)²	1.918	1.884	1.688	0.644	1.380	2.24	0.0	1.419	0.650	1.062	2.380	0.921	1.675	2.440	3.100	2.304	2.850	1.432
Total Inside Area (ft.)²	2.961	2.689	2.312	1.380	1.880	3.44	2.147	2.358	1.386	1.756	3.697	1.513	2.530	4.074	4.005	3.507	3.791	1.939
Total Area/Nominal Area	1.76	1.90	2.00	1.417	2.139	1.64	1.0	1.59	1.46	1.50	1.75	1.58	1.54	1.626	1.54	1.68	1.855	1.598
Equivalent Diameter (in.)	0.445	0.338	0.260	0.297	0.176	0.589	1.025	0.425	0.310	0.336	0.619	0.275	0.537	0.706	0.844	0.563	0.556	0.345
Fin Pitch (in.)	8	8.75	11	6.75	14	6	0	ST	ST	ST	6	ST	6	8	7.75	ST	6.75	12
Inter-Fin Spacing (in.)	0.12	0.081	0.069	0.153	0.043	0.143	0	0.122	0.153	0.117	0.080	0.086	0.066	0.306	0.061	0.120	0.067	0.040

(1) SP = Spiral Fin   ST = Straight Fin

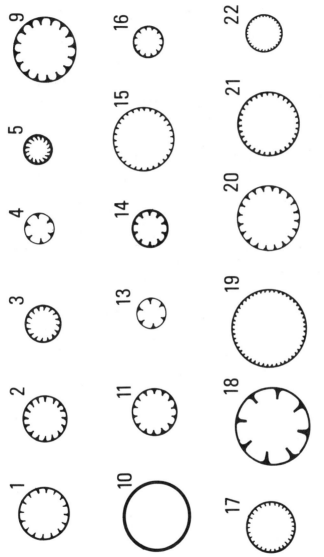

Fig. D-2. Cross-section of Inner Fin tubes tested.

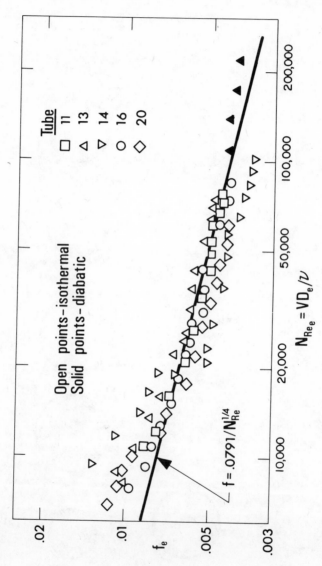

Fig. D-3. Friction factors for straight fin tubes based on equivalent diameter.

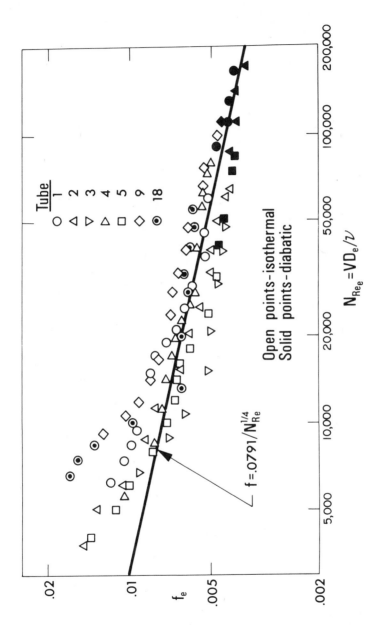

Fig. D-4. Friction factors for high spiral fin tubes based on equivalent diameter.

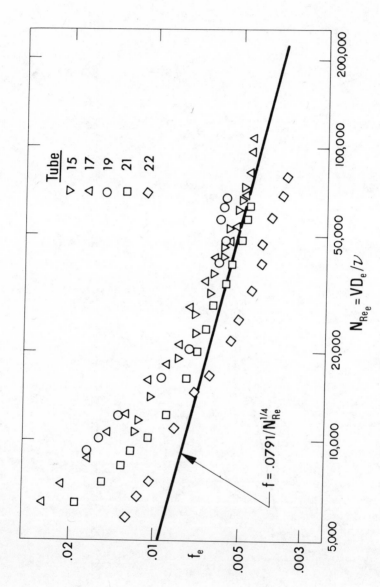

Fig. D-5. Friction factors for low spiral fin tubes based on equivalent diameter.

STRAIGHT FIN TUBES

$$f_e = \frac{0.406 \left(\frac{b}{D_e}\right)^{0.16}}{N_{Re_e}^{0.39}}$$

$$5,000 \leqslant N_{Re_e} \leqslant 75,000$$

STANDARD DEVIATION $\pm$ 10.8%

WHERE b = AVERAGE DISTANCE BETWEEN FINS

SPIRAL FIN TUBES

$$f_e = \frac{0.614}{N_{Re_e}^{0.39} \, (P/D_e)^{0.2}}$$

$$5,000 \leqslant N_{Re_e} \leqslant 75,000$$

STANDARD DEVIATION $\pm$ 12.3%

WHERE P = PITCH OF FIN (LENGTH PER TURN)

FIGURE D-6   FRICTION FACTOR CORRELATIONS

Fig. D-6. Friction factor correlations.

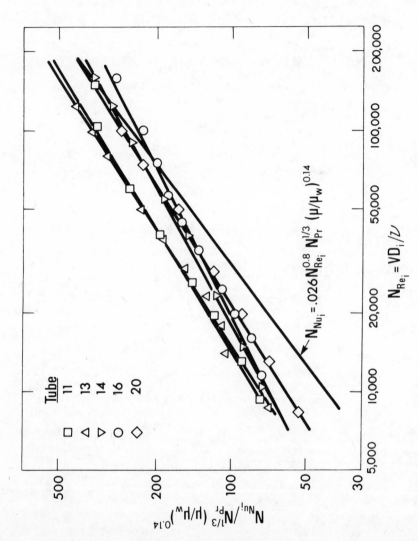

Fig. D-7. Heat transfer performance of straight fin tubes (based on nominal area and inside diameter).

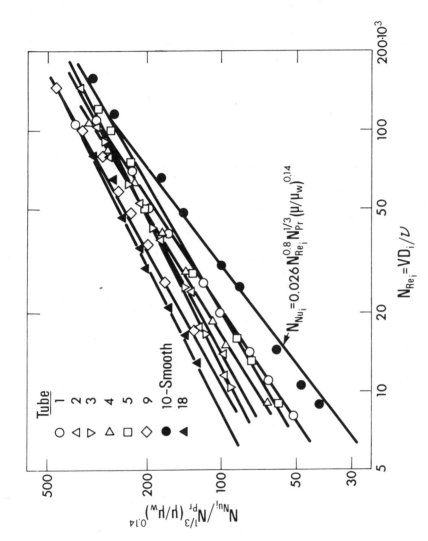

Fig. D-8 Heat transfer performance of high spiral fin tubes (based on nominal area and inside diameter).

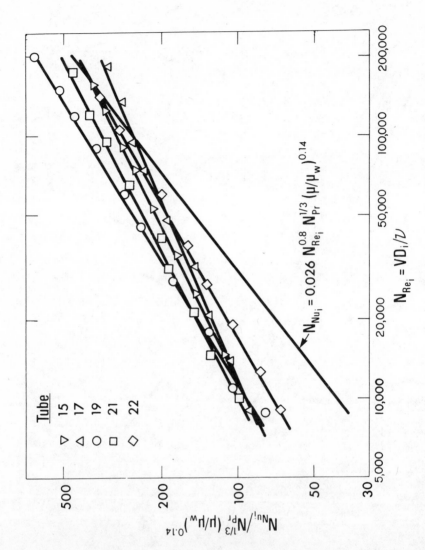

Fig. D-9. Heat transfer performance of low spiral fin tubes (based on nominal area and inside diameter).

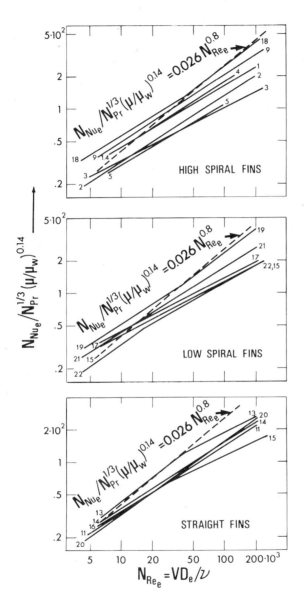

Fig. D-10. Heat transfer performance based on effective area and equivalent diameter.

STRAIGHT FIN TUBES

$$N_{Nu_e} = 0.212\, N_{Re_e}^{0.60} \left(\frac{b}{D_e}\right)^{0.34} N_{Pr}^{1/3} \left(\frac{z}{z_w}\right)^{0.14}$$

$$5,000 < N_{Re_e} < 100,000$$

STANDARD DEVIATION ± 10%

WHERE b = AVERAGE DISTANCE BETWEEN FINS

SPIRAL FIN TUBES

$$\frac{N_{Nu_e}}{N_{Pr}^{1/3} \left(\frac{z}{z_w}\right)^{0.14}} = 0.369\, N_{Re_e}^{0.63} \left(\frac{P}{D_e}\right)^{-0.27} \left(\frac{b}{D_e}\right)^{0.21}$$

$$5,000 < N_{Re_e} < 100,000$$

STANDARD DEVIATION ± 12.8%

WHERE P = PITCH OF FIN (LENGTH PER TURN)

FIGURE D-11   HEAT TRANSFER CORRELATIONS

Fig. D-11. Heat transfer correlations.

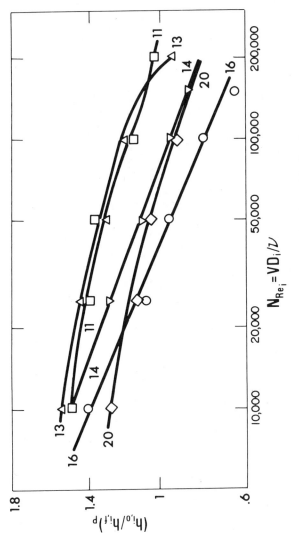

Fig. D-12. Constant pumping power comparison for straight fin tubes.

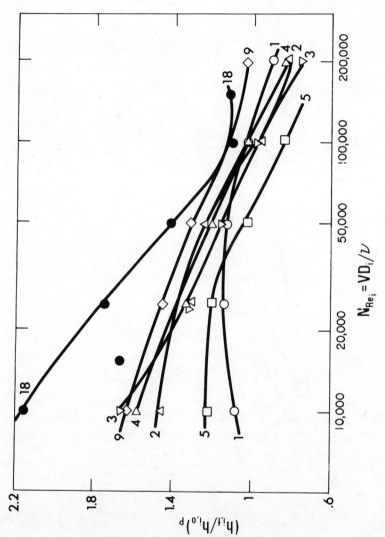

Fig. D-13. Constant pumping power comparison for high spiral fin tubes.

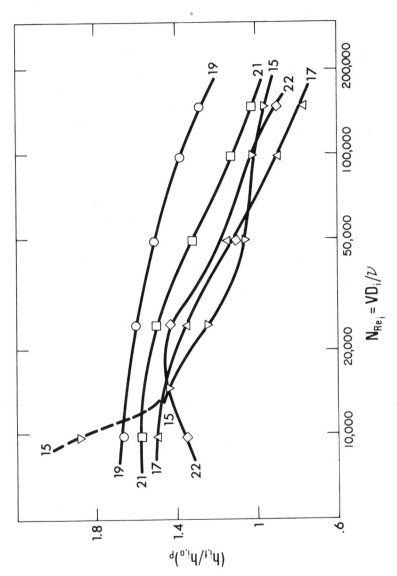

Fig. D-14. Constant pumping power comparison for low spiral fin tubes.

Chapter 18

NUMERICAL SIMULATION OF THE THERMAL BEHAVIOR OF CONVECTIVE HEAT TRANSFER EQUIPMENT

F.W. Schmidt (*)

I. INTRODUCTION

The simulation of convective heat transfer equipment has long been an area of interest to engineers. Through these techniques the steady state and transient performance characteristics of heat exchangers can be determined quickly, accurately, and economically. The steady state results are used primarily for the design and sizing of units. In addition to the temperature distributions in the fluids and the rates of heat transfer, the pressure drops and the thermal stresses in the matrix structures may be determined and the condensation or evaporation rates obtained if changes in phase occur. The transient response of the unit to changes in the flow rate and/or temperature of the entering fluids is easily determined using simulation techniques. These characteristics are extremely important in the design of control systems for chemical processing and power generation plants where the heat exchangers are integral components in the overall system.

Various levels of sophistication can be used in defining the mathematical model for the heat exchanger, several of which have been discussed by Thal-Larson [1]. The model must, however, satisfy the physcial laws, namely; the laws of the conservation of mass, momentum and energy for the fluids and of energy for the heat exchanger matrix. If these relationships are expressed in differential form a set of nonlinear partial differential equations results. Assumptions are usually made to reduce the equations to a manageable form and their solutions are obtained using analytical or numerical techniques. It is commonly assumed that the temperature of the fluid at any axial location is uniform, i.e. the temperature of the fluid varies only in the direction of fluid flow while the velocity profile remains

(*) The Pennsylvania State University, University Park, Pa. 16802 U.S.A.

N.B. ALL FIGURES QUOTED IN TEXT ARE AT THE END OF THE CHAPTER

constant through the length of the heat exchanger section. In order to calculate the rate of heat transfer and the pressure drop the concepts of a convective heat transfer coefficient and friction factor are introduced. The exact form of the energy equations for the counterflow shell and tube heat exchanger shown in Figure 1 is:

Tube fluid:
$$\rho_1 c_1 A_1 \left[\frac{\partial T_1}{\partial t} + v_1 \frac{\partial T_1}{\partial x} \right] = h_1 P_1 (T_t - T_1)$$

Tube:
$$A_t \rho_t c_t \frac{\partial T_t}{\partial t} = h_1 P_1 (T_1 - T_t) + h_2 P_2 (T_2 - T_t)$$

Shell fluid:
$$\rho_2 c_2 A_2 \left[\frac{\partial T_2}{\partial t} - v_2 \frac{\partial T_2}{\partial x} \right] = h_2 P_2 (T_t - T_2).$$

This analysis is called the distributed parameter model with the equations written in terms of a fixed coordinate system.

Several simulation methods using numerical or analog techniques employ a coordinate system which moves with the fluid. The left hand side of the fluid energy equations is then expressed in terms of the substantial derivatives:

Tube fluid:
$$\rho_1 c_1 A_1 \frac{DT_1}{Dt} = h_1 P_1 (T_t - T_1)$$

Shell fluid:
$$\rho_2 c_2 A_2 \frac{DT_2}{Dt} = h_2 P_2 (T_t - T_2).$$

The energy equation of the tube is still expressed in terms of a fixed coordinate system.

The lumped parameter approach is quite frequently used. In this model the conservation equations are integrated over the complete region so that the exchanger is represented by a set of ordinary differential equations. The temperatures are thus only a function of time. For the exchanger shown in Figure 1 the equations are:

Tube fluid:
$$\rho_1 c_1 A_1 \frac{dT_1}{dt} = h_1 P_1 (T_t - T_1) + P_1 C_1 V_1 A_1 (T_{1,i} - T_{1,o})$$

Tube:
$$A_t \rho_t c_t \frac{dT_t}{dt} = h_2 P_2 (T_2 - T_t) + h_1 P_1 (T_1 - T_t)$$

Shell fluid:
$$\rho_2 c_2 A_2 \frac{dT_2}{dt} = h_2 P_2 (T_t - T_2) + P_2 C_2 V_2 A_2 (T_{2,i} - T_{2,o})$$

Considerable inaccuracies are introduced in the results obtained with the lumped parameter model may be a satisfactory representation of heat exchanger components in a complex process control system analysis.

Many times the heat exchanger is divided into subsections and the expression for the conservation equations is obtained directly by writing mass, momentum and energy balances. These methods are especially useful in complex systems since the expressions in many instances can be expressed in terms of existing well defined relationships. As an example, the heat exchanger in Figure 1 has been divided into four sections, Figure 2. In steady state operation the solution technique can incorporate the NTU-effectiveness relationship or the LMTD method where the overall heat transfer coefficients are constant but may differ in each subsection. For a parallel flow heat exchanger the temperatures at the outlets of each sub-region can be calculated by proceeding from the subsection at the left to those on the right in a stepwise fashion using either the NTU-ϵ or LMTD methods. The final outlet temperatures can be evaluated.

Several survey papers on heat exchanger dynamics have been presented in the past notably by Williams [2] and Krüger [3]. Analytical techniques usually involving Laplace transforms have been used in determining the transient response of certain types of heat exchangers. For many units, however, analytical solutions cannot be obtained and simulation techniques must be used. Initially, heat exchangers were simulated using passive electrical circuits or hydraulic networks. As analog, hybrid and digital computers became available new simulation techniques were developed so that these facilities could be used. A survey of the application of these various methods will now be presented.

II. ANALOG NETWORKS

The analogy between the flow of heat and the flow of electrical current or the flow of heat and that of a fluid has been utilized in the simulation of heat exchangers using either a fluid flow network or a passive electrical network. The electrical network may incorporate mechanical devices to assist the simulation process through switching operations. The electrical network may also incorporate specially designed electrical components to allow the current in a branch of the network to flow only in one direction (cathode followers). The majority of the studies employing these analogies was conducted in the 1950's with flow or electrical networks actually constructed and utilized in obtaining the numerical solution for various heat exchanger configurations.

A. Hydraulic Analog

The use of hydraulic analogs for the study of heat exchangers has been reported by Juhasz and his co-workers. The application to steady state heat exchangers is reported in (4) where the liquid levels of the fluid in the hydraulic network are proportional to the temperature distribution in the heat exchanger. The application of the basic hydraulic analogy for the determination of heat exchanger transient was discussed in [5,6]. The simulation utilizes the network's capacitance and resistance to fluid flow with a sudden filling process representing a step change in temperature. The decay of the liquid levels in the various sections of the network is proportional to the transient temperature variations in the heat exchanger.

B. Passive Electrical Analog

The passive analog networks have been used for the determination of the steady state and transient response of shell and tube heat exchangers without phase changes, regenerators and multi-fluid heat exchangers. A resistance capacitance network, R-C, was used by Mozley [7] for the prediction of the response of a shell and tube heat exchanger to sinusoidal variation in the temperature of the tube fluid. A distributed parameter model was used and satisfactory agreement with experimental results was obtained. An electrical network using resistances, capacitances and inductances was proposed by Ford [8]. In subsequent discussions the inductances were replaced by electronic amplifiers and feedback techniques were employed. A passive network with buffer amplifiers and variable resistors was used by Kummel [9] to simulate a shell and tube condenser. The distributed parameter model was used.

The evaluation of the steady state performance of a heat exchanger using an R-C network was presented by Paschkis-Hlinka [10] where the length of the heat exchanger was subdivided into discrete sections. For transient studies an R-C circuit was proposed which incorporated unit gain cathode followers in the sections of the network representing the fluids. Two pure resistance networks were proposed by Wyngaard [11]. One of the units incorporated cathode followers while the other utilized the Lagrangian approach for the flowing fluids and was composed entirely of resistive elements. The derivatives with respect to time were replaced by backward finite difference expressions, thus making it possible to represent the heat capacitances by pure resistances. The Lagrangian approach was also employed by Cima and London [12] in the simulation of a counterflow heat regenerator for a gas turbine. Standard representation of the thermal resistance to the transfer of heat and the capacitance of the tube was employed while the simulation of the fluid flow was accomplished by the movement of discrete capacitance in rolling contact with the rest of the network. The speed of the contacts was proportional to the velocity of the fluids. This analog was also used in subsequent studies by London et al. [13].

All of the above are simulations of shell and tube heat exchangers. The evaluation of the steady state performance of a three fluid heat exchanger with interconnecting extended surfaces between the confining channels has been presented by Paschkis and Heisler [14]. An R-C network is used with a delta resistance circuit employed for the representation of the transfer of heat from the fluid to the passage walls directly, as well as indirectly from the fluid to the supporting extended surfaces, and then by conduction to the passage wall.

The most lasting contribution of several of these studies, exclusive of the results obtained, has been the development of the techniques which utilize the analogy between the flow of heat and electrical current in the modeling of the heat exchanger. Once the electrical network has been formed, Kirchhoff's voltage and current laws can be applied and the solution of the resulting set of equations obtained using high speed digital computers.

III. ANALOG AND HYBRID COMPUTERS

Analog and hybrid computers are commonly employed for the simulation of heat exchangers. The analog computer has proven to be extremely useful for large complex processes composed of many different components where the heat exchangers can be satisfactorily represented by a lumped parameter model. The major objectives of these studies have been to utilize the simulation for dynamic analysis to establish and verify the control systems for the processes. The analog computer is used in these analyses since it has the capability of high speed solution of ordinary differential equations. When the physical system is described by sets of partial differential equations the equations must be reduced to a set of ordinary differential equations using finite differences in order to use the analog computer. Simulation of discontinuities and logical functions can also be easily handled by an analog computer.

The hybrid computer combines the analog computer and a digital computer. The interconnecting link between these computers provides for: the high speed conversion of data from analog to digital and from digital to analog; the communication of logical information between the two computers and the monitoring and control of the analog computer. The digital computer is used for off-line digital processing, function storage and playback, analog computer set-up, checkout and control as well as many other applications. The digital computer is sometimes used for the numerical solutions of differential equations when the accuracy of the analog computer is not adequate or when the size of the problem exceeds the amount of analog equipment available. The hybrid computer has, in many cases, proven to be more economical in the solution of problems than either the pure analog or the digital computers.

A. Analog Computer Application

In a previous section several mathematical models describing heat exchangers have been presented. In the distributed parameter model, the differential equations for the fluids contain derivatives with respect to space and time. In order to solve the resulting set of equations using the analog computer the partial differential equations must be reduced to a set of ordinary differential equations using finite difference techniques. One has a choice as to which variable is left in continuous form. If the space derivative is replaced by a finite difference expression the resulting set of ordinary differential equations can be solved using the analog computer. The simulation of the counterflow heat exchanger shown in Figure 1 using an analog computer is accomplished by subdividing the exchanger into discrete sections identified by the subscript i, Figure 2. A backward finite difference is used for the approximation of the derivative with respect to space. The differential equations reduce to the following form for interior sections:

Tube fluid:
$$\frac{dT_{1,i}}{dt} = -v_1 \left[\frac{T_{1,i} - T_{1,i-1}}{\Delta x} \right] + \frac{h_1 P_1}{A_1 \rho_1 c_1} (T_{t,i} - T_{1,i})$$

Tube:
$$\frac{dT_{t,i}}{dt} = \frac{h_1 P_1}{A_t \rho_t c_t} (T_{1,i} - T_{t,i}) + \frac{h_2 P_2}{A_t \rho_t c_t} (T_{2,i} - T_{t,i})$$

Shell fluid:
$$\frac{dT_{2,i}}{dt} = v_2 \left[\frac{T_{2,i} - T_{2,i-1}}{\Delta x} \right] + \frac{h_2 P_2}{A_2 \rho_2 c_2} (T_{t,i} - T_{2,i}).$$

A forward or a central finite difference scheme could have been used also. The block analog circuit diagram for the "i"th section is shown in Figure 3. For the circuit shown each spacial subdivision would require seven integrators, three multipliers and nine potentiometers. This procedure is often called the "parallel" solution technique and although no computational instabilities arise in the solution, a considerable quantity of equipment is involved.

A comparison of a numerical and "parallel" type solution to a shell and tube heat exchanger for changes in tube fluid flow rates has been presented by Wood and Sastry [15]. Their results were compared to experimental data and satisfactory agreement was found. It is interesting to note that an analog circuit was set up for the "parallel" solution, however, an analog-digital simulator program known as Continuous System Modeling Program (CSMP) was used to obtain numerical values. A model of a parallel flow heat exchanger using a Lagrangian approach, referred to as time lumping, was proposed by Schmidt and Clark [16]. A considerable saving in the amount of necessary computational equipment was indicated. Results for step changes of inlet velocities were reported and compared with those obtained using a spacially lumped model. An analog computer simulation of a counter flow heat exchanger where a central finite difference approximation was used for the spacial derivatives was discussed by Privott and Ferrell [17]. The exchanger was divided into six spacial subdivisions and the dynamic behavior determined for variations in the flow rates of the two fluids.

The simulation of very complex systems such as steam boilers and nuclear reactors using analog computers has been reported in the literature [18 to 22]. The various sections of the steam generator; boiling channels, risers, downcomers, steam drums, etc., are handled individually. The main objective of these studies was to evaluate the response of the steam generator to various types of control systems.

B. Hybrid Computer Applications

In a distributed parameter model of a shell and tube heat exchanger the time derivatives can be replaced by a backward finite difference method. The differential equations for the counter flow heat exchanger of Figure 2 become:

Tube fluid:
$$\frac{dT_1^j}{dx} = \frac{h_1 P_1}{A_1 \rho_1 c_1 v_1} (T_t^j - T_1^j) - \left(\frac{T_1^j - T_1^{j-1}}{v_1 \Delta t} \right)$$

Tube:
$$0 = \frac{h_2 P_2}{A_t \rho_t c_t} (T_2^j - T_t^j) + \frac{h_1 P_1}{A_t \rho_t c_t} (T_1^j - T_t^j) - \left(\frac{T_t^j - T_t^{j-1}}{\Delta t} \right)$$

Shell fluid:
$$\frac{dT_2^j}{dx} = - \frac{h_2 P_2}{A_2 \rho_2 c_2 v_2} (T_t^j - T_2^j) + \left(\frac{T_2^j - T_2^{j-1}}{v_2 \Delta t} \right)$$

where j is the index for the time step with $t^j = j\Delta t$. The initial temperature distributions are known along with the inlet conditions of the fluids. The spacial variable is scaled to computer time. The set of ordinary differential equations can be integrated with respect to x if provisions can be made for the storage and recall of the temperatures as required by the differential equations. This computational technique is called the "series" method and is particularly suited for a hybrid computer. Considerable saving in equipment results when this method is employed since the analog equipment may be re-used for each successive time step.

A detailed discussion of the application of this method to the shell and tube heat exchanger has been presented by Silvey and Barker [23]. In order to facilitate the computations, the equation for the tube is rearranged to yield:

$$0 = \frac{h_2 P_2}{A_t \rho_t c_t} (T_2^{j-1} - T_t^{j-1}) + \frac{h_1 P_1}{A_t \rho_t c_t} (T_1^{j-1} - T_t^{j-1}) - \left(\frac{T_t^j - T_t^{j-1}}{\Delta t} \right) .$$

To remove computational instabilities a spacial transformation of the form $y = L-x$ was used for the shell fluid energy equation. The resulting equations are rearranged with $\overline{T}_1$, $\overline{T}_2$, and $\overline{T}_t$ representing stored temperatures.

Tube fluid:
$$\frac{dT_1^j}{dx} + \left[\frac{h_1 P_1}{A_1 \rho_1 c_1 v_1} + \frac{1}{v_1 \Delta t} \right] T_1^j = \frac{h_1 P_1}{A_1 \rho_1 c_1 v_1} T_t^j + $$

$$+ \frac{1}{v_1 \Delta t} \overline{T}_1^{j-1}$$

Tube:
$$T_t^j = \left[1 - \frac{\Delta t}{A_t \rho_t c_t} (h_2 P_2 + h_1 P_1) \right] \overline{T}_t^{j-1} + $$

$$+ \left(\frac{\Delta t h_1 P_1}{A_t \rho_t c_t} \right) \overline{T}_1^{j-1} + \left[\frac{\Delta t h_2 P_2}{A_t \rho_t c_t} \right] \overline{T}_2^{j-1}$$

Shell fluid: $\dfrac{dT_2^{\,j}}{dy} + \left(\dfrac{h_2 P_2}{A_2 \rho_2 c_2 v_2} + \dfrac{1}{v_2 \Delta t} \right) T_2^{\,j} = \dfrac{h_2 P_2}{A_2 \rho_2 c_2 v_2} \bar{T}_t^{\,j} +$

$$+ \dfrac{1}{v_2 \Delta t} \bar{T}_2^{\,j-1} \,.$$

The computer block diagram is shown in Figure 4 and computations for the values of the temperatures at time j proceed in the following manner with the storage units being already loaded with $\bar{T}_1^{\,j-1}, \bar{T}_2^{\,j-1}$, and $\bar{T}_t^{\,j-1}$. The equations for the tube fluid and the tube are integrated from $x = 0$ to $x = L$ to obtain values of $T_1^{\,j}$ and $T_t^{\,j}$. The values of $T_1^{\,j}$ and $T_t^{\,j}$ overwrite the current values in the respective stores. The shell fluid equation is then integrated from $y = 0$ to L and the new values of $T_2^{\,j}$ are placed in storage.

An analysis of the accuracy of the "series" method was made by Vichnevetsky [24] for the simplified transport equation of the fluid:

$$\frac{\partial T}{\partial t} + V \frac{\partial T}{\partial x} = 0 \,.$$

An exact solution was obtained for a step in the inlet temperature. A comparison of the "series" solution with that of the exact indicated that if a backward time difference technique was used the high frequency components moved too fast and were attenuated, while if the central time difference was used the high frequency components were not attenuated but moved too slowly. For either case the frequency response was found to be in error. A modification of the "series" method was proposed by Bosgra and Buis [25] in order to improve its frequency response. A method of characteristics was proposed by McAvoy [26] and was applied to a counterflow heat exchanger with changes in fluid inlet temperature and a ramp function change in the velocity of the fluid. In all studies the heat transfer coefficients were considered to be constant and the results indicated excellent frequency response for this method.

The hybrid simulation of a heat exchanger-reactor control system was made by Carlson [27] using the "series" method. The system studied was similar to that analyzed by Douglas et al. [28] where a pure analog computer was used. Three partial differential equations were used to describe the heat exchanger and reactor. The results of Carlson's studies indicated that the hybrid simulation used approximately 30% less hardware than required by the pure analog computer solution. A reduction in computational cost from one tenth to one twentieth was noted when comparing the results to those obtained using a digital computer.

The hybrid simulation of a complete nuclear power plant has been presented by Blake et al. [29] and Vichnevetsky [21]. A more recent study of the simulation of a natural circulation boiler using a hybrid computer was reported by Lawrence et al. [30]. The boiler was divided into preheater, mixing and boiling regions. Each of these regions was divided and the differential energy equations were solved using the "parallel" method.

IV. DIGITAL COMPUTERS

The most popular method of simulating heat exchangers is through the use of digital computers. The derivatives with respect to time and distance in the differential energy equations are replaced by finite difference expressions. Many of the methods previously presented in Section I for the mathematical modeling of the heat exchanger have been used in the development of the computer programs. Some of the programs have been very general in nature while others are for specific classes or pieces of equipment. Programs which fit into this latter category are those developed by companies for the design of their products as noted by Rissler [31]. These programs are usually classified as proprietary information and are not available or described in the open literature. The more general programs are occasionally described and made available to other users.

As the use of digital computers increases in industry the engineer is often faced with the decision of whether to develop his own program or use programs by someone else. In many cases, company policy will determine when he may purchase, lease or license programs developed outside of his organization. These external programs are available from various sources. Many are not widely publicized so it is difficult to obtain enough information to evaluate the accuracy of the methods. Several organizations have taken steps to improve the distribution of information concerning those programs which are available. A listing of computer programs for sale, lease or license was published in the July 12, 1971 issue of "Chemical Engineering". Several heat exchanger design programs were listed. A Computer Program Directory was published in 1971 by the Joint Users Group of The Association for Computer Machinery. The objective of the directory is to facilitate the exchange of program documentation among the various user groups. Provisions have been made to make this an annual publication. Algorithms are also published in periodicals such as the "COMMUNICATIONS OF THE ASSOCIATION FOR COMPUTING MACHINES", "INTERNATIONAL COMPUTER PROGRAMS QUARTERLY" and "COMPUTER AND INFORMATION SYSTEMS". Computer manufacturers have established user program sharing systems such as IBM's "SHARE" and CDC's "SWAP". NASA publishes the "COMPUTER ABSTRACT JOURNAL QUARTERLY" which is an index listing of documented computer programs developed by or for NASA, the Department of Defense and the A.E.C. and are for sale. Programs which are "sharable" within NASA and U.S. government agencies and contractors are listed in another separate journal published by NASA. A discussion of the Argonne National Laboratory Computer Code Center was presented by Butler et al. [32] with the history of the center and the documentary procedure described. The center has a working arrangement for the exchange of programs with the ENEA Computer Programme Library at Ispra, Italy and several other national computer libraries.

A. Shell and Tube Heat Exchangers

The application of finite difference techniques for the prediction of the transient temperature distribution in a heat exchanger was described by Dusinberre [33 and 34]. An energy balance method was used in the formation of the difference equations. A comparison of the transient response of a shell and tube heat exchanger using the finite difference method proposed by Dusinberre, the electrical analog network of Paschkis and Hlinka [10] and an exact solution obtained using Laplace transformations were presented by Wyngard and Schmidt [35]. Step changes in inlet temperatures and fluid velocities were studied and it was found that the finite difference methods generally yields more accurate results. Further comparison of the results obtained by these three methods for a two pass shell and tube heat exchanger was presented by Cygnarowicz [36] and substantially verified the earlier findings. A finite difference solution of the distributed parameter model shell and tube heat exchanger was presented by Privott et al. [17]. Central finite differences were used for the space derivatives and a forward difference method was used for time. A comparison of the results obtained using an analog computer and a "parallel" computational method was presented.

The transient response of heat exchangers with one fluid having an infinite heat capacity is described by Myers et al. [37]. A step change in the temperature of the infinite capacity fluid is assumed and the distributed parameter model is used for the other fluid and the tube. The two energy equations were combined to yield a second degree hyperbolic equation. The equation is rewritten in terms of its characteristic coordinates and a solution is obtained using finite difference techniques. The dynamic response of a condenser was also determined by Heidemann et al. [38]. Forward finite difference techniques can be used for the derivatives with respect to time whereas central finite difference techniques can be used for the derivatives with respect to distance. The application of these techniques to the exchanger shown in Figure 2 yields the following relationships:

Tube fluid:
$$\frac{T_{1,i}^{j+1} - T_{1,i}^{j}}{\Delta t} = -v_1 \left[\frac{T_{1,i+1}^{j} - T_{1,i-1}^{j}}{2\Delta x} \right] + \frac{h_1 P_1}{A_1 \rho_1 c_1}$$
$$\left[T_{t,i}^{j} - T_{1,i}^{j} \right]$$

Tube:
$$\frac{T_{t,i}^{j+1} - T_{t,i}^{j}}{\Delta t} = \frac{h_2 P_2}{A_t \rho_t c_t} \left[T_{2,i}^{j} - T_{t,i}^{j} \right] + \frac{h_1 P_1}{A_t \rho_t c_t}$$
$$\left[T_{1,i}^{j} - T_{t,i}^{j} \right]$$

Shell fluid:
$$\frac{T_{2,i}^{j+1} - T_{2,i}^{j}}{\Delta t} = v_2 \left[\frac{T_{2,i+1}^{j} - T_{2,i-1}^{j}}{2\Delta x} \right] + \frac{h_2 P_2}{A_2 \rho_2 c_2}$$
$$\left[T_{t,i}^{j} - T_{2,i}^{j} \right] .$$

The discretization errors are of order Δx^2 and Δt. Stability requirements impose a limit on the maximum value of Δt which may be used for a given number of subdivisions in x. The results for a step change in the coolant rate were compared with experimental data. The authors indicated that the lack of agreement found between the predicted and experimental results was due to the inaccuracies in the estimation of the convective film coefficients especially those on the condensing side. Wood [15] used the same numerical technique for the analysis of a counter flow, water to water heat exchanger and noted good agreement between the predicted and experimental results when only 10 spacial subdivisions were used.

High order approximations could have been used for the approximation of the first derivative with respect to space at all interior nodes. A backward finite difference method might also have been employed in forming the difference approximation of the derivative with respect to time and would result in an unconditionally stable set of implicit relationships being formed.

In the above analysis the temperature of the fluid was assumed to be only a function of axial distance since the radial or transverse variations in the temperature profile were neglected. These assumptions were not made in the determination of the rate of heat transfer between two non-Newtonian fluids in a counter flow plate heat exchanger by Gutfinger et al. [39]. Both flows were laminar and had fully developed velocity profiles. The thermal capacitance and the resistance of the walls between the fluids were neglected. The differential energy equation for each fluid was solved independently and an iterative method was used to match the temperature gradients at the interface.

B. Periodic Flow Regenerator

Several studies have been made of the simulation of rotary periodic flow heat exchangers. The thermal conductivity of the matrix was assumed to be zero in the direction of fluid and "metal" flow, and infinite in the direction normal to the fluid flow by Lambertson [40]. The regenerator was subdivided in the axial and circumferential directions and each element was regarded as a crossflow heat exchanger. Lumped parameter energy balances were written for the matrix and the fluids in terms of the temperatures at the surrounding elements. The resulting set of difference equations for the steady periodic operation of the regenerator was solved by using a digital computer with an iterative procedure to match the metal temperature circumferentially. The method proposed by Lambertson was extended to consider a matrix with a finite thermal conductivity in the direction of fluid flow by Bahnke and Howard [41].

A slightly different lumped parameter analysis for a radial flow regenerator was used by Mondt [42]. The procedures used in the initial set up were based on the work of Paschkis and Hlinka [10] where the metal was represented by an R-C network. A digital computer was then used to simulate the R-C network and to determine the temperature distribution in the regenerator. The computer program developed by Mondt was subsequently used by London et al. [43] to determine the transient response of periodic flow regenerators.

Willmott [44] utilized finite difference techniques for the determination of the thermal response of a stationary regenerator where hot and cool gases were alternately passed through the unit in a periodic manner. Both the heat capacity of the gas and the axial conduction in the matrix were neglected. The differential energy equations for the gas and the matrix were solved numerically using central difference approximations. For constant properties and film coefficients the integration was performed along the characteristics of the equations. The author describes a procedure which could be utilized when the thermal properties are functions of temperature and the heat transfer coefficients and mass flow rates are time dependent.

V. SPECIFIC APPLICATIONS

The discussion to this point has been concerned with a general survey of the various methods which have been used for the simulation of heat exchangers. Both the lumped and distributed parameter models have been used with the distributed parameter model more commonly used for shell and tube heat exchanger while the lumped parameter model is used for more complex systems. A full appreciation of the power of the lumped parameter analysis may be best illustrated by describing its application for the determination of the performance characteristics of a steam generator and a multifluid heat exchanger.

A. Multifluid Heat Exchanger

A study of the thermal behavior of a compact plate fin multifluid heat exchanger in steady state operation has been represented by Demetri and Platt [45]. A sketch of a section of a counterflow and parallel flow unit as well as a cross-flow unit is shown in Figure 5. The complete heat exchanger may be represented by a number of passages stacked in a repeating pattern. The channel walls are inter-connected so that it is possible for heat to be transferred from fluid "a" to fluid "c" directly by conduction as well as by convection from fluids "a" to "b" to "c".

The procedures used in these calculations are similar to those described by Paschkis and Heisler [24] in that an electrical analogy is established to represent the transfer of heat between the fluids flowing in the channels. The current flow in the various sections of the electrical network and thus the heat transferred, however, are found using the digital computer. A section of a parallel flow heat exchanger network analog is shown in Figure 6.

The expressions for the resistances to the transfer of heat in this network are determined in the conventional manner where the exchanger has been subdivided in the axial direction. The convective heat transfer resistances between the fluid and the channel "a" walls for each of these sections are expressed as:

$$R_{ca} = \frac{2}{\eta_{oa} h_a A_a} \quad \text{where} \quad \eta_{oa} = 1 - (A_f/A_a)(1 - \eta_f)$$

and A_f is the surface area of the fins, A_a is the total heat transfer area including the exposed area of channel "a" and the cross sectional area of the fins, h_a is the convective heat transfer coefficient in channel "a" and η_f is the fin efficiency. The resistance to heat transfer from channel "a" to "b" directly by conduction through the fins is expressed as:

$$R_{f,a} = \frac{l_a}{k_f A_{ta}}$$

where l_a is the length of the constant cross sectional area fins, A_{ta} is the total cross sectional area of the fins in channel "a", and k_f is the thermal conductivity of the fins. The resistance to the transfer of heat offered by the channel wall is:

$$R_{w,a-b} = \frac{l_{ab}}{k_{wab} A_a}$$

where l_{ab} is the wall thickness and k_{wab} is the thermal conductivity of the channel wall. The delta resistance circuits have been replaced by "T" circuits for simplicity. In the electrical analogy the fluid temperatures are analogous to the voltages and the current flow is proportional to the flow of heat to or from the fluids.

The orientation of the flow channels will determine the computational algorithm. For a parallel flow heat exchanger all of the fluid temperatures at one end of the heat exchanger are known. The electrical resistances for the first section are determined and Kirchhoff's voltage and current laws are used to determine the values of the currents $i_a \sim q_a$, $i_b \sim q_b$ etc. The temperatures of the fluids leaving the subsections are then found using an energy balance. For example, the temperature of the fluid leaving channel "a" is given as:

$$T_{a,i+1} = T_{a,i} \frac{q_{a,i}}{\dot{m} c_a}$$

where $\dot{m}$ is the mass rate of flow. The calculations move to the next subsection, the necessary changes are made in the values of the resistances and the new values of the fluid temperature just computed are used to determine the heat transfer in these sections. The process is repeated until the temperatures of the fluids leaving the heat exchanger are found. If the heat exchanger is counterflow or crossflow an iterative procedure must be used to obtain a solution which is consistant with all the temperatures of the entering fluids.

B. Steam Generator

The operating performance characteristics of a steam generating unit can be obtained by subdividing the unit into discrete segments and utilizing a lumped parameter analysis. The application of these techniques to obtain the outlet conditions of the steam leaving a high liquid mass flux "once-through" shell and tube heat exchanger shown in Figure 7 will be discussed. The water flows through the tubes entering as a subcooled liquid and leaving as a superheated vapor while the high temperature gas enters the unit at the left and flows perpendicular to the tubes in the direction indicated. One may appreciate the complexity of the system and realize that numerous assumptions will have to be made in order to obtain results with a reasonable amount of computational time.

Although only one tube is shown on the schematic diagram it is assumed that many water tubes are connected in parallel to the inlet and discharge manifolds. The steam generator is divided with vertical subsections each containing tubes in one vertical water pass. The vertical subsections undergo additional subdivision as shown in Figure 7. The total tube surface areas and the boundaries of each of the regions are obtained in the following manner. The mixing temperature of the hot gases and the condition of the water entering the regions are known together with the mass rate of flow of the gases and water. The temperature of the water or the steam leaving the unit is specified if the flow is single phase. If the flow is two phase, a mixture of water and steam, the exit quality or void faction is specified. This procedure has been proposed by Kiely [46] and has resulted in extremely accurate prediction of the heat transfer and pressure drop in a "once-through" steam generator. The size in the "step" in temperature, quality or superheat, is governed by the accuracy desired. An energy balance of the region, Figure 8, is used to determine the rate of heat transfer and the mixing temperature of the hot gases leaving the region. The heat capacities, $\dot{m}c$, of the two fluids can be calculated making note of the fact that when the water undergoes a change in phase the heat capacity is infinite. The total surface area for the region is obtained using the NTU-effectiveness method. The effectiveness, the rate of heat transfer divided by the smaller heat capacity multiplied by the differences in the temperatures of the hot gas and water entering the region, can be calculated. The NTU, overall heat transfer coefficient times surface area divided by the minimum heat capacity, can be determined using the relationships presented in Kays and London [47]. If the overall heat transfer coefficient is known the surface area and the length of the region can be determined.

The evaluation of the overall heat transfer coefficient will usually require an iterative procedure since the fluid properties are strongly dependent on temperature and pressure. The convective heat transfer coefficient for the gas side can be determined using a single phase cross flow relationship such as those presented by Grimson [48]. Few serious difficulties are encountered in the evaluation of the convective heat transfer coefficient if single phase flow, water or steam, is present inside the tubes. However, the evaluation of the coefficient for two phase flow is much more involved.

Before proceeding with the analysis it is worthwhile to investigate the physical phenomenon associated with the flow of water through a "once-through" unit and to note how the various flow regimes will influence the convective heat transfer coefficients. The water enters the unit as a single phase fluid and as it moves through the tubes its bulk temperature is increased due to the transfer of heat from the hot gases. The temperature of the heating surface is normally less than the local saturation temperature in this region. The temperature of the heating surface will increase also and once it has exceeded the local saturation temperature by a predictable amount nucleate boiling can and normally will occur. The bulk mixing temperature will still be lower than the local saturation temperature so the bubbles will condense soon after they are formed and the boiling is classified as subcooled on local boiling. As the fluid continues to move through the unit its bulk temperature increases and once it reaches the local saturation temperature the flow becomes classified as fully developed nucleate or saturated boiling. The majority of the bubbles do not collapse as they are entrained in the main stream of the fluid and it becomes necessary to consider the mixture to be composed of both a liquid phase and a vapor phase. Since the density of steam is much smaller than that of water, the fluid is greatly accelerated as it moves through the two phase region. Two phase flow mixtures have been classified as:

1. Bubbly — The liquid phase is continuous and the vapor phase is discontinuous.
2. Slug — The vapor bubbles combine and the flow consists of relatively large slugs of vapor and liquid following each other.
3. Annular — The liquid phase is continuous in an annulus along the wall and the vapor is continuous in the core. As the flow moves downstream discontinuities in the liquid annulus occur.

Once all the liquid has been vaporized only steam passes through the tube and the flow can again be treated as a single phase.

A schematic sketch of the flow and heat transfer regimes encountered in a "once-through" steam generator is shown in Figure 9. The mathematical expressions which are used for the determination of the rate of heat transfer and the pressure drop are dependent on the flow regime. It should be noted that in a high mass flux "once-through" steam generator the length of the sections where bubbly and slug flow occur is very short and the complete two phase flow may be considered to be annular when total evaporation is desired. A detailed discussion of these relationships is not possible at this time so the reader is referred to several books which treat the subject in considerable detail [49-52]. There is a variety of models which can be used for each of these flow regimes and one must evaluate the complexity of the model along with the effect of its inaccuracies on the overall performance of the steam generator. It should be noted that the same flow model must be used for both the hydrodynamic, pressure drop, and heat transfer calculations. Since boiling always results in a increase in the fluid velocity, the pressure drop due to the acceleration or change in momentum of the fluid is significant and must be considered.

In order to calculate the film coefficient on the water side and to obtain the overall heat transfer coefficient, U, it is first necessary to determine the pressure drop and the wall temperature. Since these items are interrelated the calculation of U must therefore employ an iterative process. An outline of the computational procedure to be followed for the evaluation of the convective heat transfer coefficients on the water side will now be presented with the flow diagrams shown on Figure 10. It is important to note that computer subroutines must be supplied in order to evaluate the thermodynamic properties and transport properties of the water and steam at the local fluid pressures and temperatures.

1. Single phase region — The temperature of the water or steam leaving the region is specified and the physical properties of the fluid are evaluated. The convective heat transfer coefficient and the overall heat transfer coefficient are determined. The surface area is determined using the NTU-effectiveness method and the average heat flux is calculated. The wall temperature at the exit of the region may be estimated and the exit pressure determined. The film temperature is found, the physical properties re-evaluated and the calculation of the convective film coefficients repeated. The procedure is repeated until the convergence of the calculated area has been obtained.

2. Nucleate boiling — Several different methods can be used in the evaluation of the convective film coefficients in the two phase region. The methods proposed by Kiely will be discussed. The heat transfer in this region is considered to be composed of two

components as indicated by Rohsenow, namely, a forced convection component and a nucleate boiling component. As noted previously, if the water experiences subcooled boiling, the size of the region is obtained by specifying the exit temperature of the water while if the bulk temperature is equal to the local saturation temperature the quality, X, or void fraction at the region's exit is specified. This is done in order to facilitate the determination of the heat transfer and pressure drop for two phase flows. The forced convective component is proportional to the temperature difference between the wall and the bulk fluid, and the nucleate boiling component is proportional to some power of the temperature difference between the wall and the local saturation temperature. An iterative procedure as shown in Figure 10 is used. The total heat flux is calculated and the overall heat transfer coefficient is determined. The NTU-effectiveness method is used to determine the surface area and the length of the region. The pressure drop is calculated using the appropriate expressions for the flows involved. The local saturation temperature as well as the exit wall temperature are determined. The calculations are repeated until the calculated surface area converges.

The complete set of calculations progresses from the water inlet to the exit in a stepwise manner following the flow path taken by the water. It may be assumed that the gas mass flux is constant in all regions of the generator. Complete mixing of the gas entering a region occurs with the resultant uniform gas temperature computed using an energy balance as shown in Figure 8. A sketch of the computational flow diagram is shown in Figure 11. Slight modification of the process must be made when one approaches the return bends in the water tube section. Once the quality exceeds 1.00 the computations proceed in the same manner as for the single phase liquid regions although different relationships are used for the evaluation of the heat transfer coefficients and the pressure drop. The calculations continue until the exit of the steam generator has been reached.

If the condition of the steam leaving the unit is not within the desired limits, adjustments in the conditions of the gas or water entering the unit can be made and the computations repeated. An iterative procedure is used to obtain the required outlet conditions.

VI. SUMMARY

A review of numerous significant contributions in the simulation of the thermal behavior of heat exchangers has been presented. These have involved passive electrical networks, analog, hybrid and digital computers. The most popular methods use the digital computer with some interest indicated in the development of techniques which utilize the hybrid computer.

The accuracy of all the methods which have been used for the determination of the performance characteristics of heat exchangers is dependent upon the ability to accurately determine the convective heat transfer coefficients. The techniques for simulating shell and tube heat exchangers are well documented in the open literature. The applications of these methods to other types of heat exchangers, however, are not as well publicized. This is due in part to the fact that computers are now used extensively for design purposes by nearly all the heat exchanger manufacturers and the details of their programs are considered by many of them to be proprietary information.

Through the use of these simulation techniques more accurate analysis can be obtained quickly and at low cost. Besides heat transfer calculations, studies can be made to determine pressure drops, stress analysis, as well as optimum cost. These programs serve as an indispensable tool for industry to use in providing an inquiry service for its customers. The economic and technical importance of obtaining accurate simulations of heat exchangers will undoubtedly stimulate the development of new and faster methods utilizing the fullest capabilities of the digital, analog and hybrid computational facilities.

NOMENCLATURE

A	area
c	specific heat
h	convective film coefficient
$\dot{m}$	mass rate of flow
P	perimeter
q	rate of heat transfer
T	temperature
t	time
v	velocity
X	quality
ρ	density

Subscripts

1	tube fluid
t	tube
2	shell fluid
i	space increment index

Superscripts

j	time increment index

REFERENCES

[1]   Thal-Larson, H., Dynamics of heat exchangers and their models, J. Bas. Engng., 82, p. 489, (1960).

[2]   Williams, T.J. and H. J. Morris, A survery of the literature on heat exchanger dynamics and control, Chem. Engng. Prog. Symp. Ser., 57, No. 36, p. 20, (1961).

[3]   Krüger, R., Wärmeübertrager und Verdampfer, VDI-Z., 112, p. 1297, (1970).

[4]   Juhasz, S. I. and F. C. Hooper, Hydraulic analog for studying steady state heat exchangers, Ind. Engng. Chem., 45, p. 1359, (1953).

[5]   Juhasz, S. I., Hydraulic analogy for transient crossflow heat exchangers, ASME paper 57-A125, (1957).

[6]   Juhasz, S. I. and J. Clark, Hydraulic analogy for transient conditions in heat exchangers, 4th Int. Instrum. Measmt. Conf., Stochholm, (1956).

[7]   Mozley, J.M., Predicting dynamics of concentric pipe heat exchangers, Ind. Engng. Chem., 48, p. 1035, (1956).

[8]   Ford, R. L., Electrical analogues for heat exchangers, Proc. Instn. elect. Engrs., 103, p. 65, (1956).

[9]   Kummel, M., Analog simulation and control of a distributed parameter system, Chem. Engng. Sci., 24, p. 1055, (1969).

[10] Paschkis, V. and J. W. Hlinka, Electrical analogy studies of the transient behavior of heat exchangers, Trans. N. Y. Acad. Sci., Ser. II, 19, p. 714, (1957).

[11] Wyngaard, J. C., Investigations of thermal transients in shell and tube heat exchangers, M.S. Thesis, Univ. of Wis., (1962).

[12] Cima, R. M. and A. L. London, The transient response of a two fluid counterflow heat exchanger -- the gas turbine regenerator, Trans. Am. Soc. mech. Engrs., 80, p. 1169, (1958).

[13] London, A. L., F. R. Biancardi and J. W. Mitchell, The transient response of gas-turbine-plant heat exchangers -- regenerators intercoolers, precoolers, and ducting, J. Engng. Pwr., 81, p. 433, (1959).

[14] Paschkis, V. and M. P. Heisler, Design of heat exchangers involving three fluids, Chem. Engng. Prog. Symp. Ser., 49, No. 5, p. 65, (1953).

[15] Wood, R. K. and V. A. Sastry, Simulation studies of a heat exchanger, Simu - a., 18, p. 105, (1972).

[16] Schmidt, R. K. and D. B. Clark, Analog simulation techniques for modeling parallel-flow heat exchangers, Simu - a., 12, p. 15, (1969).

[17] Privott, W. J. Jr. and J. K. Ferrell, Dynamic analysis of a flow forced concentric tube heat exchanger, Chem. Engng. Prog. Symp. Ser., 62, No. 64, p. 200, (1968).

[18] Chien, K. L., E. I. Ergin, C. Ling and A. Lee, Dynamic analysis of a boiler, Trans. Am. Soc. mech. Engrs., 80, p. 1809, (1958).

[19] Laübli F., The problem of simulating the dynamic behavior of steam generators in analog computers, Sulzer. Tech. Rev., 43, p. 35, (1961).

[20] McPherson, P. K., G. B. Collines, C. B. Guppy and A. Summer, Dynamic analysis of nuclear boiler, Proc. Instn. mech. Engrs., 180, p. 417, (1966).

[21] Vichnevetsky, R., Analog/Hybrid solution of partial differential equations in the nuclear industry, Simu - a., 10, p. 269, (1968).

[22] Schmidt, J. R., Computers aid in heat exchanger design, Power, 112, p. 68, (1968).

[23] Silvey, T. I. and J. R. Barker, Hybrid computing techniques for solving parabolic and hyperbolic partial differential equations, Comput. J., 13, p. 164, (1970).

[24] Vichnevetsky, R., Hybrid computer methods for partial differential equations, Seminar on Engng. Applic. Hybrid Comput., Penn State Univ., (1969).

[25] Bosgra, O. H. and J. P. Buis, Hybrid computer solution of the hyperbolic equations describing forced flow steam generating dynamics, 6th Congr. Hybrid. Comput. AICA-IFIP, Munich, (1970).

[26] McAvay, T. J., Solution of hyperbolic partial differential equations vie a hybrid implementation of the method of characteristics, Simu. - a., 18, p. 91, (1972).

[27] Carlson, A. M., Hybrid simulation of an exchanger/reactor control system, N.R.C. Conf. Process Control, Edmonton, Can., (1968).

[28] Douglas, J. M., J. C. Orcutt and P. W. Berthiaume, Design and control of feed-effluent, exchanger-reactor systems, Ind. Engng. Chem. Fundam., 1, p.253, (1962).

[29] Blake, R., M. Piggott, R. Smale and L. Lotito, Hybrid computer optimization of a nuclear power control system, Int. Symp. Analog/Hybrid Comp. applied Nucl. Energy, Versailles, (1968).

[30]   Lawrence, B. R., G. F. Crate, E. J. Wright, R. E. Gagne and J. D. Stebbing, A hybrid simulation of Pickering control, Summer Comp. Simu. Conf., Boston, (1971).

[31]   Rissler, K., Heat exchanger design by computer, Chem. Process Engng., 52, No. 10, p. 61, (1971).

[32]   Butler, M. K., C. Harrison, Jr. and W. J. Snow, The Argonne Code Center: a decade of computer program exchange, Handling of Nuclear Information, International Atomic Energy Agency, Vienna, p. 83, (1970).

[33]   Dusinberre, G. M., Calculation of transient temperatures in pipes and heat exchangers by numerical methods, Trans. Am. Soc. mech. Engrs., 76, p. 421, (1954).

[34]   Dusinberre, G. M., *Heat Transfer Calculations by Finite Differences*, International Textbook Co., Scranton, Pa., p. 108, (1961).

[35]   Wyngaard, J. C. and F. W. Schmidt, A comparison of methods for determining the transient response of a shell and tube heat exchanger, ASME paper 64-WA/HT-20, (1964).

[36]   Cygnarowicz, T.A., Transient response of two-pass shell and tube heat exchangers, M.S. Thesis, Penn State Univ., (1964).

[37]   Myers, G. E., J. W. Mitchell and C. F. Lindeman, Jr., The transient response of heat exchangers having an infinite capacitance rate fluid, J. Heat Transfer, 92, p. 269, (1970).

[38]   Heidman, R. A., C. E. Huckaba, F. S. Eisen, L. I. Weissman and G. M. Gallatig, Dynamics of convection heat exchangers, Can. J. chem. Engng., 49, p. 147, (1971).

[39]   Gutfinger, G., J. Isenberg and M. A. Zeitlin, Heat transfer to non-Newtonian fluids in countercurrent plate heat exchangers, Israel J. Technol., 8, p. 225, (1970).

[40]   Lambertson, T. J., Performance factors of a periodic flow heat exchanger, Trans. Am. Soc. mech. Engrs., 80, p. 586, (1958).

[41]   Bahnke, G. D. and C. P. Howard, The effect of longitudinal heat conduction on periodic flow heat exchanger performance, J. Engng. Pwr., 86, p. 105, (1964).

[42]   Mondt, J. R., Jr., Vehicular gas turbine periodic flow heat exchanger solid and fluid temperature distributions, J. Engng. Pwr., 86 p. 121, (1964).

[43]   London, A. L., D. F. Lampsell and J. G. McGowan, The transient response of gas turbine plant heat exchangers — additional solutions for regenerators of

the periodic flow and direct transfer types, J. Engng. Pwr., 86, p.127, (1964).

[44] Willmott, A. J., Digital computer simulation of a thermal regenerator, Int. J. Heat Mass Transfer, 7, p. 1291, (1964).

[45] Demetri, E. P. and M. Platt, A general method for the analysis of compact multifluid heat exchangers, ASME paper 72-HT-14, (1972).

[46] Kiely, D., Private communication, (1972).

[47] Kays, W. M., and A. L. London, *Compact Heat Exchangers,* 2nd ed., McGraw-Hill Book Co., New York, (1964).

[48] Grimson, E. D., Correlation and utilization of new data on flow resistance and heat transfer for cross flow of gases over tube banks, Trans. Am. Soc. mech. Engrs., 59, p. 583, (1937).

[49] Tong, L. S., *Boiling Heat Transfer and Two Phase Flow,* John Wiley and Sons, Inc., New York, (1965).

[50] Collier, J. G., and G. B. Wallis, *Two-Phase Flow and Heat Transfer,* Summer course notes, Stanford Univ., (1967).

[51] Kutateladze, S. S., *Problems of Heat Transfer and Hydraulics of Two Phase Media,* Pergamon Press, London, (1969).

[52] El-Wakil, M. M., *Nuclear Heat Transport,* International Textbook Co., (1971).

Table 1

Potentiometer Setting for Analog Circuit

1. $T_{1,i}$ at steady state

2. $\dfrac{1.0}{\Delta x}$

3. $\dfrac{P_1}{A_1 \rho_1 c_1}$

4. $T_{t,i}$ at steady state

5. $\dfrac{P_1}{A_t \rho_t c_t}$

6. $\dfrac{P_2}{A_t \rho_t c_t}$

7. $T_{2,i}$ at steady state

8. $\dfrac{P_2}{A_2 \rho_2 c_2}$

9. $\dfrac{v_2}{\Delta x}$

Table 2

Potentiometer Setting for Hybrid Circuit

1. $\left(\dfrac{h_2 P_2}{A_2 \rho_2 c_2 v_2} + \dfrac{1}{v_2 \Delta t} \right)$

2. $\dfrac{1}{v_2 \Delta t}$

3. $\dfrac{h_2 P_2}{A_2 \rho_2 c_2 v_2}$

4. $\dfrac{\Delta t h_2 P_2}{A_t \rho_t c_t}$

5. $\left[1 - \dfrac{\Delta t}{A_t \rho_t c_t} \left(h_2 P_2 + h_1 P_1 \right) \right]$

6. $\left(\dfrac{\Delta t h_1 P_1}{A_t \rho_t c_t} \right)$

7. $\dfrac{h_1 P_1}{A_1 \rho_1 c_1 v_1}$

8. $\dfrac{1}{v_1 \Delta t}$

9. $\left[\dfrac{h_1 P_1}{A_1 \rho_1 c_1 v_1} + \dfrac{1}{v_1 \Delta t} \right]$

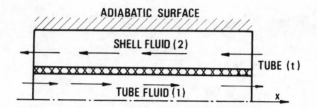

Fig. 1. Shell and Tube Heat Exchanger-Counterflow

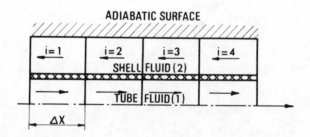

Fig. 2. Subdivision of Shell and Tube Heat Exchanger

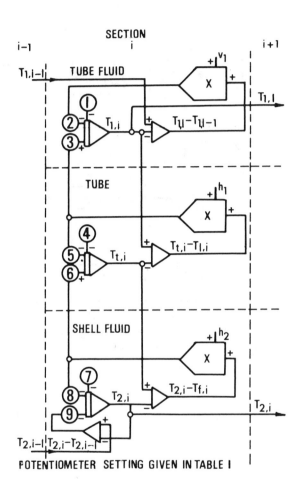

Fig. 3. Block Diagram - Analog Computer

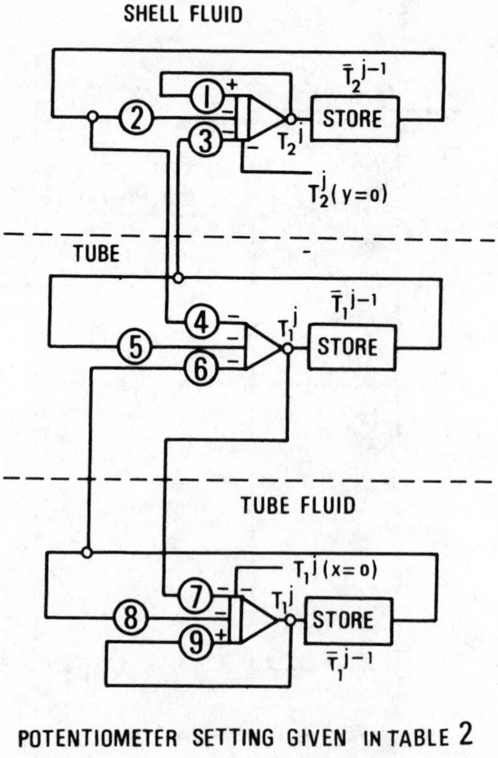

Fig. 4. Block Diagram - Hybrid Computer

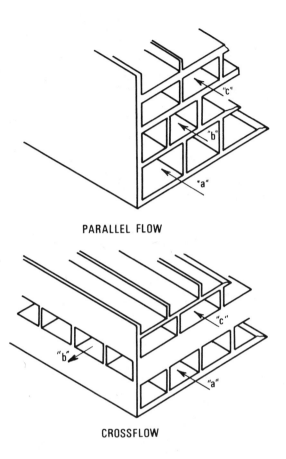

PARALLEL FLOW

CROSSFLOW

Fig. 5. Multifluid Heat Exchanger

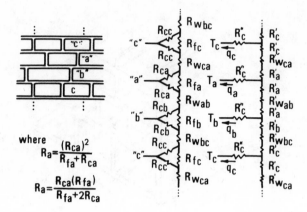

where
$$R_a = \frac{(R_{ca})^2}{R_{fa} + R_{ca}}$$

$$R_a = \frac{R_{ca}(R_{fa})}{R_{fa} + 2R_{ca}}$$

Fig. 6. Parallel Flow Multi Fluid Heat Exchanger Section

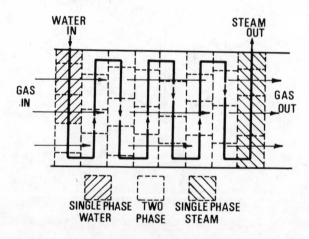

SINGLE PHASE   TWO   SINGLE PHASE
WATER   PHASE   STEAM

Fig. 7. Once Through Steam Generator

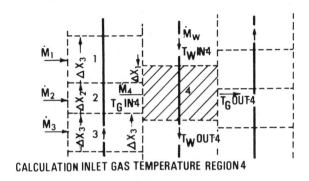

CALCULATION INLET GAS TEMPERATURE REGION 4

Fig. 8. Incremental Region of Steam Generator

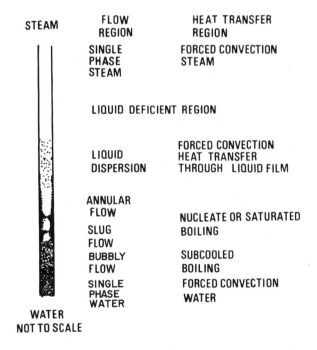

Fig. 9. Flow and Heat Transfer Regimes

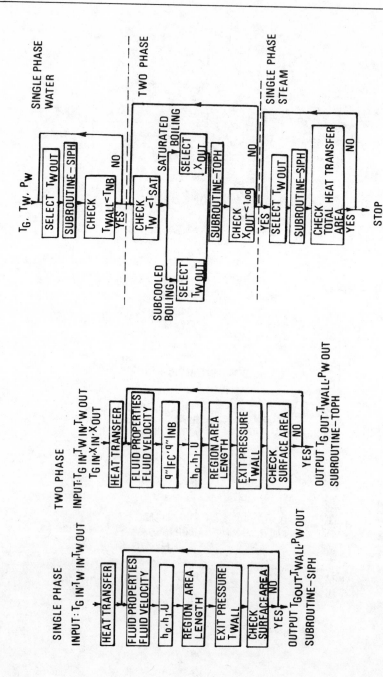

Fig. 10. Computer Flow Diagrams of Subroutines

Fig. 11. Computer Flow Diagram for Steam Generator Calculations

Chapter 19

PREDICTION OF PERFORMANCE CHARACTERISTICS OF HEAT-EXCHANGING EQUIPMENT

C.L. Spigt, F. van der Walle, H.I. Bardoux (*)

1. Introduction.

A detailed knowledge of the heat transfer and flow phenomena in heat-exchanging equipment may be considered as a prerequisite for obtaining a compatible and reliable design. For instance, the designer of such equipment may like to know whether he has not overestimated the dimensions of the heat-exchanger resulting in bad economics, or in poor control of the system. Also, he may be interested in the magnitude of the wall temperatures because of strength considerations and the occurrence of melting (burn-out) phenomena or in the possibility of encountering flow oscillations resulting in fluctuating wall temperatures which sometimes cause corrosion problems and in many cases damaging of the tubes.

In the following the stability and the transient characteristics of steam-generating units will be dealt with.

A steady state is called stable if, after a small disturbance, the system will ultimately return to the original steady state condition. The concept of small disturbances is used in this definition, as it will always be possible to destroy completely a steady state condition if the applied disturbance is sufficiently large. It is not sufficient to know whether the steady state is stable or not. Also, the response characteristics of an otherwise stable configuration, as described for instance by the amplitudes and frequencies resulting from disturbances, are important to the designer. In order to obtain information on the stability of a steady state condition of a boiling system a frequency response analysis has to be carried out from which these transfer functions can be obtained. The assumption of small disturbances in a stability analysis means in many cases that the system can be considered linear,

---

(*) DYNAFLOW—INFONET, Amsterdam, The Netherlands

N.B. ALL FIGURES QUOTED IN TEXT ARE AT THE END OF THE CHAPTER

which simplifies its analysis and makes the transfer function approach more readily applicable.

Some investigators have performed stability analysis using the Ledinegg instability [6] for explaining the onset of hydraulic oscillations. Ledinegg was amongst the first to point out that an instability can arise in a heated two-phase system having a negative slope in a part of its hydrodynamic characteristics, i.e. in the flow-pressure loss curve. At one time this was regarded as the primary criterion for instability in such systems. Flow oscillations, however, result from a dynamic interaction between the void distribution in the channel and the inlet mass flow, which is largely influenced by the dynamic aspects of the boundary conditions imposed on the channel. The Ledinegg criterium simply states that in a region with negative slope, i.e. decreasing steady state pressure drop with increasing mass flow, that any temporarily introduced displacement from the steady state condition will tend to become larger with time as a consequence of this negative slope. In a positive slope region any displacement will tend to become smaller directly after the source of the introduced disturbance has been removed.

In order to be able to evaluate the Ledinegg stability criterium in relation to the harmonic analysis approach it is desirable to define two basically different definitions, namely:

- Condition of Static Stability

  A system is considered statically stable if, directly *after* an initial displacement from a steady condition, the system *tends to move* in the direction of the original steady state condition.

- Condition of Dynamic Stability

  A system is considered dynamically stable if, upon an initial displacement from a steady condition, the system will ultimately attain again the original steady state condition.

This means that the resulting dynamic behavior after the initial displacement has to converge with time to the original condition. This convergence process may be either of a periodic or of an aperiodic nature.

It is evident that the only stability criterium that is of interest for the dynamic behavior is the dynamic stability condition. It follows from the above that a system can only be dynamically stable if it is also statically stable. However, a statically stable system need not be dynamically stable; an example of this is a damped mass-spring system. With a positive mass and a positive spring constant the spring force will always try to reduce a displacement from the equilibrium condition and the system is therefore statically stable. However, in case the damping is negative the resulting dynamic behavior afterwards will exhibit ever increasing amplitudes and the system is therefore dynamically unstable.

The Ledinegg stability criterium is in fact the static stability criterium and it follows from the above that it is neither a sufficient criterium for stability nor a necessary criterium for stability.

It is true, however, that once the Ledinegg criterium shows a statically unstable condition the systems will be dynamically unstable. A complete study of the stability characteristics, therefore, cannot be made using solely the equations describing the steady state, but has to start from equations incorporating dynamic effects.

In the following, the basic equations and the solution procedures are described. Furthermore, some examples are given of the prediction of the stability and transient characteristics of steam-generating units by means of digital computer codes.

2. Basic Equations and Correlation Functions.

Considering a coolant channel in which evaporation takes place and which is heated by a liquid or gas flow, the following conservation laws can be formulated for the evaporating part.

Conservation of mass

$$\frac{\partial}{\partial t}\left\{\rho_1 (1-a) + \rho_s a\right\} + \frac{\partial}{\partial z}\left\{\rho_1 (1-a) V_1 + \rho_s a \, SV_1\right\} = 0 \tag{1}$$

Conservation of momentum

$$\frac{\partial}{\partial t}\left\{\rho_1 (1-a) V_1 + \rho_s a \, SV_1\right\} + \frac{\partial}{\partial z}\left\{\rho_1 (1-a) V_1^{\,2} + \rho_s a S^2 V_1^{\,2}\right\}$$

$$+ \frac{\partial p}{\partial z} + g\left\{\rho_1 (1-a) + \rho_s a\right\} + F = 0 \tag{2}$$

Conservation of energy

$$\frac{\partial}{\partial t}\left\{\rho_1 (1-a) \, cT_1 + \rho_s a(cT_s + e)\right\} + \frac{\partial}{\partial z}\left\{\rho_1 (1-a) V_1 \left(cT_1 + \frac{p}{\rho_1}\right)\right. +$$

$$+ \; \rho_s \, aSV_1 \left(cT_s + e + \frac{p}{\rho_s} \right) \Big\} = \; q_w \tag{3}$$

The local heat load q_w cannot usually be given generally beforehand but has to be calculated from the given mass flow and inlet temperature of the heating fluid and the heat transfer characteristics between the two process streams. Assuming countercurrent flow of the two process streams, this equation becomes

$$\left(V_p \, \frac{\partial T_p}{\partial z} - \frac{\partial T_p}{\partial t} \right) \rho_p c_p = \; q_w \tag{4}$$

and

$$q_w = \; h \, (T_p - T_{sc}) \tag{5}$$

The heat transmission coefficient h is composed of the heat transfer coefficients at the primary and secondary side and of the material characteristics of the separating wall.

In the calculations the evaporating channel is subdivided into four regions:

(a) the non-boiling region $T_1 < T_b$ and $a = 0$ T_b is liquid temperature at which bubbles detach from the wall.

(b) the subcooled boiling region $T_b < T_1 < T_s$; for the void fraction an extra equation must be given

$$\frac{\partial}{\partial t} \left\{ \rho_s \, ae \right\} + \frac{\partial}{\partial z} \left\{ \rho_s \, aSV_1 \left(e + \frac{p}{\rho_s} \right) \right\} = \; \kappa q_w \tag{6}$$

(c) the saturated boiling region $T_1 = T_s$

(d) the superheated region $a = 1$

The outlet of the coolant channel may fall in any one of these regions. Within each region a further subdivision can be made according to an existing flow regime, e.g. slug flow, bubbly flow, froth flow, mist flow, etc., each having their own flow and heat transfer correlations.

In the three conservation laws there are four main unknowns α, V_1, T_1 and T_s. The other quantities in the conservation laws are given by the equations of state which are expressions for the physical quantities of the fluid in an analytical form and the heat transfer and fluid flow correlations. In the computer codes the physical quantities may vary along the coolant channel and with time. The equations of state can be formulated as

$$\rho_s = f\,(T_s)\,\rho_l = f\,(T_l)\ T_s = f\,(p)\ c = f\,(T_l)\ e = f\,(T_l) \tag{7}$$

Correlation functions have to be given for the slip ratio S, for the wall friction force F, for the heat division parameter k, for the liquid temperature T_b (definition of boiling region) where bubbles detach from the wall, and for the quantity h in the heat transmission equation(5). The wall friction force per unit cross-sectional area and per unit length for a two phase mixture can be expressed in terms of Fanning friction factor f and the two phase friction multiplier R, see ref. [4].

Together with the conditions to be fulfilled in each sub-region, the equations (1),(2)and (4) can be solved. The equations describing the steady-state characteristics can be derived by omitting all time-dependent terms in the equations mentioned before.

3. General Description of the Computer Programs and the Solution Procedures.

The equations that determine the thermo-hydraulic behavior of a system can be divided into:

— the conservation laws for mass, momentum and energy, describing the variations of a number of unknowns along the length of the heat-exchanger as a function of time.

— a number of heat transfer and fluid flow correlation functions.

— the equations of state.

— a number of boundary conditions e.g. prescribed conditions of natural or forced circulation, constant pressure drop across the heat-exchanger, programmed control, etc. The equations describing the boundary conditions relate the conditions at the exit of the coolant channel to those at the inlet of the coolant channel. The imposed boundary conditions are sometimes simplified by assuming, for instance, a constant pressure at the exit of the channel and a constant temperature at the inlet of the channel and an imposed pressure drop.

Regarding the study of the transient behaviour of a two-phase flow and the occurrence of hydraulic oscillations, different approaches have been used in the past for the solution of the equations. In many studies, the coolant channel is divided into a number of spatial increments for which the general equations, e.g. the conservation laws, the equations of state, etc., are formulated. In some of these studies the number of increments has been assumed to be small, e.g. only two, namely the non-boiling and the boiling region. As a rule, in these studies time delays are used to simulate the transport time of the fluid through the system and the equations are solved by means of an analogue computer. These studies are normally referred to as "lumped parameter studies" and these should be distinguished from "distributed parameter studies" in which the number of increments is large and the increments are mostly evenly distributed along the length of the channel.

In this last case the equations are solved by means of a digital computer. Another group of studies consists of those in which the occurrence of hydraulic oscillations is considered as an isolated phenomenon and in which an attempt is made to study the possible mechanism or controlling parameters. Examples of this group are the studies of Ledinegg [ref. 6] and Stenning [ref. 5].

The general solution procedure is as follows

- The differential equations (conservation laws) together with the correlation functions and equations of state are integrated by a straighforward integration technique starting with given or assumed values for the unknowns at the inlet of the heat-exchanger.

- The resulting outlet values together with the inlet values are checked for compatibility with the boundary conditions. The correct values for the assumed inlet values are determined either straightforward or by an iterative process in which successive, increasingly accurate values for the unknowns are used.

The procedure to integrate the differential equations and the solution procedure for the boundary conditions is dependent upon the type of the problem as far as the time dependence is concerned.

Two cases can be distinguished:

a. steady-state performance

b. unsteady performance, further divided into:

b.1. harmonic analysis by small perturbation techniques leading to the calculation of transfer functions

b.2. transient analysis to large amplitude disturbances.

In the steady state (a) the conservation laws become ordinary differential equations that can be integrated by well-known integration techniques as e.g. Runge-Kutta. The boundary conditions become simple algebraic expressions and the iterative solution of the correct starting values can be obtained by well-known techniques such as Newton or Regula Falsi.

In the harmonic analysis (b.1.) every unknown is written as

$$X = X_o + X_i \cdot e^{i\omega t}, \tag{8}$$

where X_o is the steady state value and

$X_i \cdot e^{i\omega t}$ is the perturbation.

Small harmonic variations are assumed which means that

$$\frac{X_i}{X_o} \ll .1 \qquad\qquad (9)$$

and X_o as well as X_i (in general complex) are only functions of the space coordinate. After linearization of the equations and after subtracting the steady state equations and dividing by $e^{i\omega t}$ a set of ordinary quasi-linear differential equations results, which can be integrated in exactly the same way as the steady state equations. The frequency ω appears as parameter in the equations, and the equations therefore have to be solved for each value of ω that is of interest. The boundary conditions are treated in the same way and result again in algebraic linear expressions for X_i with ω as parameter. The boundary conditions can be solved at once as the whole system of equations is linear, which enables the use of the superposition principle.

In the transient analysis case (b.2.) the simplifications as mentioned above are not applicable so that the complete partial differential equations have to be solved in the space-time domain. Sometimes a simplification is possible when the dynamic analysis is made for systems with large response times.

In those cases it is possible to treat the time derivatives in the partial differential equations as small perturbations, which means that it is assumed that the $\partial/\partial t$ − terms are of an order of magnitude smaller than the $\partial/\partial x$ − terms. The integration can be performed then in the space domain, only similar to the steady state procedure, while the $\partial/\partial t$ − terms appear as additional terms in the right-hand sides of the equations.

In this publication some examples of applications of the computer codes in the analysis of the performance characteristics in steady and unsteady operating conditions of heat-exchanging equipment will be given.

4. Electrically Heated Natural Circulation Boiler.

At first an example is given of the prediction of the steady state and stability characteristics of an electrically heated boiling water channel simulating a simple fuel assembly of a nuclear reactor. A simplified flowsheet is given in Fig. 1. The channel formed by the heating element and the shroud is the riser, the one formed by the shroud and the pressure vessel is the downcomer. Since the shroud is open at both ends the two channels are in direct connection with each other. When the element is heated the water in the riser starts to boil and vapor is formed. Owing to the resulting density difference between the fluid in the riser and in the downcomer the steam-water mixture in the riser flows upwards by a natural circulation process. At the water surface the steam and water are separated. The water returns to the inlet of the riser through the downcomer passing a subcooler. The steam flows to the condenser and the condensate is returned to the downcomer.

In this case the electrical power is prescribed so the heat load need not be calculated. Therefore, there is no need to consider the equations 4 and 5. For the wall friction force, F (see equation 2), the Fanning equation is used in the single phase region.

In the two-phase region use is made of the Two Phase Friction Multiplier R as evaluated by Martinelli-Nelson and formulated by Jones, Ref. [1]. For the slip ratio S (equation 1, 2 and 3) the correlation of Zuber-Findlay is used with experimentally determined values of the distribution parameter C_o and the local drift velocity, Ref.[2]. The correlation used for T_b (detachment of bubbles) was based on the criteria of Bowring, Ref. [3], while for the heat division parameter κ the correlation developed in Ref. [4] has been used.

The calculated distribution along the boiling channel of the void fraction and the liquid and the saturation temperature for two operational steady state conditions (electrical power 100 and 150 KW, subcooling .7°C, system pressure 16 atm) have been plotted in Fig. 2. Furthermore, the calculated natural circulation flow-rate V_{in} has been given with some results of experiments. As can be seen there is an excellent agreement between the measured and predicted void fraction data and a slight discrepancy between the measured and calculated values of the inlet mass flowrate.

Also the dynamic performance characteristics have been calculated. These have been performed by a harmonic analysis as outlined before leading to the calculation of transfer functions. An open loop as well as a closed loop analysis was followed. With an open loop analysis the concepts of control theory can be applied. The required cut in the closed loop has been made in the downcomer at the inlet of the riser, see Fig. 1.

A stability criterion has been derived, defined as

$$G_1 = \frac{\left(T_{s,i}\right)_{do}}{\left(T_{s,i}\right)_{in}} \tag{10}$$

G_1 determines the change in saturation temperature (pressure) at the outlet of the downcomer upon a variation in saturation temperature (pressure) at the inlet of the riser.

An instability can be expected when the modulus of G_1 is equal or larger than 1 and at the same time when the phase angle between the two temperatures in eq. (10) has reached a value of 0 or 360°C. In this case any disturbance in pressure at the inlet will be amplified.

Calculated open and closed loop characteristics have been plotted for a saturation temperature of 200°C (system pressure of 16 atm) in the intermediate frequence range of .6 to 1.4 c.p.s. for different channel powers in Figure 3. In the plot of the open loop characteristics C_1 it is shown that an unstable condition is passed when progressing from 150 to 151 kW channel power. At 150 kW the modulus of G_1 becomes larger than unity but the phase angle does not approach the value of 0° or 360°C. At 151 kW the modulus of G_1 is larger than unity and the phase angle becomes zero, which indicates that the system is unstable. In the right-hand side of the figure the transfer function from power to inlet flowrate has been plotted.

A comparison between the open and closed loop results at 150 and 151 kW leads to the conclusion that large amplitudes in inlet mass flow occur. The predicated instability threshold of 151 kW at a frequency of .947 c.p.s. corresponds with the experimentally determined value of 162 kW and a frequency of .93 c.p.s., see Ref. 4. In the calculation of the closed loop transfer functions the characteristics of the heating element were not yet taken into account, which makes it quite difficult to compare the calculated closed loop results with the experimentally determined transfer functions.

In Fig. 4. open loop characteristics are presented for the low and high frequency range. In these frequency regions G_1 also approaches or exceeds unity and the phase angle is probably close enough to 0° for oscillations to develop. By means of quantities b_v etc the dynamic characteristics of the condenser and subcooler can be varied in the computer code. As can be seen, the subcooler and condenser characteristics greatly influence the results in the low frequency range. It may be concluded that low frequency oscillations will develop over a wide range of heating powers. The channel power has a stabilizing influence in this low frequency range. These results are comparable with the observations made by Stenning, Ref. [5], and probably correspond with the instability criterion developed by Ledinegg, Ref. [6], for an aperiodic excursive stability.

In the high frequency range the phase angle of G_1 exhibits very strong variations with frequency.

As could be concluded from the calculated distribution along the length of the liquid velocity during perturbation these high frequency oscillations are characterized by near zero variations of the liquid velocity at the inlet and have the character of standing-wave oscillations, see Ref. [4].

5. Water Heated Once-Through Steam-Generator.

Secondly, an example is given of the calculated results of the heat transfer and hydraulic characteristics of a water-heated once-through steam-generator.

A simplified flowsheet is given in Figure 5. The apparatus consists of a primary water circuit working under a pressure of 2200 psi and a secondary circuit with an outlet for superheated steam and a feedwater supply. The secondary fluid enters the steam-generator in nearly saturated conditions at the bottom, undergoes a phase change and exits superheated. The once through steam-generator changes in load by exchanging boiling surface for superheating surface. For the calculations the saturated boiling region was subdivided into a bubbly flow region, each having its own heat transfer correlations.

The steady state part of the computer program calculates the fluid and wall temperatures, phase velocities, steam qualities, void fractions, pressure drops and the heat flux distribution along the heat transfer path for given values of the primary mass flow, pressure and inlet temperature and the secondary feedwater mass flow, temperature and pressure. Some results of calculations are shown in Figs. 6 and 7. Fig. 6 shows the variation of the primary fluid temperature, the wall temperature and the secondary liquid and steam-temperature. In the curve of the wall temperature the main heat transfer regions can be clearly distinguished. The subcooled boiling region with a wall temperature sharply decreasing with length, the saturated boiling region with a slightly increasing wall temperature owing to the slight change in the heat transfer capabilities of a two-phase flow mixture with increasing void fraction, followed by a sharp increase in wall temperature indicating a change in the heat transfer mechanism in that of film boiling. In the film boiling region the wall temperature increases somewhat owing to the decrease in the forced convection heat transfer. A further sharp increase indicates that the fluid has become superheated. In the superheating section the tube wall temperature approaches the primary liquid temperature since the heat transfer coefficient of the primary water inside the tube is considerably higher than the coefficient of the steam surrounding the tube.

In Fig. 7 the calculated results are plotted of steam quality, void fraction and heat transferred. As shown, most of the heat is transferred in the boiling region. Furthermore, in this figure values are given of the total heat transferred, the primary outlet temperature, the pressure loss in the boiler part and superheated part.

For the configuration outlined above the dynamic performance characteristics have been calculated. These have been performed by a harmonic analysis as outlined before leading to the calculation of transfer functions. In this case a closed loop analysis was followed while the material characteristics of the tubes have been taken into account.

By means of the transfer function from a perturbation of inlet pressure drop to the resulting change in inlet velocity (both secondary side) the dynamic performance characteristics are analyzed. Some results of the calculations are given in Fig. 8. as a function of the frequency. The plotted values (curve $k_{ci_o} = 1$) can be interpreted as the percentage change in inlet velocity relative to the steady state value due to an inlet pressure drop fluctuation of about .0025 psi. As is shown, three resonance peaks are present: the dominating one in the middle, a lower and a higher frequency peak. The system's low inlet pressure drop coefficient makes it somewhat sensitive for oscillatory behavior. In the same figure the influence is shown of the inlet pressure drop upon stability. Increasing the inlet pressure ($k_{ci_o} = 210$) drop has a large stabilizing effect on the dominating peak. This stability becomes doubtful as the two neighboring peaks become more dominant at larger values of the inlet pressure drop.

The peak values of the three resonance frequencies have been plotted as a function of load in Fig.9. From the shape of this curve it can be concluded that, for the low inlet pressure drop coefficient, between 60% and 80% load unstable conditions may develop in the middle frequency region. At lower and higher loads the peak values rapidly become less pronounced.

Comparison with unpublished results of experiments showed a good agreement regarding the variation of the temperatures along the height of the channel in the steady state conditions as well as the frequency and the amplitude of the oscillatory behavior.

6. Sodium-Sodium Heat-Exchanger.

An example will be given of the calculation of performance characteristics of a sodium-sodium heat-exchanger in steady-state conditions and of the response characteristics to a large disturbance. A simple sketch of the sodium-sodium heat-exchanger is given in Fig. 10. The primary process steam flows top to bottom between two concentric tubes. In this space the secondary process steam flows from bottom to top inside a number of tubes. In fact the computer program has been written for a sodium heated steam-generator using a description where the two-phase mixture was considered as a one-phase medium with a certain density, enthalpy, temperature, etc. Because all mathematical procedures were already developed within this program, this computer code was also used for the sodium-sodium heat-exchanger.

Owing to the fact that for this case the response of the heat-exchanger to large disturbances had to be calculated, the partial non-linear differential equations have to be solved in the space-time domain. The disturbances were, however, relatively slow : say, 3 seconds or more. In a system consisting of many components the coupling of the various components is such that the high frequency content of the response is not passed on from one component to the other. As a consequence the system as a whole responds more or less according to the time scale of the large disturbance. This characteristic makes it possible to treat the time derivatives in the partial differential equations as small perturbations, which means that it is assumed that the $\partial/\partial t$ terms are smaller than the $\partial/\partial z$ terms. As a result, the integration of the equations can be performed with rather large integration steps in the time without resulting in inaccuracies.

Some results of calculations will be given in which a 90% reduction in the secondary process stream was given in 25 seconds. In Fig. 11 the variation is given with time of the secondary process stream ϕ_s, the primary outlet temperature and the secondary outlet temperature. During the perturbation of the system the primary inlet temperature and the secondary inlet temperatures have been kept constant. As is shown, the secondary outlet temperature approaches already in a short time the primary inlet temperature, which means that in the top part of the heat-exchanger no heat is transferred. This is shown in Fig. 12 more clearly, where the primary and secondary sodium temperatures have been plotted along the length of the heat-exchanger for t = 0, t =25 seconds (end of excitation) and t = 45 seconds. It is clearly shown that at low loads the length over which the heat is transferred in the heat-exchanger is very short. The advantage of this computer code is that for these low loads an accurate temperature distribution over the length can be determined also, owing to the large flexibility with which the stepsize of the integration can be chosen.

7. Sodium-Heated Once-Through Steam-Generator.

Finally an example is given of the calculated results of the heat transfer and hydraulic characteristics in a steady state condition of a once-through countercurrent steam-generator heated by sodium as a primary fluid and of the response characteristics to a large disturbance in secondary mass flow and inlet temperature. The apparatus consists of a primary sodium circuit flow from top to bottom and a secondary water-steam circuit flowing in a number of tubes from bottom to top of the heat-exchanger. The secondary fluid enters the generator subcooled at the bottom, undergoes a phase change and exists superheated. For the calculations the total length of the steam-generator has been divided into six regions. In the calculations the two-phase mixture was considered as a one-phase medium with a certain density, enthalpy, temperature, etc.

The six regions are :
(a)   the non-boiling region
(b)   the subcooled boiling region
(c)   the saturated bubbly flow region
(d)   the saturated heat transfer crisis region
(e)   the saturated mist flow region
(f)   the superheated region.

In the non-boiling region the secondary wall temperature is smaller than the saturation temperature. For the heat transfer a single-phase heat transfer correlation can be used. In the subcooled boiling region the secondary wall temperature is equal to or larger than the saturation temperature, but the enthalpy of the fluid is smaller than the enthalpy for saturated liquid conditions. A subcooled boiling heat transfer correleation is used. In the three saturated boiling regions the enthalpy is equal to or larger than the enthalpy for saturated liquid conditions but equal to or smaller than the enthalpy for saturated steam conditions. In the bubbly flow region the same heat transfer correlation is used as in the subcooled region. When the heatflux becomes larger than a critical value, the start of the saturated heat transfer crisis region is defined. In this region the heatflux is taken according to an assumed burn-out correlation. When the burn-out heat flux becomes smaller than the heat flux calculated from a mist-flow heat transfer correlation, the start of the saturated mist flow region is defined.

The solution procedure is the same as described before. The computer code calculates the fluid and wall temperatures, pressure drop and the heatflux distribution along the heat transfer path for given values of the primary mass flow and inlet temperature and the secondary feedwater mass flow, temperature, and pressure. The system can also be perturbated with a disturbance in the foregoing independent variables.

In Fig. 13, some calculated results have been plotted of the primary sodium temperature, the wall temperature at the primary and secondary side and the secondary fluid temperature as a function of the axial coordinate along the height of the steam-generator. Also, the start of the six heat transfer regions is indicated.The calculated total of heat transferred amounts to 29.4 MW while the secondary mass flow was taken as 18.45 kg/sec. As is shown, most of the heat is transferred in the boiling region. In this case the saturated bubbly flow region (region 3) was very short, owing to the fact that at the end of the subcooled region the heat transfer rate was almost equal to the one calculated from the burn-out correlation.

In the following figures some results are given of the calculated response characteristics to a 90% reduction of the secondary process stream in about 40 seconds. In Fig. 14, the primary sodium temperature, the secondary fluid temperature and the wall temperature are given as a function of time at the bottom side and in Fig. 15, at the top side of the stream-generator. The wall temperatures show the largest variation with time at the bottom side, while the two imposed boundary conditions, constant primary sodium temperature at the top and constant secondary fluid temperature at the bottom, are clearly shown. Finally in Fig. 16, the variation of the temperatures along the length of the heat-exchanger 55 seconds after the start of the excitation are given. In the curve of the wall temperatures the main heat transfer regions can clearly be distinguished. The subcooled boiling region's wall temperature decreases with length and the saturated boiling region with an increasing wall temperature, owing to the change in the heat transfer capabilities of a two-phase flow mixture with increasing void fraction. A further sharp increase indicates that the fluid has become super-heated. In the superheating section the tube wall temperatures approach the primary liquid temperature since the heat transfer coefficient of the primary water inside the tube is considerably higher than the heat transfer coefficient of the steam film inside the tubes.

8. Concluding Remarks.

From the examples given it is clear that the computer techniques developed can be used in the analysis of the performance characteristics of heat-exchanging equipment in steady and unsteady operating conditions, and in the dimensioning and optimalization of such equipment. The result can be a more economic design or a solution of particular design difficulties.

Further development work is under way regarding the phenomena governing the onset and character of the hydraulic instabilities. Particular attention is given to the low-frequency range and to the pressure drop characteristics in order to derive analytically the relationship between the Ledinegg instability or Static Stability (S-curve of the pressure drop-mass flow characteristics of the boiling channel) and the Dynamic Stability derived from the equations given before.

NOMENCLATURE

c	heat capacity of liquid at constant pressure per unit mass, $L^2 t^{-2} T^{-1}$
e	latent heat of evaporation at constant volume, $L^2 t^{-2}$
F	wall friction force per unit cross-sectional area per unit length, $ML^{-2} t^{-2}$
f	frequency, t^{-1}
G_1	open loop transfer function, see eq. 10
g	gravitational acceleration, Lt^{-2}
H	transfer function
h	heat transmission coefficient, $ML^{-1} t^{-3} T^{-1}$
K	amplitude ratio
k	pressure loss coefficient
L	length, L
M	mass, M
P_s	system pressure, $ML^{-1} t^{-2}$
p	local pressure $ML^{-1} t^{-2}$
Q	heat input, $ML^2 t^{-3}$
q_w	heat transported per unit cross-sectional area per unit length, $ML^{-1} t^{-3}$
S	slip ratio $= V_s/V_1$
t	time, t
T	temperature, T
T_b	liquid temperature at which bubbles detach from the wall, T
ΔT_{sub}	subcooling temperature $(T_{sat} - T_{in})$
V	velocity, Lt^{-1}
x	steam quality
z	coordinate in axial direction, L
a	void fraction
ρ	density, ML^{-3}
κ	heat division parameter (defines part of heat load that is used for steam production in the subcooled region)
ϕ	mass velocity, Mt^{-1}
ϕ	phase angle

SUBSCRIPTS.

l	liquid
s	vapor or steam
sc	secondary circuit
p	primary circuit
w	wall
sat	saturated conditions in condenser
in, ci	inlet coolant channel
do	downcomer outlet
i	harmonic variation from steady state
o	steady state value
Δ	amplitude of sine
	difference between two values
mod	modulus.

REFERENCES

[1]   Jones, A.B., Dight, D.G., Hydrodynamic stability of a boiling channel, Reports Knolls Atomic Power Laboratory, Schenectady, KAPL 2170 (1961), 2208 (1962), 2290 (1963) and 3070 (1964).

[2]   Zuber, N., Findlay, J.A. Average volumetric concentration in two-phase flow systems, Transactions of the ASME, Journal of Heat Transfer, Vol. 87, p. 453-468, 1965.

[3]   Bowring, R.W., Physical model based on bubble detachment and calculation of steam voidage in the subcooled region of a heated channel, Report OECD Halden Reactor Project, HPR 10, 1962.

[4]   Spigt, C.L., On the hydraulic characteristics of a boiling water channel with natural circulation, Report Eindhoven University WW-R92, 1966.

[5]   Stenning, A.H., Veziroglu, T.N., Flow oscillation modes in forced convection boiling, Proceedings of the 1965 Heat Transfer and Fluid Mechanics Institute, Stanford University, 1965.

[6]   Ledinegg. M., Unstabilität der Strömung bei Naturlichen und Zwangsumlauf (Flow instability in natural and forced circulation), Die Wärme, Vol. 61, p. 891-898, 1938.

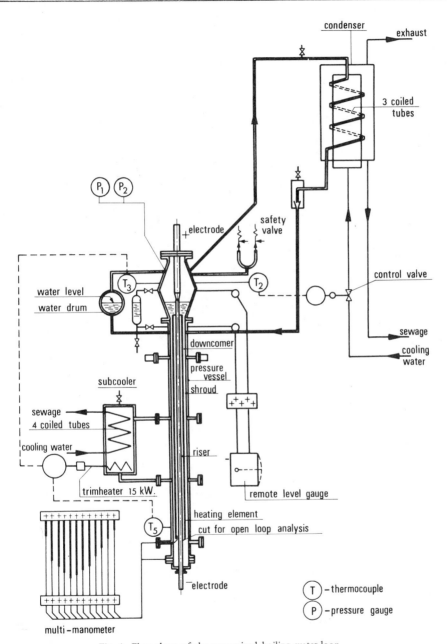

Fig. 1. Flow sheet of the pressurized boiling water loop.

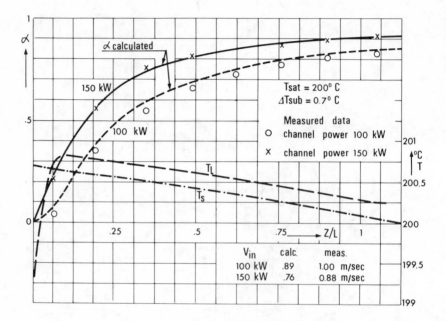

Fig. 2. Calculated and measured results of the steady-state characteristics, Test Section I.

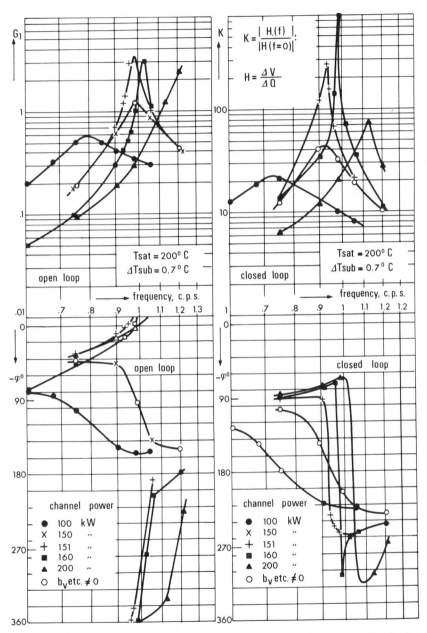

Fig. 3. Open and closed loop transfer functions in the intermediate frequency range, Test Section I.

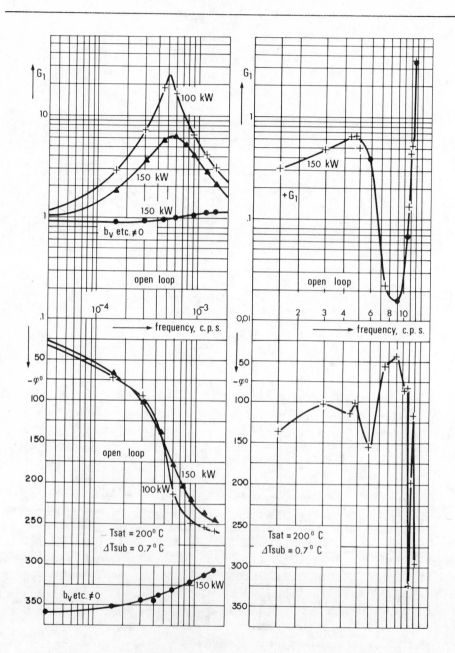

Fig. 4. Open loop transfer functions in the low-and high frequency range. Test Section I.

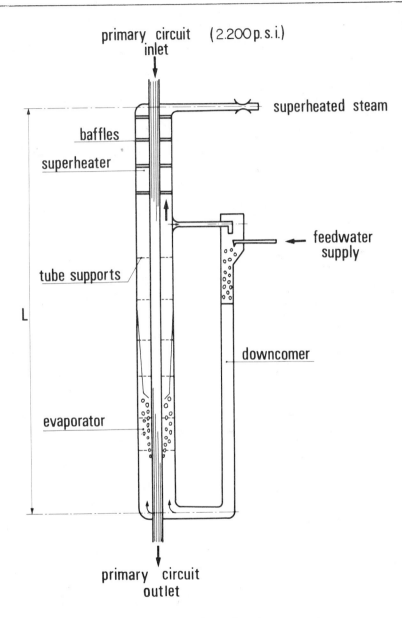

Fig. 5. Flowsheet Water-heated once-through steamgenerator.

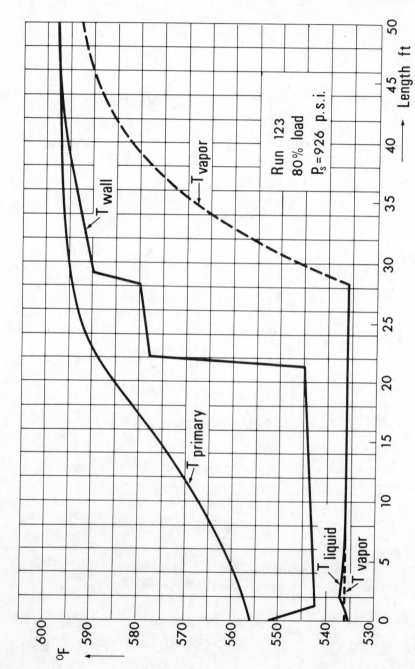

Fig. 6. Water-heated once-through steamgenerator.

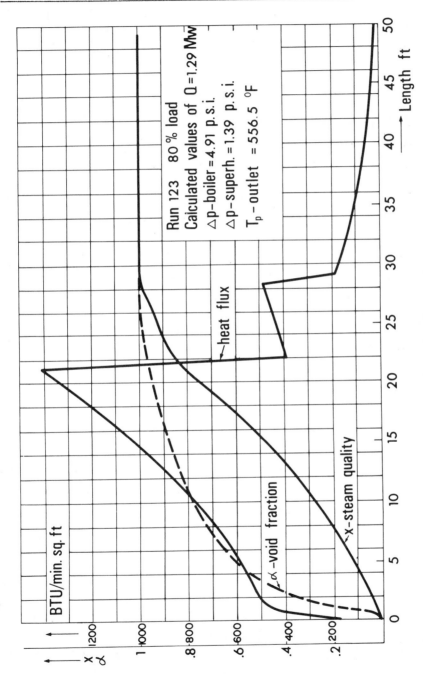

Fig. 7. Water-heated once-through steamgenerator.

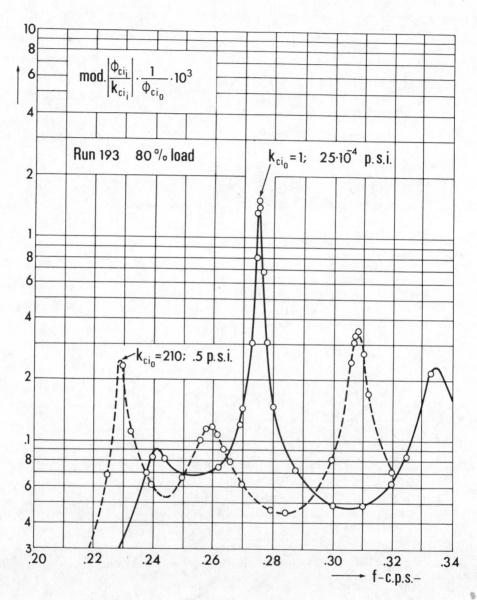

Fig. 8. Water-heated once-through steamgenerator.

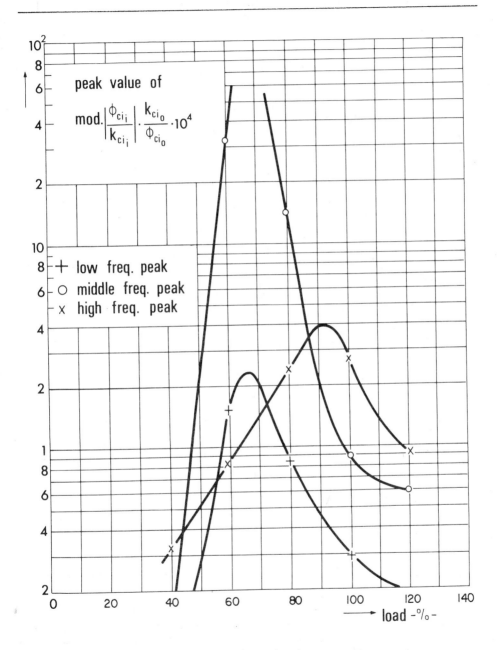

Fig. 9. Water-heated once-through steamgenerator.

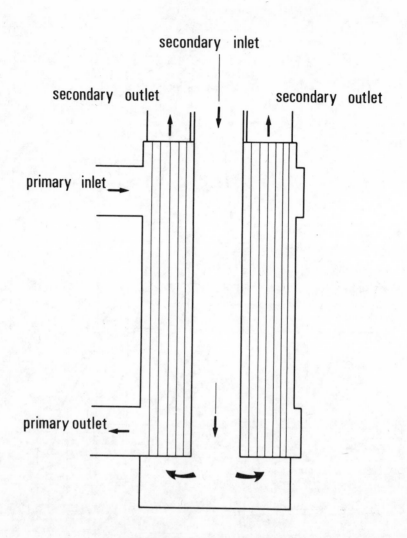

Fig. 10. Sodium-Sodium heat exchanger.

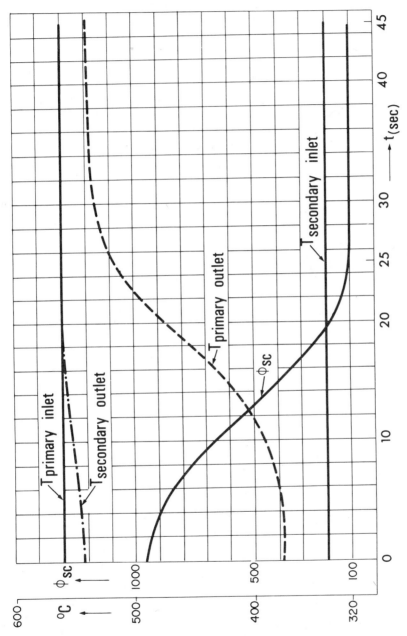

Fig. 11. Sodium-Sodium heat exchanger.

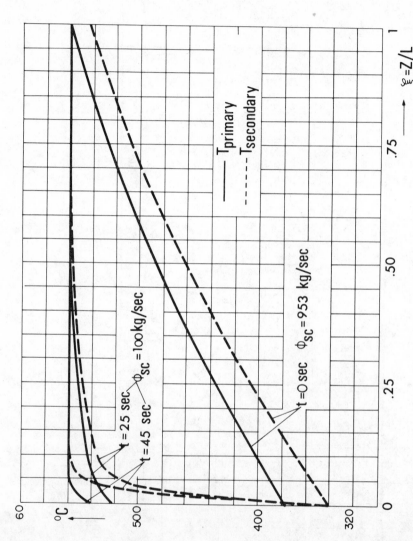

Fig. 12. Sodium-Sodium heat exchanger.

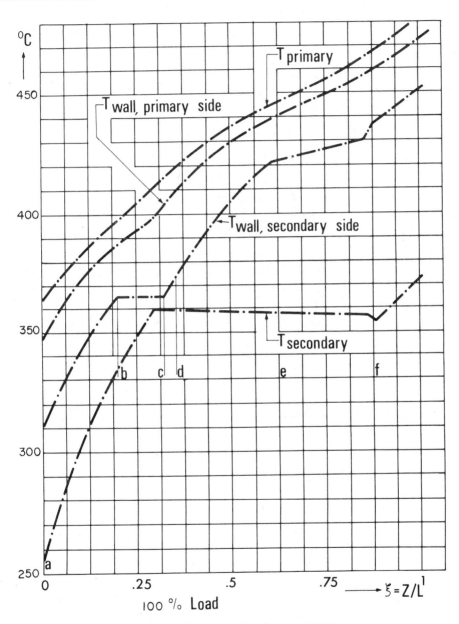

Fig. 13. Sodium-heated once-through steamgenerator.

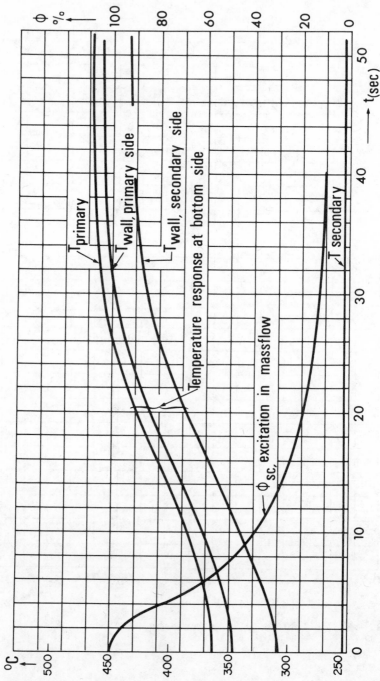

Fig. 14. Sodium-heated once-through steamgenerator.

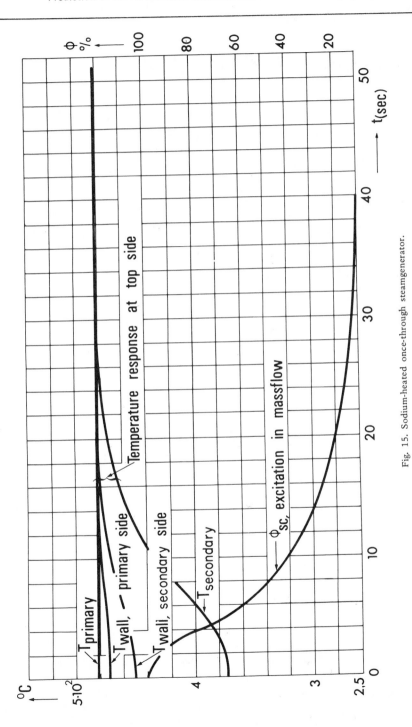

Fig. 15. Sodium-heated once-through steamgenerator.

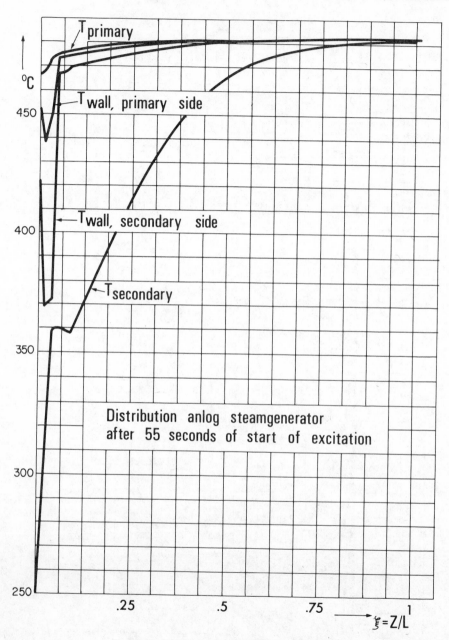

Fig. 16. Sodium-heated once-through steamgenerator.

Chapter 20

HEAT TRANSFER AND FLOW RESISTANCE CORRELATION FOR HELICALLY FINNED AND STAGGERED TUBE BANKS IN CROSSFLOW

Z. Mirković (*)

INTRODUCTION

A number of reports have been published in recent years describing attempts to improve the heat transfer from gas — cooled surface using a finned tube surface in heat exchanger designs. But even with this information there is still not enough knowledge of the effect of various geometrical and other parameters on the heat transfer performance in the heat exchanger with a finned tube surface.

In the development of any design method for heat exchangers it is necessary to know the heat transfer and pressure drop characteristics of the exchanger. When the geometry of the exchanger becomes sufficiently complicated, it is necessary to develop the experimental correlation for heat transfer and pressure drop performance in order to obtain reliable design procedure. In the exchangers with the finned tubes heat transfer coefficient on a finned surface depends on the arrangement of tubes, tube pitches, tube diameter and height and pitch of fins.

When a general relation for the Nusselt and Euler number in the function of the Reynolds number and geometry of the exchangers is derived, it is evident from the previous works (6) and (14) that it is very difficult to define the adequate equivalent diameter included in the Re number which will reduce heat transfer and pressure drop data in the narrow range around such dependence. In order to obtain single correlation for the heat transfer and pressure drop correlation with a reasonable scattering of experimental data, it is necessary to introduce correction terms, including the geometrical parameters in dimensionless form

(*) Research and Development Center for Thermal and Nuclear Technology — ENERGOINVEST Sarajevo —, Yugoslavia

N.B. ALL FIGURES QUOTED IN THE TEXT ARE AT THE END OF THE CHAPTER

effecting heat transfer and pressure drop performance.

The subject of this work was to define the functional relation of the dimensionless heat transfer and pressure drop parameters to the different geometrical parameters describing staggered arrangement of helically finned tubes. It was kept in mind that this correlation will be used in the algorithm for the calculation of heat exchangers operating in different plants when the range of dimensionless geometrical parameters was adequately selected.

ON THE PROBLEM

Generally, we can assume that there is a relation for heat transfer in the form of

$$Nu = Nu \, (\, Re, \, Pr, \, h/d, \, \delta/d, \, n/d, \, S_T/d, \, S_L/d, \, n_R \,) \qquad (1)$$

where Nu is the mean value of the dimensionless number characterizing heat transfer and is based on the total surface of the finned banks formed of n_R tube rows. It is most convenient to use the usual form

$$Nu = A_t \, Re^B_t \, Pr^m \, \phi_t \qquad (2)$$

where A_t, B_t, m and ϕ_t are the values which are to be determined experimentally. It is possible to simplify the experiment if the value of m = 0.33 will be accepted as the exponent of the Pr number.

The value $\phi_t = \phi_t \, (h/d, \, \delta/d, \, n/d, \, {}^S T/d, \, {}^S L/d, \, n_R \,)$, which is a function of geometry, can be represented in the form of

$$\phi_t \overset{N}{\underset{i=1}{\sqcap}} \phi_{ti} \qquad (3)$$

where ϕ_{ti} is the function of a dimensionless geometrical group

$$\phi_{ti} = Z_i^{a_i} \qquad (4)$$

Similar procedure can be applied for the Eu number as well

$$Eu = Eu \, (Re, \, h/d, \, \delta/d, \, n/d, \, {}^S T/d, \, {}^S L/d, \, n_R) \qquad (5)$$

i.e.

$$Eu = A_h Re^{B_t} \phi_h \qquad (6)$$

where

$$\phi_h \overset{N}{\underset{i=1}{=}} \phi_{hi} \qquad (7)$$

and

$$\phi_{hi} = Z_i^{b_i} \qquad (8)$$

For the Nu and Eu number the same dimensionless geometrical parameters are taken as follows

$$Z_1 = \frac{S_T - d}{d} \qquad (9)$$

$$Z_2 = \frac{S_L - d}{d} \qquad (10)$$

$$Z_3 = \frac{1 - n\delta}{nh} \qquad (11)$$

Accordingly, the general correlation can be written as follows

$$Nu = A_t Re^{B_t} Pr^{0.33} \left(\frac{S_T - d}{d}\right)^{a_1} \left(\frac{S_L - d}{d}\right)^{a_2} \left(\frac{1 - n\delta}{d}\right)^{a_3} \qquad (12)$$

$$Eu = A_h Re^{B_h} \left(\frac{S_T - d}{d}\right)^{b_1} \left(\frac{S_L - d}{d}\right)^{b_2} \left(\frac{1 - n\delta}{nh}\right)^{b_3} \qquad (13)$$

Based on the above, the aim of the experiment is to determine the unknown values

$$A_t, \; B_t, \; a_1, \; a_2, \; a_3$$

$$A_h, \; B_h, \; b_1, \; b_2, \; b_3$$

The values a_i and b_i are obtained by investigating how Nu i.e. Eu is changed as a function of Z_i, for a certain constant Re number. After establishing the values of a_i and b_i, the following function is formed

$$K_T = Nu \; Pr^{0.33 - 1} \phi_t = A_t Re^{B_t} \qquad (14)$$

$$K_p = \text{Eu } \phi_h^{-1} = A_h \text{Re}^{B_h} \tag{15}$$

from which it is possible to establish the values A_t, B_t and A_h, B_h.

EXPERIMENTAL EQUIPMENT

Helicoidally finned tubes in cross-flow of staggered tube banks are chosen for tests. The cross section of the test section is 570 x 600 mm. There are 40 to 48 tubes in the test section, depending on tube arrangement. The test section is 8 rows deep. The geometry of the tubes and the tube bank is given in Table No.1. The test was performed on 13 different heat exchangers.

The air rig for testing of thermal and aerodynamical characteristics is shown in Figures 1.1 and 1.2. This is an atmospheric rig accomodating a fan, air heater, test section, elements for directioning of fluid stream and measurement instruments. The device is comprised of three circulating circuits : primary circuit with air, secondary circuit with cooling water and tercial circuit with heating steam.

The measurement of air and water flow is performed with a measurement orifice (DIN 1952), and the temperature is measured with thermo--coples Chromes-Alumel. The position of thermoelements is given in Figures 1.3. and 1.4. Mass flows were within the values of 2.35 kg/sq.ms. On the basis of these mass flows and also on the basis of geometry, Re numbers were achieved in the range from $\text{Re}_h = 1600$ to $\text{Re}_h = 31000$ for aerodynamic characteristics and within the range from $\text{Re}_t = 3000$ to $\text{Re}_t = 56000$ for thermal characteristics. Measurements of pressure drop were carried out at ambient temperature, and measurements for determination of thermal characteristics were carried out at temperature range from $90°C$ to $130°C$.

The physical characteristics of the fluid are based on the mean temperature of fluids. For calculations a GAMMA−30 computer was used, and the FORTRAN language was applied.

RESULTS AND DISCUSSION

The overall heat transfer coefficient, k_u, based on inside tube surface was determined by measuring the inlet and outlet temperature of gas and water in the test section and by measuring the mass flow of the gas.

TABLE 1 – Finned Tubes and Test Section Data

Series	SECTION		TUBE		FINS			NUMB. OF ROWS IN TUBE				TUBE PITCH		
	Width a mm	Height b mm	In. d_1 mm	Out. d_2 mm	Fin d_3 mm	Thick. δ mm	Height h mm	No. of fins/m n	In long. direct n_L	In trans. direct n_{t_1}	In trans. direct n_{t_2}	In trans. direct S_T mm	In long. direct S_L mm	streched tube length mm
	2	3	4	5	6	7	8	9	10	11	12	13	14	15
1	570	600	16.6	25.4	44.45	1.27	9.525	236.22	8	6	5	100	60	570
2										5	4	120	70	
3										6	5	100	70	
4										6	5	100	70	
5										5	5	109	80	
6			42.2	50.8	69.85					6	5	100	80	
7			29.0	38.1	57.25									
8			42.2	50.8	76.20		12.70							
9					82.55	1.524	15.875							
10					76.20	1.27	12.70	118.1						
11								157.5						
12						1.575								
13						2.032								

$$k_u = \frac{G_1 c_{p1} \Delta T}{A_{cu} \Delta t_{ln} \varphi_{\Delta t}} \tag{16}$$

By assumption that the heat coefficient on the water side, α_1, can be calculated from Hausen's relation for the tube and that the thermal conductivity of the tube, λ_c, is known, it was possible to determine the heat transfer coefficient on a finned surface, α_1, for a surface efficiency reduced to unit.

$$a_1 = \frac{A_{cu}}{\epsilon A_{uk}[1/k_u - 1/a_2 - d_1/2\lambda_c \ln (d_2/d_1)]} \tag{17}$$

Now it is easy to get the Nusselt number

$$Nu_1 = \frac{a_1 de_t}{\lambda_1} \tag{18}$$

The Euler number was determined by measuring pressure drop and mass flow at the narrowest cross section. For adiabatic flow, $Q = 0$, the Euler number is given by

$$Eu = \frac{\Delta P_{st}}{(w\rho_1)^2 n_R} \rho_1 \left[1 + \frac{(w\rho_1)^2}{\Delta P_{st} \rho_1} (1 - \rho_1/\rho_2) \right] \left[1 - \frac{\Delta P_{st}}{2P_1} \right] \tag{19}$$

The Re number

$$Re = \frac{(\rho_1 w)de}{\eta_1} \tag{20}$$

is also based on the mass flow at the minimal cross section and at the equivalent diameter which is given for the thermal characteristics by the formula

$$de_t = \frac{2A_{uk}}{\pi l_k} \tag{21}$$

and for aerodynamical characteristics by the formula

$$de_h = \frac{4 V}{A_{uk}} \tag{22}$$

The results $Nu = f (Re_t)$ and $Eu = f (Re_h)$ for individual exchangers are given on diagrams, Figures 2 and 3.

Based on the results of tests on heat exchangers IT-2, IR-4 and IT-5, the values for exponents of dimensionless groups $(S_T/d-1)$ are obtained (Figures Nos. 4 and 7).

$$a_1 = 0.1$$

$$b_1 = 0.14$$

By testing the heat exchangers IT-1, IT-3 and IT-4, the values for group exponents $(S_L/d-1)$ are obtained (Figures Nos. 5 and 8).

$$a_2 = -0.15$$

$$b_2 = -0.18$$

The effect of the interfin space is tested on heat exchangers IT-6, IT-8, IT-9, IT-10, IT-11 and IT-12. Based on Figures 6 and 9, the values of the exponent group $(1 - \overset{*}{n}\delta)/nh$ are obtained

$$a_3 = 0.25$$

$$b_3 = -0.20$$

Finally, based on the experimental data for $K_T = f(Re_t)$ and $K_p = f(Re_h)$, the following was obtained

$$A_t = 0.224$$
$$B_t = 0.622$$
$$A_h = 3.96$$
$$B_h = -0.31$$

Accordingly, the general correlations are :

$$Nu = 0.224 \left(\frac{S_{T-d}}{d}\right)^{0.1} \left(\frac{S_{L-d}}{d}\right)^{-0.15} \left(\frac{1-n\delta}{nh}\right)^{-0.25} Re_t^{0.662} Pr^{0.33} \tag{23}$$

$$Eu = 3.96 \left(\frac{S_{T-d}}{d}\right)^{0.14} \left(\frac{S_{L-d}}{d}\right)^{-0.18} \left(\frac{1-n\delta}{nh}\right)^{-0.20} Re_h^{-0.31} \tag{24}$$

The above relations are obtained on the tube bank with eight rows. The Euler number refers to one row of tubes.

On IT-5 and IT-6 heat exchangers the effect of the number of rows on the Nu number is examined. The results of these tests have shown that the coefficient of heat transfer has a smaller value on the first rows. As tests were carried out with eight rows, it is necessary to use a correction factor for some other number of rows in the bank, in accordance with the diagram, Figure 12.

CONCLUSION

This experiment has shown that for the tested range of Re numbers it is possible to find the functional dependence between the dimensionless values for heat transfer and pressure drop on one side and the geometrical values of the tested finned tube banks.

— Increase of transversal tube pitch, S_T, for constant tube diameter, d, acts in such a manner that both the Nu and Eu numbers increase.

— Increase of longitudinal tube pitch, S_L, for constant tube diameter, d, acts in such a manner that both the Nu and Eu numbers decrease.

— Increase of tube diameter, d, for constant transversal and longitudinal tube pitches acts in such a manner that both the Nu and Eu numbers increase.

— Increase of width of the interfinned space, $1-n\delta$, and decrease of height of fin, h, act in such a manner that the Nu number increases, and Eu number decreases.

— All the rows do not transfer the heat equally. The first row transfers the heat in the poorest manner. From the third row and on heat transfer is more or less constant.

— General correlations for the dimensionless value of heat transfer and pressure drop over quite well the results of some other authors as well, Figures 13 and 14.

NOMENCLATURE

d_1	inside tube diameter
d, d_2	diameter of bare tube
d_3	diameter of fins
h	height of fins
δ	thickness of fins
n	number of fins per unit lenght of tube
n_R	number of rows in tube bank
S_T	transversal pitch
S_L	longitudinal pitch
de_t	equivalent diameter
de_h	volumetric equivalent diameter
z	dimensionless geometry
A_{cu}	inside tube heat transfer surface
A_{uk}	total heat transfer surface
I_k	perimeter of finned tube
v	free volume
G_1	mass flow
$w\rho$	specific mass flow
η_1	dynamic viscosity of air
λ_1	thermal conductivity of air
c_{p1}	heat capacity of gas at p = const.
Re, Re_t, Re_h	Reynolds numbers
Pr	Prandtl number
ΔT	temperature drop of gas
Δt_{ln}	mean log. temp. difference-MLTD
$\varphi_{\Delta t}$	correction factor for MLTD
ϵ	surface efficiency
k_u	overall heat transfer coefficient
a_1 a_2	Heat transfer coefficient on air side and water side, respectively
Nu	Nusselt number
Eu	Euler number
ϕ_t, ϕ_h	functions of geometry
ΔP_{st}	static pressure drop
K_T	function of Nu and ϕ_t
K_p	function of Eu and ϕ_h

REFERENCES

[1]   Brauer, H., Wärme-und Strömungtechnishe an quer angeströmten Rippenrohründeln, Teil 1 und 2, Chemie – Ing. Technik 33 Jahrg., 1961, No. 5 und No. 6.

[2]   Brauer, H., Wärmeübergang und Strömungswiderstand bei fluchtend und verstez angeordneten Rippenrohren, Dechema Monographien, Band 40, 1962.

[3]   Chatillon, M., Étude d'échangeurs de chaleur de centrales nucléaires, EDF - Bulletin du Centre de Recherches et d'essais de Chatou, No. 13, 1965.

[4]   Djordjević, R., Eksperimentalno odredjivanje lokalnih vrednosti koeficijenata prelaza toplote na površinama poprečno orebrenih cevi, Magistarski rad, Maš. Fak. Univerziteta u Beogradu, 1970.

[5]   Gardner, K.A., Efficiency of Extended Surface, Trans. of the ASME vol. 67, 1945.

[6]   Gunter, A.Y., and Shaw, W.A., A General Correlation of Friction Factors for Various Types of Surfaces in Crossflow, Trans. of the ASME, November, 1945.

[7]   Jameson, S.L., Tube Spacing in Finned – Tube Banks, Trans. of the ASME November, 1945.

[8]   Kays, W.M. and London, A.L., Compact Heat Exchangers, McGraw-Hill Book Co., Inc., London, 1964.

[9]   Krischer, O. und Kast, W., Wärmeübertragung und Wärmespanungen bei Rippenrohren, VDI - Forschungsheft 474, 1959.

[10]   Lymer, A. and Ridel, B.F., Finned Tubes in a Cross-flow of Gas, The Journal of BNEC, vol. 6, No. 4, 1961.

[11]   Mannesmann Druckschrift 5303, Rohre für Wärmeaustauscher, Berechnungsunterlagen.

[12]   Neal, S.B.H.C. and Hitchcock, J.A., A study of the heat transfer processes in banks of finned tubes in cross-flow, using a large scale model technique, Proc. 3rd Interm. Heat Transfer Conf., Chicago, 1, 11 vol. 3, 1966.

[13]   Rounthwaite, C. und Cherrett, N., Heat transfer and pressure drop performance of helically finned tubes in staggered cross-flow. The Journal of BNEC, vol. 6, No. 4, 1961.

[14]   Vampola, J., Vergleich von Rippenrohren aus Unterschiedlichen Werkstoffen für Luftkühler, Chem. Techn., 17. Jg. Heft, Januar 1965.

[15] Zhukauskas, A., Stasiulevichius, J. and Skrinska, Experimental investigation of efficiency of heat transfer of a tube with spiral fins in cross-flow, Third Intern. Heat Transfer Conf. 1966, Chicago, vol. III.

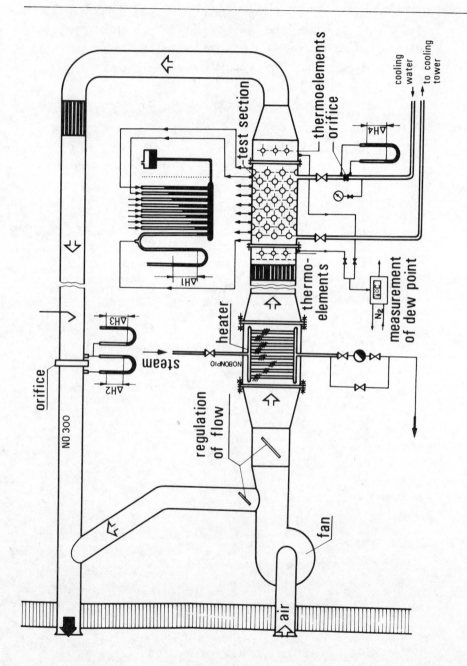

Fig. 1.1. Flow diagram of experimental rig.

Fig. 1.2. View of experimental rig.

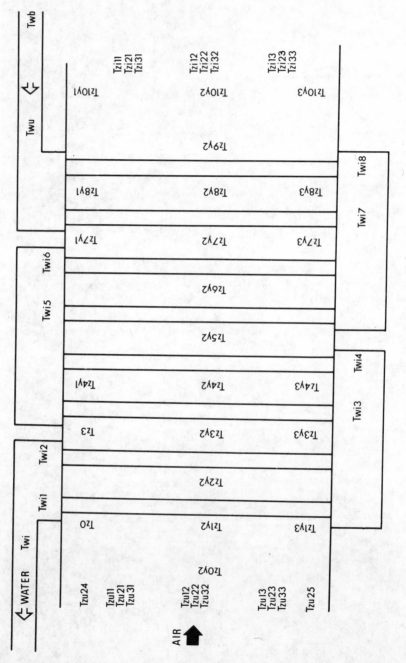

Fig. 1.3. Horizontal position of thermoelements.

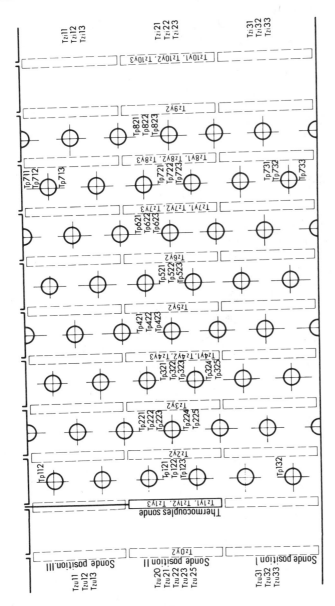

Fig. 1.4. Vertical position of thermoelements.

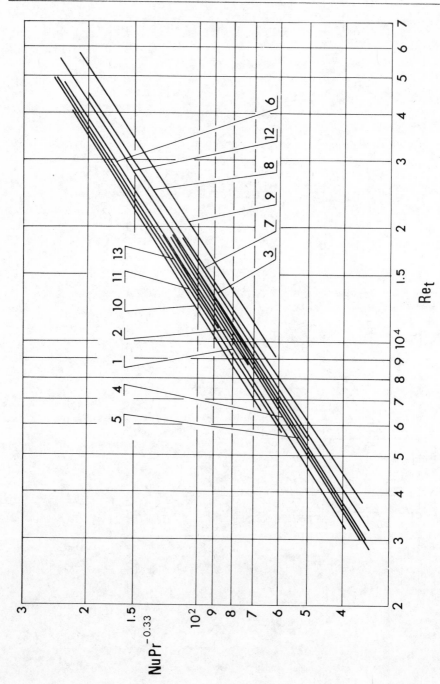

Fig. 2. NuPr$^{-0.33}$ depending on Re-number for tested heat exchangers.

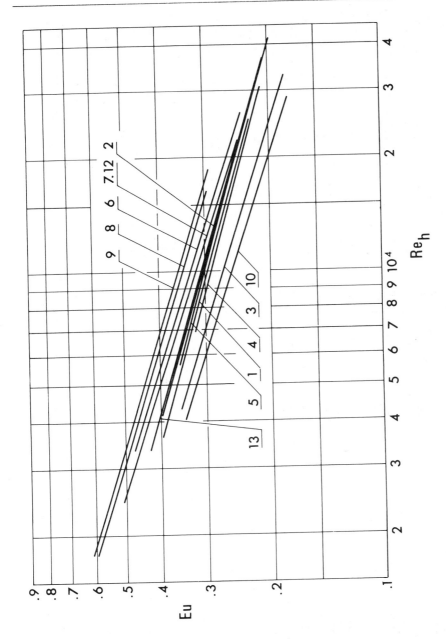

Fig. 3. Eu-number depending on Re-number for tested heat exchangers.

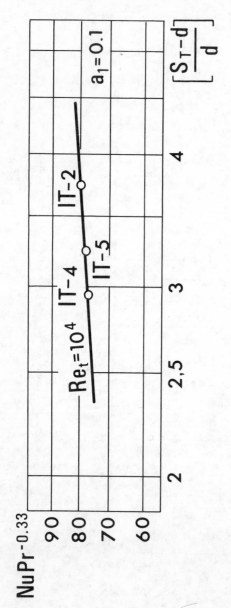

Fig. 4. Influence of group $(S_T/d - 1)$ on $NuPr^{-0.33}$.

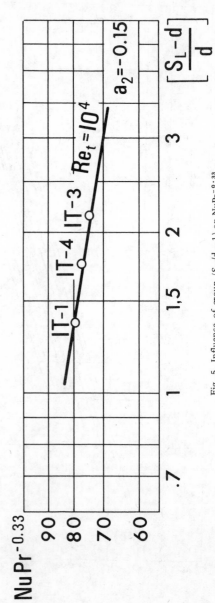

Fig. 5. Influence of group $(S_T/d - 1)$ on $NuPr^{-0.33}$.

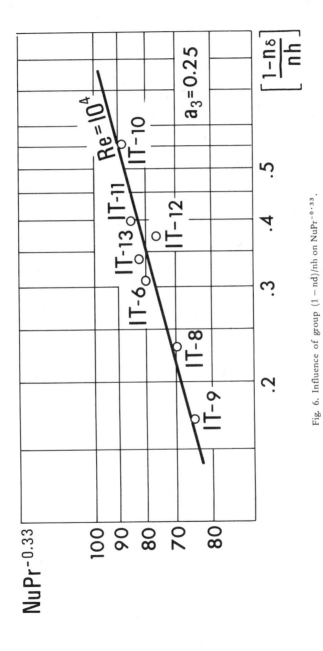

Fig. 6. Influence of group $(1 - nd)/nh$ on $NuPr^{-0.33}$.

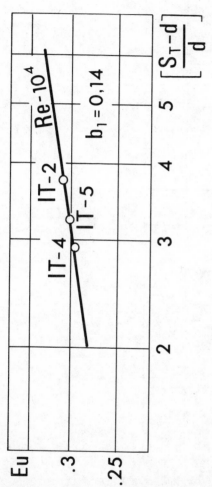

Fig. 7. Influence of group $(S_T/d - 1)$ on Eu-number.

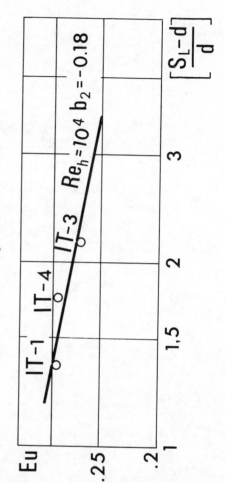

Fig. 8. Influence of group $(S_L/d - 1)$ on Eu-number.

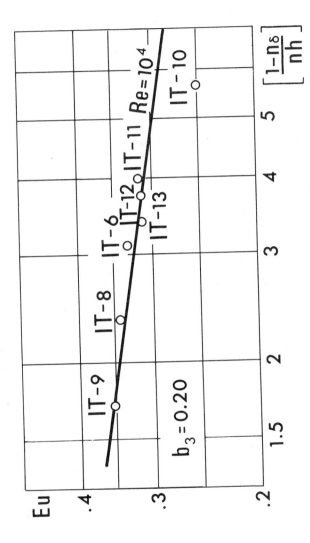

Fig. 9. Influence of group $(1 - nd)/nh$ on Eu-number.

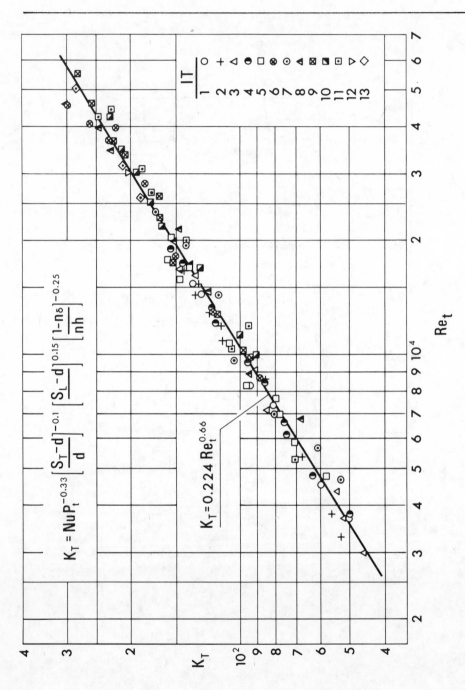

Fig. 10. K_T - function depending on Re-number.

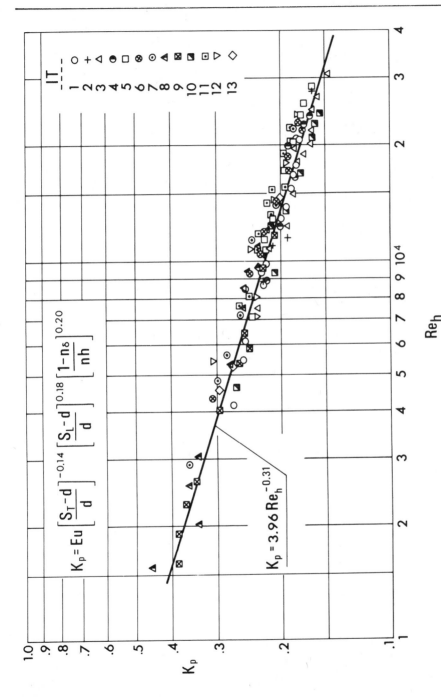

Fig. 11. K_p - function depending on Re-number.

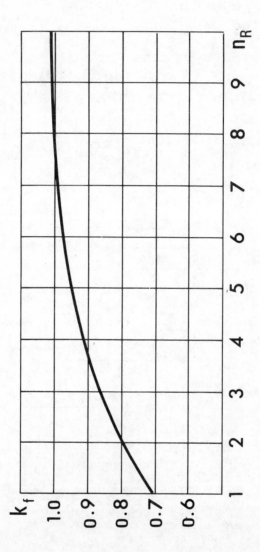

Fig. 12. Correction factor k_f for tube banks with less than 8 raws in deep.

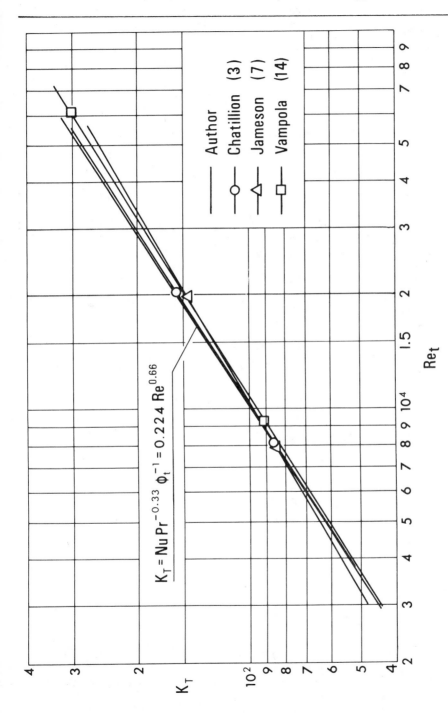

$$\frac{K_T = Nu \, Pr^{-0.33} \, \phi_t^{-1}}{} = 0.224 \, Re^{0.66}$$

Fig. 13. Comparison of K_T - function with results of other authors.

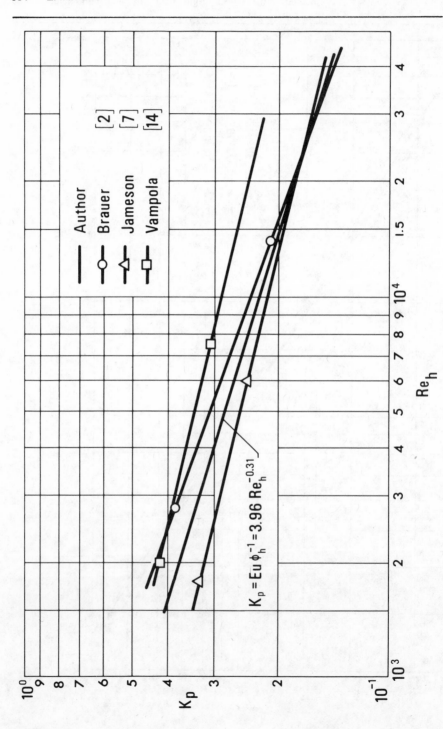

Fig. 14. Comparison of K_p - function with results of other authors.

Chapter 21

APPLICATION OF HEAT TRANSFER FINNED SURFACES IN HIGH TEMPERATURE HEAT EXCHANGERS

V.I. Tolubinsky, N.V. Zozulya (*)

The problem of creating effective heat exchangers and apparatus characterized by intensive heat transfer through a separating wall is among important scientific problems. This problem is specified in particular by the fact that in realizing energetic and technological systems based on thermal processes, at least 30–40% of metal is consumed for heat exchange equipment manufacture.

To solve this problem, complex development of the following topics is necessary:

— investigation of methods of heat transfer intensification from the heating surface, development and research of heat exchange surfaces, and constructions of new heat exchangers,

— improvement and creation of new technology for manufacturing both heating surfaces and apparatus as a whole. Work in this field has been carried out at the Institute of Engineering Thermophysics of the Academy of Sciences of the Ukrainian SSR.

The most attention has been paid to equipment operating under high pressures and temperatures, since the production of this equimpent is the traditional style (from smooth tubes) leads to high metal consumption and structural awkwardness.

The intensity of heat transfer through the separating wall in heat exchangers is unequal on different wall sides as a rule. Under these conditions metal consumption and construction dimensions are known to be determined mainly by the side where heat transfer intensity is lower. Smooth tube and plate heat exchangers have a substantial disadvantage due to the fact that the whole heat transfer surface should be able to stand the pressure gradient between heat carriers.

(*) Institute of Engineering Thermophysics, Academy of Sciences, Ukrainian SSR.

N.B. ALL FIGURES QUOTED IN THE TEXT ARE AT THE END OF THE CHAPTER

That is why extremely thick walls are needed, although they are unnecessary from the point of view of heat transfer. Improvement of the characteristics of smooth-tube heat transfer surfaces is attained by a flow velocity increase, through use of compact bundles and tubes of small diameter. However, these measures are expedient within certain limits because of an increase of hydraulic resistance, the danger of surface erosion, and a decrease of the reliability of the construction.

A process of insufficiently intensive heat transfer to the wall can be compensated by a finning of its surface extension. This method of increasing the effectiveness of heat transfer and of improving the weight and dimensions of heat exchangers is acceptable especially for heat exchangers operating under the high pressures of heat carriers inside the tubes since the main heat transfer surface, i.e. fins, (which is 5-10 times greater than that of the smooth tube) can be of lesser thickness than the tube wall and, as well, can be made from a cheaper metal. This latter point is advantageous when the separating wall is made from expensive corrosion-proof materials.

For a long time the application of finned surfaces in high-temperature heat exchangers has been obstructed by the lack of a technology for joining a thin fin with the tube — a technology which would provide reliable operation of the finned tube under high operating temperatures (several hundred degrees) and which would be characterized by high efficiency and profitability. This has encouraged the search for a solution through the joint work of the Institute of Engineering Thermophysics and other technological institutions. As a result, from the compared methods of manufacturing finned tubes – soldering with hard solder, casting, drawing from melt, pressing, knurling (fig.1) and electric welding — the last proved to be the most prospective as applied to ferrous metals and stainless steels. The complex work of the Institute of Engineering Thermophysics and E.O. Paton Electric Welding Institute, both of the Academy of Sciences, Ukr.SSR, has been completed by the development of production technology [1], the study of thermoaerodynamic characteristics, and the creation of a calculation method for heat transfer surfaces from tubes with various kinds of welded finning (Fig.2).

The results of thermotechnical and technical-economical investigation, as well as new technological solutions in the field of manufacturing finned heating surfaces, have substantially broadened the limits of their application for high-temperature energetic and technological heat exchange devices.

It is known that the use of finning reduces heat transfer intensity (calculating on the basis of the heat-absorbing surface unit) as compared with non-finned surfaces, since the fin efficiency is defined not only by connective heat transfer ($d_{equiv.}$) but also by its material thermal resistance.

The efficiency increase of finned surfaces and primarily of tubes with various kinds of welded finning has been given special attention at the Laboratory of Heat Exchange Apparatus of the Institute of Engineering Thermophysics. From this point of view, improvement of geometrical parameters and choice of materials for the manufacture of finned tubes are of great importance. However, the main way of improving such surfaces of thermoaerodynamic properties is by searching for methods of heat transfer intensification to the fin surface. This is confirmed by the well-known studies by Case and London [2] in the field of small-scale heat exchangers. Antufjev M. V. [3] and by other workers, as well as by many years of research at the Institute of Engineering Thermophysics aimed at perfecting a variety of heat exchangers for industrial purposes.

It should be noted that in finned tubes, with longitudinal finning in particular, the potential possibilities of heat transfer intensification are substantially greater than in smooth tubes. Mainly, this is due to the fact that in tubes with longitudinal finning the possibility of counterflow motion of heat carries is realized and the flow (gas, as a rule) is separated and moves along narrow channels. This results in $d_{equiv.}$ decrease.

Low hydraulic resistances cause a velocity increase in the interfin channel which, taking into consideration the counterflow of heat carriers, improves the heat transfer process.

The investigation of specimen and bundles of longitudinally finned tubes of various dimensions [4,5], carried out in laboratories and using half-industrial and industrial conditions, has shown the validity of application of tubes of usual similarity equations (at high quality welding for finned parts) :

$$\mathrm{Nu} = 0,02\ \mathrm{Re}^{0.8}\ ;\ \ \xi = \frac{0.316}{\mathrm{Re}^{0.25}}\ .$$

In this case $\quad d_{equiv.} = \dfrac{4\ F}{U}$

is the parameter with an accuracy sufficient for engineering design (t3%) and suitable for characterizing heat transfer and flow aerodynamics.

It is known that the intensity of the heat transfer process and the surface aerodynamic resistance depend mainly on the nature of the flow near the wall. Evidently, optimal solutions for the development of effective heat transfer surfaces should be searched for only in the above usage rather than in the flow turbulization at the expense of its velocity increase.

Interconnection among the surface configuration, the flow structure, and heat transfer has been studied at the Institute during experimental investigation of this problem on plates, in the channel, inside the tubes and at flows around the interfin channels of tubes with deformed fins. The solution of the question from the engineering point of view has been facilitated by the fact that the thermo-aerodynamically expedient deformation of fins can be technologically realized on fins welded to the bearing tube.

A rather simple method of flow turbulization near the wall, aimed at heat transfer intensification, can be put into practice by using roughness elements on the wall [6,7].

To find the effect of the height h, the configuration of baffles and their location (spacing S) on the flow structure and heat transfer, the investigation of air flows around the plate with one, two and many roughness elements has been carried out. Preliminary tests on the smooth plate have revealed good correspondence of the local heat transfer coefficients a_{loc}^{sm} (sm-smooth, loc-local) with the known data [8,9] and thereby have approved the experimental technique.

Under the condition of isothermal flowing around the plate with a single baffle and its heating, investigation of velocity fields and turbulent pulsations, as well as determination of $\alpha_{loc}^{r} = q(t_{f\infty} - t_{wloc})$ (r-rough), has confirmed [10] the presence of the flow breakaway and allowed specification of the flow motion pattern in the vicinity of the baffle. The following zones have been determined: stagnation zones within 1-2 h before and behind the baffle; the flow breakaway zone of the length $1 = 10$ h from the baffle edge with α_{loc}^{r} increase in this zone up to the maximum value at the point of joining (due to intense agitation in the vorted and exchange processes between the vortex and the outer flow); the flow joining zone in the section $1 > 10$ h where $\alpha_{loc}^{r}/\alpha_{loc}^{sm} > 1$ is attributed to a new boundary layer formation.

Investigation of α_{loc}^{r} in the presence of two baffles (when h changes from 10 to 2 mm and S/h changes from 4 to 12) has shown that the breakaway zone length behind the first baffle is affected by the second baffle, which reduces the former 1.5 – 2 times.

To determine optimal values of h and S/h, investigation has been carried out on the plate with many periodical roughness elements at various flow velocities. As a result, it has been stated [11] that the ratio (S/h) opt $\simeq 14$ and agrees with [12]. Optimal baffle height value h depends on U_∞. This dependence character is shown in Fig. 3 in coordinates $\alpha_{loc}^r / \alpha_{loc}^{sm} = f(h/\delta_{lam}^{sm})$ at various U_∞ where the laminar sublayer thickness for a smooth plate δ_{lam}^{sm} has been defined by the mean tangential stress value for the section corresponding to the location of the roughness elements. It has been stated that within the limits $1 \cdot 10^5 < Re_\ell < 5 \cdot 10^5$

$$\left(\frac{h}{\delta_\ell^{sm}}\right)_{opt} = 0.048 \ Re_\ell^{0.4} \ ; \ Re_r = \frac{U_\infty \ell}{\nu}$$

Periodical roughness elements in the form of baffles of rectangular and circular section have proved the independence of α_{loc}^r from the geometrical configuration of the baffles.

The data obtained from the plate with periodical roughness have been used to investigate tubes on the longitudinal finning for which elements of periodical roughness have been used.

For the intensification of heat transfer, the application of elements with small geometrical dimensions for finning is of interest [13]. In tubes with longitudinal finning such elements can be obtained by the perforation of the fins. In this case, in each working fin element exist conditions for the formation of a kind of new boundary layer on the fin. This layer can be influenced by similar formation on preceding elements.

Figure 4 presents the influence (at U = const) of the length of the perforated fin's working element on the convective heat transfer coefficient. For comparison, there is the value of $\bar{\alpha}_c$ for the tube with non-perforated finning presented in this figure.

The degree of dependence of $\bar{\alpha}_c$ on perforation parameters — the length of the working element (α 5-20mm) and the punching length (β 1-2mm) are determinded by the empirical formula [14]:

$$\overline{Nu} = 0.045 \ Re^{0.8} \left(\frac{a}{\beta}\right)^{-0.2}$$

It is of practical interest that at aerodynamic resistance growth not more than that of heat transfer, such fin deformation readily realized can increase $\bar{\alpha}_c$ 1.5 − 2 times more than with smooth tubes.

The further the dimensions of working elements decrease, the more the transition to wire finning of tubes longitudinally flown around in particular brings additional heat transfer intensity increase.

In the studies on tubular surfaces with deformed longitudinal finning, which were carried out at the Institute, one can mark surfaces with the fin dissection into petals for which the optimal angle of petal turn with respect to the flow axis has been determined.

High thermodynamic properties are characteristic for longitudinal finning with the dissection of fins into elements which are parallel to the flow axis and which have entrance and exit rims slightly bent in different directions. As visual and analytical studies have shown, due to the occurrence of a negative pressure gradient behind the rim bending and of a positive gradient from another side of the fin before the rim bending, viscous sublayer suction and blowing into the adjacent interfin channel are taking place.

A comparison between longitudinally finned tubes deformed in this way and tubes with continuous finning is given in fig. 5 at different working elements lengths t and at the rim bending height $h \simeq 0.1 \ d_{equiv}$.

It should be noted that in this case as well, heat transfer intensification is accompanied by almost linear growth of hydraulic resistance, which is much more benificial than achieving the same results at the expense of an increase of the heat carrier velocity.

The quality criterion of heat transfer surfaces is the comparison of their weight, volume and energy indices as the function of power consumed for heat carrier transport. With a certain conditionality (assuming the temperature head between heat carrier and the wall to be 1°C) some of the surfaces investigated (fig.6) have been compared (fig.7) in the coordinates of the type $\bar{\alpha}_r F_{tot}/N = f(N)$ where $\bar{\alpha}_r$ is the reduced heat transfer coefficient allowing for the thermal resistance of the fin, F_{tot} is the total heat transfer surface, and N is power consumption.

From this comparison it is seen that at N = idem energetic (as well as weight and volume) characteristics of the best kind of longitudinal profile finning with rims bending aside are 1.5 times higher than for tubes with non-deformed fins and considerably better than for smooth tubes.

Tubes with transverse finning are of great practical interest. Investigations of their thermoaerodynamic characteristics, especially with welded finning, at the Laboratory of Heat Exchange Apparatus of the Institute have been carried out during a number of years.

These studies determined the effect of leading flow turbulization on heat transfer of a bundle of tubes transversely finned [15] and showed that account should be taken of bundles with a small number of tube rows into the depth. In addition to characteristics already known [16,17] a detailed study of characteristics for tubes with strip finning has been carried out. These characteristics are evaluated as being affected by production technology of finned tubes (corrugation at cold strip coiling, flash at its welding). Heat transfer changes along tube rows have been determined together with the effect of composition and spacing of tubes located in bundles on their heat transfer and hydraulic resistance. At staggered tubes composition [18] with helical strip finning heat transfer is determined by :

$$Nu = 0,14 \left(\frac{S_1 - d_r}{S'_2 - d_r}\right)^{0,4} Re^{0,65}$$

where diagonal pitch

$$S'_2 = \sqrt{\left(\frac{S_1}{2}\right)^2 + S_2^2}$$

and Re is defined by the outer diameter of the tube.

As a result of investigations of tubes [19] with transverse finning dissected into small elements (welded petals and wire), a positive effect has been obtained, although relatively smaller than that in tubes with longitudinal, deformed finning.

On the basis of a technical-economical and operational comparison, it has been discovered that for operation under higher fluid temperatures, the tubes with welded finning are the most prospective. Constant cooperation with branch design and industrial institutions has made it possible to trace the ways of investigating the application of results and in many cases to realize them in concrete industrial apparatuses.

Surfaces with various kinds of longitudinal and transverse fins welded to tubes are widely adopted in regenerators, tap water preheaters and heat recovery units of gas-turbine power plants. They are promising also for economizers and air preheaters of steam boilers and steam-gas plants, and they are used in heaters of air preheating, an in superheated vapour temperature controllers. In a number of heat transfer devices in atomic power stations, and above all in intermediate steam superheaters, the change from smooth tubes to finned tubes offers essential gains. Such surfaces are prospective in various heat transfer apparatuses of different chemical engineering plants.

The experience in the application of tubes with welded finning confirms the data cited above. For example, substitution of smooth tubes, in a gas-turbine power plant with the capacity of 50,000kw, for tubes with welded finning allowed a decrease of consumption of the tubes by 400,000 running meters and the weight of the regenerators by 200 ton.

The production technology of tubes with longitudinal, helical strip and wire finning is mastered in a number of plants.

The wide application of such tubes has been promoted by : a) study of their work intensification methods and characteristics necessary for calculation, b) development of technology and automatic equipment for the production, of finned tubes, c) joint experimental production (of the Institute of Engineering Thermophysics and E.O. Paton Institute of electric welding) of 250–300 thousand running meters of tubes with various kinds of welded finning for experimental and pilot samples of new heat exchangers.

REFERENCES

[1] Prihod'ko, P.M.: Automaticheskaia Svarka, 1963, No. 9.

[2] Keys, V.M., London, A.L.: "Compact Heat Exchangers," Gosenergoizdat, Moskva, 1962.

[3] Antuf'ev, V.M.: "Effektivnost' razlichnyh form konvektivnyh poverhnostei nagreva," Energiia, Moskva, 1966.

[4] Kremnev, O.A., Zozulia, N.V., Havin, A.A.: Energomashinostroenie, 1961, No.1.

[5] Zozulia, N.V., Shvarc, V.A., Kalinin, V.L.: Izvestiia VUZ – Energetika, 1963, No. 8.

[6] Koshkin, V.K., Kalinin, E.K., "Teploobemenye apparaty i teplonositeli," Mashinostroenie, Moskva, 1971.

[7] Puchkov, P.I., Vinogradov, O.S.: Energomashinostroenie, 1965, No.6.

[8] Petuhov, B.S., Diatlov, A.A., Kirillov, V.V.: Zhurnal Teh. Fiz.V.24, 1954, No. 10, 1971-1972.

[9] Zhiugzhda, I.I., Zhukauskas, A.A.: Trudy AN LSSR, 1962, B, 4(31).

[10] Seban : Trans. ASME, C86, 1964, No. 2, 154-160.

[11] Zozulia, N.V., Kalinin,V.L.:"Teplomassoperenos," Vol.1, Pt.3, 1972, 244-249.

[12] Gomelauri, V.I.:Trudy Instituta Fiziki AN GSSR, Tbilisi, Vol.9, 1963.

[13] Kremnov, O.A., Duhnenko, N.T.: Energomashinostroenie, 1965, No. 6.

[14] Zozulia, N.V.: Teplofizika i Teplotehnika, 1970, No. 17, Kiev.

[15] Zozulia, N.V., Vorob'ev, Iu.P., Havin A.A.: Teplofizika i Teplotehnika, 1971, No.1.

[16] Tolubinskii, V.I., Legkii, V. M.: Teplofizika i Teplotehnika, Kiev, 1964.

[17] Iudin, V.F., Tohtarova, L.S.: Energomashinostroenie, 1964, No.1.

[18] Zozulia, N.V., Kalinin, B.L., Havin, A.A.: Teploenergetika, 1970, No.6.

[19] Zozulia, N.V., Havin, A.A.: Energomashinostroenie, 1968, No. 11.

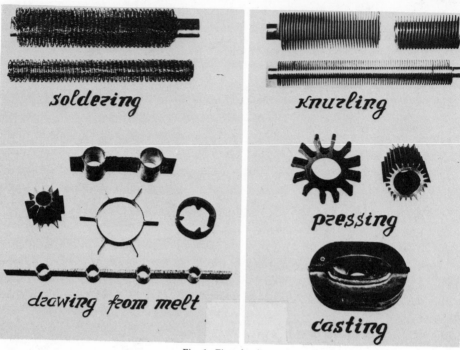

Fig. 1. Finned tubes.

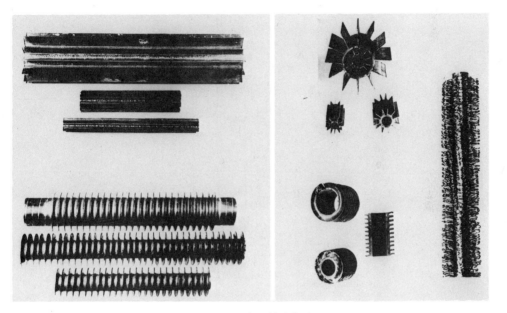

Fig. 2. Tubes with welded finning.

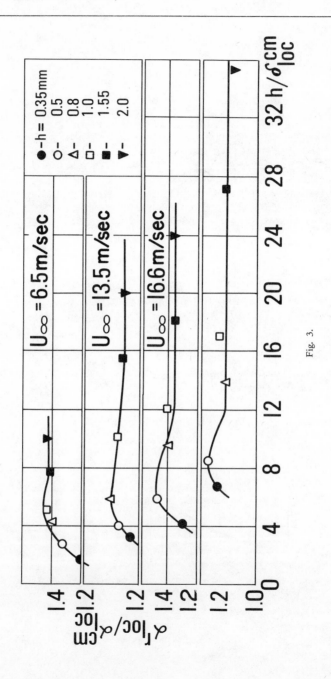

Fig. 3.

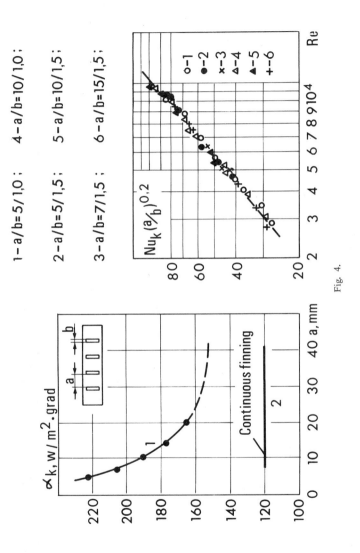

Fig. 4.

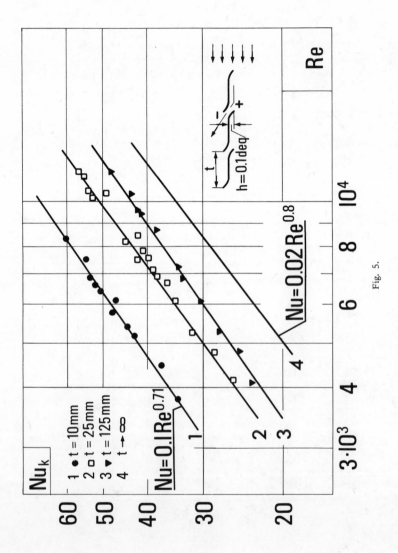

Fig. 5.

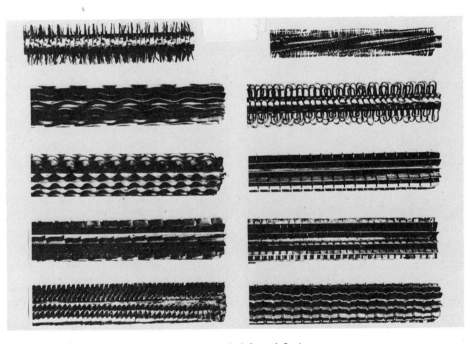

Fig. 6. Tubes with deformed finning

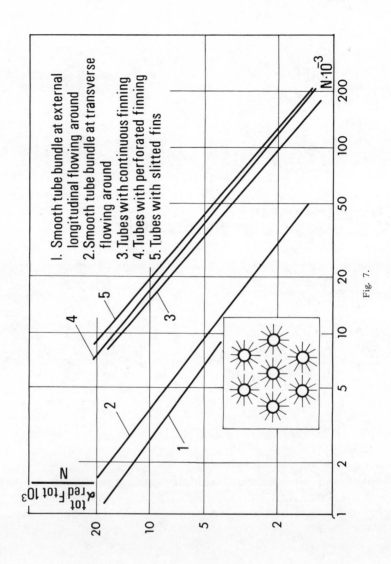

Fig. 7.

Chapter 22

EFFECTS OF TURBULENCE PROMOTERS ON THE PERFORMANCE OF PLATE HEAT EXCHANGERS

D. Pescod(*)

1. INTRODUCTION

Heat exchangers may be used in air conditioning applications for heat recovery, for dehumidification systems in conjunction with refrigeration, and for air cooling by means of the evaporation of water [1].

Cooling systems may use regenerators [2], [3], but plate heat exchangers (recuperators) possess certain advantages. They have no moving parts, the passages are easy to seal and there can be complete separation of the two fluid streams. One limitation with plate heat exchangers when used for gases is the quantity of gas which may be passed through; for this reason crossflow arrangements are frequently used even though there is some loss of efficiency when compared with counterflow arrangements.

When the fluid pressures are low it is possible to construct the heat exchanger of rigid plastic film such as polystyrene or polyvinyl chloride. Provided thin material is used, say 0.25 mm or less, the poor thermal conductivity compared with metals is not important, as the temperature gradient across the plates is low. These plastic materials are suitable for vacuum forming into various shapes, and designs were made with protrusions in the form of spikes or ripples to increase the turbulence in the air passing between the plates and thus to increase the heat transfer coefficient when compared with plane parallel passages, and economize in material used. The protruding spikes also provided physical support between

(*) Division of Mechanical Engineering, Commonwealth Scientific and Industrial Research Organization, Highett Victoria, Australia.

N.B. ALL FIGURES QUOTED IN THE TEXT ARE AT THE END OF THE CHAPTER

adjacent plates

2. DESCRIPTION

2.1. Heat Exchangers.

The heat exchangers used for the experimental investigation were made from polystyrene plates approximately 200 mm square and 0.25 thick. With later heat exchangers, rigid polyvinyl chloride was used, and the thickness was reduced to 0.20 mm. This size was chosen to provide relatively large differences in temperatures and thus the most accurate measurements of effectiveness possible with the equipment available. For practical heat exchangers of high efficiency, large plates would be necessary.

The main parameters were passage width and spike spacing, and six heat exchangers with protrusions in the form of spikes were constructed with dimensions as given in Table 1. A seventh design of heat exchanger was made by drape moulding the plastic sheets over expanded metal to create a pattern of ripples or "pockets." Adjacent plates nested together with a few protrusions to provide correct spacing.

In all cases the plates were heat sealed or electronically welded at their edges to from alternate passages at right angles to each other as in the normal cross flow arrangement. The height of the heat exchangers was approximately 150 mm.

TABLE 1

Plate type	Passage Width mm	Spike Spacing mm*	Symbol for Graphs
Spike	3.40	6.7	x
Spike	2.82	6.7	o
Spike	2.16	6.7	●
Spike	3.56	9.5	□
Spike	3.40	13.4	△
Spike	3.40	13.4**	▽
Pocket	3.56	34 +	*

*	All spikes were on a diamond pattern
**	Plates coated with rayon fibres to retain water for evaporative cooling tests
+	Pocket spacing : 9.5 mm X 28.5 mm.

2.2. Test Set-up

The arrangement of the test rig is shown in Figure 1. The experimental heat exchanger is sealed into a thermally insulated box. Air is drawn through the heat exchanger and then through a mixer to obtain a more uniform temperature. The air then flows through an electric heater and another mixer before passing through the alternate passages of the heat exchanger. The air is mixed again and blown to atmosphere through an exhaust fan. The air flow rate and electric heating could be varied. With low air flows, the pressure differences across the heat exchanger were small, and thus the measurements were less accurate. Pressure measurements were made with an inclined manometer and a "Betz" micromanometer.

A grid of nine temperature measurements was made after each mixer, using calibrated mercury in glass thermometers.

The electrical power to the heater was measured with a wattmeter, and the air flow rate could be calculated from the heat input and the rise in air temperature.

3. TESTS

The air flow rate was varied from 4 to 54 1/s, corresponding to approach velocities of 0.3 to 4 m/s. The electrical heat input was varied from 30 to 660 watts.

A total of 89 test runs were made; each one required several hours for conditions to stabilize. No distortion of the plates was detected with temperatures up to 50°C, but on one occasion when the temperature of the air was accidentally raised above 50°C, some distortion of the moulded plates was observed.

4. THEORETICAL ANALYSIS

From the measurements of temperatures, air pressures, and electrical heat input, and the physical sizes and known properties of the heat exchangers, the air flows, friction factors, heat transfer coefficients, Reynolds Numbers and Stanton Numbers could be determined.

The Reynolds Numbers and friction factors were based upon the width of the passages. In the case of the rippled plates, no corrections were made for the extended surface area compared with flat plates, because the area of the extended surface would have been difficult to measure. Values of heat transfer coefficient in all cases refer to projected area.

From the calculated values, graphs were prepared, and these are presented in Figures 2 to 7. Figures 3 to 7 are plotted on logarithmic paper. In Figures 3 to 5, curves for pane parallel walls have been included for comparison. These were obtained from Kays and London [4].

In Figure 2, some theoretical curves of effectiveness vs. approach velocity are included for comparison.

The curves have the general form

$$\epsilon = 1/(1 + A\, V_a^E) \tag{1}$$

This is based on the general equation for heat transfer coefficient

$$h = FV^b \tag{2}$$

Such curves are shown in Figure 6.

For a given temperature difference across a heat exchanger plate, B , the total heat flow $H = GV^b$, where G is a constant relating to the physical properties of the heat exchanger. $\tag{3}$

If changes in the density of the air as it passes through the heat exchanger are negligible, then the mass flow may be taken as

$$M = DV \tag{4}$$

With balanced air flows, the temperature change of air passing through,

$$(t_{co} - t_{ci}) = (t_{hi} - t_{ho}) = F\, V^b/DV = (F/D)\, V^{(b-1)} \tag{5}$$

The effectiveness

$$\epsilon = \frac{(F/D)\, V^{(b-1)}}{(F/D)\, V^{(b-1)} + B}$$

$$= 1/(1 + (BD/F)\, V^{(1-b)}) \tag{6}$$

Substituting V_a for V and using a new constant
A in place of (BD/F)
Then $\epsilon = 1/(1 + A\, V_a^E)$, where $E = (1-b)$
In Figure 8, curves in the form
Pressure gradient $Ph/L = C\, V_a^{1.75}$ were drawn for comparison with the test results.

5. RESULTS

In Figure 2 the heat exchanging effectiveness is plotted against approach velocity of air for the various heat exchangers, using symbols for the points as in Table 1. It may be seen that the pocket design has a different characteristic from the spike designs.

In Figure 3 the product Stanton Number x (Prandtl Number) 2/3 is plotted against Reynolds Number for a range of Reynolds Numbers between 93 and 1870. Again it may be seen that the pocket design is distinctly different from the spike designs and both higher Stanton Numbers than plane parallel walls.

In Figure 4 the friction factor is plotted against Reynolds Number. Again, the pocket design is different from the spike designs but the friction factors for the primary circuit is different from the friction factor for the secondary circuit because of the elongated shape of the pockets.

In Figure 5 the product Stanton Number x (Prandtl Number) 2/3 is divided by the friction factor and plotted against the Reynolds Number, to illustrate the relative efficiency of heat transfer. While there is considerable scatter of results, some designs appear to be superior to plane parallel passages. Bergles and Morton [5] in a survey of techniques to augment convective heat transfer, have shown that this is possible.

In Figure 6 the heat transfer coefficient is plotted against approach velocity of the air and again the pocket design and spike designs fall into distinct groups, with the spike designs more characteristic of turbulent flow.

In Figure 7 the air pressure gradient is plotted against approach velocity of the air. In this graph it may be seen that the different designs have their own air resistance characteristics but all tend to have the same slope corresponding to

$$Ph/L = C \ V_a^{1.75}$$

This is similar to turbulent flow characteristics. Furthermore, the effect of passage width is insignificant, indicating that the air resistance is almost entirely due to the protrusions.

6. DISCUSSION

The curves in the graphs indicate that the flow through parallel passages having small protrusions in the form of spikes is consistent with turbulent flow for

Reynolds Numbers, based on width of passage, as low as 200. The curves also indicate that the width of passage is not very significant within the range of dimensions tested; consequently the most compact heat exchanger could have plates spaced about 2 mm apart.

The high values of heat transfer coefficient show that these designs are economical in material. The poor thermal conductivity of the plastic is compensated by the use of thin films; typical values of mean temperature difference across a plate is 2% to 8% of the mean temperature difference between the air streams. The lower value could be used for practical applications.

A unit capable of handling 500 1/s with an effectiveness of 0.70 and a pressure drop of 7 mm of water could use plates of spike design, 0.2 mm thick and 400 mm square, spaced 3 mm apart to form a stack 760 mm wide. The plates could have a spike spacing of 13.4 mm. A unit capable of handling 190 1/s with an effectiveness of 0.62 and a pressure drop of 10 mm of water could use plates of pocket design 0.2 mm thick and 150 mm square, spaced 2.3 mm apart to form a stack 760 mm wide.

These types of heat exchangers have been used successfully as air coolers by wetting the plates with water on one side only. Air exhausted from a room is drawn over the wet surfaces before being discharged outside and the evaporation of the water cools the plates. Outside air is passed over the dry side of the plates where it is cooled by heat transfer without the addition of water vapour, and it is then discharged into the room.

It is possible to obtain lower room temperatures than those obtainable with simple evaporative coolers.

The results of this work have provided a basis for patents and commercial applications.

7. CONCLUSIONS

Turbulence promoters moulded into plates of heat exchangers can result in economies of material without increase in fluid power input requirements, compared with plane plates of similar proportions.

Plate heat exchangers made from vacuum formed rigid plastic films are effective for applications where the temperatures and fluid pressures are modreately low, as in air conditioning applications.

Plate heat exchangers of this type may be used as air coolers by evaporating water from one side of each plate. They do not add water vapor to the air entering the space to be cooled.

LIST OF SYMBOLS

Cp	Specific heat at constant pressure	kJ/kg °C
D	Hydraulic diameter of passage	m
f	Mean friction factor, defined by	$f = \dfrac{\Delta PD}{2\rho L\ V^2}$
h	Heat transfer coefficient	W/m² °C
k	Thermal conductivity	W/m°C
L	Total flow length through heat exchanger	m
P	Pressure	Pa
Ph	Pressure	mm of water
Pr	Prandtl number, defined by	$Pr = \dfrac{\mu Cp}{k}$
Re	Reynolds number, defined by	$Re = \dfrac{\rho VD}{\mu}$
St	Stanton number, defined by	$St = \dfrac{h}{\rho VCp}$
t_{ci}	Temperature of cold air in	°C
t_{co}	Temperature of cold air out	°C
t_{hi}	Temperature of hot air in	°C
t_{ho}	Temperature of hot air out	°C
V	Velocity	m/s
Va	Approach velocity	m/s
Δ	Difference	
ϵ	Effectiveness, defined by	$\epsilon = \dfrac{(t_{hi} - t_{ho}) + (t_{co} - t_{ci})}{2(t_{hi} - t_{ci})}$
ρ	Density	kg/m³
μ	Viscosity	Pa s

NOTE : The SI system of units is used in this paper; a table of conversions to British units is given in the Appendix.

REFERENCES

[1] Pescod, D. - Unit air cooler using plastic heat exchanger with evaporatively cooled plates. Aust. Refrig. Air Condit. Heat., 1968, 22(9) : 22.

[2] Dunkle, R.V. - Regenerative evaporative cooling systems. Aust. Refrig. Air Condit. Heat., 1966, 20(1) : 13.

[3] Morse, R.N. - A rock bed regenerative building cooling system. Mech. Chem. Engng. Trans. Instn Engrs Aust., 1968, 4(1) : 55.

[4] Kays, W.M. and London, A.L. - Compact Heat Exchangers, McGraw-Hill, 1958.

[5] Bergles, A.E. and Morton, H.L. - Survey and evaluation of techniques to augment heat transfer. Massachusetts Institute of Technology, Report No. 5382-34, February, 1965.

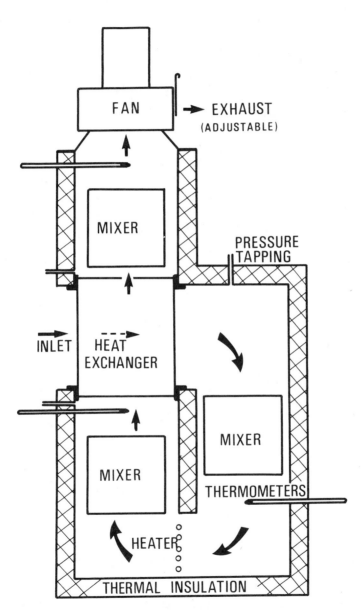

Fig. 1. Arrangement of test rig.

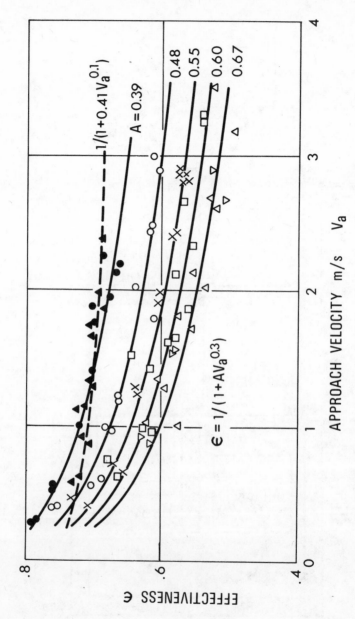

Fig. 2. Effectiveness vs. approach velocity for heat exchangers approximately 200 mm square.

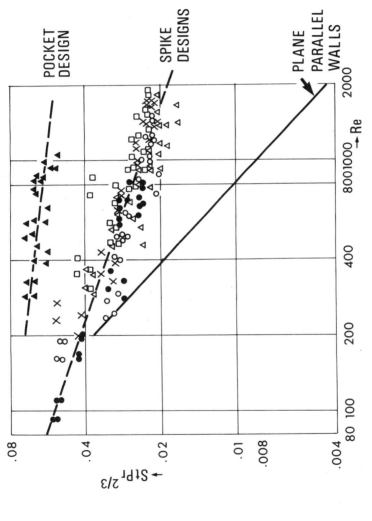

Fig. 3. Stanton No. x (Prandtl No.) 2/3 vs. Reynolds No.

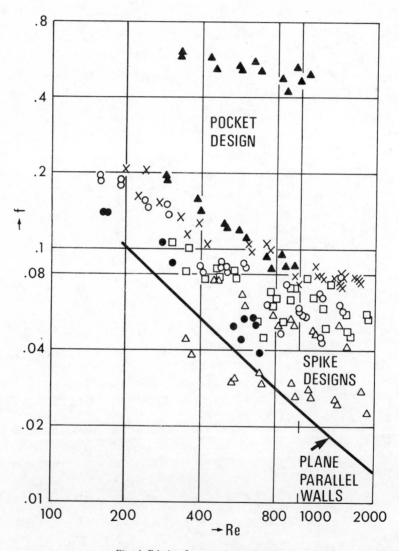

Fig. 4. Friction factor vs. Reynolds No.

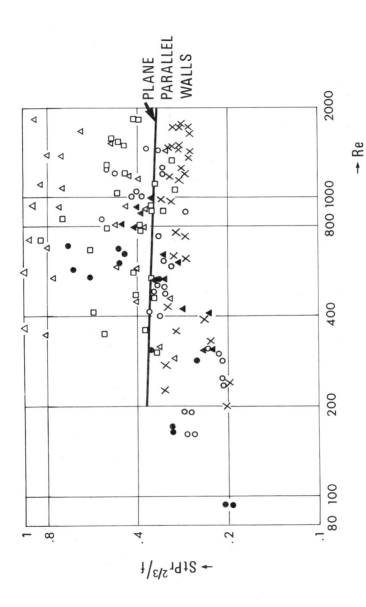

Fig. 5. Stanton No. x (Prandtl No.) 2/3 Friction factor vs. Reynolds No.

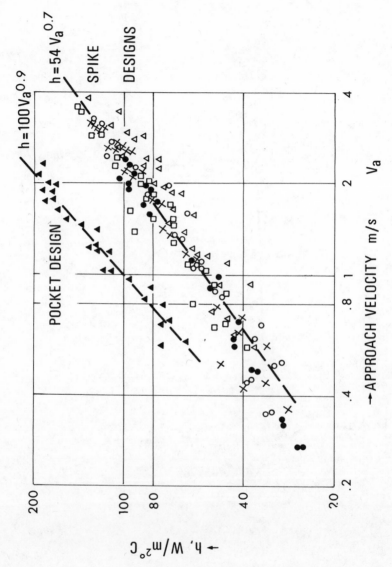

Fig. 6. Heat transfer coefficient vs. approach velocity.

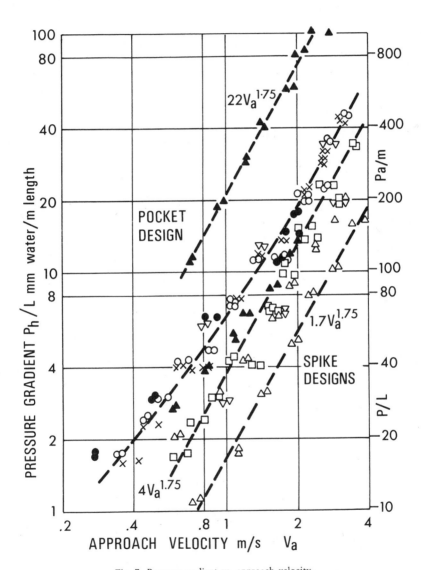

Fig. 7. Pressure gradient vs. approach velocity.

Chapter 23

INVESTIGATION OF THE HEAT TRANSFER PROCESSES
IN TUBE-BANKS IN CROSS FLOW

S. Oka, Ž. Kostić, S. Šikmanović*

1. INTRODUCTION

Heat exchangers with tube-banks in cross flow are still of great practical interest due to their wide use in thermal and chemical engineering. In the last few years appeared several very exhaustive articles which deal with mean and local values of flow resistance, pressure coefficients and Nusselt numbers, as well as with flow and heat transfer processes in tube-banks [1, 2, 3]. Besides the great number of experimental data, it is not yet possible to get a clear idea about the flow and heat transfer processes in tube-banks because of the very complicated geometry and the great number of involving parameters.

It is very difficult to make local measurements in tube-banks in order to discover response of flow and heat transfer processes to the changes of some (especially geometrical) parameters. It is also very difficult to divide the influence of different parameters.

In order to investigate real physical processes of flow and heat transfer in tube-banks and their response to the change of different parameters, it is possible to make experiments on properly choosen, simple geometrical models, in which it is possible to make measurements of local values, and to investigate the influence of only one separate parameter.

The aim of our investigations was to explore on such simple models the influence of geometrical parameters : transverse pitch, longitudinal pitch and position of tube in the bank, on the flow and heat transfer processes around the

---

(*) Thermal Physics and Engineering Department, Boris Kidric Institute, Belgrade, Yugoslavia.

N.B. ALL FIGURES QUOTED IN TEXT ARE AT THE END OF THE CHAPTER

tube in the bank.

As simple experimental models were chosen : a single transverse row of cylinders for investigation of influence of tranverse pitch, two cylinders in cross flow (one in the wake of another) for investigation of longitudinal pitch, and single cylinder in cross flow of fluid with high turbulence intensity and different macroscales for investigation of the influence of tube position in the bank. Experiments were carried out in the subcritical region of flow in a tube-bank, i.e. in the Reynolds number range from 10^3 to 2.10^5.

2. EXPERIMENTAL APPARATUS AND CONDITIONS

All experiments were carried out in the wind-tunnel with cross section 500x500 mm and with cylinders of 100 mm in diameter (except the experiment in which the shedding frequency was measured, in which cylinders of 14.5, 25 and 44.5 mm in diameter were used).

Velocity and turbulence intensity profiles in the channel were uniform [4]. It was possible to obtain velocities from 2 to 6 m/sec. Turbulence intensity in the empty channel was 0.3%.

A detailed description of the experimental apparatus, measuring tubes construction, measurement methods, and a check of the results are given in [4, 5, 6, 7].

An experiment with the single transverse row of cylinders (investigation of the influence of tranverse pitch) was carried out on Re $\div$ 30000 and for the range $S_T/D \simeq 1.066 \div 5$.

An experiment with two cylinders in cross flow (one of them in the wake of another) was carried out in the Reynolds number range from 12,000 to 40,000, and for the distance between cylinders (longitudinal pitch) in the interval $1.6 \leqslant L/D \leqslant 9$.

Measurements of the shedding frequencies for the same geometrical model behind the first and the second cylinder, was carried out for Re = 2,950 $\div$ 8,800, and for L/D = 1 $\div$ 40.

An experiment with the single cylinder in cross flow of fluid with different turbulence intensities and different macroscale of turbulence was carried out for Re $\simeq$ 20,000. Turbulence intensity was changed from 2.7% to 26% and the macroscale $\Lambda/D = 0.05 \div 0.2$.

SINGLE TRANSVERSE ROW OF CYLINDERS IN CROSS FLOW
The basic results of this experiment are given in [8].

3.1. Mean heat transfer and form drag coefficient

When the transverse pitch decreases up to 1.6 D the mean Nusselt number does not change essentially, although distributions of local Nusselt numbers and pressure coefficients already show a noticeable change in the character of flow compared to the single cylinder. For $S_T/D = 1.6$ the mean Nusselt number is greater only by about 5%. The form drag coefficient is increased threefold [8].

For $S_T/D < 1.6$ the mean Nusselt number sharply increases followed by a more severe increase of the form drag coefficient. For $S_T/D = 1.066$ in this experiment $Nu_M = 320$ (or 27% greater), and $C_{Dp} = 154$ (or 1200% greater) compared to the values for a single cylinder [8].

3.2. Distribution of local values and flow patterns

For $S_T/D \geqslant 4.4$ the character of flow is very similar to the flow around a single cylinder. On the front side of the cylinder the laminar boundary layer develops, with the separation point at $\varphi = 82°$. Pressure distribution in the rear side is also changed slightly. The distribution of the local Nu number shows a slight increase in the laminar boundary layer region, as a consequence of the increased velocities between cylinders (for the same U_0). In the rear side a significant increase in heat transfer [8] shows that the shedding frequency is increased because of the presence of other cylinders [9].

Greater differences in flow character appear for $S_T/D < 4.4$ (Fig. 1). With the decrease of S_T/D the static pressure on the front side changes more and more sharply, and the minimum static pressure on the cylinder surface decreases rapidly. As a result of the flow acceleration the separation point shifts downstream. For $S_T/D \simeq 1.6$ separation point (SL) is on $\varphi = 92°$.

Because of the increased velocities between the cylinders (for the same U_0), for $S_T/D = 2.2$ and 1.6 (Fig.1) the laminar boundary layer transition to the turbulent boundary layer occurs (transition point ST). The turbulent boundary layer separation point is at $\varphi = 120°$. With further increase of favorable pressure gradient (for $S_T/D \leqslant 1.46$), the laminar boundary layer stabilizes and turbulent boundary layer disappears.

For $S_T/D > 1.6$ heat transfer on the front side of the cylinder is only slightly greater, and on the rear side the further change of the structure of the wake takes place, with the probable increase of shedding frequency.

Up to $S_T/D = 1.25$ the character of flow is still similar to the flow around blunt bodies (fig. 2). The minimum local static pressure is already at $\varphi = 90°$. The minimum pressure coefficient slightly differs from the value obtained from the Bernouli equation for nonviscous fluid. That means that because of the great acceleration, the laminar boundary layer thickness became very small. Because of the smaller boundary layer thickness on the front side for $S_T/D = 1.6$ and 1.46 (Fig.1), heat transfer intensity sharply increases. Because of the acceleration for those values of S_T/D, the boundary layer thickness decreases downstream, and the maximum Nusselt number is no longer at the front stagnation point.

Distributions for the local static pressure coefficient and local Nusselt numbers for $S_T/D = 1.066$ show a quite different flow character. Flow through the row of cylinders with such a small transverse pitch is very similar to the flow through the convergent-divergent channels. Minimum static pressure is in the most narrow cross section ($\varphi = 90°$) and the 10 cal Nusselt number also increases up to $\varphi = 90$ and reaches maximum value at this point. The distributions of local pressure and of the Nu number in the divergent parts of those channels do not correspond further to the distributions at the rear side of the cylinders. Separated flow in this part gives very high heat transfer intensity.

4. TWO CYLINDERS IN CROSS FLOW

In this experiment one cylinder (in the text, the second cylinder) was in the wake of the another (in the text, the first cylinder). The basic results of this experiment are given in [4, 5, 6, 10, 11, 12].

4.1. Mean Nusselt number and form drag coefficient

(a) **The first cylinder.** The first cylinder form drag coefficient differs slightly from that of the single cylinder (fig. 3, L = ∞). Slight changes can be noticed at shorter distances when the vortex region is formed.

The mean Nusselt number for the first cylinder, at the distance L/D = 3, does not show any difference from those for the single cylinder. At shortened distances deviations ($3 \geqslant L/D \geqslant 1.6$) are no greater than ±8% as compared to those for the single cylinder (fig. 4), with tendency of decreasing for L/D < 2.

(b) **The second cylinder.** The form drag coefficient for the second cylinder is significantly smaller than that for the single (and the first) cylinder, even at large distances between cylinders (L/D = 9, Fig. 3). When the closed vortex region is formed between the cylinders, the form drag coefficient for the second cylinder suddenly begins to decrease, and becomes negative for L/D < 2.

For L/D > 3.8 (when the closed vortex region is not formed) the mean Nusselt number for the second cylinder is 25% higher than that for the single cylinder (Fig.4). With the decrease of the distance between the cylinders the mean Nusselt number remains constant. For distances L/D < 2.3 the mean Nusselt number for the second cylinder decreases with the decreasing of distance.

(c) **The dependence of the mean Nusselt number on the Reynolds number.** When a large part (about 50%) of the cylinder surface (either of the first or the second cylinder) is covered by laminar boundary layer, the mean Nusselt number is proportional to $Re^{0.6}$ (Fig.5).

When the largest part of the cylinder surface is in the vortex region, the mean Nusselt number depends on $Re^{2/3}$, as is the case for the second cylinder for distances less than 2.7 D (Fig.6).

4.2. Distribution of the local values and flow patter

(a) **The first cylinder.** Distribution of the local static pressure coefficient and Nusselt numbers for the first cylinder differ very slightly from those for the single cylinder. Only on the rear side at short mutual distances (1.6 ≤ L/D ≤ 3), when a closed vortex region is formed between the cylinders, the local Nusselt numbers are increased for distances L/D = 2.5 and 2, probably because of more intensive and better defined vortex flow between the cylinders. For L/D = 1.6 velocities in the closed vortex region evidently are smaller, so Nusselt numbers are again decreased [4, 5].

(b) **The second cylinder.** Distribution of the local static pressure coefficient and Nusselt numbers for the second cylinder show that, at distances L/D = 9 and 6 (when the closed vortex region is not yet formed between the cylinders) the laminar boundary layer becomes turbulent before separation [4, 5].

The heat transfer intensity within the laminar boundary layer is increased by about 20% (when compared to that of the first cylinder). This is the result of the increased turbulence level (and probably $\Lambda/D \simeq 1$) of the oncoming flow [4, 5].

Qualitatively different local static pressure coefficient and Nusselt number distributions on the surface of the second cylinder are obtained at distances L/D ≤ 3. In this case, two reattachment points are obtained at the front side of the second cylinder. They are actually the impingement points of the free shear layers separated from the first cylinder and which correspond to the maxima on the local static pressure coefficient distribution curves. The closed vortex region is already formed at these distances. The greater distance at which two pressure maxima were observed on the front side of the second cylinder (critical distance L_{cr}) is 3.8 [4, 5].

Starting from these reattachment points, laminar boundary layers are formed towards the front and rear stagnation points. The maximum heat transfer coefficient is obtained near the reattachment points.

The laminar boundary layer becomes turbulent also at these distances. Transition points are shifted downstream (compared to the L/D = 9 and 6) at $\varphi = 115° - 120°$, but the turbulent boundary layer separation points are approximately at the same place ($\varphi = 150° - 155°$) [4, 5].

The front side of the second cylinder is in the closed vortex region for these small distances. Due to the decrease of the velocities and the increase in length and thickness of the boundary layer developing towards the formed front stagnation point, the local Nusselt numbers on the cylinder front side decrease simultaneously with the decrease of L/D, but are still higher than on the rear side of the first cylinder [4, 5].

On the basis of those results, it was possible to develop an idea of the pattern and mechanism of the flow around the two cylinders (Fig.7.).

At distances between cylinders greater than a certain critical distance, (when the closed vortex region is not formed between them), the second cylinder has no influence on the processes developing on the first. The pattern and mechanism of flow in the wake of the first cylinder is the same as that for the single cylinder.

As the second cylinder is in the wake of the first, the increased turbulence level of the oncoming flow causes the transition (SL) of the laminar to turbulent boundary layer to occur before the separation (ST). The wake of the second cylinder is thus more narrow which causes significant decrease of the form drag coefficient. The flow around the second cylinder is similar to the supercritical flow (Re $> 2.10^5$) around single cylinder. Heat transfer in laminar boundary layer is increased due to the high turbulence intensity and possibly macroscale of the order of cylinder diameter.

At distances shorter than critical, the free shear layers which separate from the first cylinder impinge on the front side of the second cylinder, and the closed vortex region is formed between the cylinders. It has been found that

$(L/D)_{cr} = 3.8$.   The critical distance is greater for greater Reynolds numbers (in the range from 12,000 to 40,000).

On the second cylinder downstream of the reattachment point (R), a laminar boundary layer is formed which becomes turbulent before separation (SL).

4.3. Shedding frequency behind the first and second cylinder

The basic results of these measurements are given in [6].

For distances greater than 10 D (Fig. 8) the shedding frequency behind the first cylinder is equal to that of the single cylinder (S = 0.21). For smaller distances (but greater than the critical) the shedding frequency slightly decreases. At a critical distance the shedding frequency suddenly falls to a very small value. The vortices do not shed more.

This result confirms the assumption that for distances less than critical the closed vortex region is formed between cylinders with two stable, symmetrical vortices (free cavity, Fig. 8).

With the increase of the Reynolds number from 2,950 to 8,800, the critical distance decreases (Fig. 8). It seems that this is not in contradiction with the dependance of the critical distance obtained in the Reynolds number range 13,000 to 40,000 [4], because of the different behavior of the length of the vortex formation region in these two Reynolds number ranges [13].

For distances L/D < 20 the frequency of shedding behind the second cylinder increases up to the critical distance. Probably this is because the vortices shedded from the first cylinder flowing near the second cylinder create an effect similar to the channel blockade [9]. Besides, the higher turbulence intensity shifts the separation points downstream which cause the increase of frequency [6].

At critical distance the frequency of shedding behind the second cylinder falls to the value characteristic for distances greater than L/D = 20. For this distance there is no more shedding of vortices from the first cylinder, and influence of the first cylinder disappears [6].

5. INFLUENCE OF THE MAIN STREAM TURBULENCE STRUCTURE ON HEAT TRANSFER FROM THE SINGLE CYLINDER

The basic results of this experiment are given in [7].

5.1. Mean Nusselt number and form drag coefficient

Fig. 9 shows the change of the mean Nusselt number with the change of turbulence intensity. The Reynolds number was about 20,000. For turbulence intensity of about 16%, the mean Nusselt number is increased by about 24%. The increase of the Nusselt number becomes smaller for a greater turbulence intensity. Results obtained for five different grids and at the four distances behind them (that means for 20 different turbulence levels and 20 different macroscales) can be represented by a single curve with a maximum scatter of about ±2%. With careful analysis the influence of Λ can be noticed, but the differences are too small to find real dependence for this range of macroscale ($\Lambda/D = 0.05$ D ÷ 0.1).

Fig. 10 shows the comparison of our results for Nu_L (Re = 20,000) with the results for higher Reynolds numbers [18]. There is great difference in the turbulence level influence on Nu_L (the same case also applies to Nu_M) depending on the Reynolds number. For Re = 20,000 and Tu = 7% the increase of heat transfer in laminar boundary layer region is 15%, and for Re = 10^5 about 60%. Obviously the reason is the different boundary layer thickness for different Reynolds numbers.

The same Fig. also shows that heat transfer in the laminar boundary layer increases much more than the mean heat transfer from the cylinder. For Tu = 16% the increase of Nu_L is about 32%.

The form drag coefficient of the single cylinder is significantly decreased only when the transition to turbulent boundary layer for this Reynolds number appears on the surface of the cylinder for Tu greater than 6-8%. For Tu = 16% the form drag is decreased by about 30% [7].

5.2. Distribution of local values and flow pattern

In Fig. 11 are shown the distributions of the local static pressure coefficients for four different turbulence intensities ranging from 4% to 14%. In the region of the laminar boundary layer can be noticed a slight acceleration of the flow outside the boundary layer with an increase of turbulence, although the minimum pressure and the separation point are at almost the same place ($70°$ and $90°$). For Tu = 14% exists the laminar to turbulent boundary layer transition, and the turbulent boundary layer separation point is at about $140°$.

Fig. 12 shows the distributions of the local Nusselt numbers for the same values of turbulence level as in Fig. 11, compared with the distribution for a cylinder in the empty channel (Tu = 0.3%). The great increase of the heat transfer intensity is easily seen in the laminar boundary layer region, especially around the stagnation point where the boundary layer thickness is smaller.

An increase of heat intensity exists also in the wake region of the cylinder. The cause of this increase was believed to be the macroscale of turbulence [19] (this was the explanation of the experimental results of [15]). Recent results [16] show that the reason for the change of structure of the wake is the turbulence intensity and not the turbulence macroscale. The increase heat transfer intensity in this region certainly occurs because of the change of structure of the wake, but our results do not show the reason for this change.

For Tu = 8% there are already signs of the transition to turbulence, and the heat transfer in the rear side of the cylinder is significantly increased. For Tu = 14% the existance of the turbulent boundary layer is evident. This is the reason for great increase of heat transfer intensity in this region and in the wake region too. Of course, the value of Tu on which the transition to turbulence occurs depends on the Reynolds number.

The greatest increase of heat transfer in the laminar boundary layer region can be noticed for smaller turbulence intensities (Tu $\leqslant$ 6%), and for greater values the increase is smaller. The same case is evident after the transition to turbulence. The further increase of heat transfer in the turbulent boundary layer and wake region is smaller, too. This is the reason for the smaller increase of heat transfer for the whole cylinder at greater turbulence intensities.

6. DISCUSSION

The experimental results on simple geometrical models have shown that the characteristic changes of mean values (with the change of transverse and longitudinal pitch, and also for increased turbulence intensity) are similar to those of the tube-banks. This can help us to use the described results for explanation of the physical background of behaviour of tube-banks.

We shall note some examples of the main processes which take place in tube-banks.

With the decrease of transverse pitch the heat transfer increases in an in-line tube-bank [1], as our results (see 3) show mainly because of the acceleration of flow and significant decrease of boundary layer thickness. The decrease of heat transfer for tube-banks observed for very small $S_T/D < 1.053$ [1] and also for small longitudinal pitches, probably occurs because of laminarisation of flow in the bank due to the great favorable pressure gradient.

The decrease of the heat transfer with the decrease of S_T/D in staggered tube-banks [1] cannot easily be explained on the basis of our results. Experiments on other simple experimental models are needed.

Experiments with two cylinders in cross flow (See 4) explain very well the influence of longitudinal pitch and partially the influence of the tube position. For pitches of S_L/D greater than a certain critical value, the increase of heat transfer is the consequence of the high turbulence level. With the decrease of the longitudinal pitch the heat transfer in a tube-bank [1] increases slightly due to the increase of the turbulence level at smaller distances from the up-stream transverse row, which serves as a turbulising grid. For our experiment (and Re = 2,000) the critical pitch is about (3-4) D, but in the tube-bank it should be smaller because of the narrower wake behind tubes in inner rows.

The most favorable conditions for heat transfer is when the pitch is equal to the length of the vortex formation region, i.e. $S_L/D = 2 - 4$ in the subcritical region. With further decrease of pitches the heat transfer decreases because of slow circulation in the vortex region between the tubes [1].

The change of the exponent "m" in the dependence $Nu_M = c \cdot Re^m$ for in-line tube banks [1] with the change of transverse and longitudinal pitches is the consequence of the change of mechanism of heat transfer on the tube surface (see Sec.4). With the change of pitches (especially longitudinal) the ratio of the area under laminar, turbulent boundary layer and vortex region is changed. These changes are greater up to $S_L/D \geqslant 2$. For staggered tube-banks this ratio changes less, and the exponent "m" is constant [1].

Influence of turbulence is manifested for the pitches $S_L/D = 3 - 4$, mostly because of the increase of heat transfer in a laminar boundary layer and because of the laminar to turbulent boundary layer transition. For smaller distances influence of turbulence is of less importance because of the formation of a closed vortex region.

Although immediately behind the cylinder the turbulence intensity is about 30-40% [16], in tube-banks [1], and also in an experiment with two cylinders (Sec.5), it is not noticed that the increase of heat transfer is adequately high for this turbulence level. The reason is the very fast decrease of Tu behind the tube and the small increase of heat transfer for such high turbulence intensity.

Of course, the processes in simple geometrical models are not the same as in the tube banks. Upstream transverse rows, side longitudinal rows and other tubes are an influence on processes, so they differ quantitatively from those described for simple models. However, the main characteristics of the processes remain the same. To investigate these influences it is necessary to make experiments on more complicated geometrical models.

NOMENCLATURE

$C_{Dp} = \frac{1}{\pi} \int_0^\pi C_{po} \cos\varphi \, d\varphi$ —   form   drag coefficient

$C_{po} = (p - p_{max})/0,5\rho\, U_0^2$ —   local static pressure coefficient

$C_{Dmax} = C_{Dp} \cdot U_0^2/U^2$ max

D   tube diameter

f   frequency of vortex shedding

$Nu = \alpha d/\lambda$ —   local Nusselt number

$Nu_M = \frac{1}{\pi} \int_0^\pi Nu \cdot d\varphi$ —   mean Nusselt number

Nu_L mean Nusselt number in the laminar boundary layer region

$Nu_{Lo} - Nu_L$ for zero turbulence intensity

p   local static pressure on cylinder surface

p_{max} maximum local static pressure

q   heat flux

$Re = U_0 D/v$ Reynolds number based on U_0

Re_{max} Reynolds number based on U_{max}

S_T transverse pitch

S_L longitudinal pitch (L-in experiment with two cylinders)

$S = f \cdot D/U_0$ Strouhal number

$Tu = \sqrt{u^2}/U_o$ —   turbulence intensity

t_w local wall temperature

t_f temperature of undisturbed fluid

U_0 velocity of undisturbed fluid

$U_m = \frac{1}{\pi} \int_0^\pi U \cdot d\varphi$ —   mean velocity around cylinder

U_{max} velocity in the most narrow cross section

$\alpha = q/(t_w - t_f)$ local heat transfer coefficient

λ heat conductivity

Λ macroscale of turbulence

φ angle measured from the front stagnation point

ρ fluid density

v kinematic viscosity

$Eu = \Delta p/\rho U_0^2$ Euler number, flow resistance coefficient

REFERENCES

[1] A. Žukauskas, A Makaravičius, A. Šlančiauskas : Heat Transfer in Banks of Tubes in Cross Flow of Fluid Mintis, Vilnius, 1968 (in Russian).

[2] Žukauskas : Int. Seminar "Heat and Mass Transfer in Flows with Separated Regions and Measurement Techniques," Herceg-Novi, Yugoslavia (1969).

[3] A. Žukauskas : Int. Seminar "Recent Developments in Heat Exchangers," Trogir, Yugoslavia, 1972.

[4] Ž.G. Kostić, S.N. Oka : Int. J. of Heat and Mass Transfer, Vol.15, pp.279-299, 1972.

[5] Z.G. Kostić : MS Thesis, University of Belgrade, 1970

[6] Ž. Kostić, S. Oka, V. Pišlar, R. Rodić : XI Yugoslav Congress of Rational and Applied Mechanics, Baško Polje, 1972 (in print).

[7] S. Šikmanović : MS Thesis, University of Belgrade, 1972 (in print).

[8] D. Uščumlić, S. Oka, Ž. Kostić : I Yugoslav Congress of Chemical and Process Engineering, Beograd, 1971 (in print).

[9] E.P. Dyban, E.Ya. Epick : Heat Transfer Conference 1970, Vol. II, FC 5.7, Versailles 1970.

[10] Ž.G. Kostić : Int.Seminar "Heat and Mass Transfer in Flows with Separated Regions and Measurement Techniques," Herceg-Novi, Yugoslavia, 1969.

[11] Ž.G. Kostić, S.N. Oka : Yugoslav Congress of Rational and Applied Mechanics, Baško polje, 1970.

[12] Ž.G. Kostić, S.N. Oka: I Yugoslav Congress of Chemical and Process Engineering, Beograd, 1971 (in print).

[13] Susan Bloor : J. of Fluid Mechanics, 19, 290, 1964.

[14] J. Kestin : Advances in Heat Transfer, Vol. 3, Academic Press, New York, 1966.

[15] B.G.v. Der Hegge Zijnen : Appl. Sci.Res., A7, 205, 1957.

[16] E.P. Dyban, E.Ya.Epick, L.G. Koslova: Heat and Mass Transfer, Vol. 1, part 3, Minsk, 1972.

[17] M.C. Smith, A.M. Kuethe : The Physics of Fluids, Vol. 9, No. 12, Dec. 1966.

[18] J. Kestin, R.T. Wood : Journal of Heat Transfer, Nov. 1971, p. 321.

[19] J.O. Hinze : Turbulence, McGraw-Hill, New York, 1959.

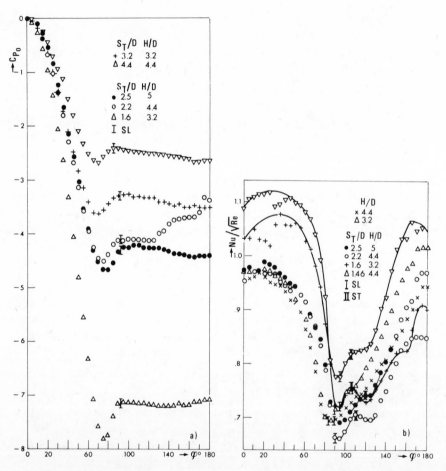

Fig. 1. Distributions of the local static pressure coefficient and Nusselt number on the surface of the tube in the single row for S_T/D = 3.2 to 1.6.

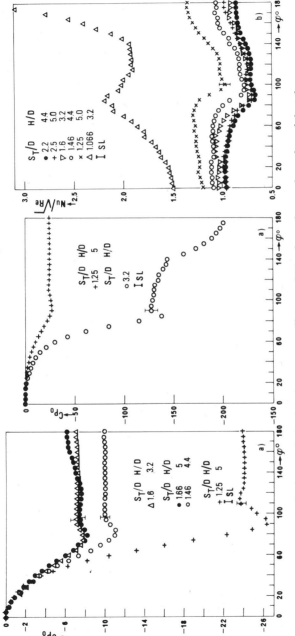

Fig. 2. Distributions of the local static pressure coefficient and Nusselt number on the surface of the tube in the single row for $S_T/D = 1.46$ to 1.006.

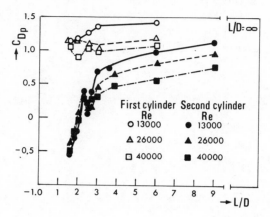

Fig. 3. First and second cylinder: form drag coefficients in dependence on distance between cylinders.

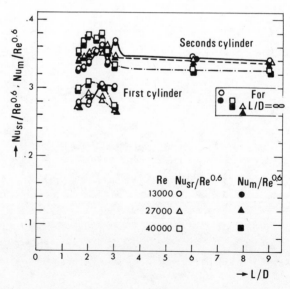

Fig. 4. First and second cylinder: dependence of the mean Nusselt number on distance between cylinders. cylinders.

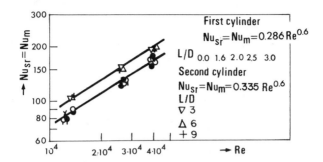

Fig. 5. First and second cylinder:mean Nusselt number in dependence on Reynolds number.

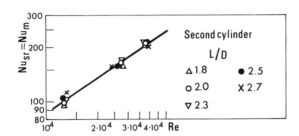

Fig. 6. Second cylinder : mean Nusselt number in dependence on Reynolds number, for $L/D \leqslant 2.7$.

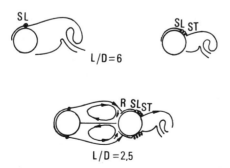

Fig. 7. Flow pattern for two cylinders in cross flow.

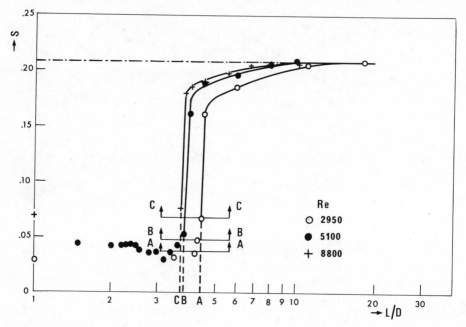

Fig. 8. Shedding frequency behind first cylinder.

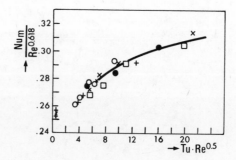

Fig. 9. The influence of turbulence intensity on mean Nusselt number.

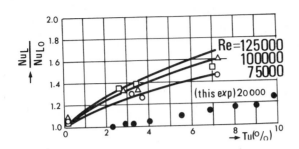

Fig. 10. The influence of turbulence intensity on heat transfer in laminar boundary layer region. (Comparison with [18]).

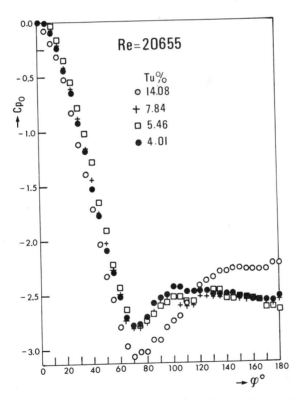

Fig. 11. Distributions of the local static pressure coefficient for different turbulence intensity.

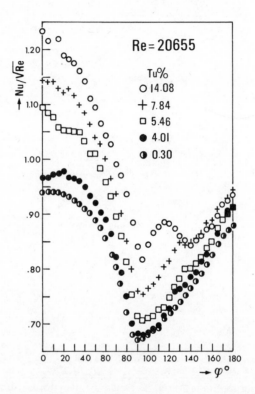

Fig. 12. Distributions of the local Nusselt number for different turbulence intensity.

Chapter 24

INVESTIGATION OF THE EFFECT OF TURNS ON TURBULENT FORCED CONVECTION HEAT TRANSFER IN PIPES

S. Kakaç, Y. Göğüş, M. R. Özgü(*)

1. INTRODUCTION

The knowledge of the heat transfer between a pipe and a fluid flowing through it is a prerequisite for a wide range of applications. The distribution of heat transfer coefficient near the thermal entrance of a circular pipe is a special but important problem of this area. The distribution of heat transfer coefficient depends not only upon the boundary conditions, but also upon the flow conditions, especially when the pipe is preceded by a bend. In many heat exchangers for a variety of reasons the pipes are not straight but include bends of various angles and radii. One of its important applications is in the design of cooling systems for chemical industry. Also heat exchangers with curved pipes can be found in power refrigeration and nuclear reactor technology. The effect of a bend in some of the cases may be very substantial and can be of importance with its influence on the over-all performance of a heat exchanger and with problems as hot spots, thermal stresses and so on.

Despite the importance of bends in heat exchangers, only a few investigations have been carried out in this particular area. Ede [1] worked on the effect of right angled bends on heat transfer in a pipe. The working fluid was water. To obtain constant heat flux boundary condition, alternating current was used in some of experiments, whereas in most of the experiments direct current was preferred. The experiments covered a range of Reynolds numbers from 700 to 50000. The heat flux employed varied from 70 m W/cm^2 to 16 W/cm^2 at the lowest and highest Reynolds numbers respectively. He found as much as sixteen times

(*) Middle East Technical University, Ankara, Turkey

N.B. ALL FIGURES QUOTED IN TEXT ARE AT THE END OF THE CHAPTER

greater heat transfer coefficient due to right angled bend if the flow was laminar. In case of turbulent flow the influence of the bend is not so strong. His experiments at the Reynolds numbers 2000 to 10000 indicate the flow behaves sometimes like turbulent and sometimes like laminar. After strongly curved bends the sudden increment of the Nusselt number is succeeded by a minimum where it gets even smaller values than the fully developed Nusselt number. The relaminarization in the curved pipe was expected to be the reason of this abnormality which continues till 50 diameters length. Even the experiments with unheated bends preceding the heated straight pipe are reported.

Mori and Nakayama [2, 3, 4] investigated forced convection heat transfer in curved pipes analytically and experimentally. Both constant heat flux and uniform temperature boundary conditions were used for laminar and turbulent flows. The velocity profile in the bend was fully examined analytically for laminar flows.

Two investigations of 180° bends were carried out by Boelter et al. [5] and Mills [6]. The investigations were for air and refer to a 180° bend of unspecified radius.

The aim of the present investigation is to determine the effect of 180° bends on the heat transfer coefficient in the region following the bend. The case of steady-state, forced convection turbulent flow is considered. The variables are:

a. Reynolds number
b. Ratio of radii of bend to pipe.

The wall heat flux is assumed to be uniform in all cases investigated. However variations in the uniformity of the heat flux is expected due to the following reasons :

i.   Variation of the resistance of the pipe material with temperature,
ii. Variation of the tube wall thickness,
iii. Nonuniformity of the pipe insulation,
iv. Axial heat conduction in the pipe material. Heat loss being greatest at the ends, due to the axial conduction.

Out of these deviations the largest one is the end heat loss due to axial conduction. But this deviation occurs also in the heat exchangers.

2. DESCRIPTION OF EXPERIMENTAL APPARATUS

A schematic diagram of the experimental set-up is shown in Fig. 1 and

2. Air from a filter-mounted fan successively flowed through a horizontal pipe 4 inches in nominal diameter, adopter, flexible hose, a calming tank, a velocity developing pipe, the test section and a mixing cup. Immediately after the fan there was a valve and a by-pass pipe with a regulating valve. By means of these two valves the flow rate and thus the Reynolds number was adjusted.

Flow measurements were made by means of an orificemeter (B.S.S. 1042) with D and D/2 tappings. Orifice plates with different plate to pipe diameter ratios were used for different desired flow rates. A copper-constantan thermocouple was inserted into the pipe measure the air temperature.

Air from the calming tank was led vertically downward through a one meter long pipe until it reached the 180° bend. Thus the air just before the entrance of the bend had a fully developed velocity profile. The inlet air temperature was measured by means of a copper-constantan thermocouple inserted into the pipe.

The test section was insulated thermally and electrically at both ends as shown in Fig. 3.

The experimental pipe, which was made of steel, had a 36 mm inside diameter, 40 mm outside diameter and 1.70 m. long. Heat was generated within the wall of the pipe by passing alternating current through it. The potential was applied between the flanges welded to the ends of the pipe by a low voltage supply from a 20 KVA 380/5-20 V single-phase tapped transformer. The condition of uniform heat flux was thereby imposed, and variations in the value of the local heat transfer coefficient were recorded as variations in the temperature of the pipe wall. This was measured at a number of stations by copper constantan thermocouples attached to the external surface of the pipe at the positions shown in the Table. Because of the expected large variation of heat transfer coefficient at the inlet, a large number of thermocouples at the inlet were used.

In the Table, the primed numbers denote thermocouples on the side of the pipe following the outside surface of the bend and the numbers without prime denote thermocouples on the opposite side. The position of each thermocouple is identified by the ratio of its distance x from the start of the test section to the diameter D of the pipe. Since the temperature of the pipe wall was expected to vary around the periphery, at each station some thermocouples were placed around the periphery of tube as well.

The rate of heat generation within the tube wall was determined from the measurements of the current by means of a current transformer and of potential difference between the two ends of the test section. The estimated error in the power input measurements was less than 1.2 percent. The test section was insulated against heat losses to the environment. This was achieved by the use of 35 mm asbestos. The whole pipe was then placed in a wooden box and the empty space of 40 mm is filled with glass wool. The bend and the straight pipe preceeding the bend were also insulated to prevent the loss of any heat, that was axially conducted from the experimental pipe to bend, to the environment.

The four bends which were used for tests were carefully constructed from straight pipe and were nearly perfect. The wall thickness, however, was substantially modified by the bending process.

A mixer was attached to the end of the test section to measure exit bulk temperature of the air. For intermediate stations along the pipe, the bulk mean temperature of the air was calculated from the measured inlet temperature, the rate of flow of air, and the measured heat input up to the station in question; corrections being made for heat losses.

3. EVALUATION OF THE HEAT TRANSFER COEFFICIENT

The local heat transfer coefficient is defined in terms of the local heat flux and the difference of wall and fluid temperatures:

$$h_x = \frac{q_{fx}}{T_{wx} - T_{bx}} \tag{1}$$

It varies by the distance from the inlet (x) and by the peripheral location. At a certain distance the following three heat transfer coefficients are of special importance: at inside with respect to the curvature of the bend (h_i), at outside with respect to the same curvature (h_0) and the peripherally averaged (or local average) heat transfer coefficient,

$$h_x = \frac{1}{2\pi} \int_0^{2\pi} h \, d\varphi \approx (h_i + h_0)/2$$

Each of these heat transfer coefficients leads to a Nusselt number according to the relation Nu = hD/k.

Similar to h, the Nusselt number may be subscribed by i, o, and x denoting local at inside with respect to bend, local at outside with respect to bend and local average at a distance x from the entrance.

The evaluation of heat transfer coefficients necessitates the measurement of heat flux to the fluid, wall and bulk temperatures. The bulk mean temperature T_{bx} at any point along the pipe is calculated from the measured inlet temperature T_{ii}, mass flow rate W and the total heat input to the third (Q_{fx}), proceeding the point in question according to the relation

$$T_{bx} = T_{ii} + \frac{Q_{fx}}{W\,C_p}$$ (2)

The small heat amount which flows to the fluid through the unheated bend which touches the heated tube is neglected. It is not practical to measure the wall temperature at the inside surface of the wall directly. First the outside wall surface temperatures are measured by means of thermocouples, then from the equation of temperature distribution in the tube wall, the inside wall surface temperatures are calculated. For this purpose the steady-state energy equation in cylindrical coordinates with constant properties, uniform heat generation and subject to the adiabatic boundary condition at outside surface is solved. It gives a relation for the temperature difference of inside and outside surfaces, $(T_{wx} - T_{0x})$, in terms of inside and outside radii (r_i, r_0) and q which is the generated heat amount per unit time and per unit volume of the tube.

$$T_{wx} - T_{0x} = \frac{q\,r_0^2}{4\,k}\left[1 - \left(\frac{r_i}{r_0}\right)^2 + 2\,\ln\frac{r_i}{r_0}\right]$$ (3a)

or

$$T_{wx} - T_{0x} = \frac{Q_t}{4\,k\,\pi L\,(r_0^2 - r_i^2)}\left[1 - \left(\frac{r_i}{r_0}\right)^2 + 2\,\ln\frac{r_i}{r_0}\right]$$ (3b)

where T_w is the inside wall temperature and Q_t is the total heat input over the length of the tube. The power input to the test section and outside wall temperature of the tube are measured, and therefore T_w can be calculated.

For the exact determination of the heat flux to the fluid (q_f) some corrections must be applied to the nominal heat generated in the tube material.

$$q_n = V.I/2\,\pi\,rl$$

These corrections are due to:

q_L : Radial heat loss through the insulation

q_a : Axial heat loss by conduction

q_p : Peripheral heat loss by conduction

q_u : Reduction in the nominally generated heat by nonuniformity of wall thickness. The desired relation for determination of qf is obtained from energy balance applied to an area element (ds) of the tube wall in the following way:

$$ds.q_n - ds.q_u = ds.q_f + ds.q_a + ds.q_p + q_L \quad ,$$

$$q_f = \frac{V.I}{2\pi r l} - q_a - q_p - (q_u + q_L). \tag{4}$$

The heat loss by conduction in the axial direction is calculated according to the following relation :

$$q_a = - k \frac{\partial^2 T}{\partial x^2} \simeq - k \frac{T_{n+1} + T_{n-1} - 2 T_n}{(\Delta x)^2} \quad . \tag{5}$$

The second relation is the finite difference form which is suitable to apply to experimentally determined temperature profile. The correction due to heat loss in the peripheral direction is not included in the following results because of missing information about the peripheral variation of temperature. It does not affect the average heat transfer coefficient; but the actual difference between the local inside and outside heat transfer coefficients with respect to bend is probably larger than those given in our results. This correction may be estimated by assuming a sinusoidal temperature variation around the periphery.

The last two corrections together as a function of wall temperature is determined by measuring the electrical heat input required to maintain the tube, which was closed at both ends and filled with glass wool to reduce natural convection and radiation, at steady-state for a given room temperature. Because $q_f \approx 0$, the generated heat amount $(q_n - q_a)$ is equal to $q_u + q_L$ according to eq. (4). If q_u is small, this correction will consist mainly of q_L which is proportional to the difference between tube wall temperature and the room temperature.

The experiments were repeated for three different heat generations and the temperature of each point of the test section is plotted against the corrected heat generation. It was found that the variation of $(q_a + q_L)$ is linear with the local temperature of each axial location.

$$q_u + q_L = C_x + B_x T_x \tag{6}$$

where C_x, B_x are constants and T_x is the temperature, all at distance x from the entrance.

One check of the correctness of this is that $-C_x/B_x$ is constant for all points along the pipe, and according to eq. (6), this constant is the ambient temperature before applying any heat flux.

4. THE RESULTS

The variation of the average local Nusselt number with x/D for the bend with curvature ratio $R/r_i = 5,7$ is plotted in Fig. 4. The succeeding Figs. 5, 6 show the variation of the local average Nusselt number along the pipe for Re numbers 9000 to 55000 with a bend to pipe radius ratio 8,3 and 17,2 respectively.

Fig. 7 shows the entry effect of bends on the average local Nusselt number approximately at Reynolds numbers 55000 and 28500 respectively. Velocity in the direction of flow inside the bend increases as we move from the inside curved surface to the outside of the bend. It is thus expected that heat transfer coefficient along the tube surface following the outside curvature of the bend will be higher. This explains the circumferential variation of heat transfer coefficient around the test section.

The local Nusselt numbers on the inside and outside of the tube with respect to the bend are shown in Figs. 10, 11, 12 and 13. Fig. 14 shows the effect of axial heat conduction in the tube wall. For fully developed turbulent flow in straight smooth tubes, the following relation is recommended by Mc Adams [7].

$$Nu = 0.023 \ (Re)^{0.8} \ (Pr)^{0.4} \tag{7}$$

The values of Nusselt number from eq. (7) are also shown in Figs. 4, 5, 6, for comparison. Flow in a circular tube whose centerline is on the arc of a circle is complicated by the fact that a secondary flow is induced, resulting in higher local conductances on the outside than on the inside of the curve. This problem has not been successfully analyzed analytically, but some experimental data are available.

Seban and Mc Laughlin conducted experiments with water and found that the average peripheral convection can be satisfactorily correlated by the analogy relation:

$$St.Pr^{0,6} = f/2 \tag{8}$$

where the friction coefficient is given by the following equation [8]:

$$\frac{f}{f \ (straight \ tube)} = \left[Re \left(\frac{r_i}{R} \right)^2 \right]^{0.05} for \ Re \left(\frac{r_i}{R} \right)^2 > 6 \tag{9}$$

where

$$f(\text{straight tube}) = 0.079 \, (\text{Re})^{-0.25} \; ; \; 5 \times 10^3 < \text{Re} < 3 \times 10^5 \qquad (10)$$

Nusselt numbers which are calculated from eq. (7) and (8) in connection with eq. (9) and (10) are also plotted in Figs. 4, 5, and 6 for comparison with experimental data obtained on a straight pipe preceded by a bend. Under the fully developed conditions, it can be shown that

$$\frac{\partial T_w}{\partial x} = \frac{\partial T}{\partial x} = \frac{\partial T_b}{\partial x} \qquad (11)$$

This explains that away from the bend the wall temperature and the bulk temperature vary linearly with x . Variation of wall temperature and bulk temperature as a function x/D for two different Reynolds numbers is plotted in Figs. 8 and 9 respectively. Variation of wall temperature and bulk temperature became nearly linear at 12 - 14 diameters from the test section inlet.

5. DISCUSSION OF RESULTS AND CONCLUSIONS

Under the conditions prevailing in this work, the thermal entrance effect is modified as a result of a change in flow conditions at the tube entrance since the bend proceeds the heated tube.

The character of the flow at the outlet of the bend is modified by secondary currents generated within the bend. The initially fully-developed velocity profile is severely distorted at the bend outlet i.e. inlet to the heated tube and there is some additional degree of turbulence in the flow stream. A higher degree of turbulence in the flow stream implies in general higher rates of heat transfer. Hence, heat transfer coefficient is higher than would be obtained in a developing turbulent boundary layer where the turbulence originates from the surface. It would appear that the effect of a distorted velocity profile is mainly revealed as a variation of the local circumferential heat transfer coefficient.

a. The Nusselt number decreases with increasing x/D as it is expected. The value of the Nusselt number became almost constant at x/D = 12 -14.

b. The local Nusselt number in the entrance region of the heated tube depends on the tube length, the ratio of bend radius to tube radius, and the Reynolds number. The value of the local Nusselt number diminishes with increase of the ratio of bend to tube and with decrease of Reynolds number.

c. At a given x/D near the entrance, the value of local Nusselt number increases with increase of Reynolds number.

d. The thermal entry length decreases with increase of Reynolds number.

e. The effect of bend on local Nusselt number is more pronounced in laminar flow conditions.

f. The minimum value of Nu_x occurs on the inside and maximum on the outside of the tube with respect to the bend. Circumferential distribution of Nu_x is nearly symmetrical about the plane of the bend. At high Reynolds numbers, the difference is larger and continues to increase with x/D values whereas at low Reynolds numbers the difference is small and diminishes at high x/D values.

g. The circumferential distribution of tube wall temperatures at several positions down stream of bends are observed. It is also observed that the circumferential temperature distribution is almost symmetrical about the plane of bend, for all four bends investigated.

Acknowledgements.

    The authors gratefully acknowledge the financial supports of the Scientific and Technical Research Council of Turkey, Scientific Affairs Division of NATO. Thanks are due to M. Serter for the additional experiments which he conducted.

NOMENCLATURE

A	Surface area, m^2
B	Constant
C	Constant
C_p	Specific heat at constant pressure, $kcal/kg°C$
D	Inside diameter of tube, m
I	Current, Amp.
L	Length, m
Nu $(= hD/k)$	Nusselt number
Pr $(= \mu Cp/k)$	Prandtl number
Q	Heat Supply per unit time, $kcal/h$
q'	Heat generation per unit volume, per unit time, $kcal/m^3 h$
q	Heat flux, $kcal/hm^2$
R	Radius of bend, m
r	Radius of tube, m
Re $(= \bar{U}D/\nu$	Reynolds number
T	Temperature, C
U	Local axial velocity component, m/s
$\bar{U}$	Average axial velocity, m/s
W	Mass flow rate, kg/h
X $(= x/D)$	Dimensionless distance
x	Axial distance, m
V	Potential difference, Volts
$\alpha = \left(\dfrac{Q/A}{Tw - Tb}\right)$	Heat transfer coefficient, $kcal/hm^2 \ °C$
k	Thermal conductivity of tube material, $kcal/hm \ °C$
k_a	Thermal conductivity of air, $kcal/hm°C$
μ	Dynamic viscosity, kg/hm
ν	Kinematic viscosity, m^2/s
ρ	Density, kg/m^3

The meaning of any other symbols are given in the text as they occur.

Subscripts

a	Axial
b	Mean bulk
i	Inside with respect to bend; inside of the tube
ii	Inlet
l	Loss
x	Local
o	Outside with respect to bend; outside of the tube
t	Total
w	Wall or surface

REFERENCES

[1]   Ede, A.J.; Int. Dev. in Heat Transfer, A.S.M.E., Part III p. 634 (1961).

[2]   Mori, Y. and Nakayama. W.; 1st. Report Int. J. of Heat Mass Transfer, 8 p. 67 - 82 (1965).

[3]   Mori, Y. and Nakayama, W.; 2nd Report, Int. J. of Heat Mass Transfer, p. 37 (1965).

[4]   Mori, Y. and Nakayama, W.; 3rd Report, Int. J. of Heat Mass Transfer, p. 681 (1965).

[5]   Boelter, L., Young, D. and Iverson, H.; Tech. Note nat. adv. Comm. Aeros Wash. No. 1451 (1948).

[6]   Mills, A.F.; J. Mech. Engng Sci. 4(1). 63(1962).

[7]   Mc Adams, W.H.; Heat Transmission 3rd edi., Mc Graw-Hill Book Company, Inc. New York, (1954).

[8]   Ito, H.; Basic Eng., Trans ASME, Ser. D, 81, 123 - 124 (1959).

[9]   Sparrow, E.M. and Siegel, R.; ASME Trans. J. of Heat Transfer, 82, 152 (1960).

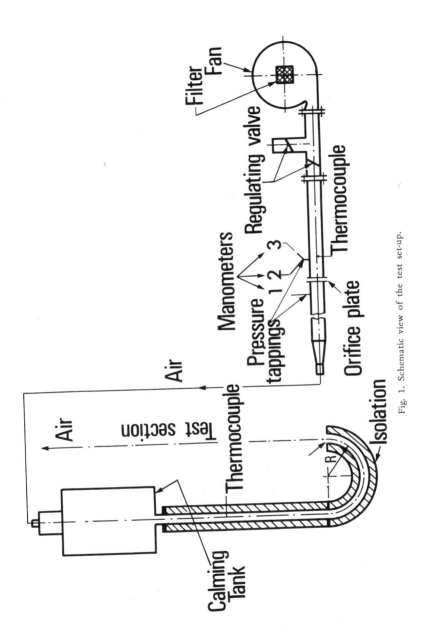

Fig. 1. Schematic view of the test set-up.

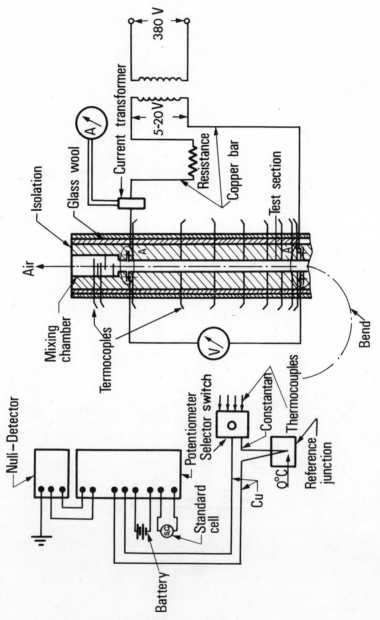

Fig. 2. Schematic view of the test Section and Instrumentation.

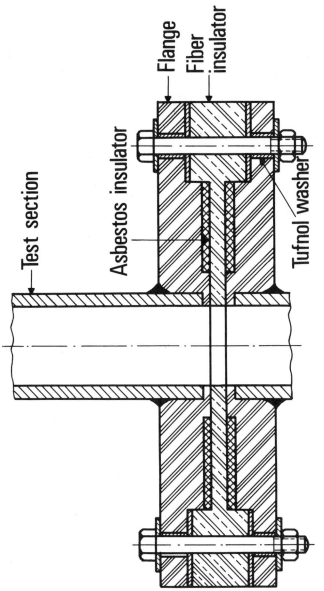

Flange

Fiber insulator

Test section

Asbestos insulator

Tufnol washer

Fig. 3. Connection of the test section to the bend.

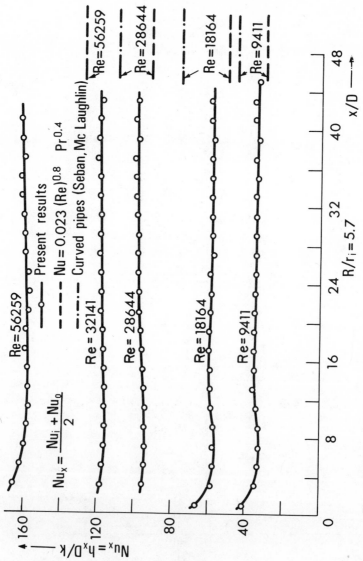

Fig. 4. Variation of local average Nusselt number along the pipe for Re numbers 9000 to 55000 with a ratio of bend to pipe radius 5.7.

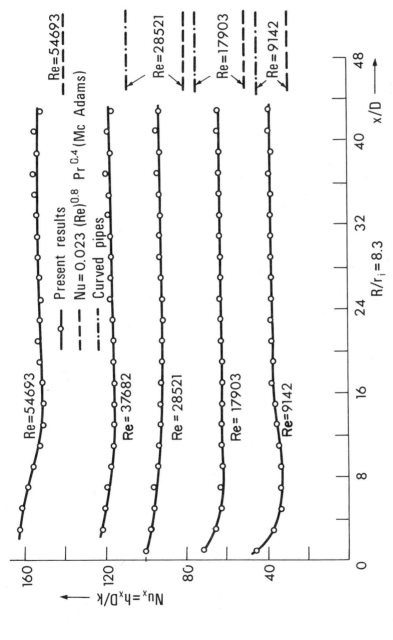

Fig. 5. Variation of local average Nusselt number along the pipe for Re numbers 9000 to 55000 with a bend to pipe radius ratio 17.2.

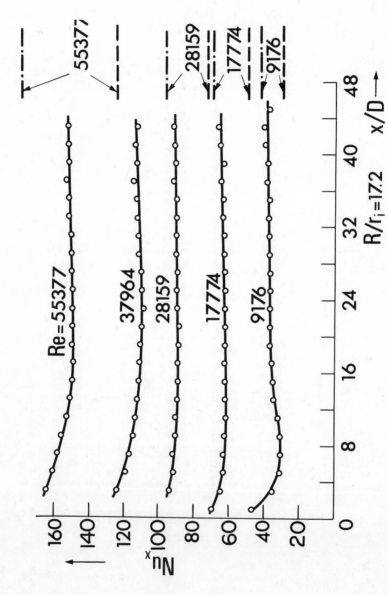

Fig. 6. Variation of local average Nusselt number along the pipe for Re numbers 9000 to 55000 with a bend to pipe radius ratio 8.3.

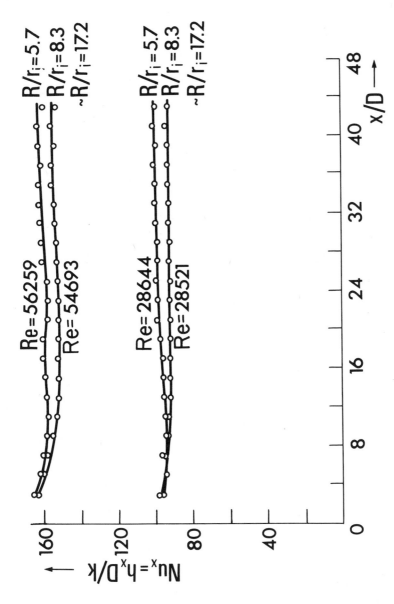

Fig. 7. Comparison of effects of bends on Nusselt number for different bend to pipe radius at different Reynolds numbers.

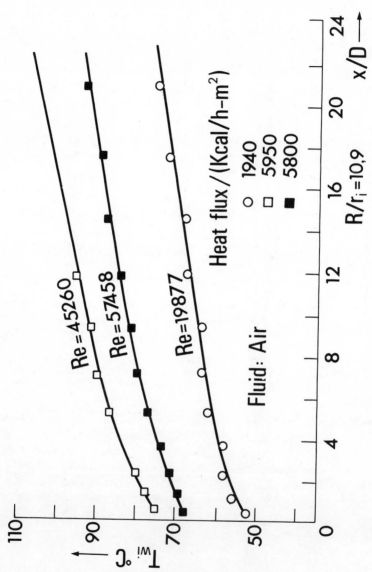

Fig. 8. Variation of wall temperature along the pipe.

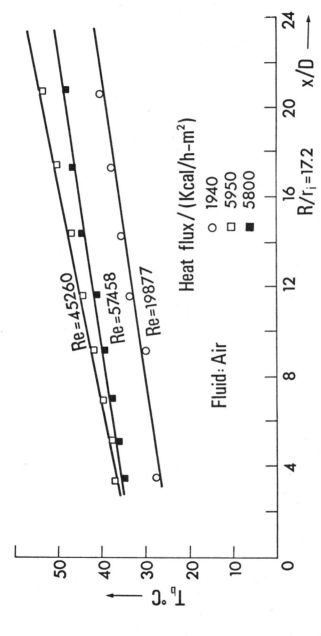

Fig. 9. Variation of bulk temperature along the pipe.

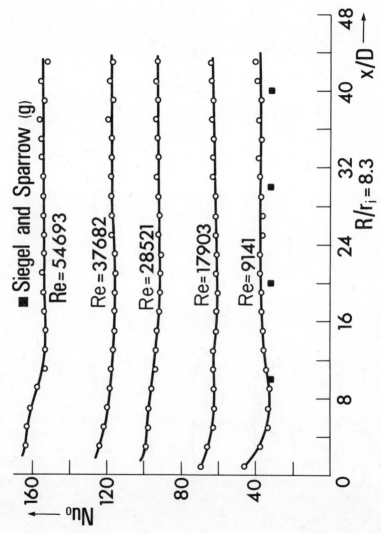

Fig. 10. Variation of outside Nusselt number along the pipe for a ratio of bend to pipe radius 8.3.

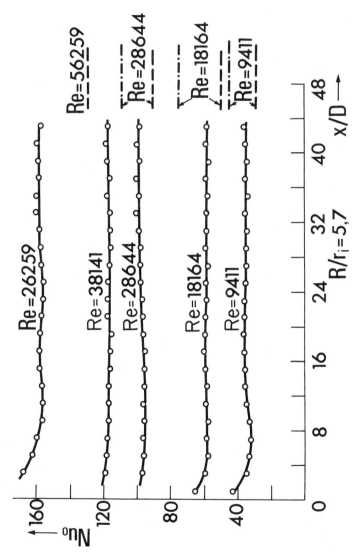

Fig. 11. Variation of outside Nusselt number along the pipe for a ratio of band to pipe radius 5.7.

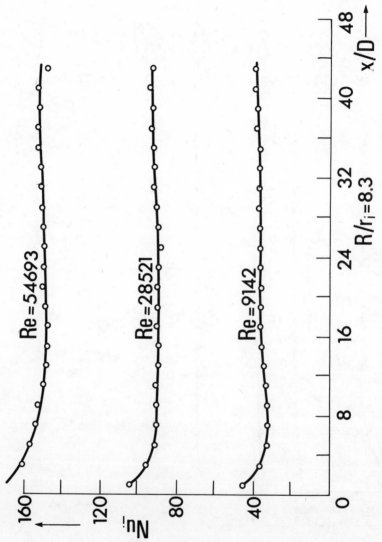

Fig. 12. Variation of inside Nusselt number along the pipe for a ratio of bend to pipe radius 8.3.

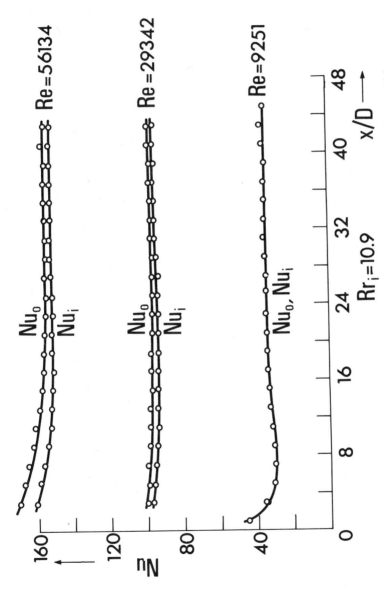

Fig. 13. Comparison of the inside and outside Nusselt number for a bend to pipe radius ratio 10.9.

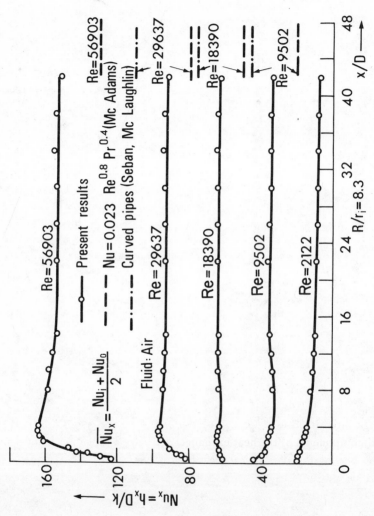

Fig. 14. Influence of axial heat conduction on the calculation of Nusselt number.

Chapter 25

STEADY AND UNSTEADY FLOW PHENOMENA IN AND BEHIND STAGGERED AND IN-LINE TUBE BANKS

E. Heinecke,   C.B. von der Decken(*)

The aim of this chapter is to give an impression of the problems with which engineers are confronted, when trying to construct and to build heat exchangers. As experience has shown it is not sufficient to know, within a certain margin, mean heat transfer and pressure loss coefficients. Numerous defects in high loaded heat exchangers in conventional as well as in nuclear power plants have shown, that detailed knowledge about instationary and stationary flow phenomena is needed, to be sure that unforeseen failure does not occur. Three items are discussed : The excitation of standing acoustical waves in heat exchangers caused by narrow band flow fluctuations, the development of nonuniform velocity profiles especially behind in-line tube banks, and the coupling of mechanical vibrations of the tubes and their wake flow.

In heat exchangers with tubes in cross flow the tubes themselves generate a wake flow with instationary components. Depending on the arrangement of the tubes (staggered or in-line, longitudinal and transverse pitches) and Reynolds number, the spectrum of the velocity and pressure fluctuations may either be of narrow band or broad band (highly turbulent) character. In any case the pressure fluctuations may have two consequences. Firstly : The fluid within the heat exchanger shell has a series of acoustical eigenfrequencies. That means that standing waves of high intensity can be excited by the flow induced pressure fluctuations. In this case all components of the heat exchanger are liable to high mechanical stresses, especially if one of the mechanical eigenfrequencies coincides with an acoustical one. In case of acoustical resonance also the pressure drop coefficient rises above

(*) Kernforschungsanlage Jülich GmbH, Jülich/Germany

N.B. ALL FIGURES QUOTED IN TEXT ARE AT THE END OF THE CHAPTER

normal level.

Secondly: Instationary flow in the wake of a tube means that drag and lift coefficient have time dependent components. The tubes are forced to vibrate mechanically also if acoustical resonance is not observed. Here one has to discern between two possibilities:

Firstly: The displacement of the tubes is so small that there is no coupling between the wake flow and the movement of the tubes (Stiff tubes). Whether mechanical failure will happen or not depends on the damping of the whole structure. Secondly : The ratio amplitude/diameter reaches a value, that a feed back coupling between the movement of the tubes and their wake flow occurs (aeroelastic effects). It needs no detailed explanation that all the above mentioned effects are the more pronounced, the more the flow fluctuations are of narrow band character. It has been the aim of many model tests to find out if any predictions can be made to help engineers to construct heat exchangers with minimum risk.

Several techniques were used to get an insight into the instationary flow phenomena. At Re-numbers $< 10^3$ flow visualisation tests with water as fluid were done, in the Re-number region $10^4 - 10^5$ experiments in air with hot wires and microphones as pickup were performed. The test rigs and measuring techniques are described in [1].

Figure 1 shows a staggered tube bundle [Re = 500]. The vortices in the wake of the tubes develop almost synchronously at the same sides of the tubes. There is obviously a strong coupling of the instationary wake flow between neighboring tubes.

The figure shows that the wake flow is still laminar. Now it would be naive to expect, that for higher Reynolds numbers the instationary phenomena would be as regular in character as in the laminar flow regime.

In an atmospherical windtunnel (cross section of test area : 0.5×0.5 m) tube banks have been investigated ($10^4 \leqslant Re \leqslant 10^5$). Microphones and hot wires were used to measure the spectral distribution of flow fluctuations. Figure 2 shows the spectral distribution of the pressure fluctuations measured with a microphone which was flush mounted to the wall of a heat exchanger between the first and second row. It is obvious that most of the energy is concentrated at a very small frequency band. (Similar curves are found if the velocity fluctuations are measured with hot wires.) Experiments have shown that for a great number of staggered arrangements such spectral distributions exist. Further downstream the spectral distribution becomes broader. However one dominating frequency can always be detected. The frequency of instationary fluctuations is usually described by the Strouhal number $Str = F \cdot D/u$. Taking that frequency at which the energy shows a peak one can define — together with tube diameter D and the mean velocity u between neighbouring tubes of one row — a Strouhal number for tube banks. The 3rd figure shows Str as a function of Re-number and longitudinal pitch. The transverse pitch is 1.86.

Obviously the longitudinal pitch has the greatest influence on the dimensionless frequency. For other transverse pitches (1.36, 2.06, 2.61) similar results were obtained. Detailed investigations on the distribution of the static pressure around the tubes and on the angle of separation of the flow explain the variation of Strouhal number with Re-number and geometrical array [1]. Now experience has shown that in tube banks acoustical resonance may happen. The flow induced pressure fluctuations may excite standing waves inside the heat exchanger shell, which as any gasfilled volume has a series of acoustical eigenfrequencies. By equating these eigenfrequencies with the frequency of the flow induced fluctuations one gets a limiting condition for the excitation of standing acoustical waves. This condition is discribed by the straight lines with 45° slope in figure 3. The first line, situated most left in the diagram, gives the condition, that the first acoustical eigenfrequency is excited. The next ones correspond to higher modes. The lines are calculated for a tube diameter of 25 mm, a velocity of sound of 345 m/s and a square test section 0.5 X 0.5 m. In figure 4 the equations are given.

Experiments with in-line arrangements showed that instationary (and stationary) flow phenomena differ in many cases from those observed in staggered tube banks. Whereas staggered arrays with small longitudinal pitches used to produce spectral distributions with well marked peaks at one frequency, in-line arrangements of comparable pitches behave quite adversely. For longitudinal pitches <2.0 the pressure fluctuations proved to be statistical in character. Figure 5 shows a typical spectrogram for in-line tube banks. The clear consequence is, that the definition of a Strouhal number for such arrays is no longer possible. That means that, contrary to staggered tube banks, no reliable prediction can be given, whether in a tube bank acoustical resonance can be expected or not. Only if the longitudinal pitch is > 2.0, the instationary flow fluctuations become regular. (Figure 6). For these arrangements the Strouhal number was found to be 0.17, a which in good agreement with results published by Borges [2]. Flow visualisation test at Re-numbers $< 10^3$ confirm the results which were gained with air as fluid.

During the investigations on the instationary flow fluctuations in in-line arrangements another striking effect attracted our attention : Tube banks with longitudinal pitches < 2.0 tend to develop velocity profiles behind the last row which are extremely nonuniform (See also [3], [4]). Figure 7 gives an example for such a velocity distribution measured 170 mm behind the last row of a tube bundle 10 rows in depth. A linearized hot wire bridge was used.

Especially in reactor technology in-line heat exchangers with very small longitudinal pitches are preferred to minimize costs. It must be feared now that in the interspaces between succeeding parts of a heat exchanger such velocity profiles develop and as a consequence may lead to high mechanical and thermal stresses.

It showed however, that it is possible by simple means to avoid the development of nonuniform velocity profiles. If the first three rows of a tube bank have a longitudinal pitch of 2.0 or greater — the following part having a longitudinal pitch $< 2.0 -$ than the velocity distribution behind the last row becomes almost uniform. The effectiveness of such a mixed array of tubes has been investigated for tube banks with 1-21 rows of tubes.

In further experiments we are trying to find a reasonable explanation for these effects. Possibly in-line arrangements with mixed longitudinal pitches have a lower pressure drop coefficient, because the equalization of a nonuniform velocity profile is accompanied by loss of energy.

At last an example of aeroelastic effects is given. In a high pressure windtunnel [5], an in-line test bundle consisting of artificially roughened copper tubes ($S_q/D = 1.7$; $S_l/D = 1.3$; length of the tubes 0.9 m; diameter 25 mm) had to be tested (figure 8). Preliminary investigations with hot wires and microphones at low pressures had given the result, that the flow fluctuations had a spectral distribution resembling the one of figure 5. So the danger of the excitation of mechanical or acoustical resonance seemed to be low.

The heat transfer tests were started at a pressure of 40 bars, with air as fluid. Seven of the tubes were destroyed within 15 minutes. Subsequent experiments with accelerometers, which were fixed to tubes, gave the result, that the tubes always vibrate in one of their mechanical eigenfrequencies with a very low amplitude, if the dynamic pressure is low enough. If the aerodynamic forces exceed a value, the exact high of which is not yet known in this case, the tubes vibrate with amplitudes in the order of 15 mm. It is surmised that a coupling of wake flow and tube vibrations similar to the flow around a single circular cylinder, occurs. Further experiments are needed to get more information on this obviously intricate problem.

REFERENCES

[1] E. Heinecke : Strömungstechnische und aeroakustische Erscheinungen in Zylindergittern. Diss. T.U. Berlin (Dezember 1971).

[2] A.R.J. Borges : Vortex shedding frequencies of the flow through two – row banks of tubes.

[3] H.G. Groehn and F. Scholz : Investigations on steam generator models of in-line tube arrangements in pressurized air and helium.
Symposium on component design for high temperature reactors using heliums as a coolant. May 1972, London.

[4] J.G. Edler von Bohl : Das Verhalten paralleler Luftstrahlen. Ing. Arch. Band XI, 1940.

[5] K. Hammeke, E. Heinecke and F. Scholz : Wärmeübergangs– und Druckverlustmessungen an querangeströmten Glattrohrbündeln, insbesondere bei hohen Reynoldszahlen. Int. J. Heat Mass Transfer Vol. 10, pp. 427 - 446, (1967).

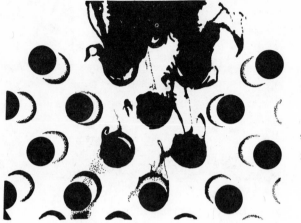

$s_q/D = 2.61$

$S_L/D = 1.5$

$Re = 500$

Fig. 1.

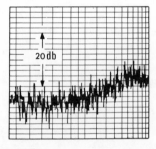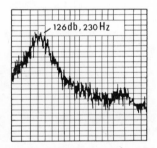

PICKUP: CONDENSER MICROPHONE

$s_q/D = 1.86$ $S_L/D = 1.5$

$D = 25$

staggered arrangement

Fig. 2.

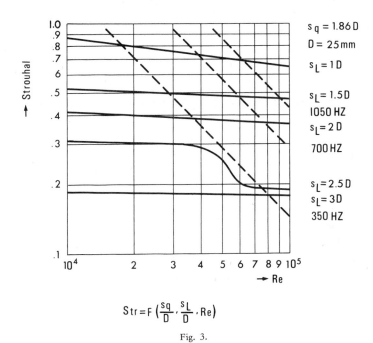

$$Str = F\left(\frac{s_q}{D}, \frac{s_L}{D}, Re\right)$$

Fig. 3.

$$Str = \text{Strouhal number}$$

1) $f_{Str} = \dfrac{Str\,U}{D}$

U = gap velocity

D = tube diameter

a = velocity of sound

2) $f_E = a \cdot F_n$

F_n = geometry factor

Resonance will occure if:

3) $f_{Str} = a \cdot F_n$

4) $Str = \dfrac{1}{Re} \cdot \dfrac{a \cdot D^2}{\nu} \cdot F_n$

Fig. 4.

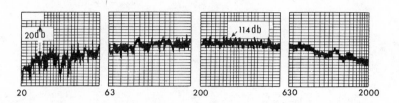

$$s_q/D = 1{,}5 \qquad s_L/D = 1{,}5$$

Pickup: Condenser Microphone

Fig. 5.

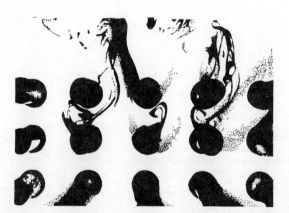

$$s_q/D = 2{.}61$$
$$s_L/D = 2{.}00$$
$$Re \;\; = 750$$

Fig. 6.

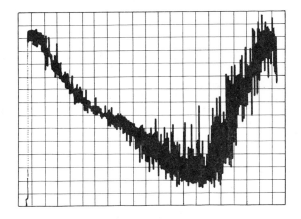

$D = 25\,mm \qquad s_q/D = 1.5 \qquad s_L/D = 1.25 \qquad Re = 4.2 \cdot 10^4$

Fig. 7

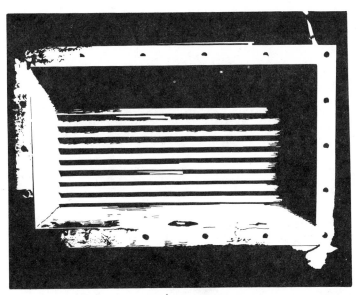

$S_g/D = 1.72 \qquad S_L/D = 1.3$

Fig. 8.

Chapter 26

SIMULTANEOUS HEAT AND MASS TRANSFER IN REGENERATORS

W.H. Granville and G. Bird (*)

INTRODUCTION

Many cryogenic chemical processes involve the purification of a high pressure gas stream before it enters the cryogenic separation process. In addition, it is usually a requirement to have efficient heat exchange between the warm high pressure feed gas and the cold products from the separation process. The removal of impurities from the feed stream may be necessary either to achieve high purity products from the separation process and/or to prevent blockages in the equipment at the cryogenic temperatures. In some cases it may be permissible to recontaminate one or more of the separation products at the higher temperature levels where blockages are no longer a problem. Such processes normally involve passing the high pressure feed gas stream through purification and heat exchange equipment.

This chapter describes a process where the purification and heat exchange are achieved in a single stage, that is, in a regenerator unit packed with adsorbent. From the point of view of heat transfer the adsorbent stores and releases heat in the usual manner but there are also some additional heat effects due to absorption and desorption. As far as mass transfer is concerned the unit acts as a short cycle pressure swing gas adsorption unit where the impurity is adsorbed at a high pressure and desorbed at a low pressure in the presense of the revert gas which may be the whole or part of the purified gas stream. This process differs from a conventional short cycle pressure swing adsorption unit using part of its purified product as a revert stream in that the adsorption beds have large longitudinal temperature gradients.

Figure 1 illustrates the essential features of the process. The hot gas at a flow rate G_A is cooled from a constant temperature T_{hi} to a mean outlet

(*) Postgraduate School of Studies in Chemical Engineering, University of Bradford England.

N.B. ALL FIGURES QUOTED IN TEXT ARE AT THE END OF THE CHAPTER

temperature T_{hom} whilst the partial pressure of the impurity is reduced from a value P_{Ai} to a mean outlet value P_{Aom}. Low pressure revert gas at a flow rate G_D enters the second regenerator at a constant temperature T_{ci} and is warmed to the mean outlet temperature T_{com}, whilst the partial pressure of the impurity is increased from the mean value p_{Dim} to the outlet mean value p_{Dom}. The difference between streams G_A and G_D will result in some other flows such as G_1 and G_2 and these may be warmed in the regenerators by using coils or tubes. After a given period time automatic changeover values reverse the gas flows through the regenerators in the normal way. The cryogenic process indicates in Figure 1 may be a gas separation plant, a closed cycle cryogenic refrigerator, a liquefaction plant, or a cooler to improve the performance of a conventional short cycle pressure swing adsorption unit.

The process is similar to that in which the impurities are frozen on a conventional regenerator packing but this process has the additional advantage that some purification is achieved at the higher temperature levels which occur during the start-up period of a cryogenic process.

DESCRIPTION OF PILOT PLANT

The process was simulated in the laboratory using a feed gas consisting of high pressure helium with nitrogen as the impurity. Purified helium leaving the regenerator was cooled in a liquid nitrogen cooled heat exchanger and returned at a lower pressure to the second regenerator. In some of the experiments part of the purified gas was removed as a product and the remainder used as revert gas for desorption and heat exchange. The regenerators did not have heat exchange coils or tubes to permit the additional heat transfer between this product stream and the feed stream.

Figure 2 illustrates the essential features of the pilot plant. Feed gas of the required composition was stored in the storage vessels (16, 17, 18). This gas was compressed using a diaphragm compressor (1) and fed to the top of a regenerator (3) via a constant temperature water bath (10). The gas flow rate and pressure were controlled. Solenoid valves and a timing device allowed the gas to pass alternately down through each regenerator in the correct manner at a predetermined cycle time. Purified gas leaving the regenerator passed through a throttling valve (6) and the liquid nitrogen cooled heat exchanger (5) before entering the second regenerator.

When operating on partial revert the product gas was removed immediately before the throttling valve. The revert and product streams passed through rotameters (14, 15) before being recycled to the storage vessels (16, 17, 18). The regenerators, throttle valve and liquid nitrogen cooled heat exchanger were enclosed in a vacuum vessel (21) which was maintained at a vacuum of 10^{-7} Torr by a mechanical vacuum pump (22) and an oil vapor diffusion pump (23). A liquid nitrogen transfer pump (25) maintained the level of liquid nitrogen in the cooler at a controlled level. Temperatures at the ends of the regenerators (T_1 T_2 T_3 T_4) were measured with copper-constantan thermocouples and recorded whilst the temperatures of the product (T_6) and revert (T_5) streams were measured with nickel resistance thermometers and recorded. Katharometers continously sampled, analysed and recorded the feed (A3), purified product (A2) and final outlet revert (A1) compositions.

The regenerators were made from thin wall cupro-nickel tubing and had brass flanges. They were 0.025 m in diameter and were packed to a depth of 0.268 m with 0.063 kg of granular activated carbon. These granules had an equivalent particle diameter of 0.00113 m and the total external surface area of the granules in each bed was 0.193m^2.

EXPERIMENTAL WORK AND RESULTS

Adsorption Isotherms

Some 60 adsorption measurements were carried out on a B.E.T. apparatus to obtain the adsorption isotherms for the activated carbon at 78°k, 195°k and 273°k. A summary of these data is given in Table 1.

TABLE 1 — Isotherm Data

Equilibrium Pressure mm Mercury	Volume of Nitrogen Adsorbed cm{(STP)}^{3} per gm solid		
	at 78° K	at 195° K	at 273° K
10	254	8.3	0.3
50	298	17.5	1.3
100	321	25.0	2.0
300	342	44.2	6.0
500	352	56.5	9.2
760	368	68.2	12.4

Pilot Plant Experiments.

The ranges in the various experimental conditions are given in Table 2.

Table 2. — Experimental Conditions

Feed Concentration	2 to 8 mole %
Adsorption Pressure	4×10^5 to 15×10^5 Nm^{-2}
Desorption Pressure	1.36×10^5 Nm^{-2}
Feed Gas Flow Rate	0.2 to 0.8 kgm^{-2} s^{-1}
Revert Ratio	0.6 to 1.0
Period Time	4 to 60 s
Product Concentration	0.005 to 4.0 mole %
Revert Concentration (outlet)	2 to 8 mole %
Hot Gas Inlet Temperature	$300°K$
Hot Gas Outlet Temperature	$100°K$ to $160°K$
Cold Gas Inlet Temperature	$90°K$ to $110°K$
Cold Gas Outlet Temperature	$260°K$ to $300°K$
Reynolds Number	10 to 70

Experimental Efficiencies

The hot heat transfer efficiencies were expressed in the form :

$$E_H = \frac{T_{hi} - T_{hom}}{T_{hi} - T_{ci}}$$

and the adsorption mass transfer efficiencies as :

$$E_A = \frac{P_{Ai} - P_{Aom}}{P_{Ai} - P_{Dim}}$$

Figures 3, 4, 5, 6, and 7 illustrate the heat transfer and mass transfer efficiency data obtained and show that the general trends in the data were :

(a) heat transfer efficiencies decreased with increasing period times as would be expected.

(b) heat transfer efficiencies decreased with increasing flow rates at the longer period times as expected, but at the shorter period times there appeared to be little effect of flow rate on heat transfer efficiencies and it should be noted that whilst this is not generally predicted there have been similar supporting data presented by Gifford et al. [3], Dermott [2] and Bretherton et al. [1] who have carried out experimental work on short period small cryogenic regenerators.

(c) mass transfer efficiencies decreased with increasing period time, the rate of decrease becoming very rapid after a particular period time which varied with operating conditions.

(d) mass transfer efficiencies decreased with increasing flow rate except at very low period times when all efficiencies were very high.

(e) mass transfer efficiencies at partial revert were lower than those at total revert as expected.

Figure 8 illustrates a peculiar effect of pressure on mass transfer efficiencies. Below pressures of 11×10^5 Nm^{-2} the mass transfer efficiencies increased with increasing pressure as would be expected. Above this pressure level there was a fall-off in efficiency. Unfortunately, it was not possible to work at higher pressures to determine whether the fall-off was localised as may be possibly deduced from figure 8. Similar pressure swing adsorption runs at ambient temperature gave the same result but later adsorption work with longer beds did not show this peculiar effect. It is thought that the fall-off with short beds may have been due to hold-up effects and to the rapid sweeping of feed gas down and out of the bed during the instantaneous changeover of the valves. It is possible that this effect would not have been noticed if the outlet valve on the regenerator had been kept closed until after pressurisation had been achieved.

SIMPLIFIED MODEL

if it is assumed that:

(a) there is plug flow of the gas through the bed.
(b) there is no heat transfer to the walls of the regenerator.
(c) there is no "hold-up" effect of the gas in the voids.
(d) there are no longitudinal heat and mass transfer effects.
(e) that only low concentrations of impurities are present so that mole fractions can be taken equal to mole ratios and that the thermal effects of adsorption are small relative to the overall heat transfer between the gas streams,

then the partial differential equations for heat and mass transfer are analogous.

During the period in which the bed is being heated and the impurity absorbed the following equations may be written for a length and time increment $d\theta$:

$$h \, (adz) \, (T - t_m) = (mdz) \, C_s \frac{\partial t_m}{\partial \theta} = - G \, Cg \left(\frac{\partial T}{\partial z} \right) \, . \, dz \quad (1)$$

$$\begin{array}{ccc}
\text{rate of heat transfer} & = & \text{rate of heat accumulation} & = & \text{rate of heat transfer} \\
\text{from gas to solid} & & \text{in solid} & & \text{from gas to solid}
\end{array}$$

$$k \, (adz) \, (p - p_m^*) = \left(\frac{mdz}{k_i} \right) \frac{\partial p_m^*}{\partial \theta} = - \frac{G}{p} \left(\frac{\partial p}{\partial z} \right) \, . \, dz \quad (2)$$

$$\begin{array}{ccc}
\text{rate of mass transfer} & = & \text{rate of mass accumulation} & = & \text{rate of heat loss} \\
\text{from gas to solid} & & \text{in solid} & & \text{from gas}
\end{array}$$

Similar equations may be written for the period during which the bed is being cooled and the impurity desobed.

It follows from these equations that the hot heat transfer and adsorption efficiencies are given by :

$$E_H = f \, (\Lambda_H, \, \Lambda_C, \, \Pi_H, \, \Pi_{C^{\circ}}) \quad (3)$$

$$E_A = f' (\Lambda_A, \, \Lambda_D, \, \Pi_A, \, \Pi_D) \quad (4)$$

where Λ = dimensionless length or reduced length

Π = dimensionless period time or reduced period

subscripts H and C refer to the heating and cooling periods (heat transfer and the subscripts A and D to the adsorption and desorption periods (mass transfer).

$$\Lambda_{H \text{ or } C} = \left(\frac{hA}{GC_g} \right)_{H \text{ or } C} \, , \quad \Pi_{H \text{ or } C} = \left(\frac{hA\theta}{MC_p} \right)_{H \text{ or } C}$$

$$\Lambda_{A \text{ or } D} = \left(\frac{kAP}{G} \right)_{A \text{ or } D} \, , \quad \Pi_{A \text{ or } D} = \left(\frac{k A K_i \theta}{M} \right)_{A \text{ or } D}$$

In these expression h and k are the overall heat and mass transfer coefficients for transfer of heat and mass from the gas into the solid and are given by the equations :

$$\frac{1}{h} = \frac{1}{h_g} + \frac{1}{h_s}$$

$$\frac{1}{k} = \frac{1}{k_g} + \frac{1}{k_s}$$

If the boundary conditions are taken the same, that is, if T_{hi}, T_{ci}, p_{Ai} and p_{Dim} are all taken as constant then the solutions to equations (3) and (4) become identical and $f = f'$. Methods of solution for equation (3) for this condition of constant inlet conditions and constant reduced parameters, have been reported by several workers such as Hausen [5], Iliffe [6] and Willmott [10] and others.

In this particular process the partial pressure of the impurity in the purified gas leaving the bottom of the one regenerator, p_{AD}, varies and hence the partial pressure entering the bottom of the second regenerator, p_{Di}, also varies. The method of solution proposed by Willmott for equation (3) may be used to introduce this varying inlet boundary condition in the solution of equation (4) for mass transfer. For the particular experimental conditions reported in this paper there is, however, little effect of taking p_{Di} as a constant and equal to p_{Dim}. This is shown in Table 3. In this table the degree of unbalance is given by β, where:

$$\beta = \frac{\Lambda_A}{\Lambda_D} \cdot \frac{\Pi_D}{\Pi_A} = \frac{p_A G_D}{p_D G_A}$$

Furthermore in this table $\Pi_A = \Pi_D$ because this condition describes the process.

COMPARISON OF PREDICTED AND EXPERIMENTAL EFFICIENCIES

Heat Transfer Efficiencies

The Reynolds number range was 10 to 70 and the gas film heat transfer coefficient, h_g, was evaluated according to the data of Littmann et.al. [7]. For the computer program this data was represented by the equation :

$$Nu = 0.36 \, Re^{0.98} \, Pr^{0.68}$$

Solid phase heat transfer coefficients, h_s, were evaluated according to the proposed method of Hausen [5]. This method is based on linear longitudinal temperature gradients and should give reasonable values when Λ/Π is large and when $(G \, C_g)_H = (G \, C_g)_C$. In the case of spheres this method gives :

$$Nu = \frac{1}{\phi}$$

where $\quad \phi = 0.1 - 0.00143 \left(\dfrac{d_p^2}{a\theta} \right)$, for $\left(\dfrac{d_p^2}{a\theta} \right) \leq 20$

and $\quad \phi = \dfrac{0.357}{\left(3 + \dfrac{d_p^2}{a\theta} \right)^{0.5}}$, for $\quad \dfrac{d_p^2}{a\theta} > 20$

TABLE 3 — Comparison of Predicted Adsorption Efficiencies

Λ_A	Λ_D	$\Pi_A, \ \Pi_D$	β	E_A^* (Variable p_{Di})	E_A^* (p_{Di} taken as constant $= p_{Dim}$)
10	5	1	2	0.955	0.949
8	4	1	2	0.925	0.923
1	0.5	1	2	0.345	0.349
10	1.67	1	6	0.986	0.985
1	0.17	1	6	0.365	0.367
16	4.0	2	4	0.998	0.999
8	2.0	2	4	0.953	0.964
1	0.25	2	4	0.316	0.320

* Values computed using method similar to Willmott's heat transfer method.

† Values computed using method similar to Iliffe.

Applying these equations it was found that the heat transfer was gas film controlled. All properties were taken at the mean temperatures of the beds.

Figure 9 shows the ratio of predicted to experimental heat transfer efficiencies for the heating period. It will be noticed that reasonable agreement is reached when Λ_H/Π_H is greater than unity. In these experiments the heat effects due to adsorption were about 10% of the heat transferred and Figure 10 shows how these results compare with data of some other workers when there have been no thermal effects due to adsorption.

Mass Transfer Efficiencies

The correlation of Gupta and Thodos [4] was used to predict the gas film mass transfer coefficient, k_g :

$$\frac{k_g \, P_{g\,f}}{\overline{G}/M_W} \cdot Sc^{2/3} = \frac{2.06}{e. \; Re^{0.575}}$$

In order to obtain solid phase mass transfer coefficients the analogy was drawn once again with heat transfer, replacing the Nusselt number by the Sherwood number giving :

$$\frac{k_s \, d_p \, RT}{D_{Km}} = \frac{1}{\phi'}$$

where : $\phi' = 0.1 - 0.00143 \dfrac{d_p^2}{De\theta}$,   for $\dfrac{d_p^2}{De\theta} \leq 20$

and: $\phi' = \dfrac{0.357}{\left(3 + \dfrac{d_p^2}{De\theta}\right)^{0.5}}$ for $d_p^2/De\theta > 20$

This expression for k_s is only approximate since in the experimental work β varied from 3 to 11, whereas it should be unity for the equations to be correct. However, Hausen has suggested that in the case of heat transfer little error is introduced for values up to about 4.

The modified Knudsen diffusion coefficient was calculated from the equation :

$$D_{Km} = D_K \cdot \frac{v}{\text{Tortuosity}}$$

and a value of 1.5. was taken for the tortuosity.

The effective diffusivity for the solid was derived on the assumption that the nitrogen was transferred into the particle by Knudsen diffusion through the pores and that all points along the pores there was equilibrium between the solid and the gas in the pores. This model leads to the following expression for D_e for this particular system :

$$D_e = \frac{v D_{Km}}{v + \frac{4 \, p \, \rho_s}{K_i \, \rho_g}}$$

where 4 = molecular weight of helium.

Comparing the values for k_g and k_s the system under consideration was shown to solid phase controlled so that $\Pi_A = \Pi_D$

Figure 11 shows a comparison between predicted and experimental mass transfer efficiencies for that data where the experimental efficiencies were greater than 0.85. This limitation was imposed because of the nature of the experimental mass transfer efficiency curves below this efficiency level where the fall-off in efficiency was very marked. The data shows that better agreement was obtained at larger Λ_A values. In predicting values of E_A it was necessary to estimate mean values for the dimensionless groups. In general properties were taken at the mean operating conditions but where the slopes of the isotherms were involved log-mean values were taken between the values at the ends of the regenerators.

DISCUSSION OF RESULTS

1. The heat transfer data in Figures 9 and 10 indicate that the results are similar to those obtained by other workers. Discrepancies between predicted and experimental efficiencies must be expected because the model does not include for the thermal effects of adsorption, maldistribution of flow, regenerative effects of the walls, hold-up of the gas in the voids in both the formulation and solution of the

partial differential equations and longitudinal conduction. In addition, there are considerable errors in estimating heat transfer coefficients at low Reynolds numbers and an error in considering a granular particle as a sphere.

2. In the case of mass transfer the hold-up effect of the gas in the voids may be more important particularly at high pressure ratios. The model predicts a zero adsorption efficiency at a zero revert gas flow rate whereas the mere reduction of pressure will give rise to some efficiency. In the experimental work there was a factor of 20 between K_i at the ends of the beds and hence depending on the mean value chosen the data in Figure 11 could have been moved nearer or further from unity. However, the data as plotted are consistent with other adsorption data obtained on the same plant at ambient temperature. The value of the product kA used in the predictions may well be questioned. Errors in k arise from the fact that the equations for ϕ' are not strictly correct because β is not unity and there are further errors in the values used for K_i and the tortuosity. The assumption of basing the external surface area A as an equivalent sphere may be unrealistic for an adsorbent when the mass transfer is controlled by diffusion within the solid.

3. The errors due to the product kA may be illustrated further. Efficiency predictions are very sensitive to Λ_A and Λ_D values when these are low. If $\Lambda_A = 2$, $\Lambda_D = 0.33$ and $\Pi_A = \Pi_D = 1$, then the predicted efficiency is about 0.6. If the value kA is multiplied by 2 then $\Lambda_A = 4$, $\Lambda_D = 0.66$, $\Pi_A = \Pi_D = 2$ and the predicted efficiency becomes 0.8 which is an increase of 33%. Increasing the initial kA value by a factor of 4 would give a predicted efficiency of 0.93 which would be an increase of 55% and this would have been within 6% of the experimental efficiency. Thus if all kA values had been increased by a factor of 2 the data in Figure 11 would have been much closer to unity, and if all kA values had been increased by a factor of 4, then Figure 11 would have shown a very good agreement between predicted and experimental efficiencies.

NOMENCLATURE

A	external surface area of particles in each bed
a	external surface area of particles per unit height
C_g	heat capacity of gas
C_s	heat capacity of particles
D_e	effective mass diffusivity
d_p	diameter of particle
D_k	Knudsen diffusion coefficient
D_{Km}	modified Knudsen diffusion coefficient
e	interstitial voidage
E_A	mass transfer efficiency (adsorption)
E_H	heat transfer efficiency (hot period)
G	gas flow rate
$\overline{G}$	gas mass flow rate
h	overall heat transfer coefficient
h_g	gas film heat transfer coefficient
h_s	solid phase heat transfer coefficient
k	overall mass transfer coefficient
k_g	gas film mass transfer coefficient
k_s	solid phase mass transfer coefficient
K_i	slope of isotherm
M	mass of particles in each bed
m	mass of particles per unit height
M_w	molecular weight
P	total pressure
P_{gf}	partial pressure of inert gas
p	partial pressure of transferable component
p^*	equilibrium pressure for isotherm
P_m^*	mean partial pressure of transferable component in particle
P_{Ai}	inlet partial pressure of transferable component during adsorption
P_{Ao}	outlet partial pressure of transferable component during adsorption
P_{Aom}	mean outlet partial pressure of transferable component during adsorption
P_{Di}	inlet partial pressure of transferable component during desorption
P_{Dim}	mean inlet partial pressure of transferable component during desorption

PD_{om}	mean outlet partial pressure of transferable component during desorption
R	universal gas constant
R_r	revert ratio
T	temperature of gas
T_{hi}	inlet temperature of gas during hot period
T_{hom}	mean outlet temperature of gas during hot period
T_{ci}	inlet temperature of gas during cold period
T_{com}	mean outlet temperature of gas during cold period
t_m	mean temperature of particle
v	intraparticle voidage
x	concentration of transferable gas in particle (isotherm)
z	distance along bed
α	thermal diffusivity
β	degree of unbalance
θ	time and period time (period time = 1/2 cycle time)
ρ_g	density of gas
ρ_s	density of particle
ϕ	penetration factor for heat transfer
ϕ'	penetration factor for mass transfer

Dimensionless Groups

Nu	Nusselt number
Re	Reynolds number
Sc	Schmidt number
$\Lambda_{H \text{ or } C}$	reduced length for heat transfer
$\Lambda_{A \text{ or } D}$	reduced length for mass transfer
$\Pi_{H \text{ or } C}$	reduced period for heat transfer
$\Pi_{A \text{ or } D}$	reduced period for mass transfer

Subscripts for symbols and groups

A	denotes absorption period
D	denotes desorption period
C	denotes cold period (bed being cooled)
H	denotes hot period (bed being heated)

ACKNOWLEDGMENTS

The authors wish to express their gratitude to the British Oxygen Company Limited, London, for the award of a B.O.C. Fellowship to carry out this work and for the cooperation received.

The authors have worked closely with Mr. D.I. Ellis of the University of Bradford who has been concerned with pressure swing adsorption and they wish to acknowledge his contribution.

REFERENCES

[1]   Bretherton, A., Granville, W.H., Harness, J.B. ; Advances in Cryogenic Engineering Vol. 16 Plenum Press, New York 1971 p. 333.

[2]   Dermott, T. ; Ph.D. thesis, University of Bradford, U.K. 1972.

[3]   Gifford, W.E., Acharya, A., Ackermann, R.A. ; Advances in Cryogenic Engineering Vol. 14 Plenum Press, New York 1969, p.353.

[4]   Gupta, A., Thodos, G. ; A.I.Chem.E. Journal 9 751 1963.

[5]   Hausen, H., Warmeubertragung in Gegestrom, Gleichstrom and Kreazstrom, Springer-Verlag, Berlin, Germany 1950.

[6]   Iliffe, E.C. ; J. Inst. Mech. Engrs, London **159** 363 1948.

[7]   Littmann, H., Barile, R.G., Pulsifer, A.H. ; Ind. and Eng. Chem. (Fundamentals) 7 554 1968.

[8]   Macdonald, W.I. ; M.Sc. thesis, University of Bradford, U.K. 1966.

[9]   Rizvi, S.N.A. ; M.Sc. thesis, University of Bradford, U.K. 1968.

[10] Willmott, A.J. ; Int. J. Heat and Mass Transfer, 7 1291 1964.

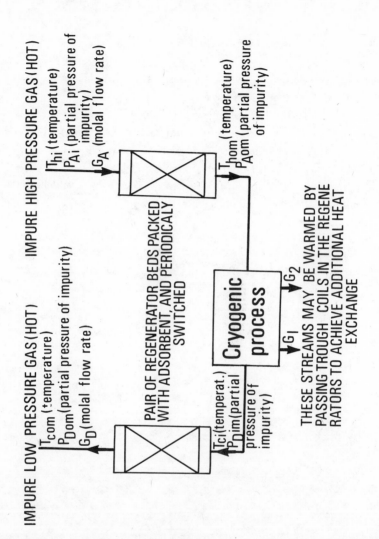

IMPURE LOW PRESSURE GAS (HOT)

IMPURE HIGH PRESSURE GAS (HOT)

T_{hi} (temperature)
P_{Ai} (partial pressure of impurity)
G_A (molal flow rate)

T_{hom} (temperature)
P_{Aom} (partial pressure of impurity)

T_{com} (temperature)
P_{Dom} (partial pressure of impurity)
G_D (molal flow rate)

PAIR OF REGENERATOR BEDS PACKED WITH ADSORBENT, AND PERIODICALLY SWITCHED

Cryogenic process

G_2

G_1

T_{ci} (temperat.)
P_{Dim} (partial pressure of impurity)

THESE STREAMS MAY BE WARMED BY PASSING TROUGH COILS IN THE REGENERATORS TO ACHIEVE ADDITIONAL HEAT EXCHANGE

Fig. 1. Schematic diagram.

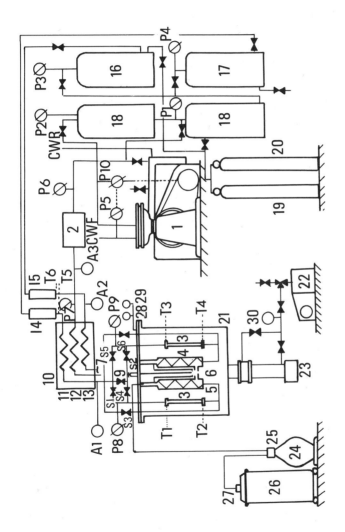

Fig. 2. Flow diagram of helium purification plant.

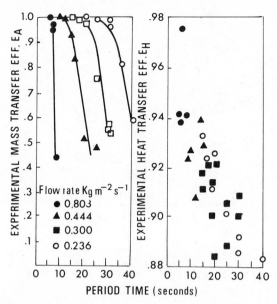

Fig. 3. The effect of period time on the mass and heat transfer efficiencies at various flow rates. (Bed length 0.286 m, adsorption pressure 8.25×20^5 Nm^{-2}, desorption pressure 1.36×10^5 Nm^{-2}, average inlet conc. 8.31% nitrogen in helium.)

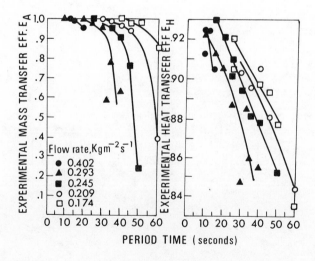

Fig. 4. The effect of period time on the mass and heat transfer efficiencies at various flow rates. (Bed length 0.286 m, adsorption pressure 11.35×10^5 Nm^{-2}, desorption pressure 1.36×10^5 Nm^{-2}, average inlet conc.

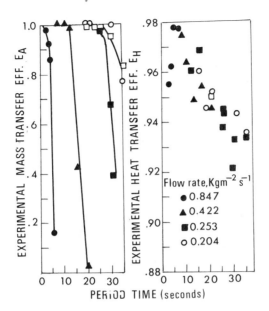

Fig. 5. The effect of period time on the mass and heat transfer efficiencies at various flow rates. (Bed length 0.286 m, adsorption pressure 6.19×10^5 Nm^{-2}, desorption pressure 1.36×10^5 Nm^{-2}, average inlet conc. 4.56% nitrogen in helium.)

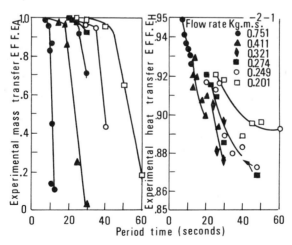

Fig. 6. The effect of period time on the mass and heat transfer efficiencies at various flow rates. (Bed length 0.286 m, adsorption pressure 8.25×10^5 Nm^{-2}, desorption pressure 1.36×10^5 Nm^{-2}, average inlet conc. 4.53% nitrogen in helium.)

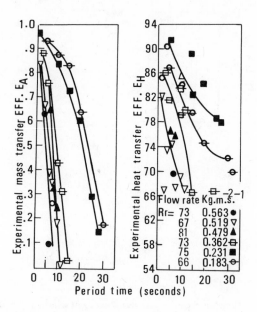

Fig. 7. The effect of period time on the mass and heat transfer efficiencies at various revert ratios and flow rates. (Bed length 0.286, adsorption pressure 8.25×10^5 Nm^{-2}, desorption pressure 1.36×10^5 Nm^{-2}, average inlet conc. 2.0% nitrogen in helium.)

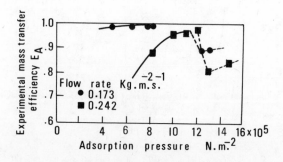

Fig. 8. The effect of adsorption pressure on the mass transfer efficiency at various flow rates. (Period time 30 s., desorption pressure 1.36×10^5 Nm^{-2}, average inlet conc. 4.30% nitrogen in helium.)

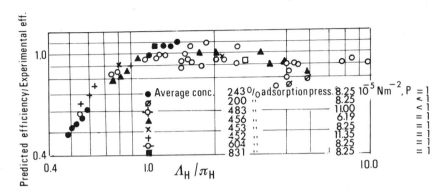

Fig. 9. The relationship between Λ_H/Π_H and the ratio of predicted to experimental efficiencies.

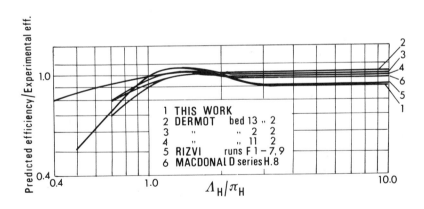

Fig. 10. A comparison of the relationships between Λ_H/Π_H and the ratio of predicted to experimental efficiencies for this and other works.

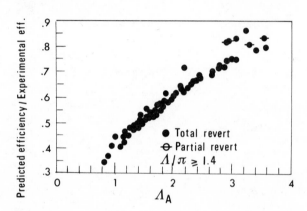

Fig. 11. The relationship between Λ_A and the ratio of predicted to experimental efficiencies for mass transfer.

Chapter 27

MOVING BED HEAT EXCHANGERS AND REGENERATORS—
A SUMMARY OF NEW DESIGN THEORY

J. Schneller, V. Hlavačka (*)

Heat exchange apparatuses in which heat is transferred between solid particles and gas represent an important part of up-to-date technological processes in many branches of industry. Among them are heat exchange systems whose active heat exchange zones are formed by moving beds. These active zones can be designed in different ways as counter-flow, cross-flow or combined flows. One of the fundamental requirements is to ensure the uniform movement of the particles through the system especially in its boundary areas (inlet, outlet), that prevents the formation of dead sections and provides the conditions for a uniform gas passage through the bed.

Some of the basic active heat exchanging zone arrangements, the greater part of which were tested in SVÚSS on dynamic particle movement characteristics, are shown in Fig. 1, laid out like a regenerative exchanger. A new concept is the use of divergent beds of solid particles, either oriented vertically or inclined enabling the designer to achieve several design simplifications, and as it is shown in (1) they even offer somewhat more convenient thermokinetic conditions. The bed extension in the direction of the gas passage contributes to the uniform temperature profile in the gas behind the bed, as is schematically shown in Fig. 2 for the cases of moving beds of constant and variable thickness respectively. From the

(*) National Research Institute for Machine Design, Behovice - Czechoslovakia.
N.B. ALL FIGURES QUOTED IN TEXT ARE AT THE END OF THE CHAPTER

thermodynamic point of view the divergent beds seem to offer greater advantages particularly a higher thermal efficiency in comparison with a constant thickness beds, the hydraulic pressure drops remaining the same. The efficiency increase, noticeable at higher values of $\alpha F/W$, varies within the range from 4 to 8 percent.

Some of the theoretical conclusions were experimentally verified by measuring the particle temperature in the bed. A new method was used based on infrared particle emission sensing in the superficial layer area by means of thermovision. This unconventional approach is part of the effort to eliminate phenomena of a stochastic character which are so characteristic for moving bed. From the experiment Fig. 3 shows the comparison between measured and calculated particle temperature curves in a bed with a oneside linear divergence. The isotherms calculated by means of a digital computer using equations according to (1) for the particular conditions of the experiment represent satisfactorily the measured curves, particularly in the central region unaffected by boundary phenomena in the neighborhood of the particle inlet into and outlet from the active zone. This work will be treated in greater detail in (2).

The contemporary effective calculating methods for simple heat exchangers, i.e. also for heat exchangers with moving beds and for the active zones of moving bed regenerators are based on their thermal efficiency which is a function of a dimensionless parameter $\alpha F/W$ and of the ratio of the gas capacity and moving particles capacity. Analytical expressions or tabular thermal efficiency values for the basic flow arrangements are known from literature. The parameter $\alpha F/W$, essentially a modification of the Stanton number, should be, for the sake of convenience, expressed for dispersion systems by the equation

$$\frac{\alpha F}{W} = 0.029 \ \frac{6\,(1-m)}{m} \ \frac{s}{d_k} \ Re^{-0,3} \tag{1}$$

which will satisfactorily approximate many experimental results of a number of authors on heat transfer in fixed beds and which, as the experiments undertaken in SVÚSS have shown, can be applied also to moving beds. Their porosity in narrow channels of active zones will be determined by means of the formula

$$m = 0.39 + 0.33 \ \frac{d_k}{s}$$

$$\text{valid for } \frac{d_k}{s} < 0.3. \tag{2}$$

The theoretical analysis of thermal conditions in moving bed regenerators and the experiences gained by the evaluation of their heat tests lead to the conclusion that the total thermal efficiency in the regeneration cycle is the decisive criterion for the evaluation of operational properties of a dispersion heat transfer system. The introduction of this magnitude into calculation leads the way to productive forms of designing such apparatuses. The method of the general derivation of the functions described below, will be treated in (3). The temperature diagram in Fig. 4 will be of service for easier orientation.

Let us consider a regenerator with heat losses; for the sake of simplification the losses are related to particle transport cycle sections. The total loss is composed, on one hand, by the heat losses caused by the regenerator surface and the conveyor heat emission into the surroundings, on the other hand by the loss of heat due to particle cooling by air leaking to the side of flue gas by the way of imperfect sealing labyrinths between both active zones. Further it is assumed that a perfect mixing of particles takes place in these sealing channels and even in the conveyor so that their temperature at the inlets can be considered to be constant. In view of the small particle dimensions (usually $d_k < 20mm$) it is not necessary in technical calculations to take the non-stationary character of their heating and cooling into consideration. The heat balance of the system is expressed by the equation

$$Q_p = Q_L + Q_{ip} + Q_{iL} \tag{3}$$

Without restriction to the general validity the partial thermal efficiencies in both the regenerator active zones can be defined by means of the relations

$$\eta_L = \frac{\Delta t_L}{\delta_{1L}} = f_L\left[\left(\frac{aF}{W}\right)_L, \frac{W_L}{W_M}\right] \tag{4}$$

$$\eta_p = \frac{\Delta t_p}{\delta_{1p}} = f_p\left[\left(\frac{aF}{W}\right)_p, \frac{W_p}{W_m}\right] \tag{5}$$

For the total thermal efficiency the following expressions can be derived

$$\eta_R = \frac{\Delta t_L}{\delta_1} = \frac{1}{\dfrac{W_L}{W_p}\dfrac{1}{\eta_p}\left(1 + \dfrac{Q_{iL}}{Q_L} + \dfrac{Q_{ip}}{Q_p}\right) + \dfrac{1}{\eta_L} - \dfrac{W_L}{W_M}\left(1 + \dfrac{Q_{iL}}{Q_L}\right)} \tag{6}$$

which is the initial relation for the complex evaluation of the thermal conditions in the regenerator.

In some cases instead of the partial thermal efficiencies in a given particle and gas flow arrangement only their relation to the thermal counter-flow efficiency will be known, i.e. the magnitudes

$$\epsilon_L = \frac{\eta_L}{\eta_{GL}} \ , \quad \epsilon_p = \frac{\eta_p}{\eta_{Gp}} \tag{7}$$

In such a case the total thermal regenerator efficiency can be defined by the expression

$$\eta_R = \frac{2\,\epsilon_L \epsilon_p}{\epsilon_p + \dfrac{W_L}{W_M}\left(\epsilon_L + \epsilon_p - 2\epsilon_L \epsilon_p\right) + \epsilon_L \dfrac{W_L}{W_p}\,\epsilon_p\, p + \epsilon_L L} \tag{8}$$

where

$$p = \left(\frac{W_L}{W_M} - \frac{W_L}{W_p}\right)\cotgh\ \frac{kF}{2W_L}\left(\frac{W_L}{W_M} - \frac{W_L}{W_p}\right)\frac{1 + \mathcal{H}}{\mathcal{H}} \tag{9}$$

$$L = \left(1 - \frac{W_L}{W_M}\right)\cotgh\ \frac{kF}{2W_L}\left(1 - \frac{W_L}{W_M}\right)(1 + \mathcal{H}) \tag{10}$$

$$\frac{1}{kF} = \frac{1}{(aF)_L}(1 + \mathcal{H})\ , \quad \mathcal{H} = \frac{(aF)_L}{(aF)_p} \tag{11}$$

Detailed tabular data for total thermal counter-flow regenerator efficiency are given in (3). For very frequently occurring, and in many respects even substantially more convenient lay-outs both side particle and gas or air cross-flow, the thermal efficiency values are given in Table 1.

Among the advantages offered by heat exchangers with a moving bed must be counted, first of all, the possibility to operate at high temperature levels and for these types of conditions these exchangers are designed either as heaters or coolers for solid particles or as special apparatuses for high temperature air heating. The high temperature regenerator model, designed for experimental purposes in the

SVÚSS laboratories, is shown in Fig. 5. The regenerator is made of ceramic material with cross-flow active zones with divergent beds and operates with alumina spheres 15 mm in diameter. The bucket conveyor, transporting particles at an inlet temperature of 800°C is cooled directly by air heated in the regenerator itself.

The bed cross-sections area on the side of the flue gases and air are 0.36 and 0.3 m respectively which permits the operation with the maximum flow rates 0.3 - 0.5 kg/s. At the flue gas temperature of 1400 - 1500°C the air was heated in this regenerator to 800 - 1000°C in dependence on the flow rate. The tests carried out under high temperature conditions up to now in the range of several hundred hours has indicated good operational properties of this apparatus and at the same time called the attention to several conditions and problems the solution of which is necessary for a successful industrial application of regenerators with moving beds.

TABLE 1

$$\frac{W_L}{W_P} = 1.0$$

$\dfrac{kF}{W_L}$	$\dfrac{W_L}{W_M} = 1.0$	0.8	0.5	0.2	0
0.5	0.312	0.315	0.317	0.317	0.316
1.0	0.443	0.448	0.448	0.441	0.432
1.5	0.517	0.521	0.515	0.495	0.475
2.0	0.565	0.569	0.556	0.522	0.491
2.5	0.601	0.603	0.582	0.536	0.497
3.0	0.629	0.629	0.600	0.543	0.498
4.0	0.662	0.666	0.624	0.551	0.500
5.0	0.680	0.680	0.638	0.554	0.500

$$\frac{W_L}{W_P} = 0.5$$

$\dfrac{kF}{W_L}$	$\dfrac{W_L}{W_M} = 1.0$	0.8	0.5	0.2	0
0.5	0.342	0.346	0.350	0.351	0.350
1.0	0.502	0.514	0.522	0.521	0.513
1.5	0.592	0.610	0.620	0.609	0.590
2.0	0.652	0.674	0.683	0.659	0.625
2.5	0.694	0.720	0.727	0.689	0.645
3.0	0.726	0.755	0.760	0.709	0.654
4.0	0.765	0.804	0.804	0.732	0.662
5.0	0.788	0.833	0.833	0.745	0.665

$$\frac{W_L}{W_P} = 0$$

$\dfrac{kF}{W_L}$	$\dfrac{W_L}{W_M} = 1.0$	0.8	0.5	0.2	0
0.5	0.373	0.378	0.384	0.386	0.388
1.0	0.560	0.584	0.604	0.607	0.604
1.5	0.658	0.692	0.734	0.736	0.718
2.0	0.712	0.765	0.814	0.823	0.785
2.5	0.747	0.804	0.867	0.870	0.828
3.0	0.771	0.832	0.902	0.899	0.853
4.0	0.796	0.868	0.942	0.945	0.888
5.0	0.810	0.882	0.964	0.967	0.909

Thermal efficiency of moving bed regenerator with both sides cross-flow arrangement.

NOMENCLATURE

$a_F = \dfrac{6(1-m)}{d_k}$ specific surface of bed $[m^2 m^{-3}]$

$d_e = \dfrac{4m}{a_F}$ effective bed diameter $[m]$

d_k particle diameter $[m]$

l   bed length $[m]$

M   mass velocity $[kg\ s^{-1}]$

m   porosity $[-]$

$R_e = \dfrac{wd_e}{mv}$ Reynolds number $[-]$

Q   transferred heat $[W]$

s   bed thickness $[m]$

t   temperature $[°C]$

W   heat capacity $[W\ deg^{-1}]$

w   superficial velocity $[ms^{-1}]$

x   longitudinal distance coordinate $[m]$

α heat transfer coefficient $[W\ m^{-2}\ deg^{-1}]$

δ inlet temperature difference $[deg]$

η thermal efficiency $[-]$

ν kinematic viscosity $[m^2\ s^{-1}]$

SUBSCRIPTS

G   counter-flow

i   loss of heat

L   air

M   moving bed

P   flue gas

R   regenerator

O   inlet thickness

1   inlet

2   outlet

REFERENCES

[1]   Schneller J., Hlavačka V.: Wärmeübertragung in gasdurchströmten beweglichen Partikelschichten verschiedener geometrischer Formen. 4th Int. Heat Transfer Conf., Versailles, Sept. 1970.

[2]   Schneller J., Hlavačka V., Struhár L.: Temperature Fields in Moving Beds with Cross-Flow of Gases. 4th CHISA Congr., Prague, Sept. 1972.

[3]   Schneller J., Hlavačka V.: Wärmeverhältnisse in Mehrstoff-Regenerativ-Wärmeaustauschsystemen. Wärme-und Stoffübertragung 5, No 3, 191 (1972).

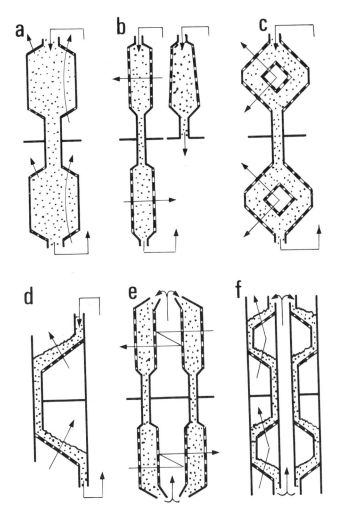

Fig. 1. Some types of arrangement of moving bed regenerative heat exchangers.

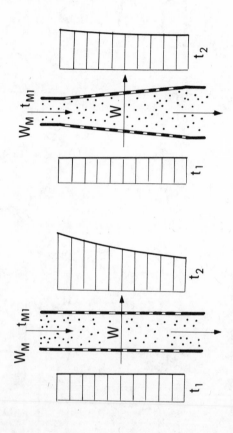

Fig. 2. Comparison of outlet gas temperature profiles for moving beds with constant and variable cross-section.

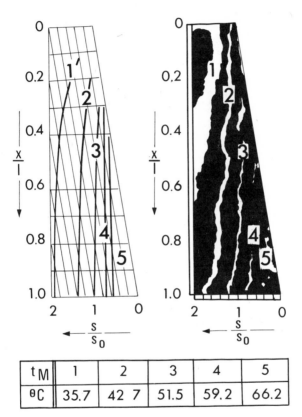

t_M	1	2	3	4	5
θC	35.7	42 7	51.5	59.2	66.2

Fig. 3. Comparison of measured and calculated temperature profiles in moving bed with variable cross-section.

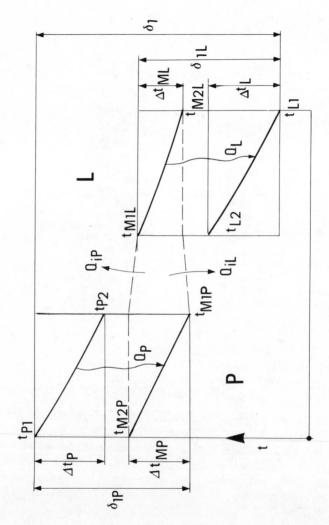

Fig. 4. General temperature diagram for moving bed regenerator.

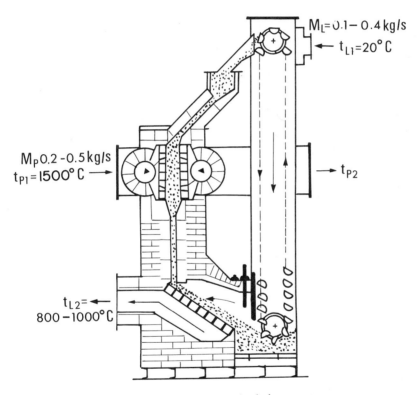

$M_L = 0.1 - 0.4\,kg/s$

$t_{L1} = 20°C$

$M_P\,0.2 - 0.5\,kg/s$
$t_{P1} = 1500°C$

t_{P2}

$t_{L2} = 800 - 1000°C$

Fig. 5. High temperature moving bed regenerator.

Chapter 28

METHODS FOR CALCULATING RADIATION TRANSFER
IN HEAT EXCHANGERS

A. Clark and E. Korybalski (*)

Introduction

This chapter deals with methods for the calculation radiation exchange in heat exchangers. Several new methods recently developed by the authors are summarized, including both steady and transient conditions. This paper is devoted to heat exchangers having surfaces that are gray-diffuse radiators without an internal participating medium.

Algebraic methods (1,2,3,4,5) are in contrast to the exact formulations in terms of a series of integral equations, the solutions for which pose extremely severe analytical difficulties (6,7,8). As a practical matter the algebraic forms represent the more meaningful approach, particularly for design calculations. The key to this method is the digital computer which makes possible not only the reduction of the enormous arithmetic operations but, when suitably programmed, an easily arranged digital simulation of the enclosures as well.

Analytical Development

A generalized enclosure representing the surfaces of a heat exchanger is shown in Figure 1. A net radiation heat transfer, q^*, is exchanged internally either directly or indirectly between all of the n-surfaces of the enclosure. Each surface, moreover, has other types of thermal interaction in addition to the internal exchange of thermal radiation. The total thermal management of a surface must

(*) Department of Mechanical Engineering, University of Michigan, Ann Arbor, Michigan, U.S.A.

N.B. ALL FIGURES QUOTED IN TEXT ARE AT THE END OF THE CHAPTER

include all of these, which involve an external thermal exchange with an ambient, q_E, an enthalpy flux across the surface (such as in a boiler tube bank), $\int wdh$, and a heat generation within the surface, P_s. These interactions are shown symbolically for surface A_i in Figure 2, where J_i is the internal radiosity flux defined as the sum of the emitted and reflected radiation heat transfer from surface A_i. Thus,

$$J_i = \epsilon_i \delta T_i^4 + \rho_i H_i \tag{1}$$

In this equation H_i is the total incident radiation heat flux on A_i received from all n-surfaces of the enclosure and originating as emission and reflection at those surfaces. The relationship of these various thermal quantities for any surface A_j, Figure 2, for the general unsteady condition is (1),

$$\frac{d}{dt} \int_{V_j} \rho c_p T dv + \int_{A_j} wdh = q_{Ej} - q_j^* + P_{sj} \cdot \quad (j = 1,2 \ldots n) \tag{2}$$

The net radiation heat transfer rate at this surface may also be written (1) as

$$q_j^* = A_j \left(\frac{\epsilon}{\rho}\right)_j \left[\sigma T_j^4 - J_j\right] \cdot \quad (j = 1,2 \ldots n) \tag{3}$$

Furthermore, for an enclosure of n-gray, diffuse, isothermal surfaces having uniform radiation properties, the radiosity flux, equation (1), for surface A_i is (1)

$$J_i = \epsilon_i \sigma T_i^4 + \rho_i \sum_{j=1}^{n} J_j F_{ij} \cdot \quad (i = 1,2 \ldots n) \tag{4}$$

Equation (4), for surface A_i, describes a general system of equations in terms of radiosity flux, J_i, and surface temperature T_i which may be written for all n-surfaces (1) as

$$
\begin{aligned}
a_{11} J_1 + a_{12} J_2 + a_{13} J_3 + \ldots + a_{1n} J_n + K_1 &= 0 \\
a_{21} J_1 + a_{22} J_2 + a_{23} J_3 + \ldots + a_{2n} J_n + K_2 &= 0 \\
\vdots \quad\quad \vdots \quad\quad \vdots \quad\quad\quad\quad \vdots \quad\quad \vdots \quad & \vdots \\
a_{n1} J_1 + a_{n2} J_2 + a_{n3} J_3 + \ldots + a_{nn} J_n + K_n &= 0 ,
\end{aligned}
\tag{5}
$$

or, in matrix form, using the conventional notation in which i designates the row and j designates the column of the resulting square matrix, the equation set (5) becomes (1).

$$(a_{ij}) \ (J_j) \ + \ (K_i) \ = \ 0 \quad \begin{matrix} (i = 1,2\ldots.n) \\ (j = 1,2\ldots.n) \end{matrix} \qquad (6)$$

In equations (5) and (6) the various matrix coefficients (a_{ij}) are

$$a_{ij} \ = \ F_{ij}, \quad (i \neq j) \qquad (7)$$

$$a_{ij} \ = \ F_{ij} \ - \ \frac{1}{\rho_i}, \, (1 = j) \qquad (8)$$

or, in a more general notation

$$(a_{ij}) \ = \ (F_{ij}) \ - \ \frac{1}{\rho_i} \ (I), \quad \begin{matrix} (i = 1,2\ldots.n) \\ (j = 1,2\ldots.n) \end{matrix} \qquad (9)$$

where, F_{ij} is the geometric view factor for surfaces A_i and A_j, ρ_i is the reflectance of A_i and (I) is the identitiy matrix. The term K_i is defined as

$$K_i \ = \ \left(\frac{\epsilon}{\rho} \right)_i \sigma T_i^4, \quad (i = 1,2\ldots.n) \qquad (10)$$

and represents an element in a column $(n \times 1)$ matrix in equation (6).

Equations (5) and (6) represent the coupled relationship between the radiosity fluxes J_j, the geometric view factors, F_{ij}, the surface temperature T_i and its diffuse radiation properties for the n-surfaces in the enclosure. These relationships are valid whether or not T_i is a known quantity, as in some cases it will not be. However, these equations are the starting points for the mathematical modelling of a large number of different physical systems. The systems evaluated in this paper involve those having the specification of surface temperature, ambient temperature or net heat flux at a surface. The purpose of the resulting calculations is the determination of either the surface temperature or the net heat flux, whichever is not initially specified. In addition, the radiosity flux is always determined at each surface.

In general, the various physical systems to be evaluated differ from one another by the nature of the external thermal interaction of the surfaces with their ambient. Additionally, a transient condition introduces another unique dimension into the analysis and calculation. The specific external interaction is described by equation (2). This expression combined with equations (3) and (4) enables a relationship to be established between the radiosity flux J_j and either the ambient temperature, $T_{\infty j}$, the surface temperature, T_j, or the net surface heat flux, $(q*/A)_j$. In a real sense equation (2) is a statement of a boundary condition between each surface and its external environment which relates the physical circumstances of the external interaction with radiation exchange within the enclosure.

In the following discussion each of the different physical systems will be presented in accordance with its class of boundary interactions.

RADTQO – The Specification of Surface Temperature and Designation of Certain Adiabatic Surfaces.

The simplest system is that in which all the surface temperatures are known. (1,2,6,7,8). For a given geometric system and known radiation surface properties, the various radiosity fluxes are found from a solution of either equation (5) or (6). While the authors' principal method of solution (1,2) is to employ computer subroutines written after the Gaussian Elimination Technique, because of economy and increased precision, it is more convenient to employ matrix algebra to show the analytical procedures. Hence, using equation (6) the radiosity fluxes, J_j are written in terms of the inversion of the (a_{ij}) matrix, as (1)

$$(J_j) = - (a_{ij})^{-1} (K_i) \quad \begin{matrix} (i = 1,2.....n) \\ (j = 1,2.....n) \end{matrix} \qquad (11)$$

The corresponding net heat transfer rates at A_j follow from equation (3) as (1)

$$q_j^* = A_j \left(\frac{\epsilon}{\rho} \right)_j \left[\sigma T_j^4 + \frac{1}{D} \sum_{i=1}^{n} A_{ij} K_i \right] \quad (j = 1,2.....n) \qquad (12)$$

In this result D is the determinant of matrix (a_{ij}) and the elements A_{ij} are its cofactors. Once the net heat flux q^* is determined it may be related to the various other thermal quantities in equation (2), depending on the type of external thermal interaction is appropriate to surface A_j. For example, should A_j be in the steady-state with the only external interaction being a heating or a cooling by a fluid flowing through the surface, the enthalpy flux of the fluid would be found to be

$$\int_{A_j} w\,dh = - q_j^* \ .$$

(13)

The restrictions imposed on the RADTQO formulation are that a minimum of 2 known surface temperatures and a maximum of (n-2) adiabatic surfaces must be used for the limits on the input data. These limits are required to satisfy the laws of thermodynamics for the steady-state.

A special provision of RADTQO is to account for a number (not exceed n-2) of surfaces A_k that are adiabatic, i.e. $q_k^* = 0$. In this case the formulation in equation (11) and the form of the (a_{ij}) matrix, equations (7) and (8) are slightly altered (1,9,10). If an adiabatic surface is designated as A_k, the elements of the corresponding (a_{ij}) matrix are written

$$a_{ij} = F_{ij}, \quad (i \neq j)$$

(14)

$$a_{ii} = F_{ii} - \frac{1}{\rho_i} \quad \begin{array}{l}(i = j) \\ (i \neq k)\end{array}$$

(15)

$$a_{kk} = F_{kk} - 1. \quad \begin{array}{l}(i = j) \\ (i = k)\end{array}$$

(16)

In other words, the principal change in the (a_{ij}) matrix for a system having adiabatic surfaces is that the diagonal coefficients corresponding to the adiabatic surface(s) are written $a_{kk} = F_{kk} - 1$. Furthermore, the term(s) K_k in the (K_i) matrix become identically zero for this case (1). The temperature of an adiabatic surface follows from equation (3) as

$$T_k = \left(\frac{J_k}{\sigma} \right)^{1/4},$$

(17)

in which J_k is computed from equation (11) with the element of the (a_{ij}) matrix as given in equations (14-16) and with K_k equal to zero. It might be of interest to note (1) from equation (16) that the numerical value of the surface radiation properties do not influence the magnitude of the temperature or radiosity flux of an adiabatic surface in the steady-state condition.

An example of the use of RADTQO will be cited later, including a comparison of results of the calculations on an enclosure with experimental data.

RADTQ – The Specification of Surface Temperatures and/or net Radiation Heat Flux at Certain Surfaces.

An improvement in the RADTQO formulation toward a more generalized approach to steady-state enclosure calculations is found in RADTQO(3). This formulation treats enclosures for which either surface temperature or net radiation heat flux may be specified arbitrarily. There are, of course, certain restrictions imposed that prevent the input of any set of temperatures or net radiation heat flux rates which would cause violations of the Laws of Thermodynamics or the creation of unreal physical circumstances, such as a negative radiosity flux on any surface. Except for these conditions, the formulation allows the input of a maximum of (n-1) arbitrary net radiation heat fluxes and a minimum of one known surface temperature or any combination of these consistent with real physical circumstances. An adiabatic condition on a surface is a special case of the net radiation heat flux (q*/A) equal to zero.

Writing equation (3) for a surface A_ℓ on which is imposed a known heat flux, q_ℓ^*/A_ℓ, the corresponding term K_ℓ, equation (10) becomes (3)

$$K_\ell = \frac{q_\ell^*}{A_\ell} + \left(\frac{\epsilon}{\rho}\right)_\ell J_\ell \ . \tag{18}$$

Hence, the row $i = \ell$ of the equation set (5), becomes (3),

$$F_{\ell 1}J_1 + F_{\ell 2}J_2 + F_{\ell 3}J_3 + \ \cdots \ + (F_{\ell\ell} - 1)\, J_\ell + \cdots\ F_{\ell n}J_n + \frac{q_\ell^*}{A_\ell}\ = 0 \tag{19}$$

The radiosity fluxes for this case may then be written as

$$\begin{bmatrix} J_1 \\ \vdots \\ J_\ell \\ \vdots \\ J_n \end{bmatrix} = -\ (a_{ij})^{-1} \begin{bmatrix} K_1 \\ \vdots \\ q_\ell^*/A_\ell \\ \vdots \\ K_n \end{bmatrix} \quad , \tag{20}$$

in which (a_{ij}) is the matrix whose coefficients are

$$a_{ij} = F_{ij} \quad (i \neq j) \tag{21}$$

$$a_{ii} = F_{ij} - \frac{1}{\rho_i} \quad \begin{array}{l} (i = j) \\ (i \neq \ell) \end{array} \tag{22}$$

$$a_{ii} = F_{\ell\ell} - 1 \quad \begin{array}{l} (i = j) \\ (i = \ell) \end{array} \tag{23}$$

For the case of specified net radiation heat fluxes the quantity (q_ℓ^*/A_ℓ) replaces K_i in the (K_i) matrix for all surfaces $(i = \ell)$ having a specified (q_ℓ^*/A_ℓ) as input data. For the special case of an adiabatic surface (q_ℓ^*/A_ℓ) equals zero and the RADTQ formulation, equations (20) - (23) becomes identical to RADTQO, equations (11), (14) - (16). In this instance also the temperature of the adiabatic surface would be computed from equation (17).

The formulation RADTQ computes the net radiation heat transfer q_j^* on all surfaces having specified input temperatures from equation (3). It also computes the temperature of all surfaces having a specified input net radiation heat flux from equations (18) and (10) as,

$$T_\ell = \left[\frac{q_\ell^*/A_\ell}{\sigma \, (\epsilon/\rho)_\ell} + \frac{J_\ell}{\sigma} \right]^{1/4} . \tag{24}$$

For an adiabatic surface q_ℓ^* equals zero and this result becomes identical with equation (17) which is written specifically for an adiabatic surface.

Calculations Involving RADTQO and RADTQ

Example calculation results and a comparison of the formulations RADTQ and RADTQO will be given using the geometric configuration shown in Figure 3 consisting of a hollow cylinder divided into six-surfaces. The geometric view factors, F_{ij},   surface areas and emittances for this configuration are summarized in Table 1. Although the cylindrical configuration in Figure 3 is of small size, the geometric view factors in Table 1 will be the same for all geometrically similar cylinders, that is, those having the same ratios r/L and L/ℓ as the cylinder in Figure 3. Further, the radiosity fluxes and net heat fluxes calculated for the cylinder of Figure 3 will be identical for all geometrically similar cylinders having the same surface temperatures and emittances. The net radiation heat transfer rate, q_j^* may be computed for other geometrically similar cylinders of length L on the basis of the similarity relation

$$q_j^* = (q_j^*)_j \left(\frac{L}{L_j}\right)^2 , \tag{25}$$

where $(q_j^*)_j$ corresponds to the net radiation heat transfer rate of the configuration in Figure 3 having length L_j and radius r_j . The net radiation heat transfer rate $(q_j^*)_j$ may be computed from the net radiation heat flux by

$$(q_j^*)_j = (q^*/A)_j A_j . \tag{26}$$

The data in Table 2 indicate that identical results are obtained from the RADTQO and RADTQ formulations for the same input data and configuration. The last two columns in this table show that with the net radiation heat flux and one surface temperature as input data, RADTQ will compute the remaining surface temperatures with a high degree of precision. In this case RADTQ used input data on $(q^*/A)_j$ corresponding to the computed output from RADTQO and RADTQ, listed in the first four columns of Table 2. By this inversion process essentially identical surface temperatures are computed by RADTQ as those which served as input for both RADTQO and the RADTQ calculation in the middle two columns.

Table 3 provides "original" and "inverted" calculations using the RADTQO and RADTQ formulations for the case of surface 4, Figure 3, being adiabatic. The comparison of the results is essentially exact.

Comparison of RADTQO and RADTQ with Experimental Data and Other Analytical Results

Calculations using RADTQO and RADTQ will now be compared with experimental data and other analytical results for systems having characteristics which may be expected to be the same as those assumed in the development of these formulations. Schornhorst and Viskanta (11) have reported some careful measurements of surface temperature and incident heat flux for both parallel and perpendicular flat plate systems. Calculated results using RADTQO and RADTQ will be compared with the physical measurements of Schornhorst and Viskanta using their reported temperatures (11) and emittance values (12) as input data. The comparison

TABLE 1

$$F_{ij}$$

Geometric View Factors for the Six-Surface Cylinder
in Figure 3

j \ i	1	2	3	4	5	6	Sum of Row i
1	0.00000	0.01105	0.79180	0.13488	0.03639	0.02588	1.00000
2	0.01105	0.00000	0.02588	0.03639	0.13488	0.79180	1.00000
3	0.24051	0.00786	0.51898	0.19853	0.03093	0.00319	1.00000
4	0.04198	0.01105	0.19853	0.51898	0.19853	0.03093	1.00000
5	0.01105	0.04198	0.03093	0.19853	0.51898	0.19853	1.00000
6	0.00786	0.24051	0.00319	0.03093	0.19853	0.51898	1.00000

Surface	Area, sq.ft.		Emittance	
	A_{j_1} inside	A_{j_2} outside	j_1 inside	j_2 outside
1	0.03221	0.03221	0.937	0.937
2	0.03221	0.03221	0.518	0.518
3	0.10603	0.11250	0.426	0.426
4	0.10603	0.10909	0.375	0.375
5	0.10603	0.10909	0.355	0.355
6	0.10603	0.11250	0.336	0.336

TABLE 2

Comparison of Results Using RADTQO and RADTQ for the Configuration in Figure 3 and Table 1

A_j	RADTQO		RADTQ		RADTQ	
	T_j	$(q^*/A)_j$	T_j	$(q^*/A)_j$	T_j	$(q^*/A)_j$
1	1400.00000*	4635.71263	1400.00000*	4635.71263	1400.00000*	4635.71260
2	400.00000*	-1249.29466	400.00000*	-1249.29466	400.00017	-1249.29466*
3	600.00000*	-1059.70026	600.00000*	-1059.70026	600.00002	-1059.70026*
4	800.00000*	- 395.00614	800.00000*	- 395.00614	800.00002	- 395.00614*
5	1000.00000*	- 69.15777	1000.00000*	- 69.15777	1000.00001	- 69.15777*
6	1200.00000*	490.93275	1200.00000*	490.93275	1200.00001	490.93275*
		- 0.22717%		- 0.22717%		- 0.22717%

*Denotes input data. All other tabular entries represent computed results based on the input data.

TABLE 3

Results Using RADTQO and RADTQ with Inversion of Input Data of the Configuration in Figure 3 and Table 1

A_j	RADTQO and RADTQ		RADTQO and RADTQ	
	T_j	$(q^*/A)_j$	T_j	$(q^*/A)_j$
1	1400.00000*	4373.99246	1400.00000*	4373.99246
2	400.00000*	-1323.16564	400.00000*	-1323.16564
3	600.00000*	-1186.05677	600.00000*	-1186.05677
4	1082.109972	0.0*	1082.109972*	- 0.00000[a]
5	1000.00000*	- 188.20980	1000.00000*	- 188.20980
6	1200.00000*	444.39817	1200.00000*	444.39817
		- 0.17970%		- 0.17970%

*Denotes input data. All other tabular entries represent computed results based on the input data.
[a]Minus sign indicates a small negative number less than 10^{-6}.

will be made for their parallel plate system which was approximated by an 11-surface enclosure (Figure 4) for the purposes of making calculations using RADTQO and RADTQ.

The results of this comparison are given in Figure 4 for surfaces of sand-blasted stainless steel and polished stainless steel. The computed results using RADTQO and RADTQ compare very favorably with both the experimental data and the diffuse analysis of Schornhorst and Viskanta (11).

Comparison of computed results will be made between RADTQ and an exact analysis for diffuse-gray surfaces by Usiskin and Siegel (13) using the cylindrical configuration of Figure 3. For this purpose the ends are considered to be black surfaces ($\epsilon_1 = \epsilon_2 = 0.99999$) at $\sigma T_1^4 = 16.0$ ($T_1 = 310.82R$) and $\sigma T_2^4 = 1.0$ ($T_2 = 155.41$ R). All other surfaces have emittances of 0.30000. A net radiation heat flux, $(q*/A)_j$, for surfaces 3,4,5 and 6 of 10.0 Btu/Hr-ft^2 is also imposed.

This calculation is well suited to the capability of RADTQ as it involves a problem having both specified temperatures and net radiation heat fluxes. Because RADTQ requires the division of the total cylindrical surface into 4 finite areas, the surface temperatures computed by the analysis of Usiskin and Siegel is spatially averaged over each area A_3, A_4, A_5, and A_6. These comparisons are given in Table 4.

TABLE 4

Comparison of Surface Temperatures Computed by RADTQ and Usiskin and Siegel (13) for the Configuration in Figure 3

A_j	Surface Temperatures	
	Usiskin and Siegel (13)	RADTQ
3	490.0	489.1
4	519.0	511.0
5	516.0	510.0
6	480.0	479.0

RADTTF – The Specification of Surface or Ambient Temperatures for External Radiative Exchange

The formulation RADTTF is written for an enxlosure or heat exchanger which may have surfaces that interact with their ambients by a radiative heat exchange. This formulation also allows for the specification of some surface temperatures where appropriate. In general, the specified input data can include n-ambient temperatures or n-surface temperatures or any combination of these. This is the first extension of these problems to include a specific type thermal exchange with the environment and is a modification of RADTQO for this purpose.

The consideration of heat exchange with an ambient requires the inclusion of the outside surface area and its radiation properties in the analysis, both of which may differ from the corresponding values on the inner surface. It is assumed that the wall of the heat exchange is sufficiently thin to permit the assumption that the inside and outside surface temperatures are the same. In the following analysis the subscripts 1 and 2 relate to the inside and outside wall conditions, respectively,

For the steady-state the relationship of q_{E_j} and q_j^* for the surface A_j, . Figure 2, which radiates to its ambient at T_{∞_j}, is found from equations (2) and (3) to be (4)

$$q_{Ej} - q_j^* = 0, \tag{27}$$

or,

$$A_{j2} \epsilon_{j2} \, \sigma \, [\, T_{\infty_j}^4 - T_j^4 \,] = A_{j1} \left(\frac{\epsilon}{\rho} \right)_{j1} [\, \sigma T_j^4 - J_j \,] . \tag{28}$$

In this case the row J $(i = j)$ of the equation set (5) may be written (4)

$$F_{j1} J_1 + F_{j2} J_2 + \ldots + (F_{jj} - E_j) J_j + \ldots + F_{jn} J_n + G_j \sigma T_{\infty_j}^4 = 0 \tag{29}$$

where

$$E_j = \left[\frac{(A_2/A_1)\, \epsilon_2 + \epsilon_1}{(A_2/A_1)(1 - \epsilon_1)\epsilon_2 + \epsilon_1} \right]_j \tag{30}$$

and

$$G_j = \left[\frac{\epsilon_2 \epsilon_1 (A_2/A_1)}{(A_2/A_1)(1-\epsilon_1)\epsilon_2 + \epsilon_1} \right]_j \tag{31}$$

The surface temperature T_j is computed (4) from J_j and T_{∞_j} by

$$\sigma T_j^4 = \frac{J_j + (1-\epsilon_1)_j\,(\epsilon_2/\epsilon_1)_j\,(A_2/A_1)_j\,\sigma\,T_{\infty j}^4}{(1-\epsilon_1)_j(\epsilon_2/\epsilon_1)_j(A_2/A_1)_j + 1} \tag{32}$$

The radiosity fluxes for RADTTF may then be written (4) for the generalized surface m having a radiative interaction with its ambient, as

$$
\begin{bmatrix}
J_1 \\
J_2 \\
J_3 \\
\cdot \\
\cdot \\
J_m \\
\cdot \\
\cdot \\
J_n
\end{bmatrix}
= - \left(a_{ij} \right)^{-1}
\begin{bmatrix}
K_1 \\
K_2 \\
K_3 \\
\cdot \\
\cdot \\
G_m \sigma T_{\infty m}^4 \\
\cdot \\
\cdot \\
K_n
\end{bmatrix}
\tag{33}
$$

in which,

$$a_{ij} = F_{ij} \qquad (i \neq j) \tag{34}$$

$$a_{ii} = F_{ii} - 1/\rho_i \quad \begin{array}{l}(i = j)\\(i \neq m)\end{array} \tag{35}$$

$$a_{ii} = F_{mm} - E_m \quad \begin{array}{l}(i = j)\\(i = m)\end{array} \tag{36}$$

Some representative calculations performed using RADTTF are given in Table 5 for the configuration in Figure 3 and Table 1.

RADTTF is also compared with the experimental measurements of Love and Gilbert (14). The results of this comparison are given in Table 6 for the emittance data and configuration in Figure 5.

TABLE 5

Computed Results Using RADTTF for the Configuration in
Figure 3 and Table 1

A_j	T_j	T_j	$(q^*/A)_j$	T_j	T_j	$(q^*/A)_j$
1	1400.00000*	NID[a]	4145.62295	1400.00000*	NID[a]	3717.27741
2	600.00000*	NID[a]	- 279.47892	600.00000*	NID[a]	- 561.79578
3	977.32248	530.00000*	- 645.65745	1007.98639	530.00000*	- 738.62286
4	845.46623	530.00000*	- 285.71110	1073.73401	1050.00000*	- 75.17715
5	753.37673	530.00000*	- 152.27165	1035.42109	1050.00000*	41.38898
6	696.20263	530.00000*	- 95.33813	789.23586	530.00000*	- 188.86534
			- 0.36889%			- 0.23967%

* Denotes input data. All other tabular entries represent computed results based on the input data.
a NID denotes "Not Input Data".

TABLE 6

Comparison of Results Using RADTTF with Experimental Measurements
of Love and Gilbert (14) for the Configuration in Figure 5

	Experimental (14)		RADTTF	
Case	T_5	$(q^*/A)_2$	T_5	$(q^*/A)_2$
A	618.0	1559.66	941.61160	1559.64840
B	591.0	926.05	853.95316	926.60476
C	569.5	349.30	713.00084	349.66936
D	563.0	219.33	669.90501	219.01269
E	555.0	170.59	649.84558	169.36160

To effect a comparison with the experimental measurements using RADTTF, the 5-surface enclosure was digitally simulated in such a way that a temperature dependent emittance could be introduced for the stainless steel surfaces with the requirement that the emittance value be fixed at 0.480000 at 500R. The simulation was conducted by varying the emittance in accordance with the measured surface temperatures of the stainless steel surfaces until the computed net heat flux on surface 2 agreed essentially with that measured.

RADTTFT − Transient form of RADTTF

The last radiation formulation to be discussed in this communication is RADTTFT, a transient form of RADTTF. This formulation is used to compute the transient temperatures, net heat flux rates and radiosities for an enclosure subjected to a step or sinesoidal (periodic) change in one or all of its ambient temperatures. A transient can also be introduced independently or simultaneously by a change in the emittance of all surfaces. Also, the influence on the thermal response of the enclosure of the density, heat capacity or thickness of any surface can be determined.

For transient conditions the thermal response of surface j, Figure 2, to a time dependent disturbance introduced into the system is found from equation (2). A disturbance can be of several kinds but for the present discussion these will include a change in the ambient temperature on any of the n-surfaces or a change in the temperature of the other (n-1) surfaces. For an external radiation exchange with the ambient and in the absence of any coolant flow through the surface or heat generation within the surface, the time rate of change of T_j is written from equations (2) and (11) as

$$\left(\frac{dT}{dt}\right)_j = \frac{A_{2j}\epsilon_{2j}\,\sigma}{(\rho c_p V)_j}\left[T_{\infty j}^4(t) - T_j^4(t)\right] -$$

$$- \frac{A_{ij}(\epsilon/\rho)_{ij}}{(\rho c_p V)_j}\left\{\sigma T_j^4(t) + (a_{ij})^{-1}[K_i(t)]\right\} \qquad (37)$$

$$(j = 1,2\ldots n \; ; \; i = 1,2\ldots n)$$

In the equation provision is made for the introduction of a disturbance by a change at time equals zero in the ambient temperature $T_{\infty j}(t)$. This change can be of any arbitrary kind. Those considered here, however, will include sinesoidal (periodic) changes only. Equation (37) also is written to include time-dependency in the radiosities, surface temperatures and surface radiation properties, if these latter vary during the transient.

The solution of the system of equations (37) for the n-interacting surfaces in the enclosure for a transient condition poses a somewhat different problem from that for the system of equations describing the steady-state. In the present case the temperature of surface j is computed for small but finite time steps starting with the conditions at time equals zero. Each succeeding calculation is dependent on the results from the preceding time step. The particular computational technique used is of the predictor-corrector type following a method developed by Krogh (15).

The response of the temperature of surface 2 of the six-surface cylinder of Figure 3 and Table 1 to a sinesoidal variation in the ambient temperatures on surfaces 2, 3, 4, 5 and 6 is given in Figure 6. For this case the ambient temperatures vary as

$$T_{\infty j}(t) = 530 + 100. \sin\left(2\pi\frac{t}{t_{oj}} + \delta_j\right), \quad (j = 2, 3, 4, 5, 6)$$

$$(38)$$

in which the period t_{oj} is equal to 0.10 hours and the phase-lag δ_j is zero. This latter puts all the ambient temperatures in phase with each other. It is a simple matter, however, to introduce both a different phase-lag δ_j or period t_{oj} for each surface.

Of some interest to note is that a sinesoidal disturbance about the initial value (530.R) of each of the ambient temperatures produces a finite change in the response of the mean temperature of surface 2. This is in marked distinction to the response of a linear system (such as heat conduction in a substance having uniform properties) which would produce an identically zero change in the mean value of its temperature for a similar transient.

Acknowledgments

The authors wish to acknowledge the assistance of the College of Engineering in providing funds for the use of the University of Michigan IBM 360/67 digital computer system with which all the data reported were obtained. The second author received Department of Mechanical Engineering fellowship support.

NOMENCLATURE

A	Area, $[\text{Ft}^2]$
A_{ij}	Cofactor of matrix (a_{ij}), Eq. (12), [1]
(a_{ij})	Geometric view factor - reflectance matrix, Eq. (6) and (9)
D	Determinant of (a_{ij}) matrix, [1]
F_{ij}	Geometric view factor, Eq. (4), [1]
(F_{ij})	Geometric view factor matrix, Eq. (9), [1]
H	Incident radiation flux, Eq. (1), $[\text{Btu/hr--ft}^2]$
h	Enthalpy, $[\text{Btu/lbm}]$
(I)	Identity matrix, Eq. (9), [1]
J	Radiosity flux, Eq. (1), $[\text{Btu/hr--ft}^2]$
K	$(\epsilon/\varsigma)\,\sigma T^4$, Eq, (10), $[\text{Btu/hr--ft}^2]$
n	Number of surfaces in enclosure, [1]
P_s	Power generation rate within surface, $[\text{Btu/hr}]$
q^*	Net radiation heat transfer rate, Eq. (3), $[\text{Btu/hr}]$
q_E	External heat transfer rate, $[\text{Btu/hr}]$
RADTQO	Name of authors' computer programs to use with formulations of this
RADTQ	paper.
RADTTF	
RADTTFT	
t	times [hours]
t_o	period of sinesoidal variation [hours]
T	Temperature, $[°R]$
V	Volume, $[\text{ft}^3]$
w	Mass flow rate, $[\text{lmb/hr}]$

Greek

ϵ	Emittance or emissivity, [1]
ρ	Reflectance or reflectivity, [1]
σ	Stefan-Boltzmann constant. $\text{Btu/hr--ft}^2\text{-}°R$

Subscripts

i	Designation of row number in matrix
j	Designation of column number in matrix
∞	Conditions in environment of enclosure
1, 2, 3, n, i, k, ℓ, n -	Designation of surfaces in enclosure
k	Adiabatic surface
ℓ	Surface having specified net heat flux
1	Inside conditions
2	Outside conditions

REFERENCES

[1] Clark, John A. and Michael E. Korybalski, "Radiation Heat Transfer in an Enclosure Having Surfaces which are Adiabatic or of Known Temperature," Paper No. HMT-15-71, Proceedings, First National Heat and Mass Transfer Confernece, Indian Institute of Technology, Madras, India, December 6-8, 1971

[2] Clark, John A. and Michael E. Korybalski, "Equivalence of the Algebraic Methods for the Calculation of Radiation Exchange in an Enclosure," Manuscript in preparation.

[3] Clark, John A. and Michael E. Korybalski, "Calculation of Surface Temperatures in an Enclosure for Radiation Heat Exchange." Manuscript in preparation.

[4] Clark, John A. and Michael E. Korybalski, "Determination of Heat Transfer Rates and Surface Temperatures in an Enclosure Radiating to Multiple Ambients." Manuscript in preparation.

[5] Clark, John A. and Michael E. Korybalski, "Radiation Heat Transfer in an Enclosure Under Transient Conditions. "Manuscript in preparation."

[6] Hottel, H.C. and A.F. Sarofim, *Radiative Transfer*, McGraw-Hill Book Co., Chap. 5, 1971.

[7] Sparrow, E.M. and R.D. Cess, *Radiative Heat Transfer*, Brooks/Cole Publishing Co., 1966.

[8] Siegel, R. and J.R. Howell, *Thermal Radiation Heat Transfer*, McGraw-Hill Book Co., 1972.

[9] Blanco-Camblor, Maria L., Private communication to J.A. Clark, Dec. 3, 1971.

[10] Lewis, John G., private communication to J.A. Clark, September 8, 1971.

[11] Schornhorst, J.R. and R. Viskanta, "An Experimental Examination of the Validity of the Commonly Used Methods of Radiant Heat Transfer, " *Journal of Heat Transfer*, Vol. 90, No. 4, Nov. 1968. p. 429-37.

[12] Viskanta, R., private communication to J.A. Clark, March 21, 1972.

[13] Usiskin, C.M. and R. Siegel, "Thermal Radiation From a Cylindrical Enclosure with Specified Wall Heat Flux," *Journal of Heat Transfer*, Vol. 82, No. 4, Nov. 1960, p. 369-75.

[14] Love, T.J. and J.S. Gilbert, "Experimental Study of Radiative Heat Transfer Between Parallel Plates," Aerospace Research Laboratories Report ARL 66-0103, June 1966, United States Air Force, Wright-Patterson Air Force Base, Ohio.

[15] Krogh, Fred. T., Computer Program for Solving a Set of n-Differential Equations of nth order. Jet Propulsion Laboratory, California Institute of Technology, Passadena, California, April, 1969.

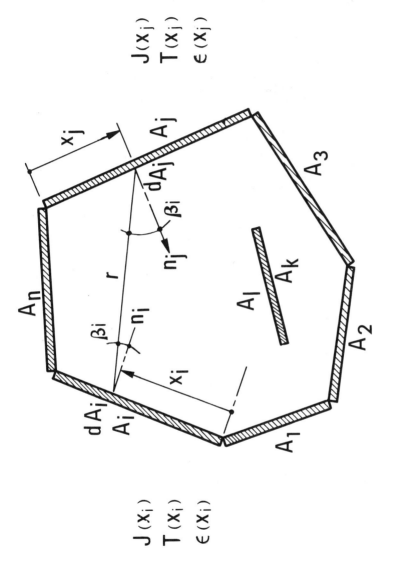

Fig. 1. A general n-surface enclosure encompassing 4π-steradians solid angle.

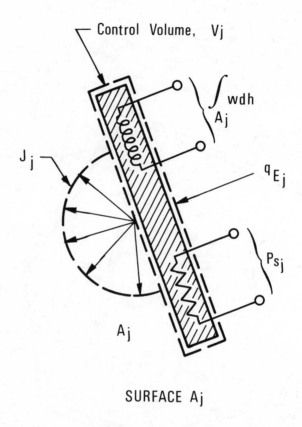

SURFACE Aj

Fig. 2. Boundary conditions on surface A_j for interactions with its surroundings.

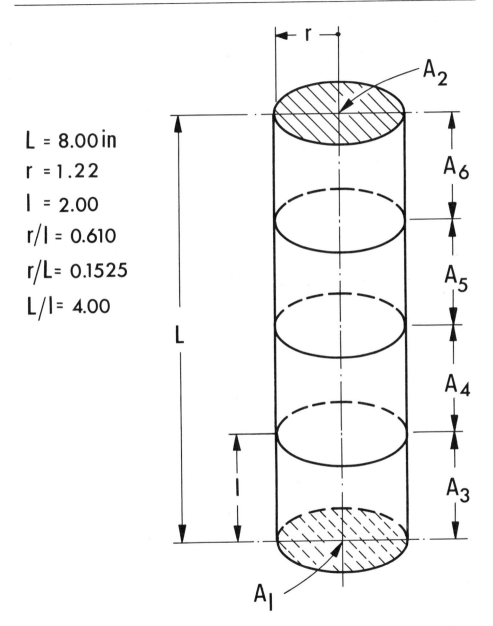

$$L = 8.00 \text{ in}$$
$$r = 1.22$$
$$l = 2.00$$
$$r/l = 0.610$$
$$r/L = 0.1525$$
$$L/l = 4.00$$

Fig. 3. Cylindrical enclosure of six-surfaces.

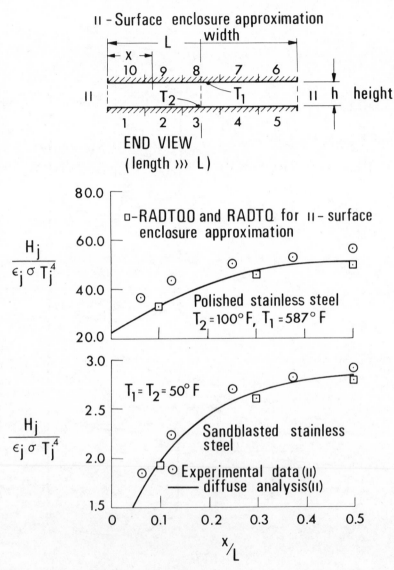

Fig. 4. Comparison of RADTQO and RADTQ with the experimental data and diffuse analysis of Schornhorst and Viskanta (11) for (H/L) = 0.125.

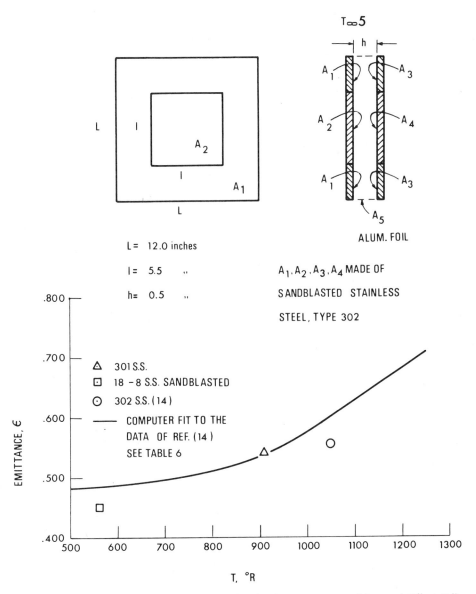

L = 12.0 inches

l = 5.5 ,,

h = 0.5 ,,

ALUM. FOIL

A_1, A_2, A_3, A_4 MADE OF

SANDBLASTED STAINLESS

STEEL, TYPE 302

△ 301 S.S.

□ 18 – 8 S.S. SANDBLASTED

⊙ 302 S.S. (14)

—— COMPUTER FIT TO THE
DATA OF REF. (14)
SEE TABLE 6

Fig. 5. Experimental configuration and emittance data for the measurements of Love and Gilbert (14).

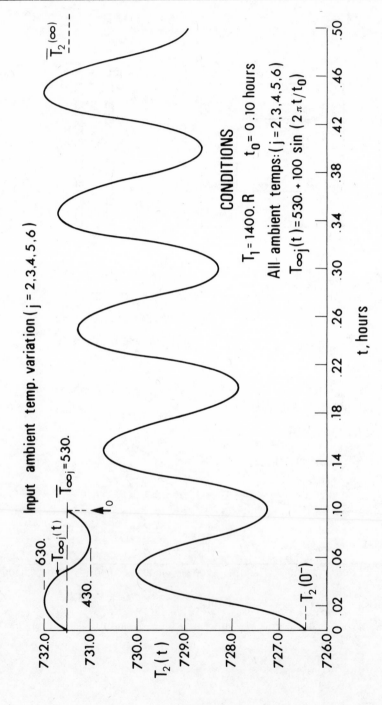

Fig. 6. Response of surface 2 to a sinesoidal variation in the ambient temperatures for the configuration in Figure 3 and Table 1. (Periodic steady-state is reached after approximately $t = 1.5$ hours at which time $\overline{T}_2 (\infty) \cong 731.58$ R).

Chapter 29

PRESSURE DROP AND HEAT TRANSFER BY VAPORIZATION OF BOILING REFRIGERANTS IN A HORIZONTAL PIPE

J. Bandel and E.U. Schlünder(*)

INTRODUCTION

In recent years heat transfer and pressure drop of vapor-liquid mixtures flowing in pipes have been the subject of a rather large number of experimental and theoretical studies.

The objective of most studies dealing with this subject has been to obtain data in the form of average pressure drop and heat transfer coefficients rather than to study systematically local pressure drop and heat transfer. Consequently, the major portion of the information available on this subject is not helpful to the understanding of the nature and mechanism of boiling refrigerants flowing in pipes. Whereas the methods of calculating single phase flow pressure drop and heat transfer are well accepted, the correlations of predicting the two phase vapor-liquid pressure drop and heat transfer show great discrepancies.

The best known calculation methods to predict frictional pressure drop are those of Lockhart-Martinelli [1], Martinelli-Nelson [2] and Chawla [3], the latter of which is complemented by the equation of Bankoff [4] at low vapor qualities. These equations are all of the same type : two phase flow pressure drop is calculated from single phase liquid and/or vapor pressure drop extended by different functions to fit the calculated or measured data.

$$\left(\frac{\Delta P}{\Delta L} \right)_{TP} = \left(\frac{\Delta P}{\Delta L} \right)_{l\,(g)} \cdot \phi \qquad (1)$$

(*) Institut für Thermische Verfahrenstechnik, Universität Karlsruhe (TH) — Germany (FR).

N.B. ALL FIGURES QUOTED IN TEXT ARE AT THE END OF THE CHAPTER

The Martinelli method is based on both liquid and vapor single phase pressure drop with the distinction of laminar and turbulent flow regimes. Chawla uses the turbulent gas flow pressure drop and Bankoff relates his equation to the single phase liquid pressure drop of the total mass flow rate. The correction functions vary from author to author. Martinellis' graphical plot of $\sqrt{\phi} = \phi_g = \sqrt{(\Delta P/\Delta L)_{TP}/(\Delta P/\Delta L)_g}$ and $\sqrt{\phi} = \phi_l = \sqrt{(\Delta P/\Delta L)_{TP}/(\Delta P/\Delta L)_l}$ versus $\chi = \sqrt{(\Delta P/\Delta L)_l/(\Delta P/\Delta L)_g}$ includes the parameters

$$\phi_{g(l)} = f\ (\dot{x}, \rho_l/\rho_g, \eta_l/\eta_g, \dot{m}, d) \tag{2}$$

The influence of mass flow rate pertains the laminar/turbulent and turbulent/laminar flow region. Chawla's correction function depends on

$$\phi = \phi_g^* = f\ (\dot{x}, \rho_l/\rho_g, \eta_l/\eta_g, \dot{m}, g, k/d) \tag{3}$$

and is given graphically.

The Bankoff correction factor depends only on the density ratio and vapor quality

$$\phi = \phi_l^{**} = f\ (\dot{x}, \rho_l/\rho_g) \tag{4}$$

The influence of the mass flow rate is different in these equations. The Lockhart-Martinelli concept leads to a dependance of pressure drop on the mass flow rate

$$\left(\frac{\Delta P}{\Delta L}\right)_{TP} \sim \dot{m}^{1 \div 1.8} \tag{5}$$

Chawla found, however, that there is a much stronger influence of mass flow rate

$$\left(\frac{\Delta P}{\Delta L}\right)_{TP} \sim \dot{m}^{1.75 \div 3} \tag{6}$$

whereas Bankoff's equation is determined by the single phase equation

$$\left(\frac{\Delta P}{\Delta L}\right)_{TP} \sim \dot{m}^{1 \text{ or } 1.8} \tag{7}$$

Heat transfer equations are commonly related to the turbulent single phase heat transfer equation of Dittus-Boelter and corrected by the turbulent pressure drop ratio of the single phase liquid to the single phase vapor flow

$$\frac{a_{TP}}{a_{l,turb}} = c \cdot \chi_{turb,turb}^n \tag{8}$$

The constants range from 2.1 to 7.55 and the exponents from -0.328 to -0.75.
Various authors introduce the so called boiling-number to this equation to allow for
nucleation effects. The type of equation is as follows

$$\frac{a_{TP}}{a_{1,turb}} = c_1 (B.10^4 + c.X_{turb,turb}^n)^{n_1} \tag{9}$$

The constants and exponents of this type of equation vary from $c_1 = 0.739$ to 2.45,
$c = 1.5$, $n_1 = 0.6$ to 1 and $n = -0.581$ to -2/3.
A somewhat different form of correlation is proposed by Chen [5], who suggests a
combined nucleation and forced convection influence

$$a_{TP} = a_{1,turb} \cdot F + a_{FZ} \cdot S \tag{10}$$

α_{FZ} is defined by the Forster-Zuber-equation for nucleate boiling and the
correction functions F and S are given in graphical form.

Chawla found from his investigation, that heat transfer to boiling
refrigerant 11 is separated into two regimes : convective boiling at low heat flux
densities and nucleate boiling at high heat fluxes. The correlation equations in the
region of convective boiling are [6]

$$Nu_1^* = \frac{1-\overset{\circ}{x}}{\overset{\bullet}{x}} \cdot \left(\frac{\rho_1}{\rho_g}\right)^{0.3} \cdot \left(\frac{\eta_1}{\eta_g}\right)^{0.8} \cdot Re_1^{0.35} \cdot Pr_1^{0.42} \cdot F_c \tag{11}$$

$$Nu_1^* = \frac{a_{TPc} \cdot d^*}{\lambda_1} = \left(1 - \frac{1}{\phi_g^{*\,8/38}}\right) \cdot \frac{a_{TPc} \cdot d}{\lambda_1} \tag{11.1}$$

The function F_c for the range of

$$Re_1 \cdot Fr_1 \leq 109$$

$$\text{is} \quad F_c = 6.6\,\dot{}\,10^{-3} (Re_1 \cdot Fr_1)^{0,475} \tag{11.2}$$

and for the range

$$Re_1 \cdot Fr_1 > 109$$

$$\text{is} \quad F_c = 1.5\,\dot{}\,10^{-2} (Re_1 \cdot Fr_1)^{0.3} \tag{11.3}$$

Nucleation data were correlated by the equation [7]

$$\frac{a_{TPn}}{a_S} = 29.Re_1^{-0.3} \cdot Fr_1^{0.2} \tag{12}$$

This equation is based on the pool-boiling equation of Stephan [8].

It should be noted from these correlations that at fixed mass flow rate and vapor quality the local heat transfer coefficient was found to be

a.) independent of the heat flux density (equ.8)

b.) dependent strongly on the heat flux (equ. 9, 10) or

c.) must be separated into two regimes according to the heat flux density (equ. 11, 12).

It is the objective of this investigation to obtain systematic experimental measurements of local pressure drop and local heat transfer coefficients for evaporating refrigerants in a horizontal tube. These measurements are to be used as a basis for the calculation of average pressure drops and heat transfer coefficients for the design of refrigerant evaporators.

EXPERIMENTAL EQUIPMENT

The experimental equipment assembly is shown schematically in Figure 1. The system consists of two circuits : a common refrigerant circuit with R 22 to condense the evaporated portion of the liquid in the test section circuit and to keep the temperature in the middle of the test section constant. In the test section circuit the refrigerant is pumped from a storage tank (1) to a 6000 W preheater (6), which is indirectly, electrically heated. Here, a portion of the refrigerant is evaporated. After a calming section of 1.5 m length the two phase flow passes through the short test section of 400 mm length (8) where the flow type is observed. The vapor-liquid mixture flows into the condenser (10) to be condensed and subcooled. The mass flow rate, adjusted by a throttle valve (5), is measured by a volumetric flow meter (11). The liquid refrigerant now flows back into the storage tank.

All cold pipes are insulated with 100 mm foam insulation. In addition to this the heated sections are insulated with 25 mm asbestos tape and 50 mm glass-wool.

Details of the electrically heated test section are shown in Figure 2. The test section pipe with 14 mm ID is heated over a length of 400 mm. Wall temperatures are measured by Ni-Cr/Ni thermocouples, which were imbedded in axial grooves and these grooves subsequently filled in with copper by electroplating methods. The depth of the grooves reached to within 0.2 mm of the inner pipe diameter. Five thermocouples are distributed around the wall in a helical fashion over a length of 50 mm while five other are distributed in the same may over 100 mm as shown in the development of pipe.

A slotted copper pipe contained the double threaded heating filament of 2 mm OD. In order to obtain good thermal contact, the heating pipe is soldered to the test section. The heating filaments are pressed into their helical grooves by a brass tube cover to which pressure is applied by five brass clamps.

Four pressure taps are connected to the test section, and each of them is used to measure absolute pressure and pressure drop over various distances along the test section.

The preheaters and the test section heating filament were connected to the secondaries of two continuously variable transformers. The original 220-volt supply was taken from a voltage stabilized net to insure constant heat input. Heat input measurements were carried out with a wattmeter of 0.1% accuracy. Bourdon gages of a 0.1% accuracy at full deflection were used to measure the pressure in the middle of the test section. Pressure drops were measured with a calibrated u-type mercury manometer of 1 mm Hg reading accuracy.

All temperature measurements were registered on an automatic, self-balancing compensator to control stationary conditions. Temperature measurements were carried out with a hand compensator combined with a mirror galvanometer accurate to 1 μvolt.

Heat losses of the test loop were determined in zero test and taken into account in the heat balance.

TEST PROCEDURE

At constant evaporating temperature, constant mass flowrate and constant vapor quality, a test run was conducted for various heat fluxes from 0 to 70 000 W/m^2. The measurements consisted of a set of readings of all instruments for a period of 20 minutes one hour after equilibrium conditions had been reached. Vapor quality was calculated from a heat balance taken between the conditions of the flowing liquid refrigerant before the throttle valve and those in the middle of the test section. The evaporating temperature was based on the saturation temperature corresponding to the absolute pressure in the middle of the test section. The average tube wall temperature was taken as an arithmetical average of all tube surface temperature readings, as no significant difference between the average of the five inner and the five outer thermocouple readings could be observed. Thus, the heat transfer coefficient was determined by the equation

$$a = \frac{\dot{q}}{\vartheta_w - \vartheta_s} \qquad (13)$$

EXPERIMENTAL RESULTS

Typical results of some heat transfer test runs at constant evaporating temperature and constant vapor quality are shown in Figure 3. From this figure it is seen that the heat transfer coefficient is independent of heat flux at low densities but strongly dependent on the mass flowrate. Increasing heat fluxes lead to nucleate boiling where the heat transfer is mainly influenced by the heat flux density, whereas the dependence on the mass flowrate is not significant. This experimental result confirms Chawla's hypothesis. All test results are therefore divided into those concerning nucleate boiling and those related to forced convective heat transfer.

The heat transfer data in the regime of nucleate boiling is compared to the above discussed equation by Chawla. Figure 4 represents the data which showed a definite dependence on heat flux density. The agreement between the calculated and measured data is very good. Without systematic errors the measured heat transfer coefficients fit the empirical curve within ± 20%.

In order to determine the influence of mass flow rate on frictional pressure drop and on convective heat transfer, the data of the refrigerant 11 measurements of Chawla and of the own investigation was plotted in Figure 5. The figure indicates that there are three regimes with different influences of the mass flow rate. This result can be related to the various flow configurations. At low mass flow rates stratified flow was observed. With increasing mass flow rates the flow becomes wavy and slug flow like and reaches annular flow at high mass fluxes. The dependence of pressure drop and convective heat transfer in the first and last regimes is the same as in single phase flow :

$$\left(\frac{\Delta P}{\Delta L}\right)_{TP} \sim \dot{m}^{1.75}$$
$$a_{TPc} \sim \dot{m}^{0.8} \qquad (14)$$

In the transition region the influence of mass flow rate is much stronger. The above discussed equations have been compared to these experimental results. Figure 6 represents the frictional pressure drop measurements compared to the equations of Lockhart-Martinelli, Chawla and Bankoff. The Bankoff method, established to predict bubble flow pressure drop, fails already compared to the measurements at low vapor qualities of $\dot{x} = 0.1$. The correlation of Chawla only fits the measured pressure drops in the transition region with refrigerant 11 whereas the Lockhart-Martinelli method gives the best results in the annular flow regime. The same result is obtained by comparing the convective heat transfer coefficients to the correlations which are based on the Lockhart-Martinelli concept (Figure 7). Best agreement between measured and calculated heat transfer coefficients in the annular flow region was found with the equation by Pujol [9]

$$\frac{a_{TPc}}{a_{l,T}} = 4 \cdot X_{turb, turb}^{-0.37} \tag{15}$$

This equation fails, when wetting effects become an influencial factor in the heat transfer. Chawla's equations predict the convective heat transfer coefficients in the transition region fairly well but fail when the density ratio ρ_l/ρ_g is low (refrigerant 12, 22) or the mass fluxes are high. Because of these discrepancies, a new concept to predict frictional pressure drop and convective heat transfer in two phase flow was established.

PROPOSED CORRELATIONS

Pressure drop

Figure 5 showed that there are two flow regimes with the same mass flow depency of frictional pressure drop as predicted for turbulent single phase flow.

The basic equations to predict frictional pressure drop in two phase flow are the modified single phase pressure drop correlations related to the liquid and the vapor phase

$$\left(\frac{\Delta P}{\Delta L}\right)_g = (\zeta_{g,w} + \zeta_{g,I}) \cdot \frac{\rho_g}{2} \cdot \bar{u}_g^2 \cdot \frac{1}{d_{h,g}}$$

$$\left(\frac{\Delta P}{\Delta L}\right)_l = (\zeta_{l,w} + \zeta_{l,I}) \cdot \frac{\rho_l}{2} \cdot \bar{u}_l^2 \cdot \frac{1}{d_{h,l}} \tag{16}$$

These equations contain, besides the wall friction factors $\zeta_{g,w}$ and $\zeta_{l,w}$, friction factors due to the momentum transfer at the interface. The pipe diameter is replaced by the hydraulic diameter of the liquid or vapor phase. To solve these equations the following assumptions have to be made :

1. Pressure drop in both phases is equal
2. As the pressure drop dependency on mass flowrate was found to be the same as in turbulent single phase flow when stratified or annular flow is present, the friction factors on the wall are determined with the Blasius equation

$$\zeta_w = \frac{0.3164}{Re_{TP}^{0.25}}$$

3. Interface friction factors are assumed to be proportional to $Re_{TP}^{-0.25}$

Then equation (16) is in rewritten form

$$\left(\frac{\Delta P}{\Delta L}\right)_{TP} = \left(\frac{\Delta P}{\Delta L}\right)_{TP,\,g} = \left(\frac{\Delta P}{\Delta L}\right)_{g,\,turb} \cdot \left(1 + \frac{F_{g,I}}{0.3164}\right) \cdot \left(\frac{d}{d_{h,g}}\right)^{1.25} \cdot \left(\frac{f}{f_{h,g}}\right)^{1.75} =$$

$$= \left(\frac{\Delta P}{\Delta L}\right)_{TP,\,l} = \left(\frac{\Delta P}{\Delta L}\right)_{l,\,turb} \cdot \left(1 + \frac{F_{l,I}}{0.3164}\right) \cdot \left(\frac{d}{d_{h,l}}\right)^{1.25} \cdot \left(\frac{f}{f_{h,l}}\right)^{1.75}$$

$$(16.1)$$

Equation (16.1) was used to determined the interphase friction functions $F_{g,I}$ and $F_{l,I}$ on the basis of measured pressure drop and calculated liquid holdup ($= f_{h,l}/f = 1 - f_{h,g}/f$) by the equations of Chawla [10] and Lockhart-Martinelli in the stratified and annular flow regime. The hydraulic diameters are defined as shown in Figure 8.

The functions which fitted both frictional pressure drop and liquid holdup are

$$F_{g,I,s} = 0 \qquad\qquad F_{g,I,a} = .6 \cdot \left(\frac{1 - \dot{x}}{\dot{x}}\right)^{.2}$$

$$(16.2)$$

$$F_{l,I,s} = 0 \qquad\qquad F_{l,I,a} = - .18 \cdot \dot{x}^{.3}$$

The prediction of frictional pressure drop is realized by trial and error, varying the diameter ratio as long as pressure drop in the gas and the liquid phase are equal.

To predict pressure drops in the transition region the stratified and annular flow regimes were limited by a critical Froude to Euler ratio. Stratified flow predominates when the ratio is

$$\left(\frac{Fr_1}{Eu}\right)_s = \frac{\rho_1 \cdot g}{\left(\dfrac{\Delta P}{\Delta L}\right)_{TP}} \geq \frac{50}{\dot{x}} \qquad (17.1)$$

and annular flow exists when the Froude to Euler ratio is

$$\left(\frac{Fr_1}{Eu}\right)_a \leq g \cdot \dot{x}_1^{.1} \qquad (17.2)$$

Pressure drop in the transition region is calculated from the critical pressure drops and mass flowrates at the critical Froude to Euler ratio for both stratified and annular flow. The equation is

$$\left(\frac{\Delta P}{\Delta L}\right)_{TP} = \left(\frac{\Delta P}{\Delta L}\right)_{TP,\,crit,\,s} \exp\left\{ \ln\frac{\dot{m}}{\dot{m}_{crit,s}} \cdot \ln\frac{\left(\dfrac{\Delta P}{\Delta L}\right)_{TP,\,crit,\,a}}{\left(\dfrac{\Delta P}{\Delta L}\right)_{TP,\,crit,s}} \Bigg/ \ln\frac{\overset{\circ}{m}_{crit,a}}{\dot{m}_{crit,s}} \right\}$$

$$(18)$$

The result of this calculation method is presented in Figure 9. Even if the agreement between measured and calculated pressure drop of our own investigation is within the measuring accuracy, this result must be examined on measurements with varied pipe diameter and other fluid properties.

CONVECTIVE HEAT TRANSFER

The analogy of the energy transport mechanism to momentum transport in convection in single phase flow leads to a relation between heat transfer and frictional pressure drop as follows :

$$\frac{\Delta P}{\Delta L} \sim \dot{m}^n \qquad\qquad a \sim \left(\frac{\Delta P}{\Delta L}\right)^{\frac{n-1}{n}} \qquad (19)$$

This result is well accepted for the turbulent single phase flow. Transferred to the two phase flow convective transfer the extended analogy leads to the equation

$$a_{TPc} = a_c \cdot c \cdot [\,(\Delta P/\Delta L)_{TP}/(\Delta P/\Delta L)\,]^{\frac{n-1}{n}} \qquad (20)$$

This result is similar to the concept of the equation (8) using the Lockhart-Martinelli parameter $\phi_{1,turb,\,turb}$, as the function $c \cdot X_{turb,\,turb}^{n1}$ may be regarded as an approximation of the parameter ϕ_1 . From experimental results the exponent n was determined in annular and stratified two phase flow to be n = 1.75. Thus, the exponent of the pressure drop ratio of the convective heat transfer equation is

$\dfrac{n-1}{n} = 3/7.$

This type of equation should be valid to predict convective heat transfer coefficients as long as tube wetting is complete. At low mass fluxes when stratified, wavy or slug-like flow is apparent, the influence of tube wetting complicates the problem, as well as at high vapor qualities. To predict convective heat transfer coefficeints in two phase flow, equation (20) was used based on the turbulent heat transfer equation of Kraussold

$$\mathrm{Nu}_l = 0.023 \cdot \mathrm{Re}_l^{0.8} \cdot \mathrm{Pr}_l^{0.42} \tag{21}$$

for the liquid flowrate, corrected by a constant and the pressure drop ratio of two phase pressure drop to single phase liquid pressure drop

$$a_{TP\,c} = a_c \cdot c \left[(\Delta P/\Delta L)_{TP} / (\Delta P/\Delta L) \right]^{3/7} \tag{22}$$

c = 1.4 refrigerant 11
c = 1.75 refrigerant 12
c = 1.8 refrigerant 22

In figure 10 the results of this investigation are compared to the proposed equation. The agreement between measured and calculated heat transfer coefficients is very good in the annular flow regime for all fluids. In the future this concept will be extended to low mass flow rates by correcting the above equation by a factor with respect to the tube wetting.

AVERAGE PRESSURE DROP AND HEAT TRANSFER COEFFICIENTS

From local pressure drop and heat transfer measurements average values can be calculated by integrating the local coefficients from inlet to outlet vapor quality. A summary of calculated average heat transfer coefficients for refrigerants is given by Schlünder et al. [11]. The heat transfer coefficients are calculated for dry expansion evaporator tubes with R 11, R 12, R 21 and R 22. For example Figure 11 shows the average heat transfer coefficient of evaporating R 11 at an evaporating temperature of + 10°C based on the equations of Chawla. With the results of the present investigation the diagrams will be extended to high flowrates.

NOMENCLATURE

B	boiling number	$= \dfrac{\dot{m} \cdot \Delta \dot{x}}{\dot{m}}$
c	constant	
d	diamter	
d_h	hydrolic diameter	
Eu	Euler-number	
f	cross section area	
f_h	hydrolic cross section area	
F	correction function	
Fr	Froude-number	
Fr_1	$\dfrac{\dot{m}^2 \cdot (1 - \dot{x})^2}{\delta_L{}^2 \cdot g \cdot d}$	
g	acceleration due to gravity	
$\dot{m}$	mass flow density	
n	exponent	
Nu	Nusselt-number	
$\dfrac{\Delta P}{\Delta L}$	pressure drop	
Pr	Prandtl-number	
$\dot{q}$	heat flux density	
Re	Reynolds-number	
Re_1	$\dfrac{\dot{m} \cdot (1 - \dot{x}) \cdot d}{\eta_1}$	
Re_{TPl}	$Re_1 \cdot \dfrac{d_{h,l}}{d} \cdot \dfrac{f}{f_{h,l}}$	
Re_{TPg}	$Re_g \cdot \dfrac{d_{h,g}}{d} \cdot \dfrac{f}{f_{h,l}}$	
S	correction function	
u	velocity	
$\dot{x}$	vapor quality $= \dfrac{\text{gas flowrate}}{\text{total flowrate}}$	

Greek letters

α	heat transfer coefficient
α_{TPc}	convective heat transfer coefficient in two phase flow
α_{TPn}	nucleate boiling heat transfer coefficient in two phase flow
ϑ	temperature
η	viscosity
ζ	density
λ	thermal conductivity
ϕ	correction function
χ	Martinelli-parameter

Subscripts

a	annular flow
crit	critical value
FZ	Forster-Zuber-equation
g	gas
I	Interphase
l	liquid
n	nucleate boiling
S	Stephan-equation
s	saturation
s	stratified flow
TP	two-phase-flow
w	wall

REFERENCES

[1] Lockhart, R.W. and R.C. Martinelli, Chem. Engr. Progr. Vol. 45, No. 39, 1949

[2] Martinelli, R.C. and B.D. Nelson, Trans. ASME, Vol 70, 1948, p. 695

[3] Chawla, J.M., Chem.Ing.Techn., Vol. 44, 1972, No. 1 + 2, p. 58-63

[4] Bankoff, S.G., Trans. ASME Ser.C., Vol. 82, 1960, p. 265-272

[5] Chen, I.C., ASME Paper No. 63-HT-43

[6] Chawla, J.M., Preprints of papers presented at the Fourth International Heat Transfer Conf. Paris-Versailles 1970, Vol. V, Session B 5.7

[7] Chawla, J.M., VDI − Forschungsheft 523, 1967

[8] Stephan, K., Abhandlungen des Deutschen Kältetechnischen Vereins (DKV), No. 18, 1964

[9] Pujol, L., Ph.D − Thesis, Lehigh University, Ann Arbor, Michigan, 1969

[10] Chawla, J.M., Chem. Ing. Techn. Vol. 41, 1969, p. 328-330

[11] Schlünder, E.U., Chawla, J.M., and E.A. Thomé, Refrigerant side heat transfer coefficients in evaporator tubes, Institut für Thermische Verfahrenstechnik, Universität Karlsruhe Germany (FR).

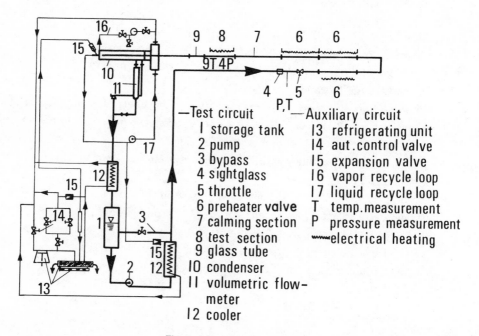

—Test circuit   —Auxiliary circuit

1 storage tank	13 refrigerating unit
2 pump	14 aut. control valve
3 bypass	15 expansion valve
4 sightglass	16 vapor recycle loop
5 throttle	17 liquid recycle loop
6 preheater valve	T temp. measurement
7 calming section	P pressure measurement
8 test section	⌇⌇ electrical heating
9 glass tube	
10 condenser	
11 volumetric flow-	
meter	
12 cooler	

Fig. 1. Schematic of equipment.

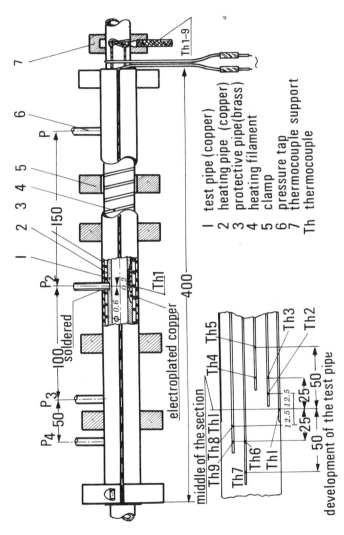

1 test pipe (copper)
2 heating pipe (copper)
3 protective pipe (brass)
4 heating filament
5 clamp
6 pressure tap
7 thermocouple support
Th thermocouple

Fig. 2. Test section.

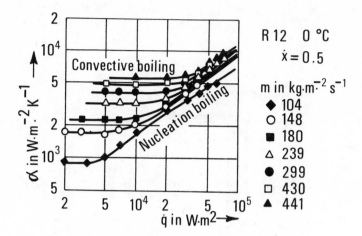

Fig. 3. Measured local heat transfer coefficients.

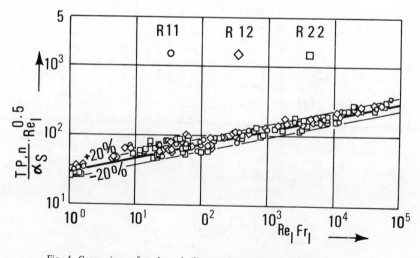

Fig. 4. Comparison of nucleate boiling results with the correlation of Chawla.

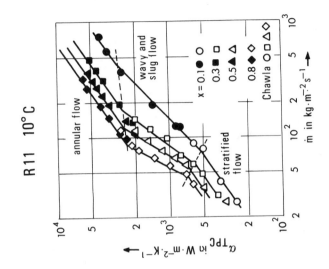

Fig. 5b. Measured convective heat transfer coefficients.

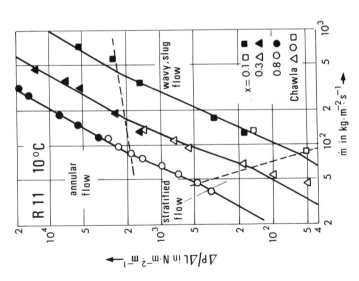

Fig. 5a. Measured frictional pressure drop.

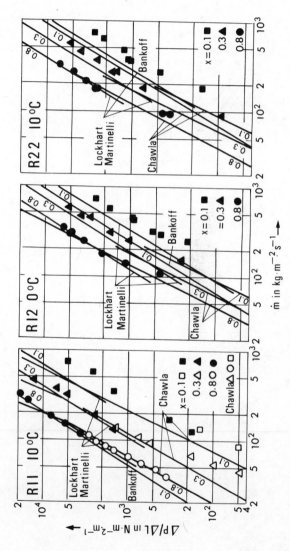

Fig. 6. Comparison of measured and predicted data.

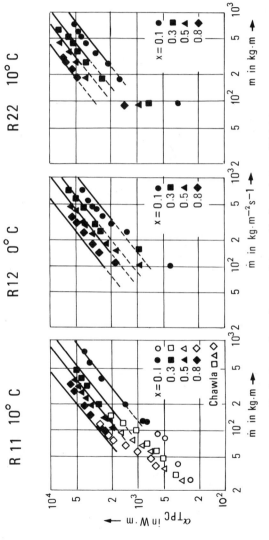

Fig. 7. Comparison of measured data with the equation of Pujol.

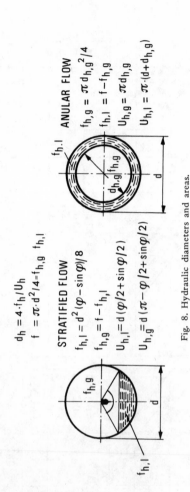

Fig. 8. Hydraulic diameters and areas.

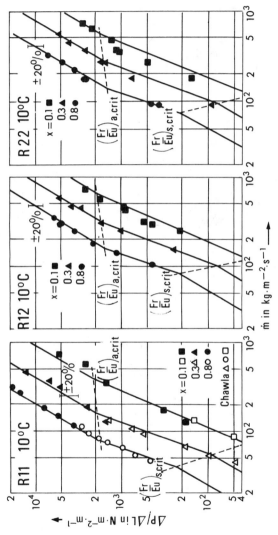

Fig. 9. Comparison of results with the proposed equation.

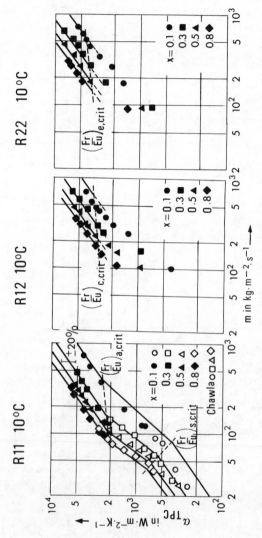

Fig. 10. Comparison of results with the proposed equation.

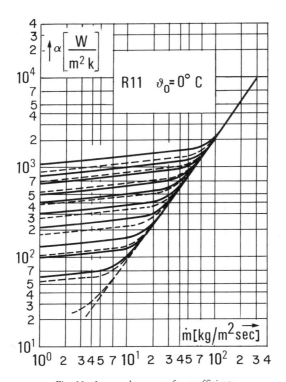

Fig. 11. Average heat transfer coefficients.

Chapter 30

ANALYTICAL MODEL OF THE EFFECT OF SEED CRYSTALS IN FALLING-FILM EVAPORATORS

R.S. Hickman and E. Marschall (*)

1. INTRODUCTION

The problem of controlling or eliminating scale formation on heat transfer surfaces is common in virtually every boiler and evaporator. The problem is especially severe in devices which evaporate brines containing high concentrations of salts and in which it is desirable to concentrate the brine to saturation conditions. In systems designed to recover minerals from brines or for closed cycle systems producing potable water, heavy scaling salt concentrations can be expected. For instance, the treatment of cooling tower blowdown liquids present such problems.

Falling film vapor compression evaporators usually operate with only a few degrees total temperature drop across the condensation-evaporation surfaces and any degradation in the thermal conductivity of the heat transfer surface will seriously limit the production capacity of the evaporator. In several desalination plants severe scaling has been encountered and then eliminated by adding small quantities of the scaling mineral as a solid phase in the brine[1, 2, 3, 4]. In each reported case it was found that mineral deposition on the heat transfer surfaces could be halted by the addition of crystals of the scaling material. The reason that scaling stopped was apparently that the crystals presented a location for deposition of the minerals of the supersaturated brine. Systematic measurements of the effectiveness of the seeding technique have not been made nor is there an analysis against which plant operation can be compared. In each reported instance the seed number density

(*) Department of Mechanical Engineering University of California — Santa Barbara — California

N.B. ALL FIGURES QUOTED IN TEXT ARE AT THE END OF THE CHAPTER

and particle sizes which prevent scaling, have been determined empirically.

The purpose of this work is to formulate a physical model of seed protection mechanism and to provide a predictive analysis which may be used to determine seed concentration and particle size for future evaporator designs.

2. MODEL FORMULATION

Figure 1 is a cross-sectional view of the falling film. The film in existing evaporators is generally turbulent and has an irregular wavy surface. Energy is being transferred from the wall and is vaporizing liquid from the film. For the seeds to prevent scaling all of the mineral originally dissolved in the evaporating fluid must find its way to a seed crystal. For our purposes we will assume that essentially all of the mineral must go to seeds which are in the laminar sublayer. We assume that the seed crystals are very small compared to both the overall film and the laminar sublayer. The first assumption may be justified on the grounds that the turbulent diffusivity will be very high and the mineral can be transferred across the turbulent layer quite easily. Also this assumption will provide a conservative estimate of the required seed content. The second assumption will place limitations on the applicability of the model especially at high Reynolds numbers and for large seeds.

For regions well away from the starting point of a falling film (on the order of 50 to 100 fim thicknesses), the flow will be steady. Waves will be present in the film but their average character and in particular the laminar sublayer will not vary with time or position. The sublayer will be at a constant temperature and supersaturated in the mineral. A slight supersaturation will be tolerated at the wall but no mass flux to the wall will be allowed since the object of the use of the seeds is to force such a condition. All variation in concentration along the wall will be taken to be zero since in practical evaporators, only 1 to 5% of the recirculating brine is evaporated in each pass over the heat transfer surface.

At equilibrium the sublayer is supplied with sufficient solute to maintain a steady concentration profile. Of course the seeds must in fact grow and these larger crystals must be removed from the sublayer or be broken up. The seed growth will not be considered here. In actual devices the seeds are continuously broken up by means of stirring paddles or the larger seeds are selectively removed. It is realizes that in any real sample of particles the particle diameters will not be constant and will have a fairly large variation in size. For tractability we assume here that only a single effective diameter particle size is present.

3. MODEL ANALYSIS

The one dimensional diffusion equation for the sublayer is

$$\partial/\partial y(D\ \partial C/\partial y) = q \tag{1}$$

where q is the volumetric rate of mineral lost to the seeds. As stated above the film is steady along the wall and convective terms do not appear in Eq. 1. If there are N seeds per unit volume and if each seed is adsorbing m mass per unit time then

$$q = N\ \dot{m} \tag{2}$$

If the gradient in supersaturation is sufficiently small then we can assume that each seed is effectively in an infinite bath whose concentration far away is constant and at the local concentration C.

The particles are small compared to the laminar sublayer thickness. The average particle center spacing is approximately

$$\ell \cong (1/N)^{1/3} \tag{3}$$

The equation for mass diffusion to the particle is

$$4\pi D\ r^2 \partial(C_L - C)/\partial r = \dot{m} \tag{4}$$

where r is the distance from a particle. $C_L - C$ is the local depression in concentration due to the presence of the particle.

The particle has a radius r_p and the concentration at the particle is taken as saturated. Then the solution to Eq. (4) for $\ell/2 \gg r_p$ yields

$$\dot{m} = 4\pi r_p D\ (C - C_s) \tag{5}$$

and Eq. (1) becomes

$$\partial/\partial y\ (D\ \partial C/\partial y) = (C - C_s)\ 4\pi r_p DN \tag{6}$$

Here C_s, the saturation concentration, is a function of y since the temperature must actually vary across the sublayer and the equilibrium saturation concentration will be a function of temperature. This variation will be small and will be taken to be zero. With this approximation and assuming D is a constant we have

$$\partial^2(C - C_s)/\partial y^2 = (C - C_s)\ (4\pi r_p N) \tag{7}$$

When the seed prevents scaling at the wall $\partial C/\partial y$ at $y = 0$ must be zero and at the edge of the sublayer, $y = \delta$, , the mass flux is

$$D \ \partial C/\partial y = \dot{M} \qquad (8)$$

Where $\dot{M}$ is the difference between the mineral released by evaporating liquid and that taken up by the seeds in turbulent region. For comparative design purposes, $\dot{M}$ will be treated as the total mineral released of $C_s \ \dot{Q}$. Here $\dot{Q}$ is the volumetric rate of evolution of fresh water per unit area, and C_s is the average concentration in the brine. The concentration solution relating the wall concentration to the saturation concentration is

$$\dot{Q} = (C_w/C_s - 1) \ D \ \sinh \ (a\delta)/[1 + (C_w/C_s - 1) \ \cosh a\delta] \qquad (9)$$

$\dot{Q}$ is the allowable vapor rate for nonscaling operation. Eq.(9) expresses the fact that for a finite vapor production rate the fluid at the wall must be slightly supersaturated. For metal surfaces not containing crystals of the scale, one would not expect scaling to occur for small levels of supersaturation.

4. RESULTS AND APPLICATION

Figure 2 shows a plot of the wall supersaturation versus $\delta \sqrt{4\pi r_p N}$. For a reduction in the supersaturation at the wall by a factor of 10 lower than at the sublayer edge, $\delta \sqrt{4\pi r_p N}$ must be greater than 3.

A relationship between r_p and N can be obtained from the mass fraction by

$$4\pi r_p^3 SN/3 = X \qquad (10)$$

where S is the specific gravity of seed crystals and X is the mass fraction of seeds. For the computations given here $S \cong 2.6$. Figure 3 is a plot of equation (10) for different particle diameters. Also α is shown for each combination of N and r_p.

Now to use equation (9) to predict the allowable yield for a given flow situation, we must know the allowable value of C_w/C_s. In general this value is not known. C_w/C_s is potentially a function of wall material, ph of the brine, seed size and seed material and other contaminate ions. Further D for many minerals is not known.

The fraction of brine evaporated during passage of the fluid over the falling film heat transfer surface is just the mass of liquid evaporated divided by the total mass flow. Using equation (9) we have

$$\phi = Q_V A\rho/\dot{W} = A\rho(C_w/C_s - 1)D\,a\,\sinh a\delta/\dot{W} \tag{11}$$

for $\qquad a\delta \ll 1$ and $C_w/C_s - 1$.

The conditions for a typical panel 5 meters high by 3 meters wide handling 1200gm/sec/meter of width have been calculated from equation (11) using $C_w/C_s - 1 = 10\%$, $D = 1.5 \times 10^{-5}$ cm^2/sec. The laminar layer is estimated to be about 40μ. The results are plotted in Figure 4.

As can be seen for vapor fraction of 3% a value of α of 325 cm^{-1} is required. From Figure 3 for 5μ particles a mass fraction of seeds of 2.5% is required. Measurements on existing plants indicate that mass fractions of from 2 to 3% are required for successful operation [3,4].

CONCLUSIONS

We have obtained an expression relating the allowable vapor production in a given flow condition to the seed slurry concentration and particle size. The Reynolds number dependency appears quite naturally since the laminar sublayer thickness is directly related to the mass flow in the falling film. The dependence upon the diffusion coefficient also appears in the formulation.

Several important parameters are currently unavailable. The maximum supersaturation at the wall and even the diffusion coefficient are unavailable in the literature in the temperature range of interest. Future work should include experimental determination of the values for polluted waters to be treated in evaporation systems.

NOMENCLATURE

A   Heat transfer surface area
C   Concentration of Mineral
D   Diffusion coefficient
$\dot{m}$ Mass flux to seed
$\dot{M}$ Total mass flux of mineral at the edge of the laminar sublayer
N   Number of particles per unit volume
$\dot{Q}$ Volumetric flow rate of liquid vaporized per unit heat transfer area
r   Radius
$\dot{W}$ Total brine flow rate
Y   Distance from the wall

Subscripts

L   Local
M   Maximum concentration
P   Particle
S   Saturation conditions
T   Total
W   Wall conditions

Greek Symbols

α $\sqrt{4\pi r_p N}$
δ laminar sublayer thickness
ρ mass density of the brine

REFERENCES

[1]   Chernazubav, V.B., Zaostravakii, F.P., Shatsillo, V.G., Golub, S.I., Navikov, E.P., and Tkach, V.I., "Prevention of Scale Formation in Distillation Desalination Plants by Means of Seeding", Proceedings of the First International Symposium on Water Desalination, Washington D.C., October 1965, Office of Saline Water.

[2]   Gainey, R.J., Tharp, C.A., and Cadwallader, E.A., "$CaSO_4$ Seeding Prevents $CaSO_4$ Scale", Industrial and Engineering Chemistry, 55, 39, 1963.

[3]   Young, K.G. and Rhoemer, W., Reading and Bates Offshore Drilling Corporation, "Private Communication", 1970.

[4]   Schmitt, R.P., Hurley, S.M., "Developments in Use of Desalting by the United States Military", Chemical Engineering Symposium, Water, 1970, pp. 178-181.

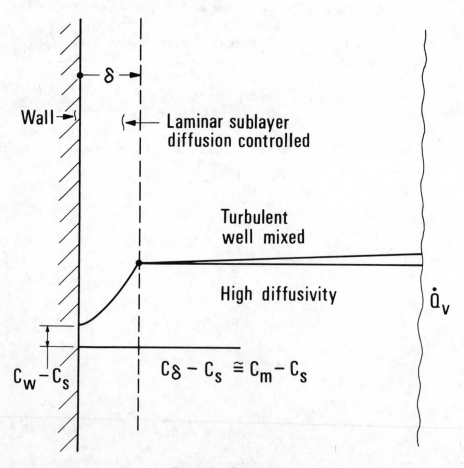

Fig. 1. Flow field geometry.

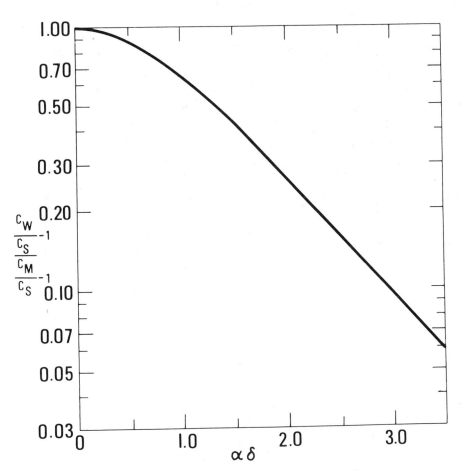

Fig. 2. Variation in the wall supersaturation as a function of $\alpha\delta$.

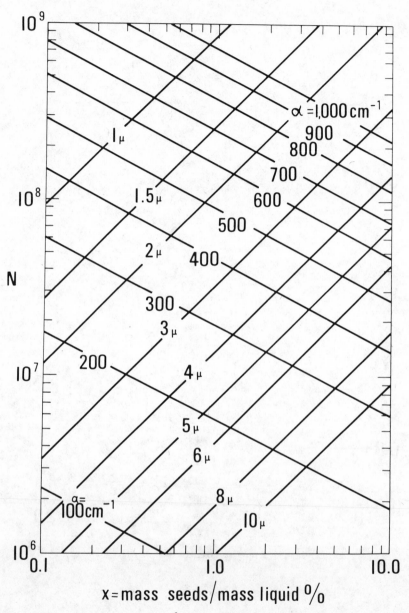

Fig. 3. Required number density (particle/cm³) versus solids mass fraction for particles of specific gravity equal to 2.7.

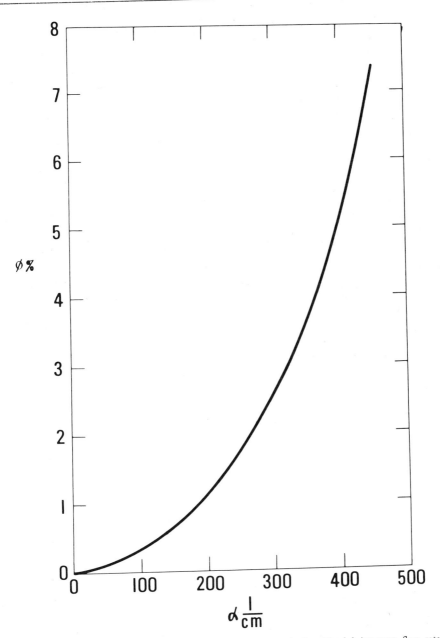

Fig. 4. Allowed vapor rate fraction from a falling film versus required α. Total brine mass flow rate at 1# m/ft-sec on an 8' wide X 15' high panel.

Chapter 31

THE WALL HEAT TRANSFER COEFFICIENT IN PACKED BEDS

W. Hennecke and E.U. Schlünder (*)

Tubular fixed-bed reactors are usually used for carrying out exothermic gas-phase reactions occurring in porous catalysts. The design of catalytic reactors requires the knowledge of the effective thermal conductivity and the heat transfer coefficient at the wall. There are reliable correlations available for estimating the effective thermal conductivity in packed beds, but there is much uncertainty in estimating the heat transfer coefficient at the wall.

A new method is developed for determining the wall heat transfer coefficient. Provided the effective thermal conductivity and the radial velocity profile are known the wall heat transfer coefficient can be calculated by this method from measured overall heat transfer coefficients.

In order to analyse the heat transfer in packed beds by mathematical methods a model is required. In the model (see Figure 1) which is used in all previous investigations it is assumed that

1. the packed tube is divided into a core region and a small wall region
2. the hydrodynamic boundary layer at the wall is disturbed at each particle and therefore, the wall heat transfer coefficient remains constant along the length of the tube
3. the radial velocity profile is uniform (rod flow)
4. the effective thermal conductivity does not vary in radial direction
5. the conduction of heat in axial direction may be neglected.

This model was also used by the authors in the beginning of the investigation. It was improved later on.

(*) Institut für Thermische Verfahrenstechnik, Universität Karlsruhe (TH) — Germany (FR)

N.B. ALL FIGURES QUOTED IN THE TEXT ARE AT THE END OF THE CHAPTER

Mathematical methods for determining the wall heat transfer coefficient

Uniform radial velocity profile

The above described model of the heat transfer in packed beds leads to the differential equation

$$C_p \rho u \frac{\partial \vartheta}{\partial \ell} = \Lambda \left[\frac{\partial^2 \vartheta}{\partial r^2} + \frac{1}{r} \frac{\partial \vartheta}{\partial r} \right] \tag{1}$$

The boundary condition at the wall is

$$r = R - \Lambda \frac{\partial \vartheta}{\partial r} = a_w (\vartheta - \vartheta_w) \tag{2}$$

It is assumed that the boundary layer ϵ is very small and that therefore $r = R$ is equivalent to $r = R - \epsilon$.

The boundary condition at the entrance is

$$1 = o \quad \vartheta = \vartheta_o \tag{3}$$

If ϑ_w is constant along the tube, integration of the differential equation leads to [5]

$$\frac{\vartheta - \vartheta_w}{\vartheta_o - \vartheta_w} = NU_w \sum_{n=1}^{\infty} \frac{J_o\left(\beta_n \frac{r}{R}\right) \cdot e^{-\frac{4\beta_n}{PE^{D/L}}}}{\left[\frac{NU_w^2}{4} + \beta_n^2 \right] J_o(\beta_n)} \tag{4}$$

with

$$\frac{NU_w}{2} J_o(\beta_n) = \beta_n J_1(\beta_n) \ . \tag{5}$$

The local overall heat transfer coefficient is defined by

$$a_x = \frac{- \Lambda \frac{\partial \vartheta}{\partial r} \Big/ r = R}{\vartheta_m - \vartheta_w} \ . \tag{6}$$

If the temperature gradient and the mean bulk temperature ϑ_m are calculated from eq. (4), it follows from eq. (6)

$$NU_x = \cfrac{\displaystyle\sum_{n=1}^{\infty} \cfrac{\exp\left(-\cfrac{4\,\beta_n^2}{PE^{D/L}}\right)}{\cfrac{NU_w^2}{4} + \beta_n^2}}{\displaystyle\sum_{n=1}^{\infty} \cfrac{\exp\left(-\cfrac{4\,\beta_n^2}{PE^{D/L}}\right)}{\left[\cfrac{NU_w^2}{4} + \beta_n^2\right]\beta_n^2}} \tag{7}$$

The average overall heat transfer coefficients follow from eq. (7) by integration

$$NU = \frac{1}{L} \int_0^L NU_x \, dl \; . \tag{8}$$

As eq. (7) does not converge for $L = 0$, the integral (8) has to be separated :

$$NU = \frac{1}{L}\left[\int_0^{L_\epsilon} NU_x \, dl + \int_{L_\epsilon}^L Nu_x \, dl \right] \tag{9}$$

L_ϵ is a small length.

The second integral can be calculated numerically. The first integral can be integrated analytically, because another solution of eq. (1) exists for short residence times [5], which leads to

$$\frac{1}{L_\epsilon} \int_0^{L_\epsilon} NU_x \, dl = NU_w\left[1 - \frac{4NU_w}{3\sqrt{\pi}\sqrt{PE^{D/L}}} + \frac{1}{2\,PE^{D/L}}\left(NU_w - \frac{3}{2} \right) \right] \tag{10}$$

Non-uniform radial velocity profile

In a packed tube the velocity in the bare cross-section is not uniform, because the porosity is not uniform. The velocity profile has a maximum near the wall, as Schwartz and Smith [25] reported. The radial velocity profiles are approximated by a parabolic equation which is useful for integrating the differential equation:

$$\frac{u}{\bar{u}} = \frac{K + \dfrac{p+2}{2}\left(\dfrac{r}{R}\right)^p}{K + 1} \; . \tag{11}$$

Eq. (11) does not show the sharp decrease very close to the wall. This effect is taken into account by the wall heat transfer coefficient.

K and P were determined from experiments made by Schwartz and Smith [25] by applying the least square method. Found to be a function of D/d were only spheres

$$K = 1.5 + 0.0006 \left(\frac{D}{d} - 2 \right)^3 \tag{12}$$

$$P = 1.14 \sqrt[3]{\frac{D}{d} - 2} \tag{13}$$

cylinders h = d:

$$K = 2.3 + 0.0002 \left(\frac{D}{d} - 2 \right) \tag{14}$$

$$P = 1.06 \sqrt[3]{\frac{D}{d} - 2} \tag{15}$$

Fig. 2 shows experimental data of Schwartz and Smith [25] and calculated radial velocity profiles according to eq. (11).

Fig. 3. shows the maximum velocity at the wall and the minimum velocity in the axis as a function of D/d.

Kubota, Ikeda and Nishimura [15] presented experimental results showing that the average velocity in the interstices does not vary with the radial position. From their experiments and the experiments of Fahien and Smith [9], who measured $u \cdot d/A$ as a function of the radial position, it was concluded that the effective thermal conductivity is uniform in the bed, although the velocity in the bare cross-section varies with the radial position.

Using (11) as the equation for the radial velocity profile the differential equation for the conduction of heat in a packed bed can be written in the form

$$\frac{K + \frac{P + 2}{2} \left(\frac{r}{R} \right)^P}{K + 1} \ \bar{U} \ \frac{\partial \vartheta}{\partial \ell} = A \left[\frac{\partial^2 \vartheta}{\partial r^2} + \frac{1}{r} \ \frac{\partial \vartheta}{\partial r} \right] \tag{16}$$

Eq. (16) cannot be integrated with the boundary condition (3), but solutions were found with the boundary condition (5) at the inlet and the boundary condition

$$r = R \quad \vartheta = \vartheta_w \tag{17}$$

at the wall for long residence times and short residence times.

Long residence times

In a dimensionless form with $\eta = \dfrac{r}{R}$, $\xi = \dfrac{1}{L}$ and $\theta = \dfrac{\vartheta - \vartheta_w}{\vartheta_o - \vartheta_w}$

Eq. (16) can be written :

$$\frac{1}{4}\, PE\, \frac{D}{L}\, \frac{\partial \theta}{\partial \xi} = \frac{K + 1}{K + \dfrac{P+2}{2}\, \eta^P}\left[\frac{\partial^2 \theta}{\partial \eta^2} + \frac{1}{\eta}\, \frac{\partial \theta}{\partial \eta}\right] \tag{18}$$

with

$$f = f_\eta (\eta) \; . \; f_\xi \; (\xi)$$

Eq. (18) can be separated in

$$\frac{1}{4}\, PE\, \frac{D}{L}\, \frac{f_\xi'}{f_\xi} = - m^2 \, . \tag{19}$$

and

$$\frac{K + 1}{K + \dfrac{P+2}{2}\, \eta^P} \cdot \frac{f_\eta'' + \dfrac{1}{\eta}\, f_\eta'}{f_\eta} = - m^2 \tag{20}$$

Integration of eq. (19) leads to

$$f_\xi = \text{const}\, . \, e^{-\dfrac{4m^2}{PE}\dfrac{D/L}{}} \tag{21}$$

Eq. (20) can be integrated by the series expansion

$$f_\eta = \sum_{n = o}^{\infty} a_n\, \eta^n \tag{22}$$

with P being restricted to natural numbers.
One of the coefficients a is given by the condition $f_\eta (0) = 0$:

$$a_1 = 0$$

A second coefficient can be chosen :

$$a_o = 1$$

The other coefficients have to be calculated with the formula of recursion

$$a_{\nu + 1} = - m^2\, \frac{K a_{\nu - 1} + \dfrac{P + 2}{2}\, a_{\nu - P - 1}}{(K + 1)(\nu + 1)^2} \tag{23}$$

From the boundary condition

$$f\eta \ (1) \ = \ 0$$

it follows that

$$a_o \ + \ a_1 \ + \ a_2 \ + \ a_3 \ + \ \ldots \ = \ 0 \tag{24}$$

The parameter m^2 has to satisfy eq. (24). There is a infinite number of solutions. The first solution is computed with a FORTRAN-program. For long residence times only the first solution is used for calculating the NU-number

$$NU_{ausg} \ = \ m^2 \tag{25}$$

Figure 3 shows NU_{ausg} for packed beds of spheres and cylinders as a function of D/d.

Short residence times

For short residence times a simple solution is found (see [11])

$$NU_{anl} \ = \ \frac{2}{\sqrt{\pi}} \ \sqrt{PE \ \frac{D}{L}} \ \sqrt{\frac{K \ + \ \dfrac{P+2}{2}}{K \ + \ 1}} \tag{26}$$

The solution for long times (25) and the solution for short times (42) are superposed

$$NU_k \ = \ \sqrt{NU^2_{ausg} \ + \ NU^2_{anl}} \tag{27}$$

Eq. (27) is a good approximation for all residence times. An approximate solution for the boundary condition (3) is

$$NU \ = \ \frac{NU_k}{1 \ + \ \dfrac{NU_k}{NU_w}} \tag{28}$$

The effective thermal conductivity is calculated by [33]:

$$\frac{\Lambda}{\lambda} \ = \ \frac{\Lambda_s}{\lambda} \ + \ \frac{Pe_d}{8} \ f_{DL} \tag{29}$$

$$f_{DL} \ = \ \frac{1 \ + \ \dfrac{L/D}{2 \ - \ (1 - 2d/D)^2}}{1 \ + \ L/D} \qquad \text{for spheres} \tag{30}$$

$$f_{DL} = 1 \quad \text{for non-spherical particles} \tag{30a}$$

$$\frac{\Lambda_s}{\lambda} = (1 - \sqrt{1 - \Psi}) \frac{\lambda_H}{\lambda} + \sqrt{1 - \Psi} \frac{\Lambda_s^*}{\lambda^*} \frac{\lambda^*}{\lambda} \tag{31}$$

$$\frac{\lambda_H}{\lambda} = \frac{1}{1 + Kn'/\Psi} + \Psi NU_{Str} \tag{32}$$

$$Kn' = \frac{2 M}{d} \frac{2 - \gamma}{\gamma} \tag{33}$$

$$NU_{Str} = \frac{0.04 \, C_s}{\frac{2}{\epsilon_{Str}} - 1} \left(\frac{Tm}{100}\right)^3 \frac{d}{\lambda} \tag{34}$$

$$\frac{\Lambda_s^*}{\lambda} = \frac{2}{N - M} \left[\frac{[N - (1 + Kn') \frac{\lambda^*}{\lambda_p^*}] B \ln \frac{N}{M}}{(N - M)^2} - \frac{B - 1}{N - M} (1 - Kn') + \right.$$

$$\left. + \frac{B + 1}{2B} \frac{\lambda_p^*}{\lambda^*} [1 - (N - M)^2 - (B - 1) Kn'] \right] \tag{35}$$

$$M = B \left[\frac{\lambda^*}{\lambda_p^*} + Kn' (1 - Bi_{Str}^*) \right] \tag{36}$$

$$N = (1 + Bi_{Str}^*)(1 + Kn') \tag{37}$$

$$\lambda^* = \lambda \sqrt{1 + (0.04 \cdot Pe_d^{0.75})^2} \tag{38}$$

$$\lambda_p^* = \frac{\lambda_p}{1 + \frac{S_{ox}}{\lambda_{ox}} \frac{\lambda_p}{d}} \tag{39}$$

$$\frac{\lambda_{ox}}{S_{ox}} = 5 \cdot 10^3 \div 10^4 \quad \text{W/m grd}$$

$$Bi^*_{Str} = NU_{Str} \frac{\lambda}{\lambda^*_p} \tag{40}$$

$$B = C \left[\frac{1 - \Psi}{\Psi} \right]^{1.11} \tag{41}$$

The coefficients C are listed in table 1

TABLE 1

	C
spheres	1.25
granular materials	1.4
cylinder, Raschigrings	2.5

In [11] a suitable equivalent diameter for nonspherical particles was found :

cylinders and hollow cylinders $h \geqslant d$:

$$d^* = \left[\sqrt{\frac{1}{2} + \frac{h}{d}} + \sqrt{\frac{1}{2} + \frac{h}{d_i} \left(\frac{d_i}{d_a} \right)^2} \right] d \tag{42}$$

granular material :

$$d^* = 1.23 \, d \tag{43}$$

(d is the arithmetic mean diameter)

If overall heat transfer coefficients are reported in the form

$$Nu = f \, (Re_D, \, Pr, \, L/D, \, D/d)$$

one gets

$$NU = Nu \Big/ \frac{\Lambda}{\lambda} \tag{44}$$

where Λ/λ is calculated by eq. (45) and

$$PE \, \frac{D}{L} = Re_D \cdot Pr \, \frac{D}{L} \bigg/ \frac{\Lambda}{\lambda} \tag{45}$$

If a uniform velocity profile is assumed,

$$NU_w = f\left(NU, PE \, \frac{D}{L}\right)$$

is calculated with a FORTRAN-Program.

If a non-uniform velocity profile is assumed, one gets

$$NU_w = \frac{NU}{1 - NU/NU_k} \tag{46}$$

with NU_K according to eq. (27).

Finally one gets

$$Nu_w = NU_w \, \frac{\Lambda}{\lambda} \, \frac{d}{D} \tag{47}$$

Experiments

Experimental data of many authors were analysed during this investigation. As there were many discrepancies, it seemed to be necessary to measure overall heat transfer coefficients again. The experimental equipment is described in [11]. Figure 4 shows the main features of the experimental set up.

The test unit consisted of an inner tube of clay and an outer tube of plexiglass and was made up of several sections. Water evaporated at the inner wall and the evaporation rate was measured with a burette. The temperatures at the wall and at the entrance of the tube were determined by thermoelements. The exit temperatures were calculated from the flow rate of air and the rate of the evaporating water.

This experimental method of evaporating water permits the measurement of heat transfer coefficients and mass transfer coefficients simultaneously. The coefficients were calculated for each section. In this way the influence of the tube length could be investigated throughly. The ratio L/D was varied between 0.3 and 4.3.

Packings of spheres, cylinders (h = d and h = 2 d), and Raschigrings made from styropore, polystyrene, glass and copper, were investigated.

Determination of wall heat transfer coefficients

5000 experimental data of overall heat transfer and mass transfer coefficients of 14 investigations, including our own experiments [6, 8, 11, 13, 14, 16, 17, 18, 19, 20, 21, 24, 29, 30, 32], were analysed. Figures 5 and 6 show the scope of the investigation.

In the beginning the wall heat transfer coefficients were calculated on the assumption of a uniform velocity profile. Figures 7 and 8 show series of experiments for spheres of ceramic and metallic materials. There is much scattering, but one can see that the Nu_w-numbers for short tubes are much higher than the Nu_w-numbers for long tubes. Generally the Nu_w-numbers decrease with decreasing Pe_d-numbers, but in some cases there is a sharp rise. For a number of experiments negative Nu_w-numbers are calculated.

Figures 9 and 10 show the same experiments. The Nu_w-numbers are now calculated on the assumption of a non-uniform radial velocity profile. Positive Nu_w-numbers are calculated for all experiments.

The Nu_w-numbers for short tubes are much lower, whereas the Nu_w-numbers for long tubes are only a little lower. Therefore, the Nu_w-numbers do not depend on the length of the tube to such a high degree as they had before, but there is still a marked dependence. It was assumed, that the wall heat transfer coefficient was constant along a tube. If Nu_w-numbers depend on the tube length, this may result from errors in measurements or from failures of the method by which they were determined. If wall heat transfer coefficients are determined from measured overall heat transfer coefficients, small errors in measurements lead to great errors in the calculation of the wall heat transfer coefficients.

If the tubes are very long $(L/D > 10)$, it is very difficult to find the exact temperature of the flowing medium at the exit of the tube, when the medium is heated with constant wall temperature. Systematic errors in temperature measurements occurred in the investigations of Chu and Storrow [6] Koch [24] and to some extent of Leva [17, 18, 19, 20] and Kling [13].

It is seen from our own experiments (for example, figure 11) that there is a systematic influence of the tube length, if Pe_d-numbers are low. It is possible that there are special effects at the entrance, for instance free convection. But the large influence of the tube length cannot result from these effects. Probably this influence is caused by failures of the method. The influence of the tube length is more pronounced if the Nu_w-numbers are determined on the assumption of a uniform velocity profile. If the Nu_w-numbers are calculated with the more complicated model of a non-uniform velocity profile, they are not dependent on the tube length for high Pe_d-numbers.

It is shown, in a separate investigation [11], that it is permissible to neglect axial conduction. But at low Pe_d-numbers it is probably not allowed to assume that the hydrodynamic boundary layer is independent of the tube length. If the boundary layer varies along the tube length, the wall heat transfer coefficient also varies. In this case it is not possible to calculate wall heat transfer coefficients from measured overall heat transfer coefficients.

Formula for wall heat transfer coefficients

In order to develop a correlation for wall heat transfer coefficients, a simple model is assumed (Figure 12).

There is a laminar boundary layer at the wall, which is determined by the particles. In the distance ϵ the area is partly bored by the particles. In the free area ϕ heat is conducted through the laminar boundary layer by convection, in the area $(1 - \varphi)$ heat is conducted through the fluid medium and the particle. Assuming constant temperatures in the fluid and in the solid phase and neglecting radiation, one can formulate :

$$\alpha_w = \zeta \, \alpha_{convection} + (1 - \zeta) \alpha_{conduction} \qquad (48)$$

Temperature gradients being large, one may set

$$\alpha_{convection} = \frac{\lambda}{\epsilon} \qquad (49)$$

and

$$\alpha_{conduction} = \frac{\Lambda_w}{\epsilon} \qquad (50)$$

Λ_w is the effective thermal conductivity at the wall near the point of contact. If the boundary layer is determined by the particles it is convenient to bring eq. (49) in a dimensionless form

$$\frac{d}{\epsilon} = \frac{\alpha_{convection} \cdot d}{\lambda} = Nu_p \qquad (51)$$

Nu_p is calculated in the same way as the Nusselt-number of a plate with the length d. The area φ is determined by geometrical relations, as if the particles were spheres in a cubic arrangment

$$\zeta = 1 - \pi \left[\frac{\epsilon}{d} - \frac{\epsilon^2}{d^2} \right] = 1 - \pi \left[\frac{1}{NU_p} - \frac{1}{NU_p^2} \right] \tag{52}$$

Finally, one gets the formula for spheres, taking also in account radiation and the effect of the tube length:

$$Nu_w = \left[Nu_p + c_a \cdot \pi \left(\frac{\Lambda_w}{\lambda} - 1 \right) \left(1 - \frac{1}{NU_p} \right) + NU_{Str} \right] \frac{1}{1 + \frac{L/D}{Pe_p} c_d} \tag{53}$$

$$NU_p = \frac{c_b}{Pr^{1/6}} \sqrt{0.194 \, Pe_p^2 + \frac{0.34 \cdot 10^{-4}}{Pr^{2/3}} \, Pe_p^3} \tag{54}$$

$$Pe_p = \sqrt{\left(\frac{u_w d}{a} \right)^2 + c_c} \tag{55}$$

$$u_w = \frac{\bar{u}}{\Psi} \frac{K + \frac{P+2}{2}}{K+1} \tag{56}$$

$$K = 1.5 + 0.0006 \left(\frac{D}{d} - 2 \right)^3 \tag{12}$$

$$P = 1.14 \sqrt[3]{\frac{D}{d} - 2} \tag{13}$$

$$c_c = c_c' \frac{34.5}{D/d} \tag{57}$$

$$\frac{\Lambda_w}{\lambda} = (1 - \sqrt{1 - \Psi_w}) \frac{\lambda HW}{\lambda} + \sqrt{1 - \Psi_w} \frac{\Lambda_w^*}{\lambda} \tag{58}$$

$$\Psi_w = 0.5 - \frac{1/6}{NU_p - 1} \tag{59}$$

$$\frac{\lambda HW}{\lambda} = \frac{1}{1 + Kn_w'/\Psi_w} \tag{60}$$

$$Kn_w' = \frac{2M}{d} \frac{2 - \gamma}{\gamma} Nu_p \tag{61}$$

$$\frac{\Lambda^*_w}{\lambda} = \frac{2}{N-M} \left[\frac{\left[N - (1 + Kn'_w) \frac{\lambda}{\lambda^*_p}\right] B \ell n \frac{N}{M}}{(N-M)^2} - \frac{B-1}{N-M} (1 - Kn'_w) \right] -$$

$$- \frac{B+1}{2B} \frac{\lambda^*_p}{\lambda} \left[1 - (N-M)^2 - (B-1) Kn'_w \right] \qquad (62)$$

with

$$M = B \left[\frac{\lambda}{\lambda^*_p} + Kn'_w \right] \qquad (63)$$

$$N = (1 + Kn'_w) \qquad (64)$$

$$\lambda^*_p = \frac{\lambda_p}{1 + \frac{S_{ox}}{\lambda_{ox}} \frac{\lambda_p}{d}} \qquad (39)$$

$$\frac{\lambda_{ox}}{S_{ox}} = 5 \cdot 10^3 \div 10^4 \ W/m^2 \ grd$$

$$B = C \left(\frac{1 - \Psi_w}{\Psi_w} \right)^{0.75} \qquad (65)$$

The coefficients C are listed in table 1 (page 9)

$$Nu_{Str} = \frac{0.04 \ C_s}{\frac{2}{\epsilon_{Str}} - 1} \left(\frac{T_m}{100} \right)^3 \frac{d}{\lambda} \qquad (34)$$

The parameters c_a, c_b, c'_c and c_d had to be determined in such a way that the deviation was minimal. The results are shown in table 2 (page 18)

The formula for mass transfer is

$$Sh_w = \left[Sh_p + c_a \pi \left(\frac{\Delta w}{\delta} - 1 \right) \left(1 - \frac{1}{Sh_p} \right) \right] \frac{1}{1 + \frac{L/D}{Pe'_p} c_d} \qquad (67)$$

Sh_p has to be calculated with eq. (54) using Sc instead of Pr and Pe'_p instead of Pe_ρ.

$$Pe'_p = \sqrt{\left(\frac{u_w \cdot d}{\delta}\right)^2 + c_c} \qquad (68)$$

$$\frac{\Delta w}{\delta} = 1 - \sqrt{1 - \Psi_w} \qquad (69)$$

Eq. (53) and (67) are also valid for granular materials if the equivalent diameter d*(eq.43) is used instead of the diameter of the sphere.

Eq. (67) is also valid for mass transfer in packed tubes of cylinders and hollow cylinders using the equivalent diameter d*(eq.42).

For heat transfer in packed beds with cylinders h = d the formula is

$$Nu_w = \frac{a_w d^*}{\lambda} = \left[Nu_p + c_a \pi \left(\frac{\Lambda_w}{\lambda} - 1\right)\left(1 - \frac{1}{\sqrt[3]{Nu_p}}\right) + Nu_{Str} \right] \frac{1}{1 + \dfrac{L/D}{Pe_p} c_d} \qquad (70)$$

and for cylinders h = 2 d and hollow cylinders (raschigrings)

$$Nu_w = \frac{a_w d^*}{\lambda} = \left[Nu_p + c_a \pi \left(\frac{\Lambda_w}{\lambda} - 1\right)\left(1 - \frac{1.5}{\sqrt[3]{Nu_p}}\right) + Nu_{Str} \right] \frac{1}{1 + \dfrac{L/D}{Pe_p} c_d} \qquad (71)$$

Nu_p has to be calculated with eq. (54) using the equivalent diameter and calculating K and P with eq. (14) and (15). In eq. (54) the parameter c_b is a function of D/d : Cylinders, $D/d \geqslant 5$

$$c_b = c_b' \left(1 + \frac{3.5}{D/d}\right) \qquad (72)$$

Raschigrings, $D/d \geqslant 5$

$$c_b = c_b' \left(1 + \frac{7}{D/d}\right) \qquad (73)$$

In case of $D/d < 5$: $c_b = c_b'$

Figure 13 shows wall heat transfer and wall mass transfer coefficients for spheres of styropor, glass and copper. The equations for the Nu_w-numbers are based on experiments in the range of

$$10 < Re_d < 10^5$$

$$0.6 < Pr < 7$$

$$2 < D/d < 50$$

$$0.3 < L/D < 58$$

The average deviation is less than 20% for the own experiments and less than 30% for the experiments of the other authors.

TABLE 2

		c_a	c_b	c_c'	c_d
spheres	heat transfer	0.5	1.3	1500	40
	mass transfer	0.5	1.3	6000	40
	heat transfer water	0.5	1.3	30000	40
granular mat.	heat transfer	0.5	1.3	1500	40
		c_a	c_b'	c_c'	c_d
cylinders h = d	heat transfer	1.0	0.8	1000	30
	mass transfer	1.0	0.8	250	10
cylinders h = 2d	heat transfer	1.0	1.1	250	30
	mass transfer	1.0	0.8	250	10
raschigrings	heat transfer	1.0	1.2	250	30
	mass transfer	1.0	1.2	250	10

Comparison with other authors

Figure 14 shows correlations for Nu_w-numbers. Beek [1] recommends the equation of Thoenes and Kramers [27]. Hlavacek [16] recommends the equation of Yagi and Kunii [30].

For large Pe_d-numbers these are in agreement with our own equations. For small Pe_d-numbers they deviate because there is an influence of the tube length. Valstar, Bik and van den Berg [28] revealed a striking discrepancy between the correlation of Yagi and Kunii and their own experiments.

The equation found in this investigation is corroborated by experiments of Brötz [2] (Figure 15).

Valstar, Bik and van den Berg [28] measured temperature profiles in a pilot reactor. They could verify these temperature profiles by theoretical calculations using wall heat transfer coefficients which are in accordance with the authors own equation.

Calculation of overall heat transfer calculations

Overall heat transfer coefficients can be calculated with the equation

$$Nu_D = \frac{NU_k}{1 + NU_k/NU_w} \frac{\Lambda}{\lambda} \tag{73}$$

The effective thermal conductivity has to be calculated with eq. (29).

$$Nu_w = NU_w \frac{D}{d} \bigg/ \frac{\Lambda}{\lambda} \tag{74}$$

The Nu_w-number has to be calculated with equation (53), (70) or (71) taking into account the form of the particles. In case of non-spherical particles the equivalent diameter eq. (42) or (43) has to be used instead of the particle diameter.

The Nu_K-number has to be calculated with eq. (27). Mass transfer coefficients can be calculated analogously. With a comparison of calculated and experimental values a good agreement is found (Figure 16). Large deviations are found only with Chu and Storrow [6], whose experimental results are doubtful because of their method of measuring the outlet temperature.

NOMENCLATURE

a) **Latin Letters**

a	m^2/s	temperature conductivity of the fluid
A	m^2/s	radial effective thermal conductivity $(A = \Lambda/gc_p)$
c_a	—	coefficients listed in table 2 (page
c_b	—	coefficients listed in table 2 (page
c_b'	—	coefficients listed in table 2 (page
c_c'	—	coefficients listed in table 2 (page
c_d	—	coefficients listed in table 2 (page
c_p	J/kg grd	heat capacity of the fluid phase
c_{Str}	$W/m^2 grd^4$	radiation coefficient of a black body
d	m	diameter of a particle
$d*$	m	equivalent diameter (eq. 51 or 52)
d_i	m	inside diameter of a hollow cylinder
D	m	tube diameter
h	m	length of a cylinder
J_o	—	Bessel-function of zero Order and first kind
J_1	—	Bessel-function of first Order and first kind
K	—	coefficient in eq. (17)
l	m	length
L	m	tube length
m	—	separation parameter
P	—	coefficient in eq. (17)
r	m	radius
R	m	tube radius
s_{ox}	m	thickness of a layer of oxide
T	K	temperature
u	m/s	velocity in the bare cross-section
$\bar{u}$	m/s	medium velocity in the bare tube

b) Greek letters

α	W/m^2 grd	overall heat transfer coefficient
α_w	W/m^2 grd	wall heat transfer coefficient
β_w	m/h	wall mass transfer coefficient
γ	—	accomodation coefficient
δ	m^2/s	diffusivity
Δ	m^2/s	radial effective diffusivity
ϵ_{Str}	m^2/s	emissivity
ϑ	grd.	temperature
ϑ_o	grd	inlet temperature
λ	W/m grd	thermal conductivity of the fluid phase
λ_{ox}	W/m grd	thermal conductivity of a layer of oxide
λ_p	W/m grd	thermal conductivity of a particle
Λ	W/m grd	radial effective thermal conductivity
M	m	free path of a molecule
ν	m^2/s	kinematic viscosity
ζ	kg/m^3	density
Ψ	—	porosity

c) Subscripts

d	particle diameter
D	tube diameter
m	average
r	radius
W	wall
X	local

e) dimensionless numbers

Nu_D	=	$\alpha \cdot D/\lambda$
Nu_w	=	$\alpha_w d/\lambda$
NU	=	$\alpha \cdot D/\Lambda$
NU_w	=	$\alpha_w D/\Lambda$
Pe_d	=	$\bar{u}d/a$

$Sc = v/\delta$

$Sh_w = \beta_w d/\delta$

Pe_p		Peclét number defined by eq. 55
Pe'_p		Peclét number defined by eq. 68
Nu_{Str}		Nusselt-number defined by eq. 34

$PE\, D/L = \bar{u}D^2/AL$

$Pr \quad = \quad v/a$

$Re_D \quad = \quad \bar{u}D/v$

$Re_d \quad = \quad \bar{u}d/v$

REFERENCES

[1]   Beek, J., Advances in Chemical Engineering, Vol. 3, Academic Press, New-York, (1962)

[2]   Brötz, W., CIT 23, 408/416 (1951)

[3]   Calderbanks, P.H. and Pogorski, L.A., Trans. Inst. Chem. Engrs. 35, 195/207 (1957).

[4]   Campbell, J.M. and Huntington, R.L., Petr. Ref. 31, 123/131 (1952)

[5]   Carslaw, H.S. and Jaeger, J.C., Conduction of Heat in Solids, Oxford Press (1948)

[6]   Chu, Y.C. and Storrow, J.A., Chem. Eng. Sc. 1, 230/237 (1952)

[7]   Coberly, C.A. and Marshall, W.R., Chem. Eng. Progr. 47, 141 (1951)

[8]   Colburn, P., Ind. Eng. Chem. 23, 910/913 (1931).

[9]   Fahien, R.W. and Smith, J.M., A.I.Ch.E.J. 1, 28/37 (1955)

[10]   Felix, J.R., Ph. D. thesis, University of Wisconsin (1951)

[11]   Hennecke, Fr.—W., Diss. Universität Karlsruhe (TH) (1972)

[12]   Hlavacek, V., Ind. Eng. Chem. 62, 8/26 (1970)

[13]   Koch, R., VDI—Forschungsheft Nr. 469 (1958)

[14]   Kubota, H., Ikeda, M. and Nishimura, Y., Kagaku Kogaku, 4, 58/61 (1966)

[15]   Kunii, D., Suzuki, M. and Ono, N., J. of Chem. Eng. Jap. 1, 21/26 (1968)

[16]   Leva, M., Ind. Eng. Chem. 38, 415/419 (1946)

[17]   Leva, M., Ind. Eng. Chem. 39, 865/882 (1947)

[18]   Leva, M., Ind. Eng. Chem. 42, 2498/2501 (1950)

[19]   Leva, M. and Grummer, M., Ind. Eng. Chem. 40, 415/419 (1948)

[20]   Leva, M., Weintraub, M., Grummer, M. and Clark, E.L., Ind. Eng. Chem. 40, 747/752 (1948)

[21]   Plautz, D.A., Ph.D. thesis, University of Illinois (1953)

[22]   Quinton, J.H. and Storrow, J.A., Chem. Eng. Sc.5, 245/257 (1956)

[23]   Schlünder, E.U., CIT 38, 1161/68 (1966)

[24]   Schwartz, C.E. and Smith, J.M., Ind. Eng. Chem. 45, 1209/18 (1953)

[25]   Seidel, H.P., CIT 37, 1125/32 (1965)

[26]   Thoenes, D., Jr. and Kramers, H., Chem. Eng. Sc.8, 271 (1958)

[27]   Valstar, J.M., Bik, J.D. and van den Berg, P.J., Trans. Inst. Chem. Eng. 47, 136 (1969)

[28]   Verschoor, H. and Schuit, G.C.A., Appl.Sc. Res.A2, 97/119 (1949)

[29]   Yagi, S. and Kunii, D., International Development in Heat Transfer, Part. IV, 750 (1961)

[30]   Yagi, S., Kunii, D. and Endo, E., Int. J. Heat Mass Transfer 7, 333/339 (1964)

[31]   Yagi, S. and Wakao, N., A.I.Ch.E.J. 5, 1, 79/85 (1959)

[32]   Zehner, P., Dissertation, Universität Karsruhe (1972).

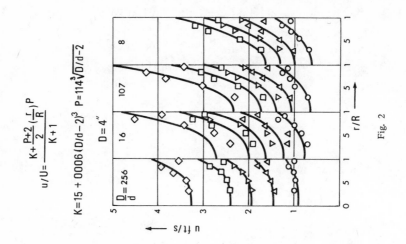

$$u/U = \frac{K + \frac{P+2}{2} \left(\frac{r}{R}\right)^P}{K+1}$$

$K = 1.5 + 0.006 (D/d - 2)^3$ $P = 1.14 \sqrt[3]{D/d} - 2$

$D = 4''$

Fig. 2

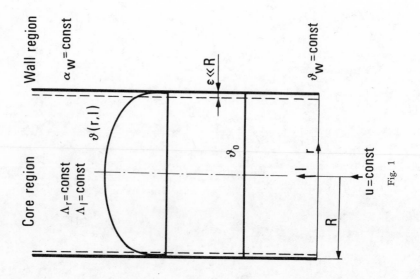

Wall region $\alpha_W = \text{const}$

$\vartheta(r,l)$

$\varepsilon \ll R$

$\vartheta_W = \text{const}$

Core region $\Lambda_r = \text{const}$
$\Lambda_l = \text{const}$

ϑ_0

R

$u = \text{const}$

Fig. 1

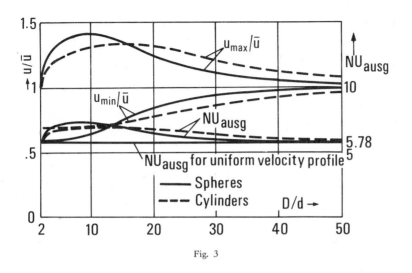

Fig. 3

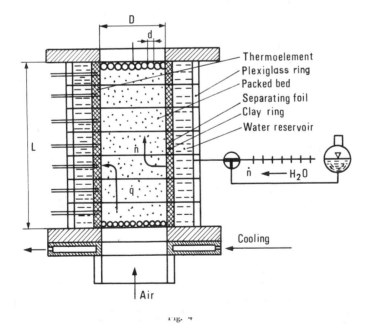

Fig. 4

author	D mm	L/D	D/d	material	
Chu & Storow [6]	25.4	12;24;30;36;42;48	3.5-25.8	glass,lead,steel	
Colborn [8]	78.8	17.8	3.4;3.1	ceramic zinc	
Kling [13]	100;50;47	2-12.6	1.6-30.8	glass,steel,aluminium	
Koch [14]	50	19.8	3.9;4.8	ceramic	
Kunii et al.[16]	140	4.3	3.3.5	cellite	
Leva [18]	52.7;15.8	174;57.9	4.1-12.1	glass,clay porcelain	also N_2,CO_2
Leva [19]	52.7;20.9;15.8	16.7;17.2	1.5-3.6	glass,porcelain	
Leva & Grummer[20]	52.7;20.9	17.4;43.8	3.5-9	glass,lead,cast iron,alum,copp.	
Leva et al.[21]	52.7;20.9	11-17.4	3.7-12	glass,porcelain	also CO_2
Schlunder [24]	69;43;25	0.2-10.7	2.5-36	glass	
Verschoor u Schuit [29]	50;43;30	4.3-8.8	3.9-14.3	glass,lead,steel	also H_2
Yagi & Wakao [32]	36	5.6-10	11.6-47	glass,lead,steel	also Sh_w
Yagi et al.[30]	60	8.3	9.4-26.7	glass,steel	water
Own experiments	69	0.3-4.3-12.3	6.9-34.5	styrop.,polyst,glass,copper	also Sh_w

Fig. 5

author	form	d mm	D mm	L/D	material
Colborn [8]	granula mat.,pellets,pebbles	5.5-24.9	78.8	17.8	ceramic
Kling [13]	cylinder	12×12	250; 100	2-30	carbon
Koch [14]	Raschigring	5.4×5.6×0.8×16.7×15.4×2.3	50	19.6	glass
Leva et al [19,20]	granula mat.	4.3	20.9	17;43.8	aloxite
"	cylinder	10.2×8.2-5×3.8	52.7;20.9	17	katal.mat.alumin.copper
"	cylinder,lg	3.2×6.4	20.9	17	katalysator mat.
"	Raschigring	6.3 9.5	52.7 20.9	17;174	clay, brass
Veschoor & Schult [29]	granula mat.	2.7	50;29.7	6;8,8	pumice
Yagi Wakao [32]	cylinder	5.2×5.2	50;43	4.2;7	ceramic
Own experiments	granula mat.	13-43	36	5,6;10	cement,clinker
"	cylinder	2×2;5×5;10×10	69	0,3-4,3-12,3	glass,copper
"	cylinder,lg	2×4;5 10	69	0,3-4,3-12,3	glass,copper
"	Raschigring	2.7×2.7×0.6 5.2×5.2×0.7	69	0,3-4,3-12,3	glass

Fig. 6

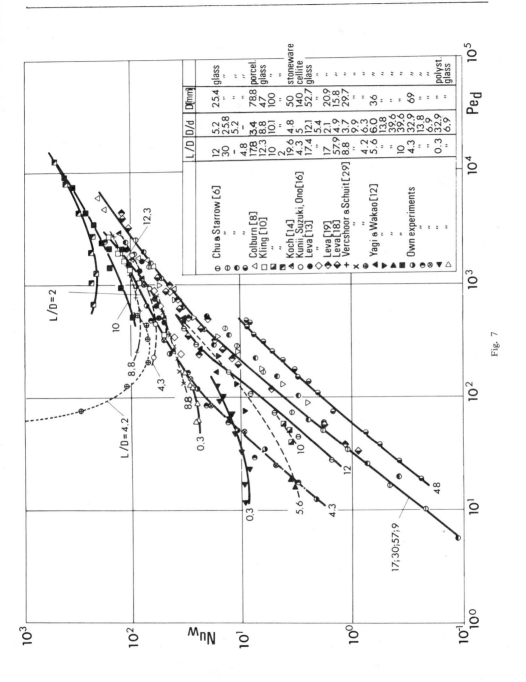

	L/D	D/d	D[mm]	
Chu a Starrow [6]	12	5.2	25.4	glass
"	30	25.8	"	"
"	-	5.2	"	"
Colburn [8]	4.8	3.4	78.8	porcel.
Kling [10]	12.3	8.8	47	glass
"	10	10.1	100	"
Koch [14]	2	5.4	50	stoneware
Kunii, Suzuki, Ono[16]	19.6	4.8	140	cellite
Leva [13]	4.3	4.5	52.7	glass
"	17.4	12.1	"	"
Leva [19]	17	5.4	"	"
Leva [18]	57.9	2.1	"	"
Verschoor a Schuit [29]	8.8	4.9	"	"
"	"	3.7	"	"
Yagi a Wakao [12]	4.2	9.9	36	"
"	5.6	6.3	"	"
"	"	6.0	"	"
Own experiments	10	13.8	69	"
"	"	39.6	"	"
"	4.3	32.9	"	"
"	"	13.8	"	"
"	0.3	6.9	"	polyst.
"	"	32.9	"	glass
"	"	6.9	"	"

Fig. 7

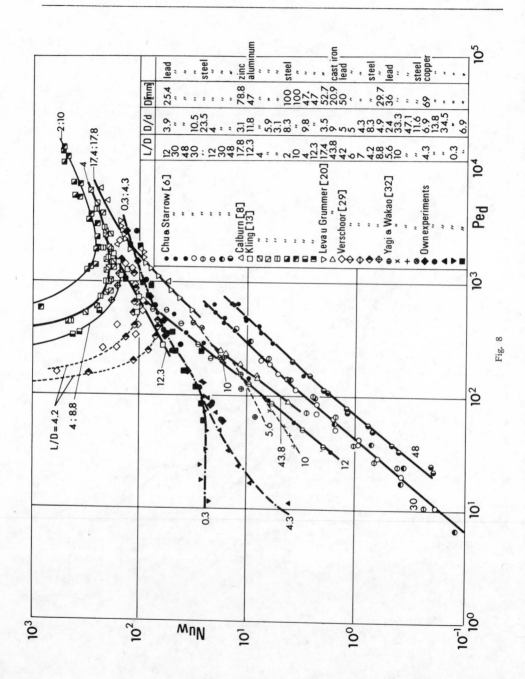

Fig. 8

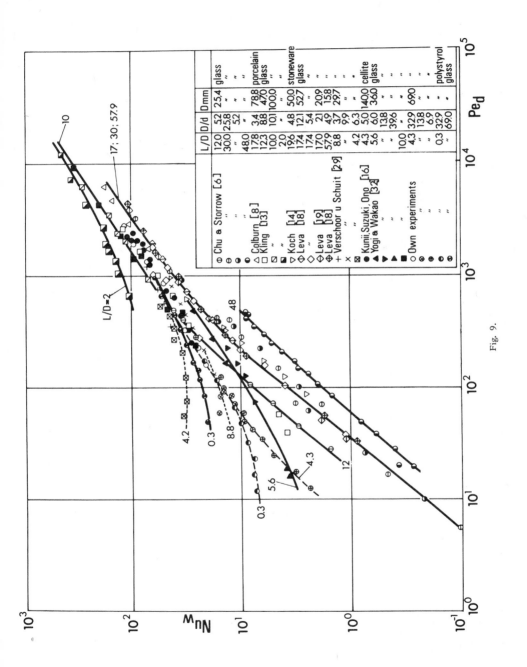

Fig. 9.

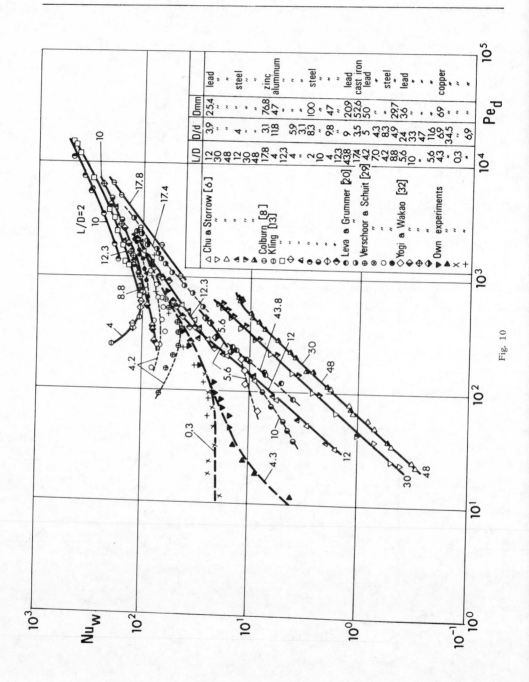

Fig. 10

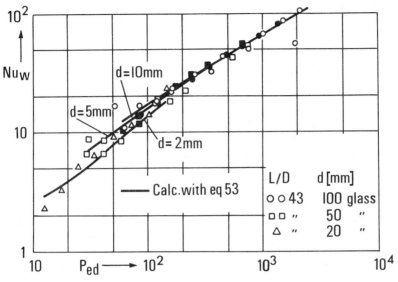

Fig. 11

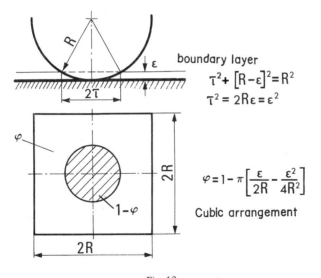

Fig. 12

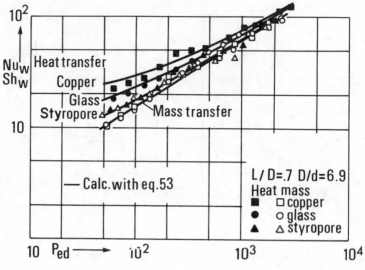

Fig. 13

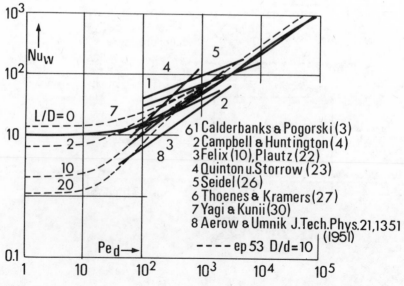

Fig. 14

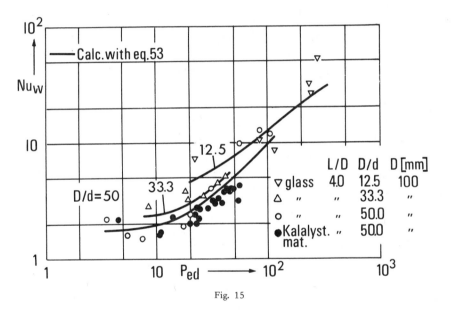

Fig. 15

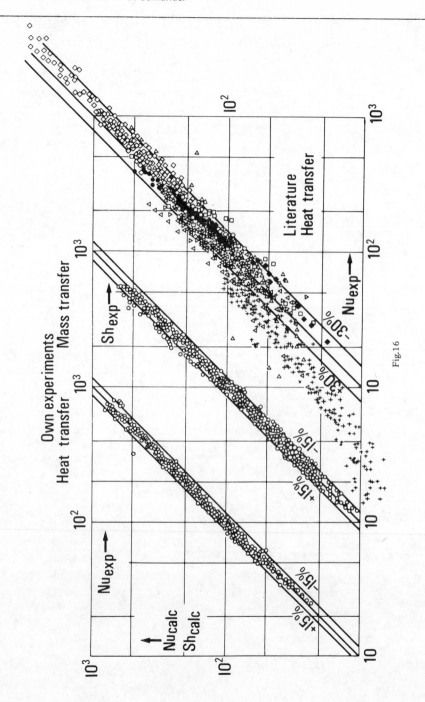

Fig.16

Chapter 32

PERFORMANCE OF A MOVING–BED HEAT–EXCHANGER

J. Dul (*)

The pebble-bed heater described below was built as the joint effort of the Institute of Nuclear Research — Świerk and CEA — Saclay in the frames of the Franco-Polish research cooperation in 1966-69. It was put into operation in December, 1969, and then tested during the next two years. The heat-exchanger is used to preheat 0.5 kg/s air for the magneto-hydro-dynamic installation. The described type can be used for different industrial applications using technical gases preheated to high temperature.

1. General description

The heater shown in fig. 1 consists of two vertical sections of the 760 mm inner diameter (upper part) and 735 mm i.d. (lower part), connected by the throat of 127 mm i.d. The differential pressure controller maintains equal pressure across the throat between both sections of the exchanger. The pebbles are heated by combustion gases discharging from two kerosene-fired combustion chambers, working with an excess amount of air. The combustion gases pass into the upper section through 12 inlets, filled by immobiles alumina spheres 70 mm dia. This solution provides the circumferential gas channel for the introduction of the pebbles.

The waste gases after cooling in the upper section are discharged to the chimney through the automatic valves system. The air passes into the lower part through the perforated cone made of 18/8 stainless steel. After heating, the air flows out through the perforated roof to the insulated pipeline of special construction, shown in the fig. 2.

The spheres are scattered down in counter flow through the throat

---

(*) Institute of Nuclear Research, Świerk, Poland.
N.B. ALL FIGURES QUOTED IN THE TEXT ARE AT THE END OF THE CHAPTER

performing the valve function. The pebbles at the nominal rate of 2000 kg/h are transported through the feeder and automatic valve by a pneumatic elevator with variable output.

2. The pebbles

The heat-exchanger is filled actually by the 4000 kg pebbles manufactured of 89.9% alumina with the diameter between 8.75 and 10 mm and average weight density 3.1 g/ccm. Numerous examinations of the pebbles and ceramic lining materials were carried out.

Among them the most important were :
— chemical analysis;
— high temperature strength under load;
— impact resistance;
— thermal shock resistance;
— abrasive wear resistance.

The pebble wear rates, due to continuous pebble replacement, depend in general on the structural properties of the pebble and on correct functioning of the lifting gear. The results of the wear rates for different tested pebbles after work in the heat-exchanger are presented below :

Test number	pebble	charge kg	wear rate		wear rate
			kg/day	per mil day	kg ton sift
I.3.A. + I.4	type I	4115.9	3.89	0.945	0.138
I.5	type II	4515.4	6.5	1.44	0.197
I.6	type I	3963.1	1.64	0.413	0.0651
	type III	35.98	——	0.23	0.0357

The pebble of the type I was manufactured by reeling method, type II by injection and type III - like type I, but with improved chemical composition.

Characteristic properties of the pebbles

pebble	alumina contents %	average density g/ccm	max.temp.utilization without sintering under load 1 kp/cm^2	Impact resistance by energy 10 kpcm /% destroyed spher.	
Type I	89.9	3.1	1470	⊖	80
Type II	99.5	3.9	1550		87.5
				⊕	65
Type III	95.8	3.6	1550		45

(remark: 74% spheres of the type II had production fissures).

3. Controlling and measurement system

The heat-exchanger is controlled by the control-board and its operation is observed on the measurement panels located in the separate control-room. The comprehensive measurement equipment allowed for examination of the performance characteristics, heat transfer coefficients and flow resistance. It consists of :

— Data logger for 100 channels, equipped with thermocouples-linearization unit.

— Acquisition data system with perforator for 30 channels.

— Digital-computer for on-line operation used for measurement data handling during and after experiments.

— 6-points recorders.

— Visual observation of the moving elements of the pebble-bed transport system.

— Panels for remote water cooling pump control, fuel pumping system and water cooling circuits measurement.

The following instruments for particular measurements are applied : Thermocouples Platinum — Platinum 10% Rhodium and Chromel-Alumel — for gases and ceramic lining temperature measurements.

Sucking off Pt-Pt 10% Rh thermocouples — for hot air temperature measurement in the pipeline ;

Total radiation pirometer with recorder — for pebbles temperature measurement at the throat outlet ;

Resistance thermometers for pebble temperature measurement in front of the feeder and for water cooling circuits ;
Pressure transducers for pressure and gas flows measurements;
Turbine — flow — meters with recorders for fuel flow measurements.

4. Experimetnal results

During the particular programs, the moving-bed heater was operated in a range of air-flows through the lower section between 570 and 1600 kg per hour and adequately in a range of combustion gases through the upper section between 610 and 1740 kg per hour. The maximum outlet air temperature attained was approximately $1300^{+25}°C$. The thermal power given from the waste gases to the pebbles in the upper section and regenerated by the air in the lower section is shown in fig. 3. Figure 4 shows the thermal efficiency of the heater. Both characteristics are presented as a function of the modified Reynolds' number

$$Re' = \frac{v \cdot d}{\nu(1 - \epsilon)} = \frac{G_m \cdot d}{\mu(1 - \epsilon)}$$

Packing density $1 - \epsilon = 0.636$ has been obtained by measurement in a hundred liter tank and compared to the bed filling in the heater. The average diameter of the alumina spheres amounted to 9.3 mm.

Pressure drop through the pebble bed is calculated by :

$$\Delta P = \xi \frac{\rho \cdot v^2}{2g} \cdot \frac{6(1-\epsilon)}{\epsilon^3} \cdot \frac{L}{d}$$

where the flow resistance coefficient ξ versus modified Reynolds number Re_m compared with some another authors is shown in fig. 5. Consequently, the modified Reynolds number is here defined as :

$$Re_m = \frac{v \cdot de}{\epsilon \cdot \nu}$$

where equivalent diameter of pebbles

$$d_e = \frac{4\epsilon}{a} = \frac{2}{3} \cdot \frac{\epsilon \cdot d}{1-\epsilon} \cdot$$

Thus

$$Re_m = \frac{2}{3(1-\epsilon)} \cdot \frac{\rho \cdot v \cdot d}{\mu} \cdot$$

On the basis of experience, within the Reynolds number Re_m ranging from 100 to

1500, the friction factor can be correlated with :

$$\xi = \frac{36}{Re_m} + 0.29.$$

Fig. 6 presents the run of the long-term experience, carried out in November 1971.

5. Conclusions

Actual efforts are directed towards the improvement of the heat transfer coefficient and efficiency.

Consequently, reduction of temperature difference between the gases and pebble, reduction of dimensions and increase of preheat temperature should be obtained.

We hope at present, after the experiences we have gained so far, that the temperature of the heated gas may be increased to 1400°C using high-alumina pebbles and refractories.

NOMENCLATURE

v m/s   —   gas velocity based on empty cross sectional area

ρ $\frac{kg}{m^3}$ —   density of gas

υ $\frac{m^2}{s}$ —   kinematic gas viscosity

μ $\frac{kg}{ms}$ —   dynamic gas viscosity

G_m $\frac{kg}{m^2 s}$ —   mass flow of gas

d m   —   spheres diameter

d_e m   —   equivalent pebble diameter

ϵ —   bed porosity

$1 - \epsilon$ —   packing density

a $\frac{m^2}{m^3}$ —   characteristic surface of the bed

L m   —   length of pebble bed

ξ —   flow resistance coefficient (friction factor)

ΔP $\frac{kp}{m^2}$ —   gas pressure drop

g $\frac{m}{s^2}$ —   gravitational constant

$Re' = \dfrac{G_m \cdot d}{\mu(1 - \epsilon)}$ —   modified Reynold's number

$Re_m = \dfrac{2}{3} \cdot Re'$ —   modified Reynold's number

η_o —   thermal efficiency coefficient

N_t $\frac{kW}{m^3}$ —   thermal power regenerated in the bed.

REFERENCES

[1]   Glaser, Thodos : Heat and momentum transfer in the flow of gases through packed beds (A.I.Ch.E. Journal 4, 1958).

[2]   Ergun : Fluid flow through packed columns(Chem.Eng.Prg. 48 No. 2, 1952)

[3]   Freund, Pinon : Etude des coefficients de perte de charge linéique et de transfert de chaleur dans un lit fixe de boulets.(R.G.T. No. 63 et 64, Mars, April 1967).

[4]   Żaworonkow, Aerow : Gidrawliczeskoje soprotiwlenija i płotnost upakowki żernistego sloja (ZTF, t.XXIII, 3, 1949).

[5]   Pták : Tlakové ztráty proudiciho plynu statickou vrstvou keramických kuliček (Zpráva SVVSS 66-05120, 1966).

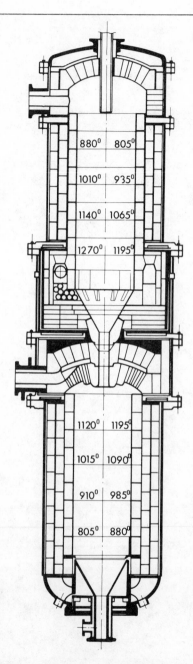

Fig. 1. Cross-section of the heat exchanger

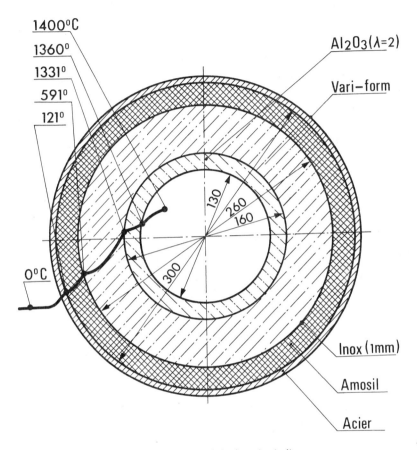

Fig. 2. Cross-section of the hot air pipeline

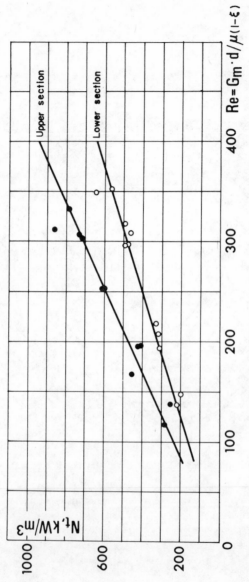

Fig. 3. Thermal power regenerated in the pebble-bed

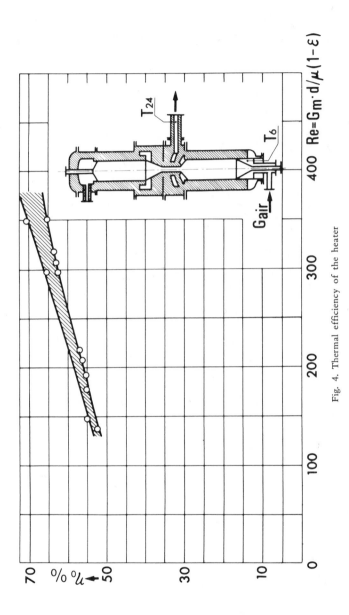

Fig. 4. Thermal efficiency of the heater

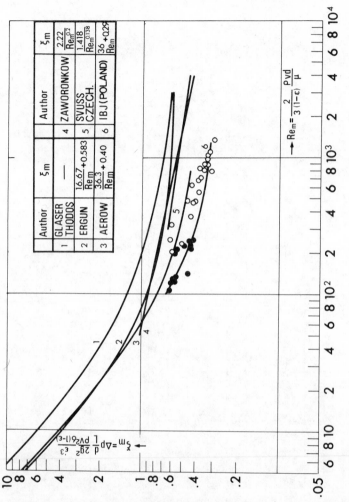

Fig. 5. Flow resistance through the bed

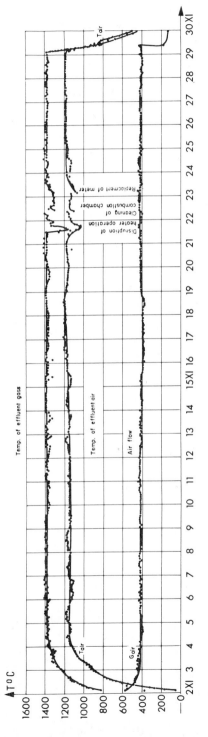

Fig. 6. Diagram of the long-term experience, Nov. 1971

Chapter 33

SCALING UP OF A DIRECT CONTACT HEAT EXCHANGER

M. Perrut and B. Paules (*)

INTRODUCTION

In classical exchangers, heat transfer takes place through metallic surfaces, leading to fouling problems and drastic decreases of the heat transfer coefficients with time. These problems reach a dramatic range when heating sea water in desalting plants.

With direct contact heat transfer, no metallic surfaces are used and two immiscible fluids are contacted as in liquid-liquid extraction operations.

The spray tower is the simplest device for counter-current flow : it is a vertical empty tower where one fluid moves, by gravity, dispersed in droplets within a continuous immiscible phase. It is known that such contactors have a low efficiency and high capacity.

As evidenced in the early 1960s, the dense packing flow allies this high capacity with a good efficiency. It consists of a dense dispersion of droplets of 40% to 90% hold-up.

Heat transfer and mass transfer experiments [4, 9] have shown that the dense packing flow presents two advantages :
— A high interfacial area,
— A low value of axial dispersion.

SPRAY TOWER HYDRODYNAMICS

In a spray tower, for a pair of flow rates of each phase, two flows can exist in the general case :
— Loose packing flow with a low dispersed phase hold-up ($<$ 40%) (figure 1),

(*) Centre de Cinétique et Physique Chimique C.N.R.S. Département de Génie Chimique NANCY (FRANCE)

N.B. ALL FIGURES QUOTED IN THE TEXT ARE AT THE END OF THE CHAPTER

— Dense packing flow with a high dispersed phase hold-up (between 40% and 90 %) where droplets are packed like glass beads (figure 2).

Dense dispersions have been obtained for 15 liquid-liquid pairs covering a large range of physical properties, dispersing either the light phase or the heavy phase [9]. Measurements of mean drop size and dispersed phase hold-up have been made for each liquid-liquid pair.

1. Mean drop size

Numerous measures lead to the following conclusions [11]:
— The perforated plate wettability is very important and the plate material will be chosen so as to be wetted by continuous phase ;
— No coalescence appears on the whole column ;
— The continuous flow rate is of no importance, except for very high values ;
— The way of formation of the droplets depends on the injection velocity value :

$$U_N = \frac{U_D}{N \; \Phi^2 /4} \tag{1}$$

with N and Φ number and diameter of holes. For values of the injection velocity higher than the jetting velocity, drops result of jet break up and the mean drop diameter $\bar{d}$ is no longer depending on U_N and can be correlated by :

$$\frac{\bar{d}}{\Phi} = 2.06 \left(1 - 0.185 \; g \; \frac{\Delta\rho}{\sigma} \; \Phi^2 \right) \tag{2}$$

2. Relative velocity ·

The spray tower hydrodynamics can be characterized by the relative velocity of the two phases :

$$V_r = \frac{U_D}{\beta \; \Omega} + \frac{U_C}{(1 - \beta) \; \Omega} \tag{3}$$

All our experimental results [9] lead to two relations (figure 3) :
— In loose packing flow :

$$\beta < 0,30 \qquad \frac{V_r}{V_r^\infty} = 1 - \beta^{0,6} \qquad\qquad (4)$$

— In dense packing flow :

$$\beta > 0,45 \qquad \frac{V_r}{V_r^\infty} = p\,(1-\beta) \qquad\qquad (5)$$

where V_r^∞ is the terminal velocity of a single drop of $\bar{d}$ diameter moving in a stagnant continuous phase.

The value of p depends on the viscosity ratio $\dfrac{\mu_C}{\mu_D}$ between 0.65 and 1 (figure 4)

HEAT OR MASS TRANSFER IN SPRAY TOWERS

Heat transfer and mass transfer investigations [4, 9] have shown that the model of piston flow with axial mixing in the continuous phase can give a right interpretation of the experimental temperature or concentration profiles..

This model has 3 parameters :

$$r = \frac{U_C}{U_D}\,\frac{(\rho C_p)_C}{(\rho C_p)_D} \qquad\qquad (6)$$

$$N_C = \frac{h_C\, a\, L\, \Omega}{U_C (\rho C_p)_C} \qquad \text{real transfer unit} \qquad (7)$$

$$P_C = \frac{U_C\, L}{\Omega a_C} \qquad \text{Peclet number} \qquad (8)$$

and the efficiency E_C is given by a relation given in the annex :

$$E_C = f\,(r,\, N_C,\, P_C) \qquad\qquad (9)$$

1. Heat transfer coefficient :

Evaluation of the interfacial area $a = \dfrac{6\beta}{\bar{d}}$ is easy and permits the calculation of h_C from the N_C value.

The heat transfer coefficient h_c was found to be a function of V_r and the Margoulis (or Stanton) number :

$$Ms_C = \frac{h_C}{(\rho C_p)_C V_r} \tag{10}$$

is constant for each liquid-liquid pair and can be measured easily with a single droplet [6, 10].

2. Axial dispersion coefficient

The axial dispersion coefficient α_C was also found to be a function of V_r (figure 5) :

$$a_C = A \cdot V_r^2 \tag{11}$$

with A increasing strongly with the ratio $D/\bar{d}$ of tower diameter to droplet diameter (figure 6).

OPTIMAL DESIGN OF A DIRECT CONTACT HEAT EXCHANGER

A flowrate U of a fluid is to be heated with an other immiscible fluid with a fixed efficiency E_C in a direct contact heat spray tower.
Several choices are to be made :

1. Dispersed phase :

The higher viscosity fluid is to be dispersed because it leads to higher values of V_r^∞ and so, to a higher capacity of the tower.

2. Flowrates ratio :

The V_c/V_D ratio needs to be acceptable with the existence zone of flowrates of the dense packing flow.

3. Droplet diameter - injector characteristics :

In order to obtain a given droplet diameter $\bar{d}$, with drop formation by jet disruption, the injector characteristics (N, Φ) are to be calculated with :

$$
\begin{cases}
\dfrac{\bar{d}}{\Phi} = 2,06\left(1 - 0,193\ g\ \dfrac{\Delta\rho}{\sigma}\ \Phi^2\right) & (2)\\[3mm]
U_N = \dfrac{U_D}{N\ \Phi^2/4}\ \#\ 2\,U_J & (12)
\end{cases}
$$

U_j can be obtained by the correlation given by Meister and Scheele [7] :

$$U_J = 1.73 \left[\frac{\sigma}{\rho_D \Phi} \left(1 - \frac{\Phi}{d} \right) \right]^{1/2} \tag{13}$$

4. Dispersed phase hold-up — Tower dimensions :

The choice of drop diameter $\bar{d}$ leads to the value of V_r^{∞} and the choice of dispersed phase hold-up β leads to :

Tower diameter D : $(\Omega = \frac{\Pi D^2}{4})$ is obtained from:

$$V_r = \frac{U_D}{\beta \Omega} + \frac{U_C}{(1 - \beta) \Omega} = p \ V_r^{\infty} (1 - \beta) \tag{14}$$

Tower height L: V_r is known from above relation, $D/\bar{d}$ also; so h_c is known from V_r and Margoulis number ; α_C can be calculated from V_r and $A = f(D/\bar{d})$.

From

$$\frac{N_C}{L} = \frac{h_C \ a \ \Omega}{U_C (\rho C_p)_C} \ , \ \frac{P_C}{L} = \frac{U_C}{\Omega a_C} \ , \ r \ \text{and} \ E_C \ , \ L$$

L can be obtained by iterative calculation from equation (9).

5. Laboratory measurements:

Easy laboratory measurements are needed :

. Physical properties of the two fluids,
. Terminal velocity V_r^{∞} which is a function of the droplet diameter d,
. Margoulis number Ms_C of a single droplet.

OPTIMAL DESIGN

The 3 parameters : r, $\bar{d}$ and β may vary on certain ranges and optimal values minimizing the cost that is to be sought leading to optimal design of the column.

DISCUSSION

The scale up is now possible up to 50-60 cm diameter towers and is certainly interesting on the economical point of view [17].

A problem is not yet resolved : viability of the dense packing flow.

NOMENCLATURE

a	interfacial area
C_p	specific heat capacity
$\bar{d}$	mean drop diameter
E	efficiency
h	heat transfer coefficient
L	tower height
N	number of holes of the perforated plate
N_C	transfer unit number
p	constant (equation (5))
P_C	Peclet number
r	ratio $= \dfrac{U_C(\rho C_p)_C}{U_D(\rho C_p)_D}$
T	temperature
U	flowrate or velocity
V_r	relative velocity
V_r^∞	terminal velocity of a single drop in a quiescent continuous phase
α	axial mixing coefficient
β	dispersed phase hold-up
Ω	tower section
μ	viscosity
ρ	density
Φ	hole diameter

Indices

C	relative to continuous phase
D	relative to dispersed phase
E	relative to tower entrance
S	relative to tower issue

REFERENCES

[1]   Hazlebeck D.E., Geankoplis C.J., Ind. Eng. Chem. Fund. 2, 1963, 310

[2]   Kehat E., Letan R., A.I.Ch.E. J., 17, 1971, 984

[3]   Letan R., Kehat E., A.I.Ch.E. J., 14, 1968, 398

[4]   Loutaty R., Thèse Doctorat ès Sciences, Univ. de Nancy, 1968

[5]   Loutaty R., Vignes A., Chem. Eng. Sci., 25, 1970, 201

[6]   Loutaty R., Vignes A., Le Goff P., Chem. Eng. Sci., 24, 1969, 1795

[7]   Meister B.J., Scheele G.F., A.I.Ch.E. J., 14, 1968, 15

[8]   Mixon F.O., Whitaker D.R., Orcutt J.C., A.I.Ch.E. j., 13, 1967, 21

[9]   Perrut M., Thèse Doctorat ès Sciences, Univ. de Nancy, 1972

[10]   Perrut M., Loutaty R., Le Goff P., (to be published)

[11]   Perrut M., Loutaty R., (to be published)

[12]   Yeheskel J., Kehat E., Chem. Eng. Sci., 26, 1971, 1223

[13]   Henton J.E., Cavers S.D., Ind. Eng. Chem. Fund., 9, 1970, 384

[14]   Cavers S., Ewanchyna J.E., Can. J. Chem. Eng., 35, 1957, 113

[15]   Loutaty R., Le Goff P., Communication Journées Génie Chimique Toulouse, 5-6 nov. 1970

[16]   Mouton J.E., Rapport interne, 1971

[17]   Letan R., Kehat E., Brit. Chem. Eng., 14, 1969, 803.

APPENDIX

The model equations can be solved and lead to following values of the efficiency E_C.

If $r = 1$

(9 bis)
$$E_C = \frac{(1 + P_C)(N_C + P_C) + P_C \dfrac{P_C(N_C + P_C) - 1}{N_C(N_C + P_C)} - N_C \, e^{-(N_C + P_C)}}{2(N_C + P_C) + P_C \left[\dfrac{P_C}{N_C} + N_C + P_C\right] - N_C \, e^{-(N_C + P_C)}}$$

$$\# \quad \frac{(1 + P_C)(N_C + P_C) + \dfrac{P_C}{N_C(N_C + P_C)}[P_C(N_C + P_C) - 1]}{2(N_C + P_C) + P_C \left[\dfrac{P_C}{N_C} + N_C + P_C\right]}$$

If $r \neq 1$

(9 ter)
$$E_C = 1 - \frac{(1 - r)(\lambda_2 - \lambda_3)P_C}{\Delta}$$

$$\lambda_{2,3} = \frac{-(P_C + rN_C) \pm \sqrt{\delta}}{2}$$

where
$$\Delta = -r\,P_C(\lambda_2 - \lambda_3) + \lambda_2\,\lambda_3(e^{\lambda_3} - e^{\lambda_2}) + P_C(\lambda_2\,e^{\lambda_3} - \lambda_3\,e^{\lambda_2})$$

With
$$\delta = P_C + (r\,N_C)^2 - 2\,r\,N_C P_C + 4 N_C P_C$$

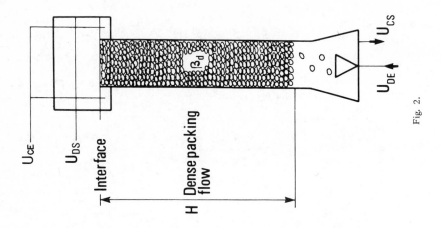

U_{CE}

U_{DS}

Interface

H Dense packing
flow

β_d

U_{CS}

U_{DE}

Fig. 2.

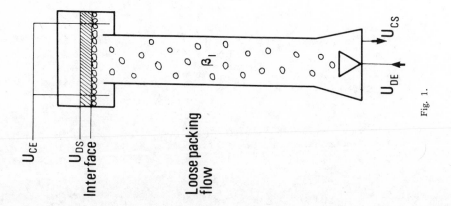

U_{CE}

U_{DS}
Interface

Loose packing
flow

β_l

U_{CS}

U_{DE}

Fig. 1.

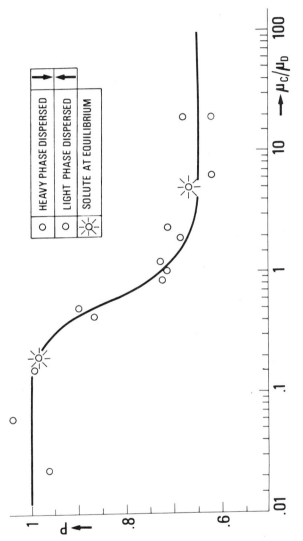

Fig. 3. Variation of V_r/V_r^∞ versus dispersed phase hold-up.

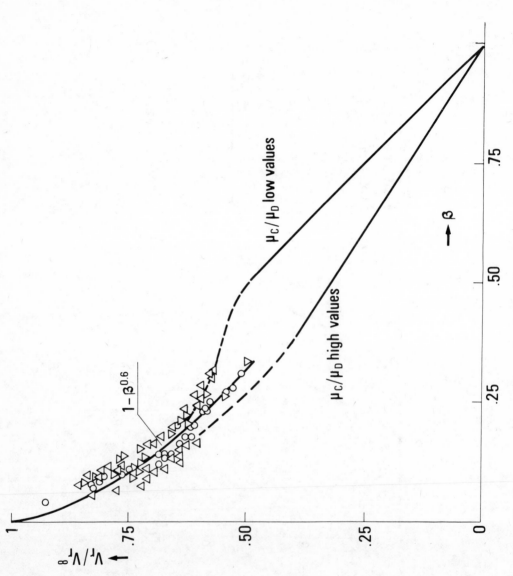

Fig. 4. Variation of ρ versus viscosity ratio (μ_C/μ_D).

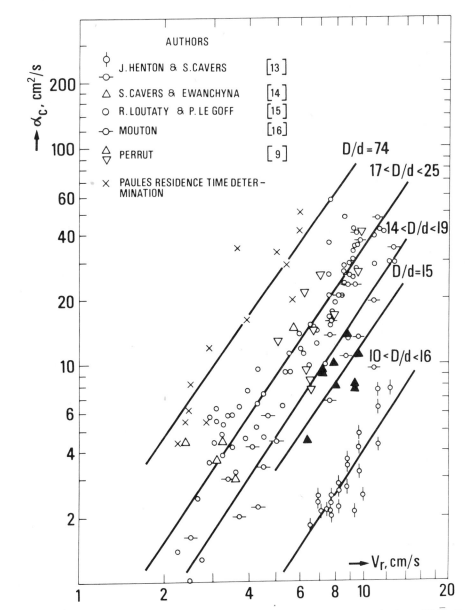

Fig. 5. Variation of axial mixing coefficient versus relative velocity V_r for several values of D/d̄ ratio.

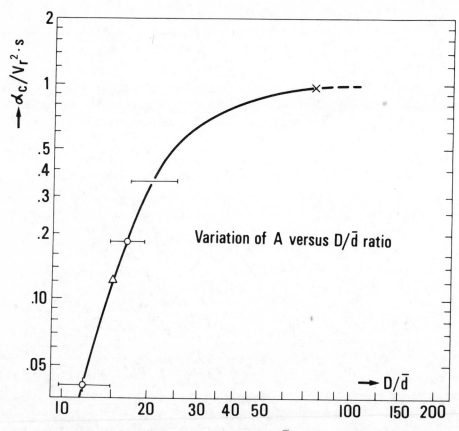

Fig. 6. Variation of A versus D/d̄ ratio.

Chapter 34

ANALYSIS OF DIRECT CONTACT CONDENSERS:
SINGLE- AND TWO–PHASE SYSTEMS

D. Moalem and S. Sideman (*)

INTRODUCTION

Direct contact exchangers, in which heat is transferred between a volatile disperesed fluid and an immiscible or miscible liquid media, are extremely efficient. High transfer coefficients and relatively small flow rates of the transfer fluid, evaporating or condensing, assure compactness. Moreover, the absence of solid transfer areas eliminates resistance due to scale and affords operation at very low temperature driving forces. Other advantages, as well as a concise review of the pertaining literature, are given elsewhere [1, 2] and will not be repeated here. The practical applications range from water desalination plants to in situ cooling of computer components [3, 4].

The present work, motivated by the quest for more efficient 3-phase heat exchangers as well as for additional insight into the transfer mechanism, deals with condensation of gravity-driven vapor bubbles. The vapor enters the column through a distribution plate- orifices or nozzles. The bubbles, with an average initial radius R_o, rise in a subcooled liquid column, condense along their way upwards to their final finite radii, R_f, and leave the column. As will be shown below, the value of R_f depends on the system (single or two-phase) and inerts contents.

Under steady state conditions the heat released by the bubbles is removed by replacing the liquid in the column continuously. The height of the liquid column corresponds to the rising time required for 99% of total possible condensation. Whereas a lumped-parameter approach yields, via an overall energy balance, the inlet and outlet conditions, the height depends on the transfer rates. These, in turn, depend on interaction between the bubbles, i.e., the effect of bubble frequency (vertical distance between the bubbles) and the number of nozzles

---

(*) Department of Chemical Engineering Technion, Israel Institute of Technology — Haifa — Israel

N.B. ALL FIGURES QUOTED IN THE TEXT ARE AT THE END OF THE CHAPTER

(horizontal distance) on the flow and temperature fields. The analysis of these effects on the condensation rate (or time) is the object of this study.

Condensation of single bubbles freely rising in subcooled miscible and immiscible liquid media has been studied experimentally [5, 6, 7] and theoretically [6, 7]. Exact numerical [7] and approximate analytical [8, 9] solutions have been suggested, the latter including the effect of non-homogeneous distribution of the ever present non-condensable gases inside the bubble. These studies yielded the required insight into the mechanism of bubble condensation. Particularaly interesting is the case of the 3-phase (2-component) system (pentane-water) where the condensate remains within the confines of the bubble walls. Recent studies of condensation of a single bubble train [10] provide the tools for the present analysis. Simultaneous numerical solution of the interrelated energy and momentum equations, while possible for a single train [11], is obviously out of the question for a multi-train system. Hence, an approximate solution, an extension of the one successfully utilized in the single train studies, is considered here [10]. Basically, the following questions are answered : 1) What is the collapse history of a single bubble in multi-bubble system, and 2) How are the field parameters which control the collapse rate - the rise velocity and temperature field - affected by bubble frequency and spacing (or number of nozzles) along the bubble's path.

In pursuing the solution, use is made of the quasi-steady state solution for a single bubble in potential, or modified-potential, flow field. Thus, in addition to its simplicity, this approach leads to a general solution, encompassing single and two-component systems (single phase and two-phase bubbles, respectively) and including non-condensables, whether homogeneously or non-homogeneously distributed inside the bubble.

THE CONDENSATION COLUMN

Under steady state conditions, the vapor (say pentane) enters saturated, at T^*, as bubbles of radii R_o, at a frequency F, through n equally spaced nozzles (per unit area). The cross section area is S.

The continuous phase, either liquid pentane or water, enters with a superficial velocity U_ℓ and a temperature T_∞. The vapor may contain non-condensable inerts, with an initial molar concentration, $Y_o \geqslant 0$. The bubbles rise at a velocity (relative to the continuous liquid) U_∞^M. The latter is a function of the gas holdup, ϵ, which in turn, is a function of F, n and H, the condensation height, or

$$U^M_\infty = f_1(\epsilon) = f_2(F, n, N) \tag{1}$$

where N is the number of equally spaced bubbles, from R_o to R_f, which determine column height. H and N are related by

$$H = \frac{1}{F}(N-1)U_b = t_f U_b \tag{2}$$

where U_b is the bubble rise velocity, related to the wall, and t_c is the final condensation time.

Assuming no heat losses to the surroundings, all the latent heat of condensation is transferred to the continuous phase, and a longitudinal temperature gradient is formed. The local temperature is T^M_∞ and the driving force, between the bubble and the continuous median, is thus a function of H (or N), n, and F or

$$\Delta T^M \equiv T_w - T^M_\infty = f_3(F, n, N) \tag{3}$$

where T^M_∞, the local field temperature, is given by

$$\delta T^M \equiv T^M_\infty - T_\infty = f_4(F, n, N) \tag{4}$$

Before proceeding with the analysis, it is noted that whereas F and n are independent variables, N depends upon the system (single or two-phase), inerts contents and the nominal (inlets) driving force $\Delta T^* = T^* - T_\infty$.

THE GENERAL COLLAPSE EQUATIONS

For a single bubble rising freely in an unconfined liquid column at a velocity U_∞, the average heat flux in a potential flow field under quasi-steady state conditions ($P_e > 1000$, and $\partial R/\partial t \ll U_\infty$) is given by [10, 12]:

$$q = k\frac{(T - T_\infty)}{\sqrt{\pi}}\frac{1}{R}\left[\frac{2RU_\infty}{a}\right]^{1/2} \tag{5}$$

where $T = T^*$ for pure vapor and $T = T_w'$ the wall temperature, in the presence of inerts $T_w = T^*$ when $Y_0 = 0$.

Equating (5) with the energy balance at wall of the collapsing bubble, i.e. $(-\lambda \dot{R} \rho_v)$ yields:

$$\dot{R} = -\frac{k\Delta T}{\rho_v \lambda} \left(\frac{2U_\infty}{\pi a\, R}\right) \quad ; \quad \Delta T = T_w - T_\infty \qquad (6)$$

where R is the instantaneous radius. Defining

$$\theta_w \equiv \frac{T_w - T_\infty}{T^* - T_\infty} \quad ; \quad \tau \equiv \text{Ja Pe}^{1/2}\, \text{Fo} \quad ; \quad \beta = R/R_0$$

$$\text{Ja} = \frac{\rho C_p (T^* - T_\infty)}{\lambda\, \rho_v} \quad ; \quad \text{Pe} = \frac{2 R_0 U_\infty}{a} \quad ; \quad \text{Fo} = \frac{ta}{R_0^2}$$

Eq. (6) reduces to

$$\dot{\beta} = -\frac{1}{\sqrt{\pi}}\, \frac{1}{\beta^{1/2}}\, \theta_w \quad ; \quad \beta = 1 \ e \ \tau = 0 \qquad (7)$$

The application of Eq. (6), or (7), for the case at hand requires redefining the operating parameters in accordance with the conditions prevailing in a bubble column. Thus, U_∞^M replaces U_∞ and T_∞^M replaces T_∞. However, for ease of comparison with single bubble studies, the Ja and Pe numbers are left unchanged as defined above.

The modified Eq. (7) now reads :

$$\dot{\beta}^M = \left(\frac{\partial \beta}{\partial \tau}\right)^M = -\frac{1}{\sqrt{\pi}} \left(\frac{U_\infty^M}{U_\infty}\right)^{1/2} \theta_w^M\, \frac{1}{\beta^{1/2}} \qquad (8)$$

where

$$\theta_w^M \equiv \frac{\Delta T^M}{\Delta T^*} = \frac{T_w - T_\infty}{T^* - T_\infty}$$

or, introducing $T_\infty^M = T_\infty + \delta T^M$ from Eq. (4),

$$\theta_w^M = \theta_w - \frac{\delta T^M}{T^* - T_\infty} \equiv \theta_w - \delta\theta^M \qquad (9)$$

Note that θ_w is the dimensionless temperature driving force for a single bubble system, and $\delta\theta^M$ represents the fractional decrease of the temperature driving force affected by the multi-bubble system. From Eqs. (8) and (9)

$$-\left(\frac{d\beta}{d\tau}\right)^M = \left(\frac{k_v}{\pi}\right)^{1/2}\left(\frac{U_\infty^M}{U_\infty}\right)^{1/2}\frac{1}{\beta^{1/2}}\left[\theta_w - \delta\theta^M\right] \tag{10}$$

where K_v, the velocity factor by which the potential flow solution for flow around a sphere is "transformed" to yield the average heat flux that would be obtained in a viscous flow field, is now introduced for generality. The velocity factor is given by [7, 8]:

$$k_v = 0.25\ Pr^{-1/3}$$

for a two-component system (two-phase bubbles) and $K_v = 1$ for a single phase bubble.

Defining

$$k = (k_v/\pi)^{1/2} \ ; \quad A = (U_\infty^M/U_\infty)^{1/2} \ ; \quad B = A\delta\theta^M \tag{11}$$

equation (10) reduces to

$$-\frac{1}{k}\left[\frac{d\beta}{d\tau}\right]^M = A\ (\theta_w - \delta\theta^M)\frac{1}{\beta^{1/2}} = \frac{A\theta_w}{\beta^{1/2}} - \frac{B}{\beta^{1/2}} \tag{12}$$

The solution of Eq. (10), or (12), depends on the relation between θ_w and β, which is given by [7, 9]:

$$\theta_w = \frac{\beta^3 - \beta_f^3}{\beta^3 - 1/G^*} \tag{13}$$

where $\beta_f = R_f/R_o$ and $G^* = \rho_L/\rho_v$.

The term $1/G^*$ in Eq. (13) is due to the condensed liquid which accumulates in the "two-phase" bubble (say pentane condensing in water) in contrast to the single-phase bubble (say pentane condensing in pentane). In the absence of noncondensables $\beta_f = 0$ for the single-phase bubble and $\beta_f = G^{*-1/3}$ for the two-phase bubble. The dependency of β_f on y_o, the initial mole fraction of inerts, and the temperature driving force is given by [7, 9]

$$\beta_f = \left[\frac{\hat{R} T^{*2} y_0}{\lambda (T^* - T_\infty)} + \frac{1}{G^*} \right]^{1/3} \tag{14}$$

where $\hat{R}$ is the specific gas constant. The term $1/G^*$ in Eqs. (13) and (14) vanishes for single-phase bubbles.

Note that Eq. (13) was derived for a two-phase single bubble, assuming an homogeneous distribution of noncondensables and accounting for the volume of the condensed liquid which remains within the confines of the bubble. The relation between θ_w and β for nonhomogeneous distribution of the inerts within the bubble is given elsewhere [9] and can be used to extend this problem, if desired.

The relationship between β_f (single bubble) and β_f^M (multi-bubble), evolved during the integration of Eq. (12) while utilizing (13), is given by :

$$(\beta_f^M)^3 = \frac{A}{A - B} \beta_f^3 - \frac{B}{A - B} \frac{1}{G^*} \tag{15}$$

or

$$(\beta_f^M)^3 = \left(\frac{1}{1 - \delta\theta_{AVE}^M} \right) \left[\beta_f^3 - \frac{\delta\theta_{AVE}^M}{G^*} \right] \tag{16}$$

for a single bubble system $\delta\theta^M = 0$ ($T_\infty^M = T_\infty$) and $\beta_f^M = \beta_f$. It is evident by inspection of Eq. (15) that $\beta_f^M > \beta_f$, indicating that for identical starting conditions (T_∞, R_0) and condensation times the single bubble will leave the system smaller — more condensed — than a comparable bubble in a multi-bubble column. This is as expected in view of the fact that $\Delta T^M < \Delta T^*$.

THE TIME DEPENDENT RADIUS

For a pure vapor, $T_w = T^*$ and $\theta_w = 1$ ($\beta_f = G^{*-1/3}$ for the two-phase bubble and $\beta_f = 0$ for the single phase bubble). Equation (12) reduces to:

$$- \frac{1}{K(A - B)} \left(\frac{d\beta}{d\tau} \right) = \frac{1}{\beta^{1/2}} \quad ; \quad \beta = 1 \text{ at } \tau = 0$$
$$A - B = A (1 - \delta\theta^M) \tag{17}$$

integration of Eq. (17) yields :

$$\tau_o^M = \frac{1}{A-B} \; \frac{\sqrt{\pi}}{\sqrt{k_v}} \; \frac{2}{3} \; (1 - \beta^{3/2}) \tag{18}$$

or

$$\tau_o^M = \frac{1}{A-B} \; \hat{\tau}_o \tag{18a}$$

where

$$\tau_o = \frac{\sqrt{\pi}}{\sqrt{k_v}} \; \frac{2}{3} \; (1 - \beta^{3/2}) \tag{19}$$

Equation (19) is the $\tau - \beta$ relationship for the inert-free single bubble, directly obtainable by integration of Eq. (6). Obviously, for a single bubble $A = (U_\infty^M / U_\infty)^{1/2} = 1$, $B = 0$ (since $\delta\theta^M = 0$) and $\tau_o^M = \tau_o$.

For unpure vapors containing permanent gases, $\theta_w \neq 1$. The relationship between τ^M and β, obtained by integrating Eq. (12), utilizing Eqs. (13) and (14) and a transformation variable $X = \beta^3 - (\beta_f^M)^3$, is:

$$\tau^M = \frac{1}{A-B} \; (\tau_o(\beta) + \tau_1(\beta, \beta_f^M)) \tag{20}$$

where τ_o is given by Eq. (19) and

$$\tau_1(\beta, \beta_f^M) = \frac{1}{3} \; \frac{(\beta_f^M)^3 - 1/G^*}{(\beta_f^M)^{3/2}} \; \ell n \left[\frac{1 - (\beta)^{3/2}}{1 + (\beta)^{3/2}} \cdot \frac{\beta^{3/2} + (\beta_f^M)^{3/2}}{\beta^{3/2} - (\beta_f^M)^{3/2}} \right] \tag{21}$$

At the limit, $A = 1$, $B \to 0$, $\delta\theta^M \to 0$, $\beta_f^M \to \beta_f$ and Eq. (20) reduces to the relation between τ and β for a single bubble [8, 9]:

$$\lim_{F \to 0} \tau^M = \tau = \tau_o(\beta) + \tau_1(\beta, \beta_f) \tag{22}$$

where $\tau_1(\beta, \beta_f)$ is identical in form with Eq. (21), with β_f replacing β_f^M

THE CONTROLLING PARAMETERS

Equation (18) to (20) represent close-form solutions for the time dependent radius of one bubble, (pure or contaminated with inerts) entering and leaving the bubble column. However, a quantitative presentation requires the knowledge of the "constants" A and B, or rather U_∞^M and δT^M.

The relative velocity

The relative velocity between the bubbles and the continuous phase in counter-current flow is given by

$$U_\infty^M = U_\ell + U_b \qquad (23)$$

where U_b is the bubbles' rise velocity, related to the wall and U_ℓ is the superficial velocity of the continuous phase in the downwards direction. Eq. (23) is substantiated by the experimental conclusion of Baker and Chao [14] that the relative (gas-liquid) velocity of bubbles with $R > 0.3$ cm are practically independent of the continuous phase velocity.

The rise velocity of a bubble in a bubble swarm as a function of the porosity, or fractional hold-up, ϵ, is given by Marrucci [15]:

$$U_\infty^M = U_\infty \frac{(1 - \epsilon)^2}{1 - \epsilon^{5/3}} \qquad (24)$$

where, here

$$\epsilon = \frac{S n (4\pi/3) \sum_{i+1}^{N} R_i^3}{S H} = \frac{n(4\pi/3) \sum_{i=1}^{N} R_i^3}{(1/F) U_b (N - 1)} \qquad (25)$$

Obviously, the correct value of ϵ requires the knowledge R_i, U_b, F and N (or t_c), i.e. the complete collapse history. Hence, the utilization of Eq. (24) requires an iterative procedure (coupled with the "external" iteration (see below) which yields the R_i values).

For a given set of operating conditions (F, n, R_o etc.), the solution starts with assuming $U_\infty^M = U_\infty$; evaluating U_b by Eq. (23); calculating ϵ by Eq. (25) with R_i taken from the external iteration loop; evaluating U_∞^M by Eq. (24) and repeating till convergence of U_∞^M.

As a first approximation ϵ can be estimated by taking $\sum_{i=1}^{n} R_i^3 = N [(R_o + R_f)/2]^3$ and $U_b \simeq U_\infty - U_\ell$, or

$$\epsilon = \frac{Fn (\pi/6) N (R_o + R_f)^3}{(N - 1) (U_\infty - U_\ell)} \qquad (26)$$

note that for small driving forces, $N \gg 1$ and $N/(N - 1) \approx 1$.

The fractional temperature decrease, $\delta\theta^M$

Assuming a periodic steady state the outlet liquid temperature is constant at each $1/F$ time interval, and an overall energy balance yields

$$(U_\ell S)\rho_\ell \; Cp_\ell \; \frac{1}{F} \; \delta T_o^M \; = \; (n.s) \; \frac{4\pi}{3} \; (R_o^3 - \; R_f^3)\rho_v \lambda \qquad (27)$$

where $\delta T_o^M = T_{out} - T_{in} = T_{out} - T_\infty$, in the continuous liquid phase. Defining

$$\delta\theta_o^M \equiv \frac{\delta T_o^M}{\Delta T^*} = \frac{T_{out} - T_\infty}{T^* - T_\infty}$$

and combining with Eq. (25) yields, in dimensionless parameters :

$$\delta\theta_o^M = \frac{Q_{v\ell}(1 - \beta_f^3)}{Ja} \qquad (28)$$

where

$$Q_{v\ell} \equiv \frac{Q_v/S}{Q_\ell/S}$$

is the volumetric ratio of the vapor to liquid flow rates (= the ratio of their superficial velocities, in an empty column of cross section S).

Assuming a linear temperature variation in the continuous phase, the average fractional decrease (hence the average driving force) is taken as

$$\delta\theta_{ave}^M = \frac{1}{2} \; \delta\theta_o^M \qquad (29)$$

It is noted that Eq. (29), together with Eqs. (26) and (24), can now be used to solve for A and B in Eq. (17), to yield an analytical, albeit approximate, solution of the collapse history, Eq. (20).

The solution can be refined to yield a more accurate solution of the condensation rate. Consider again the bubble column. Once a thermal equilibrium is reached, heat is released and removes continuously. Here, however, we consider a periodic operation. During the time interval $\Delta t \; (= 1/F)$ the successive bubbles move within the (assumed) well mixed thermal field, left behind the preceeding bubbles, which is at a temperature $T_{\infty,i}^M$, corresponding to $\delta T_i^M \equiv (T_{\infty,i}^M - T_\infty)$. Thus, rather than assuming that the bubble is affected by an overall average driving force (δT_{ave}^M), we evaluate the temperature along the bubbles' pass in $(N{-}1)$ discrete sections, each (U_b/F) high. Consequently, as the limitation of a constant driving force is removed, a more realistic bubble collapse history can now be evaluated. The energy balance, Eq. (25), is now applied sectionally. Starting with R_i

we proceed to solve for the local $\delta T^M (\equiv \delta T_i^M)$ while at the same time searching for the correct R_{i+1}. Since n, F, T*, δT_i^M, N and R_{i+1} are interrelated, an iterative procedure must be adopted. The first approximation, based on an overall average driving force, can be used to obtain the initial values of R_{i+1}.

In pursuing the solution, recall that R_f (or $\beta_1 = \beta_f$, if we count the bubbles downwards) as well as T_∞, the approach temperature for the upper bubble, are known. Starting with $\beta_1 = \beta_f$ we seek a solution in terms of δT_i^M and N, so that $\beta_N = 1$.

With reference to Eq. (27), the enrgy balance on section i yields :

$$\delta T_{i+1}^M = \delta T_i^M + \frac{nF\,\rho_v\,\lambda(R_{i+1}^3 - R_i^3)(4\pi/3)}{U_\ell\,\rho_\ell\,C_{p\ell}} \tag{30}$$

where, counting i downwards, δT_{i+1}^M is the temperature of the continuous phase leaving section i, related to T_∞, and $\delta T_{i+1}^M - \delta T_i^M = T_{i+1}^M - T_i^M$.
In dimensionless terms, Eq. (30) becomes, analogous to Eq. (28) :

$$\delta\theta_{i+1}^M = \delta\theta_i^M + \frac{Q_v\,\ell}{Ja}(\beta_{i+1}^3 - \beta_i^3) \tag{31}$$

THE COLLAPSE HISTORY

As already indicated, the collapse history, i.e. β vs. τ, can be solved analytically from Eq. (20) by approximating A by Eqs. (24) and (26) and B by Eq. (29), derived by averaging the characteristic parameters over the whole column. A more accurate result is obtained by evaluating $\delta\theta^M$ and R_i locally, along the bubbles paths.

The average collapse rate in section i is obtained by applying Eq. (10)

$$-\dot\beta_{ave}^M = \left(\frac{k_v}{\pi}\right)^{1/2}\left(\frac{U_\infty^M}{U_\infty}\right)^{1/2}\left|\theta_{w\,ave} - \delta\theta_{i,i+1}^M\right|\frac{1}{\beta_{ave}^{1/2}} \tag{32}$$

where

$$\dot\beta_{ave}^M = \frac{\beta_{i+1} - \beta_i}{\Delta\tau} \;;\; \Delta\tau = Ja\,Pe^{1/2}\frac{a}{R_o^2}\cdot(1/F)$$

$$\beta_{ave} = (\beta_i + \beta_{i+1})/2$$

$$\delta\theta^M_{i,i+1} = (\delta\theta^M_{i+1} + \delta\theta^M_i)/2$$

and

$$\theta_{w, ave} = (\theta_{w,i} + \theta_{w,i+1})/2, \text{ evaluated for } \beta_i \text{ and } \beta_{i+1} \text{ by Eq. (13).}$$

The value of β_{i+1} is obtained by solving Eq. (32), now written as :

$$(\beta_{i+1} - \beta_i)(\beta_{i+1} + \beta_i)^{1/2} + C_1 (\beta^3_{i+1} - \beta^3_i) - C_2 = 0 \qquad (33)$$

where :

$$C_1 = \sqrt{\frac{k_v A}{2\pi} \frac{Q_v \ell}{Ja}} \Delta\tau$$

$$C_2 = \sqrt{\frac{2k_v A}{\pi}} (\theta_{w,av} - \delta\theta^M_i) \Delta\tau$$

It is convenient to start the solution of Eq. (33) by estimating β_i by assuming a linear relationship between R_f and R_o, hence

$$\beta_i = \beta_f + (1 - \beta_f) \frac{i-1}{N-1} ; \quad \begin{array}{l} \beta_1 = \beta_f \\ \beta_N = 1 \end{array} \qquad (33a)$$

Starting with $\beta_i = \beta_1 = \beta_f$ counting downwards, the solution proceeds as follows:

a. Utilizing Eq. (33a) get the first approximation for $\beta_{i+1} (= \beta_2^{(1)})$
b. Using $\beta_2^{(1)}$, calculate $\theta_{w,av}$ (Eq. 13), C_1 and C_2 (note $\delta\theta^M_1 = 0$).
c. Solve Eq. (33) for $\beta_{i+1} = \beta_2^{(2)}$.
d. Calculate $\theta_{w,av}, C_1, C_2$.
e. Calculate $\beta_2^{(3)}$ by Eq. (33).
f. Continue steps d. and e. until $\beta^{(m)} = \beta^{(m-1)} \sim$

The first value of β_{i+1} $(=\beta_2)$ is then used to calculate $\delta\theta^M_{i+1}$, by Eq. (31). The above procedure is then repeated to evaluate all β_{i+1} ($i = 1, 2 \ldots$, N−1).

Evaluation of N and ΔT^*

At a given bubble density, n and F, the number of bubbles, N which constitute the bubble column, depends on the temperature driving force. Low N represents small condensation time due to high temperature driving force, while high values of N (at the same n and F) represent low condensation rates.

Utilizing the design date which consists of T^*, R_o, R_f n and F, and following the above procedure, the value of N must satisfy the condition $R_N = R_o$ (or $\beta_N = 1$). A somewhat simpler procedure is obtained by reversing the problem and evaluating ΔT^* for integer values of N. The above procedure is used, with ΔT^* now being an assumed rather than a given value. The procedure continues normally (a to g), to evaluate β_2 to β_N. However, if $\beta_N \neq 1$, ΔT^* is corrected and the procedure is started again. Note that U_∞^M must be calculated for each new value of N, since $U_\infty^M = f(\epsilon)$ and ϵ varies with N (Eq. 25). A summary of the general procedure used to evaluate the effect of the various parameters follows:

Summary of the General Procedure

a. Set n.
b. Set F.
c. Set N, and estimate $\beta_{i+1}^{(1)}$, the initial values of β_{i+1}, Eq. (33a).
d. Determine U_∞^M, by solving Eq. (24) according to outlined procedure
e. Assume ΔT^*.
f. Calculate β_{i+1} by Eq. (33) and the associate iteration procedure.
g. Use calculated value of β_{i+1} to calculate δT_{i+1}^M by Eq. (31)
h. If $\beta_N \neq 1$ change ΔT^* and repeat f till $\beta_N = 1$.
i. Take new N and repeat stages d to h.
j. Take new F and repeat c to i.
k. Take new n and repeat b to j.

RESULTS AND DISCUSSION

The effects of the basic parameters (n and F) at various operating conditions (Q_v, Q_1, ΔT^*, N and β_f) are best demonstrated by evaluating the characteristic "constants" A and B in Eqs. (18) and (20). Only counter-current flow is considered here, and $R_o = 0.25$ cm for all runs considered. The latter implies a constant-pressure vapor supply chamber.

Fig. 1. represents the velocity ratio $U_\infty^M/U_\infty \equiv A^2$ as a function of bubble frequency at different values of horizontal spacing of active injection parts. It is to be noted that A was found to be independent of ΔT^*, or the condensation rates, except at very low (up to $.0.5°C) \Delta T^{*\prime}$s. At these small driving forces the value of A decreases with ΔT^* due to relatively high value of ϵ (low condensation rate), Eq. (24). Fig. 1 may thus be used only for $\Delta T^* > 0.5°C$, single and two-component systems.

As is to be expected, A decreases as bubble density, F and/or n increases. The effect of non-condensables in the bubbles on the velocity ratio is comparatively small, and is not shown here. It was, however, incorporated in the subsequent calculations.

The variation of $\delta\theta^M (= B/A)$ along the column was calculated step by step by Eq. (31) and is presented in Fig. 2 for a single component (pentane-pentane) system and in Fig. 3 for a two component (pentane-water) system. In all cases $\delta\theta^M$ increases in the direction of flow of the continuous phase. The increase is more pronounced as bubble density, vertical and / or horizontal, increases. Not included in Figs. 2 and 3 is $\delta\theta_o^M$, the dimensionless overall temperature rise of the continuous phase, Eq. (28). The deviations between the values of the temperatures of the continuous phase evaluated by the overall heat balance (eq. 28) and the stepwise numerical method (Eq. 31) is within 2%. This excellent agreement validates the consistancy of the numerical method. Included in Figs 2 and 3 are θ_w and θ_w^M, (Eq.9), representing the effective local driving force along the column under various operating conditions. As seen from Figs 2 and 3, the effect of non-condensibles on $\delta\theta^M$ is relatively small when plotted against β. (It will be more pronounced when plotted against column height, for instance). Note that for identical flow rates, the presence of inerts will affect the outlet temperature (or $\delta\theta_o^M$) by approximately (1 $- \beta_f^3$). Thus, for $\beta_f = 0.4$, $\delta\theta_o^M$ will be some 6% smaller than for a pure system with $\beta_f = 0$. However, by Eq. (8) or (10), the collapse rate is determined by $\theta_w^M = \theta_w - \delta\theta^M$ and θ_w (Eq. 7) is strongly affected by inerts (or β_f). Consequently θ_w and θ_w^M, the dimensionless local driving force, vary along the column, exhibiting a shift in the maximum value from the bottom of the column where $\beta = 1.0$, to some point inside the column. This effect is particularly pronounced at high bubble densities.

Figs. 4 and 5 represent $(A-B) = A (1 - \delta\theta_{ave}^M)$ as a function of the nominal driving force, ΔT^*, for various operating conditions. Note that for pure vapors $A - B = \tau_o/\tau_o^M$ i.e. the ratio of single bubble to multi-bubble condensation times corresponding to the same β. This relationship does not hold in the presence of non condensables, as seen by Eq. (20).

For a given bubble spacing (n = const.) the value of (A—B) decreases with frequency i.e. the complete condensation time increases with F. The effect of frequency is more significant as F increases and is stronger as n increases. For the same F, (A—B) decreases as n increases. These results are at variance with those obtained for a single bubble-train in an infinite expense where (A—B) approaches unity as F increases [10]. This is due to the fact that frequency in a single train affects an increase in the rise velocity, thus enhancing condensation through stronger convection effects. Here, however, the rise velocity decreases as the vapor hold-up fraction increases.

The effect of non condensables is to increase (A—B) relative to the corresponding pure systems, and thus to reduce the effects of n and F. This is consistant with the larger condensation times (or height) required for un-pure vapors.

As seen in Fig. 6, (A—B) increases as the continuous flow rate is increased and the condensation rate increases accordingly. This effect is more pronounced at low bubble frequencies, where the required ΔT^* decreases by some 30% with a two-fold increase of the continuous phase velocity. Consistant with other studies [13], the effect of the continuous phase flow rate decreases as the dispersed phase flow rate increases.

The dimensionless bubble collapse history is presented in Fig. 7. As expected in view of the above results the collapse rate decreases as the spatial density of the bubbles increases. This effect is more pronounced in the single component (pentane-pentane) system, consistent with the corresponding larger $\delta\theta^M$ values exhibited by this system, as compared with those of the two component (pentane-water) system. Clearly, this is due to the difference in the volumetric heat capacities of the two continuous phases. The collapse history of a single bubble in an infinite, constant temperature, expance is also included in Fig. 7 for comparison.

Figs. 8 and 9 represent the column height required for complete condensation i.e. the height required to condense 99% of the volume of the vapor that can condense at the given operating conditions. In general, the closer the horizontal and/or vertical bubble spacing at identical nominal temperature driving forces, the higher the column required. This effect is particularly noted at low temperature driving forces and, consistant with single train-studies [10], is much more pronounced in the presence of inerts. For a given n, flow rate and inlet continuous phase temperature, T_∞, the frequency (i.e. the vapor flow rate) determines the exit temperature and the corresponding temperature driving force at the bottom of the column. The assymptotic minimum possible ΔT^* noted in Figs. 8 and 9 correspond to $\theta_w^M \to 0$ and represent the minimum nominal ΔT^* which may still yield complete condensation. In practical terms this means that the inlet temperature T_∞, should be at least $(\Delta T^* \text{-min})$ lower than the saturation temperature of the vapor ; otherwise, condensation will be partial, and limited to the upper part of the column. As can also be seen from Figs. 4 and 5, the assymptotic value of ΔT^* increases with n and F, i.e. the dispersed phase flow rate.

As can be seen from Figs. 8 and 9, the effect of frequency on column height is stronger at higher horizontal bubble densities (higher n). Similarly, the effect of n is more pronounced at the higher frequencies. This is understandable in view of the fact that the local temperature driving force decreases noticeably at high bubble densities. Since the change in the temperature driving force depends on the volumetric specific heat, the pentane-pentane system should be more sensitive to change in F and n. This is indeed confirmed by comparing the single and two component systems, and noting that the column height required for the pentane-water system is less sensitive to change of F and n.

Also included in Figs. 8 and 9 are the corresponding condensation heights for single bubbles in a quiescent infinite expanse of the continuous phase maintained at various values of T_∞. It is noteworthy that the effect of the counter-current continuous-phase flow is to decrease the column height even below that required for a single bubble. This is particularly noted at low bubble densities of pure systems, where the apparent height for the multibubble column is lower than that of a single bubble. This is easily understood by realizing that the collapse rate is determined by the relative vapor liquid U_∞^M (which at low bubble densities is roughly equal to U_∞) while the column height is evaluated by reference to the bubbles' velocity relative to the wall, $U_b (= U_\infty^M - U_\ell)$. Since for a single bubble in a still column $U_b = U_\infty$ the condensation height ratio (multi to single bubble) is given by

$$\frac{H^M}{H} = \frac{(U_\infty^M - U_\ell) \ t_f^M}{U_\infty \cdot t_f} = \frac{\tau_f^M}{\tau_f} \left(A^2 - \frac{U_\ell}{U_\infty} \right) \qquad (34)$$

For low n and F the values of A and τ^M/τ are close to unity, and the condensation height ratio will strongly depend on U_ℓ, the continuous phase (superficial) velocity.

It is interesting to compare the instantaneous interfacial heat transfer coefficient for the multi-bubble system with that of a single bubble in an infinite medium. The instantaneous coefficient is defined by $h^M = q^M/s\Delta T_w^M$ where q^M, s and ΔT_w^M are the instantaneous heat flux, surface area and temperature driving force, respectively. For identical values of β, Eqs. (5) (7) and (8) yield:

$q^M/q = \beta^M/\beta = (\theta_w^M/\theta_w)A$ and

$$\frac{h^M}{h} = A = (U_\infty^M/U_\infty)^{1/2} \tag{35}$$

Thus, one can easily determine h^M by utilizing Eq. (5) and Fig. 1.

A somewhat more useful information is gained by defining the volumetric heat transfer coefficient

$$U_v = \frac{Q/t_f}{V \cdot \Delta T_{ave}} \tag{36}$$

where Q/t_f represents the average heat flow rate, V denotes the optimal volume (based on height for complete condensation as defined above) and ΔT_{ave} is the (arithmetic) average temperature driving force along the column. Figs. 10 and 11 represent the calculated values of U_v which are plotted, for ease of reference, against the nominal driving force. In general, the volumetric transfer coefficient increases with increasing the dispersed phase flow rate (F and n), consistant with earlier spray column studies of evaporating drops [13]. Also noted is the large effect of increasing n as compared to that of F. As already noted in our single bubble studies [7], the pentane-water system exhibits transfer coefficients which are some 50% above those of the pentane-pentane system. Again, this is due to the higher heat capacity of water as compared to pentane.

The values of U_v range from $1.6 \cdot 10^3$ Kcal/hr–m^3 – °C [$\simeq 10^2$ BTU/m–ft^3 – °F] at low F and n to 4.3×10^5 Kcal/hr–m^3–°C [$\simeq 2.7 \times 10^4$ BTU/hr–ft^2 – °F] at n = 4 and F = 26. (Note that n = 4 represents a square pitch of 4 nozzles per cm^2, which with R_o = 0.25 cm, represent the case where the incoming bubbles practically touch one another in the horizontal plane). The corresponding vapor superficial velocities, 0.63 cm^3/sec–cm^2 to 6.5 cm^3/sec–cm^2, were kept low, so as to maintain the identity of each bubble, as suggested by Fair's experiments with air-water system [16]. In this sense, this study is limited to the "streamline" flow region, where (for air-water system) the hold-up is linearly proportional to the dispersed flow rate. Here, however, the dispersed phase hold up decreases due to condensation as the bubbles rise along the column.

The numerical values for the overall heat transfer coefficients are in general agreement with those realized in these laboratories and others reported in the literature. Direct-contact condensation in a venturi mixed co-current pipe flow of steam in Aroclor yielded values in the range of 1.5×10^5 to 4×10^5 BTU/hr–ft^3 – °F [17]. Order of magnitude smaller values were reported when a co-current spray column was used [17, 18]. Values up to 4×10^5 BTU/hr–ft^3 – °F were reported by Harriott and Wiegandt [19] for a co-current, turbulent, downflow sieve plate condenser. However, these values were arbitrarily based on the exit temperature driving force and a 3" height. Condensation of methyl-chloride in water in co-current flow through a packed bed yielded U_v between 6.5×10^4 to 1.5×10^5 BTU/hr–ft^3 – °F [19]. Wilke et al. [18] reported U_v of about 6×10^3 BTU/hr–ft^3 – °F for steam condensing Aroclor in a counter current packed bed. Again, meaningful comparison is difficult since arbitrary values were used for column heights.

CONCLUSIONS

1. An analysis of the collapse history of a multi-bubble system in the streamline region was obtained by assuming quasi-steady state and solving for the local driving forces along the bubble column. The solution allows to evaluate the independent effects of bubble frequency F, horizontal spacing n and inerts contents in single and two component systems.
2. In general, increasing the horizontal and/or vertical bubble density affects a decrease in the rise velocity and the local temperature driving force. Bubble collapse time, and the required column height, increases accordingly.
   It is noteworthy that for a single train in an infinite expance the frequency affects an increase in the rise velocity and a decrease in the temperature driving force, yielding, at high frequencies, a collapse history approaching that of a single bubble.
3. The effect of frequency is more pronounced at higher horizontal bubble densities.
4. The effect of increasing horizontal densities is appreciably stronger than that of increasing the frequency.
5. The general effect of increasing the continuous phase flow rate is to "moderate" the effects of n and F, i.e. to increase the condensation rate, and decrease the required column height.

6. The presence of non-condensables strongly affects the temperature driving force, and the maximum driving force is shifted towards the interior of the column. These effects are particularly noted high n and F values.

7. Collapse rates are some 50% higher in the two-component pentane-water, system than in the single component pentane-pentane system, consistant with earlier single bubble studies. The practical advantages of utilizing a two-component, three-phase heat exchanges are thus demonstrated. The calculated overall volumetric heat transfer coefficients are in general agreement with available experimental data.

NOMENCLATURE

A	velocity ratio $(\sqrt{U_\infty^*/U_\infty})$
B	operational variable $(A\delta\theta^M)$
C_1	constant, Eq. (33)
C_2	constant, Eq. (32)
Cp_ℓ	heat capacity, continuous phase
F	bubble frequency
F_o	Fourrier number $(\alpha t/R_o^2)$
G^*	density ratio of volatile fluid (ρ_L/ρ_v)
H	condensation height
h	instantaneous heat transfer coefficient
i	index of a bubble in a row
Ja	Jakob number $(\rho_\ell Cp_\ell(T^* - T_\infty)/\lambda\rho_v)$
K	constant $(\sqrt{K_v/\pi})$
K_v	velocity factor
k	thermal conductivity, continuous phase
N	number of bubbles in a row
n	number of nozzles per unit area
P^*	system pressure
Pe	Peclet number $(= 2R_o U_\infty/\alpha)$
Q	heat released by bubbles going from R_0 to R_f
Q_ℓ	volumetric flow rate of continuous phase
Q_v	volumetric flow rate of dispersed vapor
$Q_{v\ell}$	flow rate ratio (Q_v/Q_ℓ)
q	instantaneous heat flux
R	instantaneous radius of bubble
R_o	initial radius of bubbles
R_f	final condensation radius
$\overset{\circ}{R}$	radial velocity (dR/dt)
$\hat{R}$	specific gas constant
S	cross section area of column
s	instantaneous area of bubble
T	temperature
T_{out}	outlet temperature, continuous phase
T_w	bubble wall temperature

T^*	saturation temperature, at P^*
T_∞	approach (inlet) temp., continuous phase
T_∞^M	local temperature of nonlinear phase
ΔT	temperature driving force $(T_w - T_\infty)$
ΔT^*	temperature driving force $(T^* - T_\infty)$
ΔT^M	temperature driving force $(T_w - T_\infty^*)$
δT^M	local temperature increase $(T_\alpha^M - T_\alpha)$
δT_o^M	overall temperature increase $(T_{out} - T_\infty)$
t	time
t_f	time, final condensation
U_b	bubble rise velocity, relative to wall
U_ℓ	superficial velocity of liquid phase (Q_ℓ/S)
U_∞	rise velocity of single bubble, stagnant medium
U_∞^M	rise velocity in a multi-bubble system, relative to liquid
U_v	volumetric heat transfer coefficient
V	optimal column volume
x	transformation variable
y_o	initial concentration of noncondensables (mole fraction)

Greek Letters

α	thermal diffusivity, continuous phase
β	dimensionless radius (R/R_o)
β_f	final dimensionless radius (R_f/R_o)
λ	latent heat
θ	dimensionless temperature, $(T - T_\infty)/(T^* - T_\infty)$
θ_w	dimensionless temperature, $(T_w - T_\infty)/(T^* - T_\infty)$
θ_w^M	dimensionless temp. driving force $((T_w - T_\infty^M)/(T^* - T_\infty))$
$\delta\theta^M$	local fractional decrease in temp. driving force $((\delta T^M)/(T^* - T_\infty))$
$\delta\theta_o^*$	overall fractional decrease in temp. driving force $((\delta T_o^M/(T^* - T_\infty))$
ρ	density, continuous phase $(=\rho_\ell)$
ρ_L	density of liquid, dispersed phase
ρ_v	density of vapor, dispersed phase
ϵ	gas hold-up volume fraction
τ	dimensionless time $(Ja\ Pe^{1/2}F_o)$
τ_o	dimensionless time, pure vapor condensation
τ_1	dimensionless time, correction due to inerts

Subscripts

ave	average value
f	final
i	index of bubble in a row (i = 1 . . . N), counting downwards
in	inlet
ℓ	continuous phase
L	liquid dispersed phase
N	last bubble in a row
out	outlet
v	vapor
w	wall
o	initial (bubble), overall
∞	approach value, inlet.

Superscripts

M	multi-bubble system
(m)	iteration index
*	saturation

REFERENCES

[1] S. Sideman, "Advances in Chemical Engineering", 6, 207, Academic Press (1966).

[2] E. Kehat and S. Sideman, "Recent Advances in Liquid-Liquid-Liquid Extraction", Chap.13, 455 (1971).

[3] S. Oktay, IBM J. of Res. and Develop., (1971), 15, No. 5, 342.

[4] S. Oktay and H.G. Elrod, IBM Technical Rept. TR—22982 (1970).

[5] S. Sideman and G. Hirsch, AIChE J. (1965) II, 1019.

[6] O.D. Wittke and B.T. Chao, ASME J. Heat Trans. (1967), 89, 7.

[7] J. Isenberg and S. Sideman, Int. J. Heat Mass Trans. (1970) 19, 945.

[8] J. Isenberg, D. Moalem and S. Sideman, Proc. 4th Int. Heat Trans. Conf. Paris (1970), Vol. V. 25B.

[9] D. Moalem and S. Sideman, Int. J. Heat Mass Trans. (1971), 14, 2152.

[10] D. Moalem, S. Sideman, A. Orell and G. Hetsroni, "Condensation of Bubble Trains : An Approximate Solution", Proc. Int. Symp. Two-Phase Systems, (1971), Haifa, Israel, paper 1-11 ; Progress in Heat Mass Transfer Series, Vol. 5, Pergamon Press, (1972).

[11] D. Moalem, S. Sideman, A. Orell and G. Hetsroni, "Direct Contact Heat Transfer with Change of Phase : Condensation of a Bubble Train", (1972). Submitted for publication, Int. J. Heat Mass Trans.

[12] M.J. Boussinesq, J. Math. Pures App. Serv., 1905, 1.

[13] S. Sideman and Y. Gat, AIChE J. 12, 296, (1966).

[14] S. Sideman, G. Hirsch and Y. Gat, AIChE J., 11, 1081 (1965).

[15] G. Marrucci, Ind. Eng. Chem. Fund. 4, 224, (1965).

[16] J.R. Fair, A.J. Lambright and J.W. Anderson, Ind. Eng. Chem. Proc. Des. Develop. 1, 33, (1962).

[17] D.L. Lackey, M.S. Thesis, Univ. Calif. Berkeley, (1961).

[18] C.R. Wilke, C.T. Cheng, V.L. Ledesma and J.W. Pórter, Chem. Eng. Prog 59, 69, (1963).

[19] P. Harriott and H. F. Wiegandt, AIChE J., 10, 755, (1964).

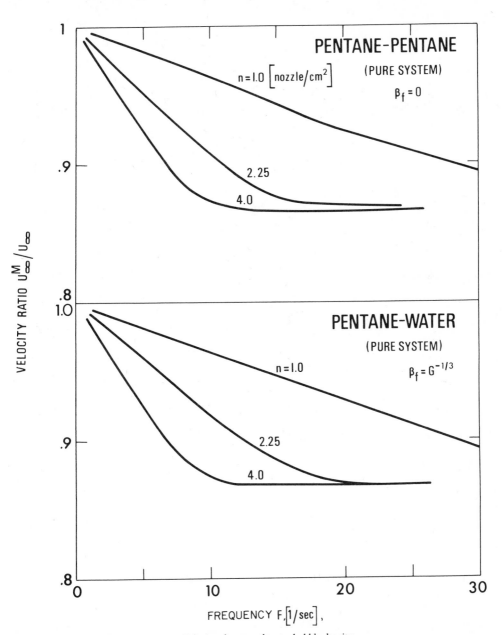

Fig. 1. Velocity decrease due to bubble-density.

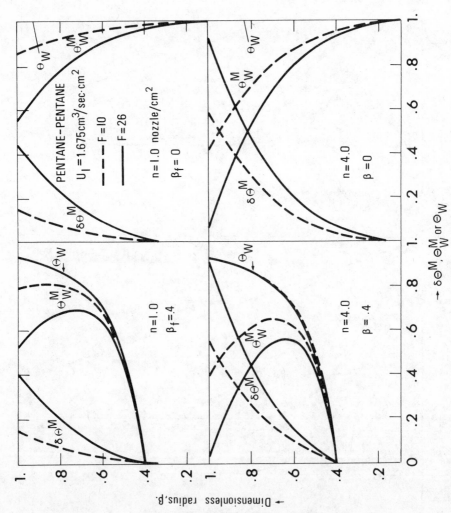

Fig. 2. Temperature decrease, bubble wall temperature and temperature driving force along the column. Pentane-Pentane system.

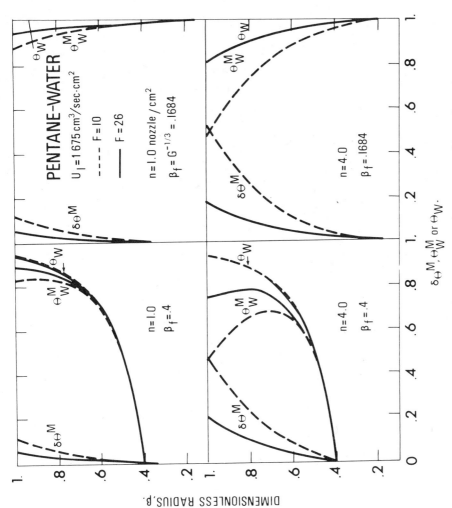

Fig. 3. Temperature decrease, bubble wall temperature and temperature driving force. Pentane-Water system.

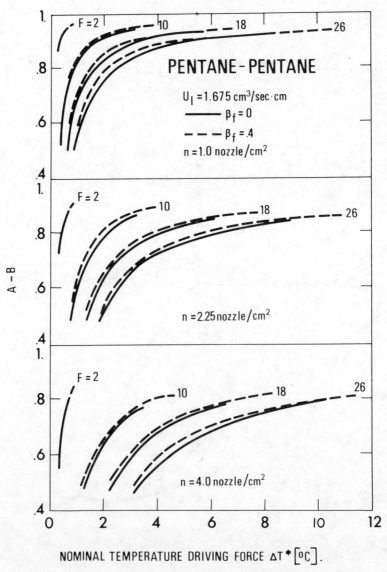

Fig. 4. Effects of frequency and horizontal bubble spacing on A-B as a function of nominal driving force.

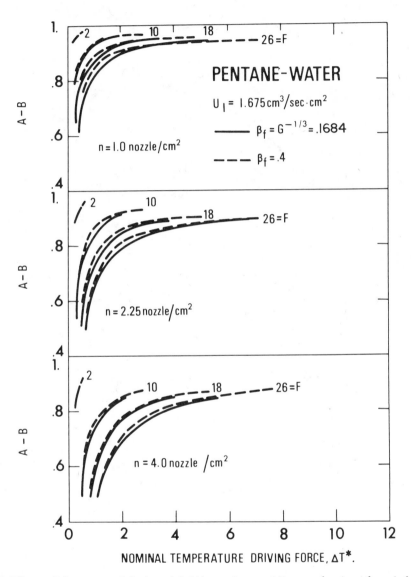

Fig. 5. Effects of frequency and horizontal bubble spacing on A-B, as a function of nominal driving force [°C].

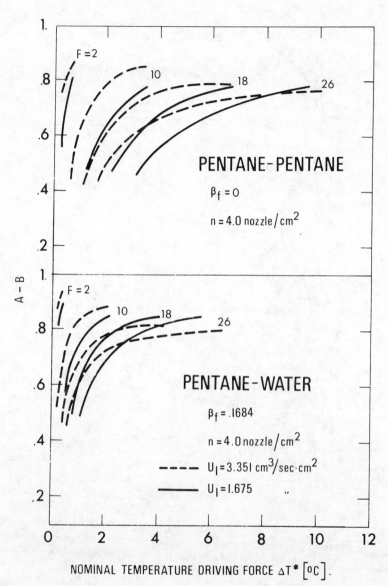

Fig. 6. Effect of the continuous phase flow rate on A-B.

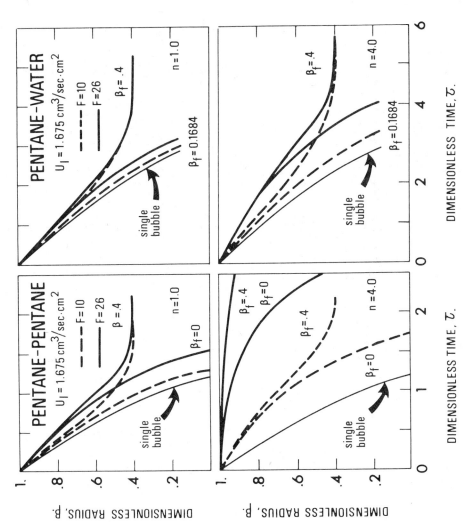

Fig. 7. Collapse history in multi-bubble system—comparison with single-bubble.

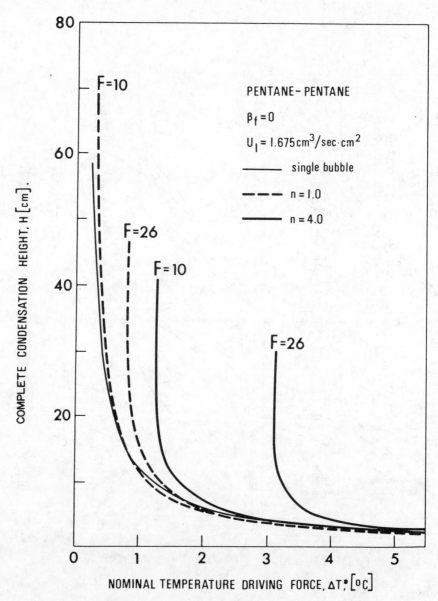

Fig. 8a. Effect of bubble-density on the complete condensation height.

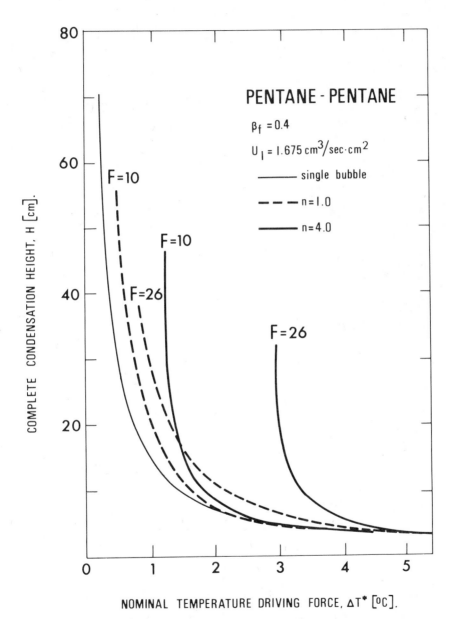

Fig. 8b. Effect of bubble-density on the complete condensation height.

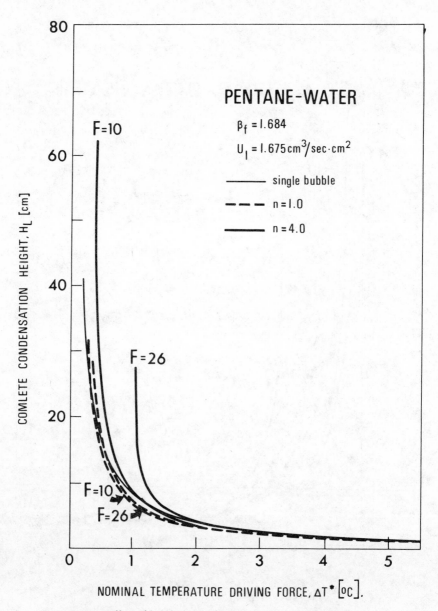

Fig. 9a. Effect of bubble-density on the complete condensation height.

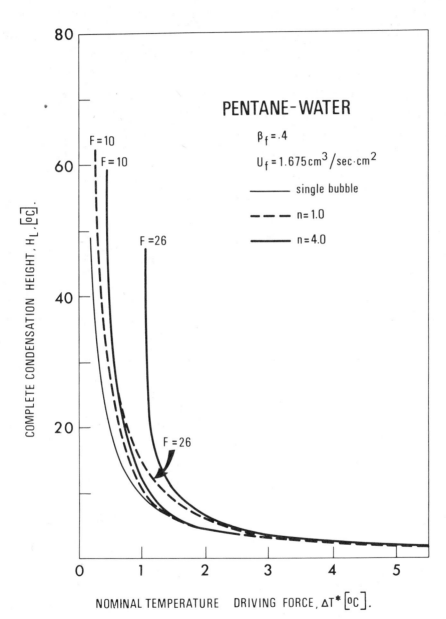

Fig. 9b. Effect of bubble-density on the complete condensation height.

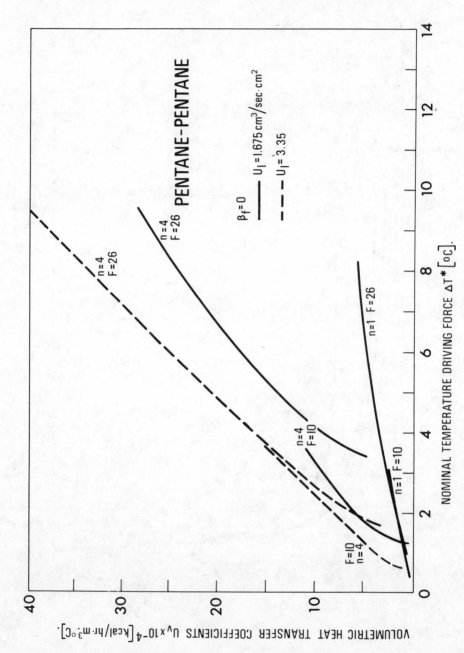

Fig. 10. Overall volumetric heat transfer coefficients at various operating conditions. Pure Pentane-Pentane system.

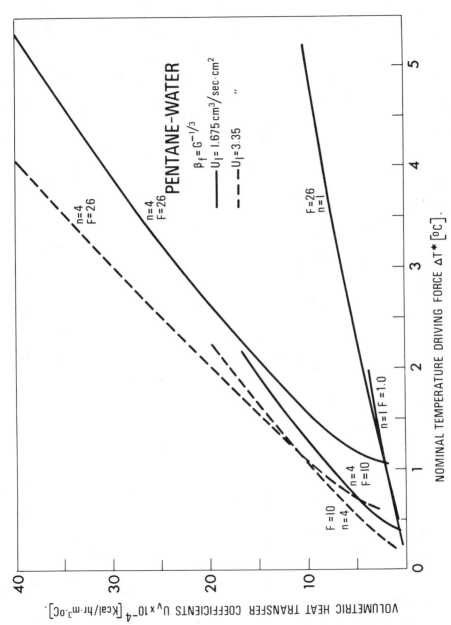

Fig. 11. Overall volumetric heat transfer coefficients at various operating conditions. Pure Pentane-Water system.

Chapter 35

PREDICTION OF LOCAL HEAT TRANSFER IN COMPACT TUBULAR HEAT EXCHANGERS WITH VARIOUS BODY FORCES

A. V. Luikov, O. G. Martynenko, V. L. Kolpashchikov (*)

High heat fluxes and a high degree of regeneration in current heat exchangers require extensive heat transfer surfaces and large heat-transfer coefficients. As the convective heat transfer coefficient for gases is small, the heat transfer coefficient is small, too. As a consequence, one has to enlarge heat transfer surfaces, thus increasing weight and overall dimensions of heat exchangers. Reduction in their size, and thus in the amount of metal used for their production, may be achieved by intensification of heat transfer. However, higher heat transfer coefficient entails an increase in the hydraulic resistance of the apparatuses which sometimes exceeds the magnitude of the heat transfer coefficient by tens of times.

Besides applying different methods for intensifying heat transfer, there is a general tendency to minimize the hydraulic diameters of channels. This gives great compactness numbers and transition to a laminar flow in channels. Here, a decrease in the intensity of heat transfer is compensated by an extended working surface, spaced in a limited volume. For example, through-channels in fillers used for heat exchangers of gas-turbine installations are about 1-2 mm in diameter. Flow velocities of gas and air in such channels do not exceed 10-30m/sec. Thus, the Reynolds number is less than 2000, and the flow is laminar under all operating conditions, The Grashof number expressing the effect of natural convection on the operating conditions is within $50 - 500$. The relative length of channels e/d is usually rather great, and for the existing specimens it is $100 \div 500$.

Below, the effect of non-uniform distribution of heat flux over the perimeter of a horizontal channel is analysed. For the purpose of generality, account is made of the influence of mass forces on the structure of velocity and temperature fields. Non-uniform distribution of thermal boundary conditions is observed both in

(*) Heat and Mass Transfer Institute, B.S.S.R. Academy of Sciences, Minsk, B.S.S.R., U.S.S.R.

N.B. ALL FIGURES QUOTED IN THE TEXT ARE AT THE END OF THE CHAPTER

tubular heat exchangers (for a heat-transfer agent inside tubes due to their position in a cross-flow) and in different plate exchangers. A tube has been chosen as the object of this investigation, because tubular heat transfer surfaces are in wide use and the main concern of the analysis was external heat transfer from tube bundles in longitudinal and transverse flows.

The paper considers the influence of non-uniform distribution of heat flux over the tube perimeter on the internal heat transfer when there is mixed laminar convection and :

1 — a section in the upper part of the tube 2β is heated by a constant heat flux, while over the remaining part of the tube the heat flux is zero ;

2 — the heat flux depends on the angular coordinate in the following way :

$$q = q_0 (1 + S \cos \varphi + c \sin \varphi)$$

To solve the problem stated, it is necessary jointly to consider the Navier-Stokes continuity and energy equations under appropriate boundary conditions. This system of equations is analysed under the following restrictions : it is assumed that there is a fully-developed velocity profile in the tube ; a portion of the tube with stabilized heat transfer is considered ; physical properties of gas (except density) are assumed to be independent of temperature ; variation of density with temperature is taken into account only in the gravitational term ; dissipation is neglected ; axial diffusion in comparison with convection is not accounted for in the Z-direction.

The initial system of the Boussinesque equations in a dimensionless form can be written as

$$W \frac{\partial W}{\partial z} + U \frac{\partial W}{\partial r} + \frac{V}{r} \frac{\partial W}{\partial \varphi} = - \frac{\partial P}{\partial z} - \frac{Gr}{Re^2} \frac{2r}{Re^2 Pr} \cos \varphi \cdot \cos a +$$

$$+ \frac{Gr}{Re^2} \sin a . \theta + \frac{1}{Re} \Delta W, \tag{1}$$

$$W \frac{\partial U}{\partial z} + U \frac{\partial U}{\partial r} + \frac{V}{r} \frac{\partial U}{\partial \varphi} - \frac{V^2}{r} = - \frac{\partial P}{\partial r} + \frac{Gr}{Re^2} \cos \varphi \cos a \theta +$$

$$+ \frac{1}{Re} \left(\Delta U - \frac{U}{r^2} - \frac{2}{r^2} \frac{\partial V}{\partial \varphi} \right), \tag{2}$$

$$W \frac{\partial W}{\partial z} + U \frac{\partial V}{\partial r} + \frac{V}{r} \frac{\partial V}{\partial \varphi} + \frac{UV}{r} = - \frac{1}{r} \frac{\partial P}{\partial \varphi} - \frac{Gr}{Re^2} \sin \varphi \cos a \theta +$$

$$+ \frac{1}{Re} \left(\Delta U - \frac{2}{r^2} \frac{\partial U}{\partial \varphi} - \frac{V}{r^2} \right), \tag{3}$$

$$\frac{\partial}{\partial z}(rW) + \frac{\partial}{\partial r}(rU) + \frac{\partial V}{\partial \varphi} = 0, \tag{4}$$

$$W\frac{\partial \theta}{\partial z} + U\frac{\partial \theta}{\partial r} + \frac{V}{r}\frac{\partial \theta}{\partial \varphi} - \frac{1}{Pe}\left(\Delta\theta - 2\frac{\beta}{\pi}W\right) \tag{5}$$

where the dimensionless variables are

$$\theta = \frac{t - t_m}{q_{cm}\frac{R}{\lambda}}, \quad W = \frac{W}{W_{cp}}, \quad U = \frac{U}{W_{cp}}, \quad V = \frac{V}{W_{cp}},$$

$$P = \frac{P}{\rho_m W_{cp}^2} \quad z = \frac{z}{R}, \quad r = \frac{r}{R}$$

and the boundary conditions are

$$\frac{\partial \theta}{\partial r}\bigg|_{r=1}; \; -\beta < \varphi < \beta = 1, \; \frac{\partial \theta}{\partial r}\bigg|_{r=1}, \; \beta < \varphi < 2\pi - \beta = 0,$$

$$W, U, V\bigg|_{r=1} = 0 \tag{6}$$

For a developed flow and a stabilized heat transfer the sought variables are independent of the longitudinal coordinate Z, which considerably simplifies the system (1) – (5).

Let us introduce the stream function Ψ from the relationships

$$U = \frac{1}{r}\frac{\partial \Psi}{\partial \varphi} \tag{7}$$

$$V = -\frac{\partial \Psi}{\partial r} \tag{8}$$

Substituting (7) and (8) into (2) and (3) and eliminating pressure from these equations, we arrive at

$$\frac{1}{Re}\nabla^4\Psi + \frac{1}{r}\left(\frac{\partial \Psi}{\partial r}\frac{\partial}{\partial \Psi} - \frac{\partial \Psi}{\partial \varphi}\frac{\partial}{\partial r}\right)\nabla^2\Psi =$$

$$= \frac{Gr}{Re^2}\left(\frac{\partial \theta}{\partial r}\sin\varphi\cos a + \frac{1}{r}\frac{\partial \theta}{\partial \varphi}\cos\varphi\cos a\right) \tag{9}$$

Equation (9) should be solved together with (1) and (5) under boundary conditions (6).

The problem is restricted to small heat fluxes. Therefore, a free convective flow is considered as a secondary problem. This is manifested by the small value of the parameter $\epsilon = Gr/Re^2$, which defines the effect of free convection on forced convection, and hence the desired functions can be expanded with respect to this parameter :

$$W = W_o(r) + \sum_{n=1}^{\infty} \epsilon^n W_n(r,\varphi), \quad \theta = \theta_o(r,\varphi) + \sum_{n=1}^{\infty} \epsilon^n \theta_n(r,\varphi)$$

$$\Psi = \sum_{n=1}^{\infty} \epsilon^n \Psi_n, \quad P = P_o(z) + \sum_{n=1}^{\infty} \epsilon^n P_n(r,\varphi,z) \tag{10}$$

In zero approximation with respect to ϵ, free convection is absent, and therefore there are no terms with the zero power ϵ in the expansion for Ψ.

Substituting expansions (10) into the system of equations and grouping the terms with the zero power ϵ, we obtain the system of equations in zero approximation

$$\frac{1}{Re} \Delta W_o = -\frac{\partial P_o}{\partial z}, \quad \Delta \theta_o = 2 \frac{\beta}{\pi} W_o$$

with the boundary conditions

$$W_o \Big|_{r=1} = 0,$$

$$\frac{\partial \theta_o}{\partial r}\Big|_{r=1,-\beta<\varphi<\beta} = 1, \quad \frac{\partial \theta}{\partial r}\Big|_{r=1,\beta<\varphi<2\pi-\beta} = 0 \tag{11}$$

For W_o the known parabolic distribution of velocities is obtained

$$W_o = 2(1 - r^2) \tag{12}$$

Now we pass to a solution of the energy equation of zero approximation

$$\Delta \theta_o = 4 \frac{\beta}{\pi}(1 - r^2) \tag{13}$$

The complete solution will have the form

$$\theta_o = \frac{\beta}{\pi}\left(r^2 - \frac{r^4}{4} - \frac{7}{24}\right) + \sum_{n=1}^{\infty} \frac{2}{n^2\pi} \sin n\beta r^4 \cos n\varphi \tag{14}$$

Now, write out the system of equations in the first approximation with respect to ϵ:

$$U_1 \frac{\partial W_o}{\partial r} + \frac{V_1}{r} \frac{\partial W_o}{\partial \varphi} = - \frac{\partial P_1}{\partial z} - \frac{2r}{RePr} \cos \varphi \cos a + \sin a\, \theta_o + \frac{1}{Re} \overline{\Delta W_1} \tag{15}$$

$$\frac{1}{Re} \nabla^4 \Psi_1 = - \left(\frac{\partial \theta_o}{\partial r} \sin \varphi \cos a + \frac{1}{r} \frac{\partial \theta_o}{\partial \varphi} \cos \varphi \cos a \right) \tag{16}$$

$$U_1 \frac{\partial \theta_o}{\partial r} + \frac{V_1}{r} \frac{\partial \theta_o}{\partial \varphi} = \frac{1}{Pe} \left(\Delta \theta_1 - 2 W_1 \frac{\beta}{\pi} \right). \tag{17}$$

Substituting θ from (14) into (16), we obtain

$$\frac{1}{Re} \nabla^4 \Psi_1 = \left[\left(\frac{\sin 2\beta}{\pi} - \frac{2\beta}{\pi} \right) r + \frac{\beta}{\pi} r^3 \right] \sin \varphi \cos a +$$
$$+ \sum_{n=2}^{\infty} \frac{2}{n\pi} \sin (n+1) r^4 \sin \overset{\circ}{n}\varphi \cos a \tag{18}$$

Now we seek again a particular solution of a nonuniform equation. It has the form:

$$\Psi_1^{4acT} = \frac{Re \cos a}{1152} \left[\frac{\beta}{\pi} r^7 + 6 \left(\frac{\sin 2\beta}{\pi} - \frac{2\beta}{\pi} \right) r^5 \right] \sin \varphi +$$
$$+ \sum_{n=2}^{\infty} \frac{Re \sin (n+1)}{16(n+1)^2 (n+2)\pi} \frac{\beta}{} r^{n+4} \sin n\varphi \cos a \tag{19}$$

Next, introduce a new function F, for which

$$\Psi_1 = \Psi_1^{4acT} + F \tag{20}$$

then obtain the equation for F

$$\Delta\Delta F = 0 \tag{21}$$

with the boundary conditions

$$\frac{\partial F}{\partial \varphi} \bigg|_{r=1} = - \frac{\partial \Psi_1}{\partial \varphi} \bigg|_{r=1} , \qquad \frac{\partial F}{\partial r} \bigg|_{r=1} = - \frac{\partial \Psi_1}{\partial r} \bigg|_{r=1} \tag{22}$$

But any limited biharmonic function F may be presented in the form $F = \varphi_1 + r^2 \varphi_2$, , where φ_1 and φ_2 are flat potential functions which may be presented inside a circle by series of the form 4

$$\varphi = \sum_{n=0}^{\infty} r^n (a_n \cos n\varphi + b_n \sin n\varphi)$$

Therefore, we assume that

$$F = \sum_{n=0}^{\infty} r^n (a_n \cos n\varphi + b_n \sin n\varphi) + \sum_{n=0}^{\infty} r^{n+2} (a_n' \cos n\varphi + b_n' \sin n\varphi)$$

(23)

Coefficients a_n, b_n, a_n', b_n', are determined from the conditions (26). Substituting the series (27) into (26), we obtain

$$B_1 = \left[\frac{Re}{192} \left(\frac{\sin 2\beta}{\pi} - \frac{2\beta}{\pi} \right) + \frac{Re}{576} \frac{\beta}{\pi} \right], \qquad b_n = \frac{Re \sin (n+1)\beta \cos a}{16(n+1)^2 (n+2) \pi},$$

$$B_1' = \left[\frac{Re}{96} \left(\frac{\sin 2\beta}{\pi} - \frac{2\beta}{\pi} \right) + \frac{Re}{384} \frac{\beta}{\pi} \right], \qquad b_n' = - \frac{Re \sin (n+1)\beta \cos a}{8 (n+1)^2 (n+2) \pi}.$$

The complete solution for Ψ_1 can be written in the form :

$$\Psi_1 = \Psi_1^{4acT} + F = \frac{Re \cos a}{1152} \left[\frac{\beta}{\pi} r^7 + 6 \left(\frac{\sin 2\beta}{\pi} - \frac{2\beta}{\pi} \right) r^5 - 3 \left[\frac{\beta}{\pi} + \right. \right.$$

$$\left. \left. + 4 \left(\frac{\sin\beta}{\pi} - \frac{2\beta}{\pi} \right) \right] r^3 + 2 \left[\frac{\beta}{\pi} + 3 \left(\frac{\sin 2\beta}{\pi} - \frac{2\beta}{\pi} \right) \right] \right] \sin \varphi +$$

$$+ \sum_{n=2}^{\infty} \frac{Re \sin (n+1)\beta \cos a}{16(n+1)^2 (n+2) \pi} (r^n - 2r^{n+2} + r^{n+4}) \sin n\varphi$$

(24)

From (24) U_1 and V_1 are determined by using formulas (7) and (8).

Solution of the equation for the axial velocity component in the first approximation is

$$W_1 = \frac{Re}{288} \frac{\beta}{\pi} \sin a (2 r^6 - 18 r^4 + 21 r^2 - 5) +$$

$$+ \left\{ - \frac{Re^2 \cos a}{23040} \frac{\beta}{\pi} r^9 - \frac{Re^2 \cos a}{2304} \left(\frac{\sin 2\beta}{\pi} - \frac{2\beta}{\pi} \right) r^7 - \right.$$

$$- \frac{Re^2 \cos a}{2304} \left(\frac{4 \sin 2\beta}{\pi} - \frac{7\beta}{\pi} \right) r^5 + \left[\frac{1}{4Pr} \cos a - \right.$$

$$\left. - \frac{Re^2 \cos a}{1152} \left(\frac{3 \sin 2\beta}{\pi} - \frac{5\beta}{\pi} \right) - \frac{Re}{4\pi} \sin \beta \sin a \right] r^3 +$$

$$+ \frac{Re}{4\pi} \sin\beta \sin a - \frac{1}{4Pr} \cos a + \frac{Re^2 \cos a}{23040}\left(- 49\frac{\beta}{\pi} + \right.$$

$$\left. + 30 \frac{\sin 2\beta}{\pi}\right)\Bigg] r^3 \cos\varphi - \sum_{n=2}^{\infty} \Bigg\{\Bigg[n \frac{Re^2 \sin(n+1)\beta \cos a}{16 \ (n+1)^3 (n+2)\pi} - \tag{25}$$

$$- \frac{Re \ \sin n\beta \sin a}{2 \ n^2 (n+1)\ \pi}\Bigg] r^{n+2} - \frac{n \ Re^2 \ \sin(n+1)\cos a}{16(n+1)^2 (n+2)^2 \pi} r^{n+4} +$$

$$+ \frac{n \ Re^2 \ \sin(n+1)\beta \cos a}{48(n+1)^2 (n+2)(n+3)\pi} r^{n+6} -$$

$$- \Bigg[\frac{(n^3 + 6n^2 + 11n)Re^2 \sin(n+1)\beta \ \cos a}{48 \ (n+1)^3 (n-2)^2 (n+3)\ \pi} -$$

$$- \frac{Re \ \sin\beta \sin a}{2n^2 (n+1)\ \pi}\Bigg] r^n \Bigg\} \cos n\varphi \ .$$

Finally, solution of the energy equation (17) is found under the specified boundary conditions.

Such a form of boundary conditions is explained by the fact that the tube material is considered to be sufficiently heat-conducting and all non-uniformities in the wall temperature due to convection are smoothed. Nonuniform wall temperature, due to non-uniform distribution of a heat flux on the boundary, is taken into account in zero approximation.

Solving equation (17) by the method, used for the solution of (13), we obtain

$$\theta_1 = \frac{Re}{288.32} \frac{\beta^2}{\pi^2} \sin a \ (r^8 - 16r^6 + 42r^4 - 40r^2 + 13) \ +$$

$$+ \sum^{\infty} \frac{PrRe^2 \sin n\beta \sin(n+1)\beta \cos a}{16\pi^2 (n+1)^2 (n+2)} \left(\frac{r^{2n}}{2n^2} - \frac{r^{2n+2}}{n(n+1)} + \right.$$

$$\left. + \frac{r^{2n+4}}{2(n+2)n} - \frac{1}{n^2 (n+1)(n+2)}\right) + \Bigg\{ - \frac{Re^2 \cos a\beta^2}{1152\cdot120 \ \pi^2} \frac{1+Pr\cdot20}{20} r'' +$$

$$+ \frac{Re^2 \cos a}{1152\cdot80}\left[\frac{\beta^2}{\pi^2}(1 + 14 \ Pr) - \frac{\beta\sin 2\beta}{2\ \pi^2}\frac{1+12 \ Pr}{6}\right] r^9 + \frac{Re^2 \cos a}{1152\cdot48} \times$$

$$\times \left[- \frac{\beta^2}{\pi^2}\frac{7+90 \ Pr}{2} + \frac{2\beta\sin 2\beta(1+12 \ Pr)}{\pi^2}\right] r^7 + \left[\frac{\beta\cos a}{96 \ Pr\ \pi} + \frac{Re^2 \cos a}{1152\cdot24}\right] \times$$

$$\times \left(\frac{\beta^2}{\pi^2} (5 + 52 \text{ Pr}) - \frac{3}{2} \frac{\beta \sin 2\beta}{\pi^2} (1 + 20 \text{ Pr}) \right) - \frac{\text{Re} \beta \sin \beta \sin a}{96 \pi^2} \Bigg] r^5 +$$

$$+ \left[\frac{\text{Re} \beta \sin \beta \sin a_n}{32 \pi^2} - \frac{\beta \cos a}{32 \text{ Pr}\pi} + \frac{\text{Re}^2 \cos a}{1152 \cdot 8} \left(- \frac{\beta^2}{\pi^2} \frac{49 - 400 \text{ Pr}}{20} \right. \right.$$

$$+ \frac{3}{2} \frac{\beta \sin \beta}{\pi^2} (1 + 8 \text{ Pr}) \Bigg) \Bigg] r^3 + \left[\frac{\text{Re}^2 \cos a}{1152 \cdot 40} \left(\frac{\beta^2}{\pi^2} \frac{409 + 1325 \text{ Pr}}{30} \right. \right.$$

$$- \frac{\beta \sin 2\beta}{\pi^2} \frac{57 + 444 \text{ Pr}}{12} \Bigg) + \frac{\text{Re}\beta}{48\pi^2} \sin \beta \sin a - \frac{\beta \cos a}{48 \pi \text{Pr}} \Bigg] r \Bigg\} \cos \varphi +$$

$$+ \sum_{n=2}^{\infty} \left\{ \frac{n \text{Re}^2 \sin(n+1)\beta \cos a\beta}{64(n+1)^2(n+2)\pi^2} \left[- \frac{1 + 3(n+3)\text{Pr}}{12(n+3)(n+4)} r^{n+8} + \frac{1 + 4\text{Pr}(n+2)}{3(n+2)(n+3)} r^{n+6} - \right. \right.$$

$$- \frac{1 + 5(n+1)\text{Pr}}{2(n+1)(n+2)} r^{n+4} + \frac{n^2 + 6n + 11 + 6(n^3 + 11n + 6)\text{Pr}}{3(n+1)^2(n+2)(n+3)} r^{n+2} -$$

$$- \frac{n^3 + 12n^2 + 57n + 118 + 2(n+1)(3n^3 + 36n^2 + 113n + 16)\text{Pr}}{12(n+1)^2(n+2)(n+3)(n+4)} r^n \Bigg] +$$

$$+ \frac{\beta \text{Re} \sin \beta \sin a}{8n^2(n+1)\pi^2} \left[\frac{r^{n+4}}{2(n+2)} - \frac{r^{n+2}}{n+1} + \frac{n+3}{2(n+1)(n+2)} r^n \right] \Bigg\} \cos n\varphi +$$

$$+ \sum_{n=0}^{\infty} \frac{\text{Re}^2 \text{Pr} \cos a \sin(n+1)\beta}{1152(n+1)\pi} \left\{ \frac{1}{2(n+4)} \frac{\beta}{\pi} r^{n+8} + \frac{3}{n+3} \left(\frac{\sin 2\beta}{\pi} - \frac{2\beta}{\pi} \right) r^{n+6} - \right.$$

$$- \frac{3}{2(n+2)} \left(\frac{4\sin 2\beta}{\pi} - \frac{7\beta}{\pi} \right) r^{n+4} + \frac{1}{n+1} \left(\frac{3\sin 2\beta}{\pi} - \frac{5\beta}{\pi} \right) r^{n+2} - \left[\frac{1}{2(n+4)} \frac{\beta}{\pi} + \right.$$

$$+ \frac{3}{n+3} \left(\frac{\sin 2\beta}{\pi} - \frac{2\beta}{\pi} \right) - \frac{3}{2(n+2)} \left(\frac{4\sin 2\beta}{\pi} - \frac{7\beta}{\pi} \right) + \frac{1}{n+1} \left(\frac{3\sin 2\beta}{\pi} - \frac{5\beta}{\pi} \right) r^n \Bigg\} \cos n\varphi$$

$$\sum_{k=1}^{\infty} \sum_{n=2}^{\infty} \frac{\text{Re}^2 \text{Pr} \sin(n+1)\beta \cos a \sin k\beta \cdot n}{16(n+1)^2(n+2)k\pi^2} \left(\frac{1}{2kn} r^{k+n} - \frac{1}{k+1} r^{k+n+2} + \frac{3}{2(k+2)} r^{k+n+4} - \right.$$

$$- \frac{2 + 3k + k^8 - kn + k^2 n}{2kn(k+1)(k+2)} r^{|n-kl|} \Bigg) \cos(n-k)\phi + \sum_{n=2}^{\infty} \frac{\text{Re}^2 \cos a \text{Pr} \sin(n-1)\beta}{1152(n-1)\pi} \Bigg\{ -$$

$$\frac{\beta}{2\pi(n+3)} r^{n+6} -$$

$$- \left(\frac{\sin 2\beta}{\pi} - \frac{2\pi}{\pi} \right) \frac{3r^{n+4}}{n+2} + \left(\frac{4\sin 2\beta}{\pi} - \frac{7\beta}{\pi} \right) \frac{3r^{n+2}}{2(n+1)} - \left[- \frac{\beta}{2\pi(n+3)} - \left(\frac{\sin 2\beta}{\pi} - \frac{2\beta}{\pi} \right) \frac{3}{n+2} + \right.$$

$$+ \left(\frac{4\sin 2\beta}{\pi} - \frac{7\beta}{\pi} \right) \frac{3}{2(n+1)} r^n \Bigg\} \cos n\varphi + \sum_{n=2}^{\infty} \frac{\text{Re}^2 \text{Pr} \sin n\beta \sin(n+1)\beta \cos a}{8\pi^2(n+1)^2(n+2) n} \left(- \frac{r^{2n+4}}{4(n+2)} + \right.$$

$$+ \frac{r^{2n+2}}{4(n+1)} - \frac{r^{2n}}{4(n+2)(n+1)} \Bigg) \cos 2n\varphi + \sum_{k=1}^{\infty} \sum_{n=1}^{\infty} \frac{\text{Re}^2 \text{Pr} \sin(n+1)\beta \cos a \sin k\beta}{8(n+1)^2(n+2)\pi^2 \cdot 4} \times$$

$$\times \left(\frac{r^{n+k+2}}{n+k+1} - \frac{r^{n+k+4}}{n+k+2} - \frac{r^{n+k}}{(n+k+1)(n+k+2)} \right) \cos \phi(n+k)$$

Now we shall write the results of solution of the problem for heat flux distribution according to the law

$$q = q_o (1 + s \cos \phi + c \sin \varphi) , \tag{26}$$

$$W_o = 2 (1 - r^2), \tag{27}$$

$$\theta_o = (c \sin \varphi + s \cos \varphi) r - \frac{r^4}{4} + r^2 - \frac{7}{24}, \tag{28}$$

$$\Psi_1 = \frac{Re}{1152} (r^7 - 12r^5 + 21r^3 - 10r) \sin \varphi - \frac{Rec}{64} (r^4 - 2r^2), \tag{29}$$

$$W_1 = \left[- \frac{Pe}{288 \cdot 80} (r^9 - 20r^7 + 70r^5 - 100r^3 + 49r) + \frac{1}{4Pr} (r^3 - r) \right] \cos \varphi \tag{30}$$

$$\theta_1 = \frac{Re^2 PrS}{32 \cdot 576} (r^8 - 16r^6 + 42r^4 - 40r^2 + 13) + \left[- \frac{Re^2}{1152 \cdot 1200} (r^{11} - \right.$$

$$- 30r^9 + 175r^7 - 500r^5 + 735r^3 - 381r) -$$

$$- \frac{Re^2 Pr}{1152 \cdot 24} (2r^{11} - 42r^9 + 225r^7 - 520r^5 + 600r^3 -$$

$$\left. - 625r) + \frac{8 + Re^2 Pr^2 c}{Pr \cdot 384} (r^5 - 3r^3 + 2r) \right] \cos \varphi - \tag{31}$$

$$- \frac{Re^2 Prc \cdot s}{384} (r^5 - 3r^2 + 2r) \sin \varphi - \frac{Re^2 Pr}{384 \cdot 60} (r^8 -$$

$$- 15r^6 + 35r^4 - 21r^2)(c \sin 2\varphi + s \cos 2 \varphi).$$

Solution of the energy equation with regard for the first approximation and with respect to the expansion parameter is built in the form

$$\theta = \theta_o + Gr/Re^2 \theta$$

where θ_1 is determined by expression (40).

Taking into account (40), (41) and (42), we get

$$\theta = - \frac{r^4}{4} + r^2 - \frac{7}{24} + br \cos(\varphi + \gamma) + \frac{PrGrb \cos \gamma}{576 \cdot 32} (r^8 -$$

$$- 16r^6 + 42r^4 - 40r^2 + 13) + \left[- \frac{Gr}{11520 \cdot 120} \left(r^{11} - \right. \right.$$

$$- 30r^9 + 175r^7 - 500r^5 + 735r^3 - 381r) -$$

$$- \frac{Gr}{1152 \cdot 240} (2r'' - 42r^9 + 225r^7 - 520r^5 + 600r^3 -$$

$$- 625r) + \frac{Gr}{Re^2 Pr \cdot 48} (r^5 - 3r^3 + 2r) \Big] \cos\varphi +$$

$$+ \frac{GrPr\, b^2 \sin\gamma}{384} (r^5 - 3r^3 + 2r) \sin(\varphi + \gamma) -$$

$$- \frac{GrPr\, b}{384 \cdot 60} (r^8 - 15r^6 + 35r^4 - 21r^2) \cos(2\varphi + \gamma)$$

$$(32)$$

where

$$\gamma = \operatorname{arctg} \frac{c}{s} \; ; \quad b \sin\gamma = c; \qquad b \cos\gamma = s$$

The function θ was numerically calculated by formula (32) for various values of the parameters entering into it.

The calculation results are graphically presented in Figs. 1 and 2. For Gr = 500, Re = 1000, Pr = 0.73, θ is shown as a function of the radius in a vertical plane $(0, \pi)$ at different values of B for the cases :

    a) $\gamma = 0$ (maximum upward heat flux)

    b) $\gamma = \pi$ (maximum downward heat flux)

We have found that owing to regard for the effect of free convection, a term appears in the solution which describes the downward displacement of the temperature minimum.

In the present case this downward displacement of the temperature minimum will be summed up with the displacement (found earlier for a zero approximation) caused by a nonuniform heat flux distribution at the tube boundary.

In the case (a) both displacements have the same direction (downwards). Therefore, the total displacement of the temperature minimum relative to the axis has the form depicted in Fig. 1.

In the case (b) the effect of nonuniform heat flux distribution and that of free convection are reverse. Thus the resulting displacement is determined by the largest one. At large B values the temperature minimum lies above the axis due to the dominating effect of a non-uniform heat flux distribution. It is clear from (Fig. 2) that at B values lying near 0.5, the temperature minimum coincides with a geometric axis of the tube.

Figs. 3 and 4 give the comparison of the effect of non-uniform heating and the effect of free convection on the temperature distribution inside a tube. The temperature graph in a zero approximation, compared with that with regard for the first approximation with respect to Gr/Re^2, shows that the temperature minimum displacement downwards caused by free convection depends considerably on the parameter γ. This displacement increases with heating from below $(\gamma = \pi)$ and decreases with S growth at heating from above $(\gamma = 0)$. This phenomenon may be treated as follows : a displacement due to free convection from a variable heat flux component which, depending on B or S, is imposed on that of the temperature minimum due to heat transfer at non-uniform heat flux and free convection caused by a constant heat flux component. The physical sense of the last effect is explained as follows : hot layers of gas from a heated wall section are transferred by free convection to the opposite wall so that the temperature minimum shifts in the direction of the heat flux maximum.

The system of equations for the second approximation is of the form

$$\nabla^4 \Psi_2 = -\frac{Pe}{r}\left(\frac{\partial \Psi_1}{\partial r}\frac{\partial}{\partial \varphi} - \frac{\partial \Psi_1}{\partial \varphi}\frac{\partial}{\partial r}\right)\nabla^2 \Psi_1 - Re\left(\frac{\partial \theta_1}{\partial r}\sin\varphi + \frac{1}{r}\frac{\partial \theta_1}{\partial \varphi}\cos\varphi\right)$$

$$(33)$$

$$\Delta W_2 = Re\frac{\partial P}{\partial z} + \frac{Re}{r}\left(\frac{\partial \Psi_1}{\partial \varphi}\frac{\partial W_1}{\partial r} - \frac{\partial \Psi_1}{\partial r}\frac{\partial W_1}{\partial \varphi}\right) + \frac{Re}{r}\frac{\partial \Psi_2}{\partial \varphi}\frac{\partial W_0}{\partial r} \qquad (34)$$

$$\Delta \theta_2 = 2W_2 + \frac{RePr}{r}\left(\frac{\partial \Psi_1}{\partial \varphi}\frac{\partial \theta_1}{\partial r} - \frac{\partial \Psi_1}{\partial r}\frac{\partial \theta_1}{\partial \varphi}\right), +$$

$$+ \frac{RePr}{r}\left(\frac{\partial \Psi_2}{\partial \varphi}\frac{\partial \theta}{\partial r} - \frac{\partial \Psi_2}{\partial r}\frac{\partial V_0}{\partial \varphi}\right)$$

$$(35)$$

By substituting first-approximation solutions into this system, we may show that the solutions of this system are of the form

$$\Psi_2 = \sum_{j=1}^{7}\sum_{i=1}^{9} P_{ij}^2 \xi_j r^{(2i-2+\eta i)}$$

$$W_2 = \sum_{j=1}^{9}\sum_{i=1}^{9} V_{ij}^2 \xi_j r^{(2i-2+\eta i)}$$

$$\theta_2 = \sum_{j=1}^{9} \sum_{i=1}^{10} T_{ij}^2 \, \xi_j \, r^{(2i-2+\eta i)} \tag{36}$$

$$\xi_1 = 1; \quad \eta_1 = 0$$

$$j = 2,4,6,8; \quad \xi_j = \cos \frac{j}{2} \varphi, \qquad \eta_j = \frac{j}{2}$$

$$j = 3,5,7,9; \quad \xi_j = \sin \frac{j-1}{2} \varphi, \qquad \eta_j = \frac{j-1}{2}$$

Coefficients P_{ij}^2, V_{ij}^2, T_{ij}^2 were calculated on the "Minsk-22" computer using a specially composed program. The same program provides temperature and velocity profiles from the solutions obtained and calculation of the integral Nusselt number. The integral number Nu is determined by the expression

$$\text{Nu} = \frac{2 \left(\dfrac{\partial \theta}{\partial r} \right)_{r=1}}{E_{cm} - t_{cp}} \tag{37}$$

where: $(\partial\theta/\partial r)\, r = 1$ is a perimeter-average temperature gradient on a wall ;

E_{cm} is a perimeter-average wall temperature ;

t_{cp} is a mean mass gas temperature.

Since integrals of $\sin n\varphi$ and $\cos n\varphi$ $(n \neq 0)$ equal zero, then only the terms with no trigonometric functions T_{i1}^1 and T_{i2}^2 will contribute to average values.

The Nu may be presented in the form

$$\text{Nu} = \frac{2 \left(1 + \dfrac{\text{Gr}}{\text{Re}^2} S_1 + \left(\dfrac{\text{Gr}}{\text{Re}^2} \right)^2 S_2 \right)}{\left(\dfrac{11}{24} - \dfrac{\text{Gr}}{\text{Re}^2} S_3 - \left(\dfrac{\text{Gr}}{\text{Re}^2} \right)^2 S_4 \right)} \ , \tag{38}$$

where

$$S_1 = \sum_{i=1}^{5} T_{i1}^1 (2i-2);$$

$$S_2 = \sum_{i=1}^{10} T_{i1}^2 (2i-2);$$

$$S_3 = \sum_{i=1}^{5} T_{i1}^1 \left(\frac{1}{2i} - \frac{1}{2i+2} \right);$$

$$S_4 = \sum_{i=1}^{10} T_{ii}^2 \left(\frac{1}{2i} - \frac{1}{2i+2} \right).$$

Fig. 5 presents temperature profiles for Gr = 500 and different values of γ and B built with regard for a second approximation. A comparison with profiles built with regard for only the first approximation reveals that convergence of solutions depends on the parameter B. It gets considerably worse with B growth. Convergence for $\gamma = \pi$ is also worse than that for $\gamma = 0$.

Fig. 6 depicts Nu versus Gr, B and γ. It is seen that Nu may decrease or increase depending on the Grashof number, this being determined by the angle γ. For $\gamma = \pi$ the dependence of Nu on Gr is minimum within Gr = 300, for $\gamma = \pi/2$ this dependence decreases and for $\gamma = 0$ increases. An increase in the parameter B makes Nu increase or decrease (curves 1, 2, 6, 7).

NOMENCLATURE

q	heat flux
ρ	gas density
ρ_m	average gas density over the tube section
W	longitudinal velocity component
U	radial velocity component
P	pressure
g	free fall acceleration
μ	gas viscosity
ν	kinematic viscosity
t	gas temperature
t_m	weighted-mean temperature
W_{cp}	mean gas flow velocity
R	tube radius
λ	thermal conductivity of gas
$\beta = 1/\rho\,(\partial\rho/\partial t)_p$	gas expansion coefficient
C_p	gas heat capacity at constant pressure
B	amplitude of heat flux non-uniformity
$Re = W_{cp}/\nu$	Reynolds number
$Gr = g\beta R^4 q/\lambda\nu^2$	Grashof number
$Pr = \mu C_p/\nu$	Prandtl number
$Pe = RePr$	Peclet number

$$\Delta = \nabla^2 = \partial^2/\partial r^2 + 1/r\,\partial/\partial r + 1/r^2\,\partial^2/\partial\varphi^2 \qquad \text{Laplacian operator}$$

$$\Delta\Delta = \nabla^4 = \partial^4/\partial r^4 + 2/r\,\partial^3/\partial r^3 - 1/r^2\,\partial^2/\partial r^2 + 1/r^3\,\partial/\partial r + 2/r^2\,\partial^4/\partial\varphi^2\,\partial r^2 + 4/r^4\,\partial^2/\partial\varphi^2 + 1/r^4\,\partial^4/\partial\varphi^4 - 2/r^3\,\partial^3/\partial\varphi^2\,\partial r \qquad \text{Laplacian of Laplacian}$$

REFERENCES

[1]   Morton B.R. Laminar convection in uniformly heated pipes at low Rayleigh numbers, A. J. Mech. Math., 1959, 12, No. 4.

[2]   Martynenko, O. G. and Kolpashchikov, V. L. Distribution of refractive index in a gas lens with account of free convection. (In a collection: Study of thermo-hydrodynamic light guides, convection. Ed. by A. V. Luikov), Minsk, 1970.

[3]   Petuklov, B. M. Heat Transfer & Resistance in Laminar Flow of Fluids in Pipes, Moscow, "Energy Press", 1967.

[4]   Frank, I. and Mizes, P., Differential integral equations of mathematical physics, Moscow—Lenningrad, 1937.

[2]   Мартынеко О.Г., Колпащиков В.Л., Распределение показтеля преломления в газовой линзе с учётом влияния свободний конвекции. Сб. "Исследование термогидродинамических световодов", под редакцией акад. А.Н.БССР А.В.Лыкова, Минск, 1970 г.

[3]   Петухов Б.М., Теплобмен и сопотивле при ламинаром течении жидкости в трубах.М, "Энергия", 1967 г.

[4]   Фяанк Ф., Мизес Р., Дифференлиальные интегральные уравнения математнческой физики, М.-Л., 1937 г.

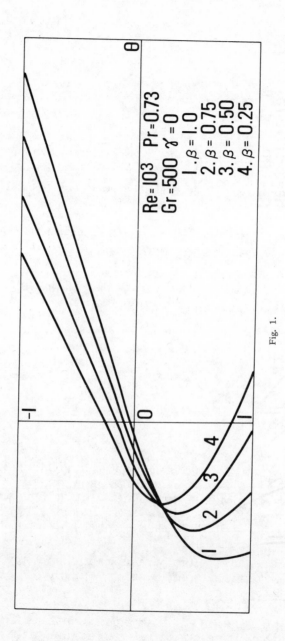

Fig. 1.

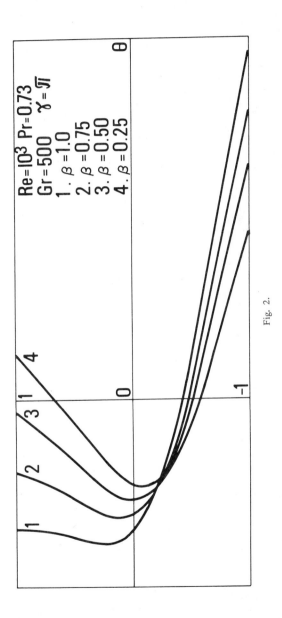

Fig. 2.

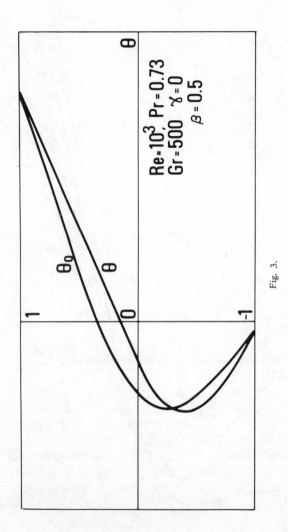

Fig. 3.

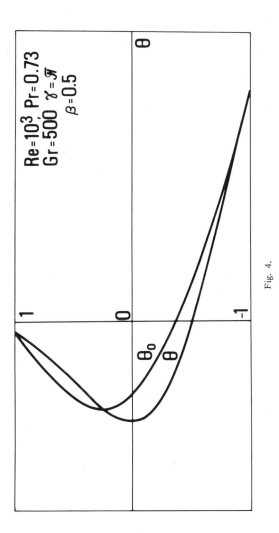

Fig. 4.

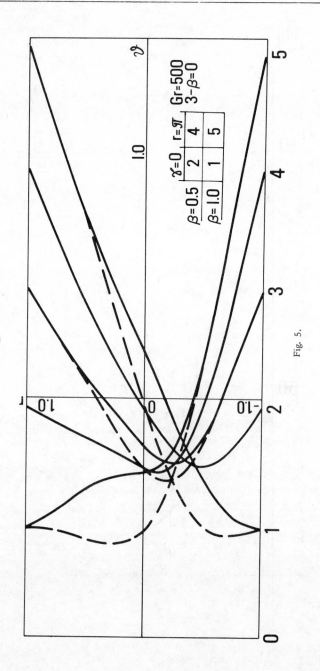

Fig. 5.

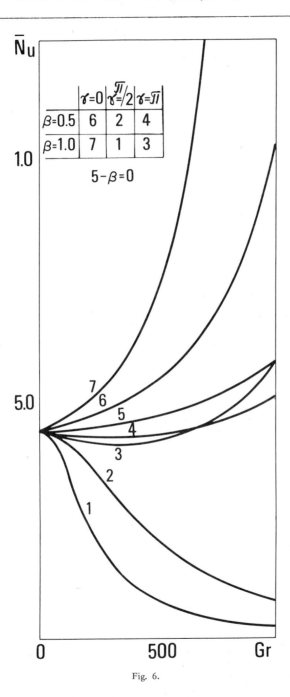

Fig. 6.

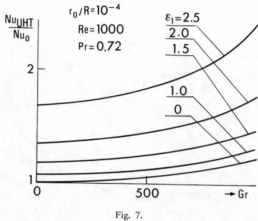

Fig. 7.

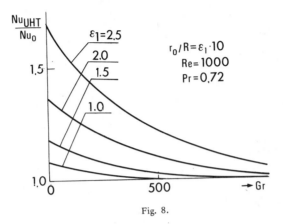

Fig. 8.

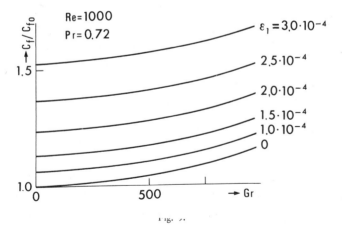

Fig. 9.

INDEX

Design & Calculation Worknotes

Design & Calculation Worknotes

Design & Calculation Worknotes

Design & Calculation Worknotes

Design & Calculation Worknotes

Design & Calculation Worknotes

Design & Calculation Worknotes

Design & Calculation Worknotes

Design & Calculation Worknotes

Design & Calculation Worknotes